Multiphase Flow and Fluidization

Continuum and Kinetic Theory Descriptions

MULTIPHASE FLOW AND FLUIDIZATION

Continuum and Kinetic Theory Descriptions

DIMITRI GIDASPOW
Department of Chemical Engineering
Illinois Institute of Technology
Chicago, Illinois

Academic Press
An Imprint of Elsevier
Boston San Diego New York
London Sydney Tokyo Toronto

A more complete description of the front cover art can be found on pages 325 and 326.

ACADEMIC PRESS, INC.
525 B Street, Suite 1900, San Diego, California 92101-4495

United Kingdom Edition published by
ACADEMIC PRESS LIMITED
24-28 Oval Road, London NW1 7DX

Library of Congress Cataloging-in-Publication Data

Gidaspow, Dimitri, 1934–
Multiphase flow and fluidization: continuum and kinetic theory descriptions with applications / Dimitri Gidaspow
p. cm.
Includes bibliographical references and index.
ISBN 0-12-282470-9 (alk. paper)
1. Multiphase flow--Mathematical models. I. Title.
TA357.5.M84G53 1993
620.1'064--dc20 93-9115
CIP

Printed in the United States of America
04 05 06 07 08 BB 9 8 7 6 5 4 3 2

CONTENTS

Preface *ix*
Nomenclature *xiii*

Chapter 1 TRANSPORT EQUATIONS **1**

1.1 Basic Approach 1
1.2 Mass Balances 2
1.3 Momentum Balances 3
1.4 Energy Balances 7
1.5 Entropy Balance 10
1.6 Mixture Equations 11
1.7 Multicomponent Multiphase Flow 20
Exercises 21
Literature Cited 28

Chapter 2 ONE-DIMENSIONAL STEADY GAS–SOLID FLOW **31**

2.1 One-Dimensional, Steady Mixture Momentum Balance 31
2.2 One-Dimensional, Steady Gas Momentum Balance: Pressure Drop in Both Phases — Model A 33
2.3 Fluid Particle Drag 35
2.4 Buoyancy 37
2.5 Drag for Models B and C 39
2.6 Entrance Length in Pneumatic Transport 40
2.7 Pressure Drop 45
2.8 Pressure Drop Correlation in a Dilute Lift Line 51
Exercises 52
Literature Cited 58

Chapter 3 DRIFT FLUX **61**

3.1 Introduction 61
3.2 Hold-Up in Homogeneous and Slip Flow 61
3.3 Single Particle Analysis 62
3.4 Ergun Equation Prediction 63
3.5 Two Regimes in a Bubble Column 66

3.6 Other Applications 68
Exercises 69
Literature Cited 71

Chapter 4 CRITICAL GRANULAR FLOW 73

4.1 One-Dimensional Granular Flow Momentum Balance 73
4.2 One-Dimensional Statics: Jannssen Equation 75
4.3 Incompressible, Frictionless Flow and Discharge 77
4.4 Thermodynamics of Powders: Compressibility 79
4.5 Critical Flow Theory for Granular Materials 82
4.6 Example of Critical Granular Flow 86
4.7 "Sound Speed" from Kinetic Theory 88
Exercises 89
Literature Cited 95

Chapter 5 THE FLUIDIZED STATE 97

5.1 Introduction: Fluidization Regimes 97
5.2 Minimum Fluidization Velocity 98
5.3 Geldart's Classification of Powders 104
5.4 Kinetic Energy Dissipation Analysis 106
Exercises 109
Literature Cited 113

Chapter 6 ON THE ORIGIN OF BUBBLES 115

6.1 Introduction 115
6.2 Void Propagation in Incompressible Fluids 116
6.3 Shocks and Dispersion with No Solids Stress 119
6.4 Bubbling Criterion for Small Particles 121
Acknowledgment 123
Exercises 123
Literature Cited 125

Chapter 7 INVISCID MULTIPHASE FLOWS: BUBBLING BEDS 129

7.1 Basic Equations 129
7.2 Compressible Granular Flow 131
7.3 Well-Posedness of Two-Phase Models 134
7.4 Homogeneous Flow and Pressure Propagation 139
7.5 Solids Vorticity and Void Propagation 141
7.6 Davidson Bubble Model 146
7.7 Computer Fluidization Models 149
7.8 Comparison of Computations to Observations 154
Exercises 183
Literature Cited 192

Chapter 8 Viscous Flow and Circulating Fluidized Beds **197**

8.1 Introduction 197
8.2 Multiphase Navier–Stokes Equation Model 198
8.3 Dimensional Analysis: Scale Factors 201
8.4 CFB or Riser Flow: Experimental 207
8.5 Need for Clusters in One-D Modeling 214
8.6 Computation of Cluster Flow 216
8.7 Computation of Core–Annular Regime 224
8.8 Radial Profiles and Turbulence 228
Exercises 233
Literature Cited 235

Chapter 9 Kinetic Theory Approach **239**

9.1 Introduction 239
9.2 Maxwellian Distribution for Particles 240
9.3 Properties of the Maxwellian State 242
9.4 Dynamics of an Encounter between Two Particles 244
9.5 The Frequency of Collisions 247
9.6 Mean Free Path 252
9.7 Elementary Treatment of Transport Coefficients 253
9.8 Boltzmann Integral–Differential Equation 256
9.9 Maxwell's Transport Equation 259
9.10 Conservation Laws with No Collisions 261
9.11 Second Approximation to the Frequency Distribution 263
9.12 Integral Equation Solver Strategy 272
9.13 Viscous Kinetic Stress Tensor 274
9.14 Dense Transport Theorem 276
9.15 Particulate Momentum Equation 279
9.16 Fluctuating Kinetic Energy Equation 281
9.17 Viscosity — Collisional Momentum Transfer 284
9.18 Granular Conductivity 290
Literature Cited 294

Chapter 10 Applications of Kinetic Theory **297**

10.1 Granular Shear Flow 297
10.2 Flow down a Chute 307
10.3 Bubbling Bed: Flow Patterns 311
10.4 Liquid–Solid Fluidization 318
10.5 Circulating Fluidized Bed Loop Simulation 324
10.6 Maximum Solids Circulation in a CFB 330
Literature Cited 334

Chapter 11 KINETIC THEORY OF GRANULAR MIXTURES 337

11.1 Empirical Input: Restitution Coefficients 337
11.2 Boltzmann Equations for a Mixture 339
11.3 Dense Transport Theorem 339
11.4 Granular Temperatures and Applications 341
11.5 Particle-to-Particle Drag 343
11.6 Summary 345
Exercises for Kinetic Theory (Chapters 9 – 11) 348
Literature Cited 353

Chapter 12 SEDIMENTATION AND CONSOLIDATION 355

12.1 Conservation of Particles 355
12.2 Settling in a Sedimentation Column: Introduction 357
12.3 Free Settling 360
12.4 Compression Settling 366
12.5 Consolidation: Relation to Osmotic Pressure 370
12.6 Electrokinetic Phenomenon: Zeta Potential 376
12.7 Effect of Zeta Potential on Sedimentation 379
Exercises 382
Literature Cited 388

Appendices 391

FORMULATION OF CONTINUUM PROBLEMS: INTRODUCTION

Appendix A: Overall (Macroscopic Balances) 393
Appendix B: Eulerian Approach: One-D Energy Balance 405
Appendix C: Leibniz Formula and Relation to Transport 411
Appendix D: Lagrangian Approach: One-D Conservation of Species and Population Balance 413
Appendix E: Reynolds Transport Theorem 423

Appendices 429

THE METHODS OF CHARACTERISTICS: INTRODUCTION

Appendix F: First Order Partial Differential Equation 431
Appendix G: Solution of a Hyperbolic System of First Order Partial Differential Equations 441

Index 457

PREFACE

Multiphase flow occurs in many operations in the chemical, petroleum, and power generation industries. The majority of chemical engineering unit operations, such as boiling and condensation, separations and mixing, involve multiphase flow. It is the central unsolved problem in the safety analysis of nuclear reactors, both light water and breeders. It is the principal engineering science problem resulting in the poor performance of multiphase processing, when compared with the design, of chemical plants processing solids. Poor solids processing plant performance and erosion caused by particle impacts also threaten the economic viability of future clean combustion of coal processes using fluidized beds.

Yet the basic texts on the subject, such as Bird, Stewart and Lightfoot's *Transport Phenomena*, do not treat multiphase flow. This is, in part, due to the fact that until recently it was impossible to obtain predictive solutions using multiphase flow theory. The development of computer work stations and the refinement of multiphase flow theory has changed this situation. A number of organizations and national laboratories have now written computer codes that can be useful as valuable research tools. Unfortunately, many of these codes cannot be used by anyone who hasn't taken special courses in their use and in multiphase theory, principally because of numerical instabilities caused by ill-posedness. It was pointed out at many multiphase flow meetings that a book on the subject would help to speed up the progress in this emerging engineering science.

I visualize this book as an advanced text to be used in transport phenomena and fluidization courses, as well as by industrial researchers. It is based on my AIChE Kern award lecture published in *Applied Mechanics Reviews* in January 1986. The kinetic theory approach is only 10 years old and has not appeared in any book on multiphase flow or fluidization. The kinetic theory approach clarifies many physical concepts, such as particulate viscosity and solids pressure. It also, naturally, introduces the new dependent variable, not found in single phase fluid mechanics: the volume fraction of the dispersed phase.

Unlike traditional treatment of the subject, fluidization is described as a branch of transport phenomena, based on the principles of conservation of mass, momentum and energy for each phase.

The equations in the first chapter are derived using the well-known conservation principles of single phase flow. The appendices present a review of the Eulerian (shell balance) and the Lagrangian approaches used in the text in order to make the book suitable for use in senior level and first year graduate courses.

The second chapter treats steady one-dimensional pneumatic transport from the point of view of conservation of mass and momentum to demonstrate how the equations developed in Chapter 1 can be used in a practical problem. The concept of fluid-particle drag, different models leading to ordinary differential equations solvable on a personal computer, and simple design procedures for estimating pressure drop are also discussed in this chapter.

Chapter 3 shows how hold-up, the volume fraction of the dispersed phase, can be estimated by the use of algebraic balances of mass and momentum.

Chapter 4 gives a traditional development of granular flow balances using continuum theory. A new explanation of the maximum discharge rates of powders based on the concept of critical flow is presented. The knowledge of the maximum permissible flow is the key to the design of safety and feeding systems in two-phase flow. For example, Apollo 13 barely returned to earth because of a maldesign of a relief valve. The designer used single-phase flow theory to over-predict the flow rate, causing an excessive rise of pressure of oxygen in the tank feeding a fuel cell, which was followed by an explosion.

Chapter 5 is an introduction to fluidization. It presents the concepts of minimum fluidization, Geldart's classification, and bed behavior in its attempt to minimize energy losses.

Chapter 6 shows that bubbles can be interpreted as shocks and that Geldart's classification can be rationalized based on hydrodynamic theory.

Chapter 7 discusses well-posedness of initial value problems, reviews older fluidization models from the point of view of multiphase flow, and shows the tremendous predictive capacity of multiphase models with zero viscosity.

Chapter 8 presents a simple derivation of multiphase Navier–Stokes' equations and the scale factors that arise from these balances. It is shown that flow regimes observed in dense vertical gas-solid transport can be predicted using these equations.

Chapters 9, 10, and 11 together provide an extensive review of kinetic theory of gases as applied to granular flow by Professor Savage and collaborators. Starting with a derivation of the Boltzmann equation for particle distribution, Chapter 9 presents the details of the derivation of conservation equations for mass, momentum, and the oscillating kinetic energy, called granular temperature. It shows how the solids pressure and the solids viscosity are obtained using the well-known kinetic theory of dense gases. Chapter 10 reviews various applications of kinetic theory, including simulation of flow in a complete loop of a circulating fluidized bed. Chapter 11 provides a molecular theory of gases approach to flow and segregation of mixtures of particles.

Chapter 12 discusses sedimentation and consolidation from an elementary point of view. The first half of the chapter can be used as a introduction to multiphase flow.

Exercises at the end of each chapter test the reader's comprehension and lead into a variety of applications not covered in the text (exercises for Chapters 9, 10, and 11 are grouped together at the end of Chapter 11).

A great deal of the work reported in this book was carried out at Illinois Institute of Technology by about a dozen Ph.D. students, starting with our present department chairman, Professor Hamid Arastoopour (1978), who is continuing independent research in fluidization. Dr. Bozorg Ettehadieh (1982), working with Dr. Robert Lyczkowski of Argonne National Laboratory, first modified an ill-posed computer code developed for gas-liquid flow at Los Alamos and predicted bubble formation in fluidized beds. He also extended the code to include gas phase reactions. Dr. M. Syamlal (1985) extended the computer program to *n* phases and has continued the research at Morgantown Energy Technology Center under the direction of Dr. Tom O'Brien. Dr. Jacques Bouillard (1986) worked on an analytical solution of bubble formation and with Dr. Yang-Tsai Shih (1986) computed settling of particles in a lamella settler following earlier work of Dr. Firooz Rasouli (1981). Dr. David F. Aldis (1987) applied the code to detonation of dust. Dr. Yuam Pei Tsuo (1989) computed the flow regimes in circulating fluidized beds. Dr. Jianmin Ding (1990) extended the code to include the granular temperature equation obtained from kinetic theory, generalized it to three dimensions and briefly considered gas phase reactions. Dr. Umesh K. Jayaswal (1991) improved the code and briefly showed its applications to gas–liquid–solid fluidization. His Ph.D. thesis contains the latest version of the code. The three experimental Ph.D. theses by Dr. Y. C. Seo (1985), Dr. K. Luo (1987) and Professor Aubrey Miller (1991) were essential for providing guidance and data for modeling. Dr. Isaac K. Gamwo (1992), working with Dr. Robert Lyczkowski, applied the code to tube erosion.

Several chapters in this book and particularly the appendices on problem formulation were used in our freshman graduate course in transport phenomena over the last 10 years.

Earlier versions of the kinetic theory chapters were used in a course in fluidization. Clarifications, in the form of tables, were added as a result of this teaching experience.

During the last several years the research reported in this book was partially funded by the National Science Foundation and by the U.S. Department of Energy University Coal Research Program with contributions from Exxon and Amoco Corporations. The typing was done by two first-rate secretaries, Eleanor Thompson and our present Administrative Associate, Marilyn McDonald. The author is particularly grateful to the former Academic Press associate editor Vicki Jennings who had the manuscript reviewed, retyped it, and saw it through till the end. The author also thanks other Academic Press staff for editing the manuscript.

Dimitri Gidaspow

NOMENCLATURE

A	Surface area, m^2
A_k	Constant related to phase k pressure
A_κ	Constant, used to evaluate conductivity
$\mathbf{a}$	Area
$\mathbf{A}$	Constant vector, Chapman and Cowling's (1961) notation
a	Cross-sectional area
a_v	Volumetric compressibility, $1/G$
B_k	Constant related to phase k viscosity
B_μ	Constant, used to evaluate viscosity
$\mathbf{B}$	Constant vector, Chapman and Cowling's (1961) notation
b_{ij}	Mobilities, related to Onsager friction coefficients
$\mathbf{C}$	Peculiar particle velocity = $\mathbf{c} - \mathbf{v}$
C	Propagation velocity in Chapter 6
C_D	Drag coefficient
C_s	Critical or sonic velocity
C_v	Granular specific heat
C_σ	Specific heat at a constant stress
$\mathbf{c}$	Instantaneous particle velocity
$\mathbf{c}_{12}$	Relative velocity $= \mathbf{c}_1 - \mathbf{c}_2$
D	Diameter

d	Dielectric constant
D	Diffusion coefficient or consolidation coefficient
D_{iss}	Dissipation of energy, J/s
D_o	Orifice diameter
D_t	Tube diameter
d_p	Diameter of particle
E	Electric field strength
EM	Electrophoretic mobility
e	Restitution coefficient
e	Void ratio
e_{yx}	Unit shearing stress in the x–y plane
F	Flux
$\mathbf{f}$	External force per unit mass acting on the particle
$\mathbf{f}_i$	Force acting on phase i
$\dot{\mathbf{f}}_i$	Force per unit mass acting on phase i
f	Frequency distribution function of particle velocities
f_g	Gas wall friction factor
$f^{(0)}$	Maxwellian distribution function
$f^{(2)}$	Pair distribution function
G	Particle–particle modulus $(\partial P_s/\partial \varepsilon_s) = (\partial \sigma/\partial \varepsilon_s)$, $\mathrm{N/m^2}$
$\mathbf{G}$	Center of mass velocity, Eq. (9.31)
g	Acceleration due to gravity
g_0	Radial distribution function
$\mathbf{g}$	Gravitational acceleration
$\tilde{g}$	Buoyancy group, $g(\rho_s - \rho_f)/\rho_s$
H	Height of interface
H_0	Initial slurry height
$\overline{H}$	Dimensionless height of interface
h_i	Enthalpy of phase i per unit mass, J/kg

h_i^n	Enthalpy of phase i at non-equilibrium
h_v	Volumetric heat transfer coefficient, $\mathrm{kW/m^3}$
I	Current
I(A)	Integral of A, Chapman and Cowling's (1961) notation
I(B)	Integral of B, Chapman and Cowling's (1961) notation
$\mathbf{I}$	Unit matrix or unit tensor
J	Bracket integral, Chapman and Cowling's (1961) notation
J_m	Mass flux
J_s	Maximum solids flux
j_{gs}	Drift flux of gas $= \varepsilon\left(v_g - \bar{v}\right)$
K	Effective friction coefficient defined by Eq. (4.11)
k	Thermal conductivity, J/sec-m^2-K
k	Permeability
k_c	Cohesive force per unit area
k_B	Boltzmann constant
$\mathbf{k}$	Unit vector along the line from center of particle 1 to 2
L	Length
ℓ	Mean free path
M	Molecular weight
M_i	Molecular weight of species i, kg/mol
m	Mass of particle
m_i	Mass of phase i, kg
m_i'	Mass of rate of production of phase i, $\mathrm{kg/m^3}$total-s
m_{is}'	Specific rate of production of phase i, $\mathrm{kg/m^3}$"i"-s
m_k'	Rate of phase k production, $\mathrm{kg/s\text{-}m^3}$
N_c	Source like contribution defined by Eq. (9.192)
N_{12}	Number of binary collisions per unit time per unit volume
n	Number of particles per unit volume
n_i	Number of particles of type i per unit volume

$\mathbf{n}$	Normal, outward drawn
P	Pressure
P_c	Collisional pressure like contribution defined by Eq. (9.193)
P_k	Pressure of phase k
P_i	Pressure of phase i, Pa
$\mathbf{P}$	Particle stress or pressure tensor
$\mathbf{P}_c$	Collisional stress tensor
$\mathbf{P}_k$	Kinetic stress tensor
$\mathbf{p}_i$	Momentum supply or phase interaction, N/m^3
Q	Quantity like mass, momentum or energy
Q_i	Volumetric flow rate of phase i
Q_i	Heat input into phase i, J
q	Charge, electrical
q_i	Net rate of heat outflow out of phase i, kW/m^3
$\mathbf{q}$	Conduction like flux vector of fluctuation energy
R	Ideal gas law constant, J/mol-K
Re	Reynolds number
r	Radius of particle or radial coordinate
r_B	Bubble radius
r_i	Rate of reaction of component i, mol/m^3-s
$\mathbf{r}$	Position vector
$\mathbf{r}_c$	Center of mass position vector
S	Entropy
S_L	Slip ratio $= v_s/v_g$
S_i	Entropy of phase i, J/K-kg
$\mathbf{S}$	Rate of shear tensor $= \dot{\nabla}^s \mathbf{v}$
T	Temperature
$\mathbf{T}$	Mixture stress tensor
$\mathbf{T}_i$	Stress tensor of phase i

$\mathbf{T}_k$	Stress tensor of phase *k* (traction)
t	Time
$\bar{t}$	Dimensionless time
U	Internal energy
U_g	Superficial gas velocity $= \varepsilon v_g$
U_o	Inlet superficial gas velocity, flow rate divided by pipe area or reference velocity
U_{mg}	Minimum bubbling velocity
U_{mf}	Minimum fluidization velocity
U_i	Internal energy of phase *i* per unit mass, J/kg
U_i'	Energy supply, kW/m^3
u	Velocity component in the *x* direction
u_B	Bubble velocity
u_i	Velocity of phase *i* in the *x* direction, m/s
V	Volume, m^3
V_b	Bulk specific volume, m^3/kg
V_i	Specific volume of phase *i*, m^3/kg
v	Particle velocity
v_r	Relative velocity, gas minus solid
v_t	Terminal velocity
v_y	Velocity component in the *y* direction
$\mathbf{v}_i$	Velocity of phase *i*, m/s
$\mathbf{v}_k$	Velocity of phase *k*
W	Mass flux, kg/m^2-s
W_i	Weight fraction of phase *i*
W_s	Solids flux
x	Coordinate in the direction of flow
x_i	Coordinate in the *i*th direction
x, y, z	Cartesian coordinates

$\mathbf{x}$	Position vector
Y_i	Weight fraction of component i in the gas phase

Greek Letters

α	Volume fraction of dispersed phase (bubble)
β	Fluid-particle friction coefficient, kg/m^3-s
β_A, β_B	Fluid-particle friction coefficient for Model A, Model B, kg/m^3-s
γ	Dissipation due to inelastic collisions, Eq. (10.5)
Γ	Any differentiable function such as U_i
δ	Internal angle of friction
ε	Fluid volume fraction
ε_i	Volume fraction of phase i
ε_k	Volume fraction of phase k
ε_s	Solids volume fraction
ζ	Zeta potential
ζ_s	Solids vorticity
θ	Solids loading
θ_i	Granular temperature for phase $i = \frac{1}{3} m_i < \mathbf{C}_i^2 >$
Θ	Granular temperature $= \frac{1}{3} < \mathbf{C}^2 >$
κ	Conductivity of granular temperature
λ	Characteristic velocity, eigenvalue
λ_k	Bulk viscosity of phase k
μ	Viscosity
μ_k	Viscosity of phase k
μ_{sc}	Viscosity coefficient of solid phase, $\mu_s = \varepsilon_s \mu_{sc}$
μ_w	Coefficient of friction in Eq. (2.3)
ξ	Bulk viscosity
ρ	Density
ρ_b	Bulk density$= \varepsilon_s \rho_s$, kg/m^3

ρ_i	Density of phase i, kg/m^3
ρ_k	Density of phase k
ρ_m	Density of the mixture
σ	Normal solids stress
σ^2	Variance
σ_i	Rate of entropy production in phase i, J/K-s-m^3 i
σ_r	Normal stress in the r direction
σ_s	Applied stress
σ_t	Total normal stress, fluid plus solids pressure
τ	Shear
ϕ	Sphericity
ϕ_E	Electric potential
Φ	Potential function
χ	Radial distribution function of Chapman and Cowling (1961)
ψ	Single particle quantity, such as mass, momentum, or energy
Ψ	Stream function

Operators and Identities

Traceless Tensor

$$\dot{\mathbf{W}} = \mathbf{W} - \tfrac{1}{3}\left(W_{xx} + W_{yy} + W_{zz}\right)\mathbf{I}; \text{ where } I \text{ is the unit tensor}$$

Rate of Shear Tensor

$$\mathbf{S} = \dot{\nabla}^s \mathbf{v} = \tfrac{1}{2}\left(\nabla \mathbf{v} + \nabla^T \mathbf{v}\right) - \tfrac{1}{3}(\nabla \mathbf{v}:\mathbf{I})\mathbf{I}$$

Convective Differentiation Moving with Phase i

$$\frac{d}{dt}^{i} = \frac{\partial}{\partial t} + \mathbf{v}_i \nabla$$

Substantial Derivative

$$\frac{D}{Dt} = \frac{\partial}{\partial t} + \mathbf{v}\nabla$$

Chapman and Cowling's Substantial Derivative

$$Df = \frac{\partial f}{\partial t} + \mathbf{c}\frac{\partial f}{\partial \mathbf{r}} + \mathbf{F}\frac{\partial f}{\partial \mathbf{c}}$$

Symmetric Tensor

$$\mathbf{W}^s = \tfrac{1}{2}\left(\mathbf{W}^T + \mathbf{W}\right)$$

where $\mathbf{W}^T$ is the transpose of $\mathbf{W}$ obtained by interchanging rows and columns of $\mathbf{W}$

∇	Gradient
$\nabla\cdot$	Divergence
∇^T	Transpose of the gradient
∇^S	Symmetric gradient $= \tfrac{1}{2}\left(\nabla + \nabla^T\right)$
$\nabla^S\mathbf{v}$	Deformation tensor $= \tfrac{1}{2}\left(\frac{\partial v_i}{\partial x_j} + \frac{\partial v_j}{\partial x_i}\right)$
$\nabla\mathbf{v}$	$\tfrac{1}{2}\left(\nabla\mathbf{v} + \nabla^T\mathbf{v}\right) + \tfrac{1}{2}\left(\nabla\mathbf{v} - \nabla^T\mathbf{v}\right)$

Identity

$$\nabla\nabla^S\mathbf{v} = \tfrac{1}{2}\nabla^2\mathbf{v} + \tfrac{1}{2}\nabla\nabla\cdot\mathbf{v}$$

Subscripts

A, B, C	Refers to Model A, B, or C in Table 2.1
g	Gas
gF	Fanning's friction factor for gas
s	Solid
sF	Fanning's friction factor for solids phase
sM	Well mixed solid
wg	Wall with gas
ws	Wall with solid

1

TRANSPORT EQUATIONS

1.1 Basic Approach 1
1.2 Mass Balances 2
1.3 Momentum Balances 3
1.4 Energy Balances 7
1.4.1 Enthalpy Representation 9
1.4.2 Entropy Representation 9
1.5 Entropy Balance 10
1.6 Mixture Equations 11
1.6.1 Mixture Conservation of Mass 14
1.6.2 Mixture Identities 15
1.6.3 Mixture Momentum 16
1.6.4 Mixture Entropy Restriction 18
1.7 Multicomponent Multiphase Flow 20
Exercises 21
1. Concept of Volume Averaging 21
2. Conservation of Total Energy for Phase *i* 22
3. Energy Equation Using Bulk Density State Equation 23
4. Population Balance for a Fluidized Bed 23
5. Population Balance for a Fluidized Bed with Changing Particle Size and Density 24
6. Population Balance for a Riser 25
7. Local Population Balance for a Fluidized Bed 26
8. Balances in Cylindrical and Spherical Coordinates 27
Literature Cited 28

1.1 Basic Approach

It is assumed that the system on which balances are made consists of a sufficient number of particles so that discontinuities can be smoothed out; therefore, derivatives of various properties exist and are continuous, unless otherwise specified. Thus, for a property per unit volume ψ, the Reynolds transport theorem is used. For a volume V that may change with time t, there is, for a system bounded by a closed surface, the following mathematical identity (Aris, 1962, and Appendices C, D, and E):

$$\frac{d^i}{dt} \iiint_{V(t)} \psi \, dV = \iiint_{V(t)} \left(\frac{\partial \psi}{\partial t} + \nabla \cdot \psi \mathbf{v}_i \right) dV. \tag{1.1}$$

In Eq. (1) the system moves with velocity $\mathbf{v}_i$. Hence, differentiation with respect to time carries the superscript i to emphasize this fact.

In the derivation of the multicomponent mass balances for single phase flow, ψ is the partial density of the component i. The volume V is the same for all species. Equating the rate of change of the mass of species i to its rate of production by chemical reaction and performing the usual mathematical operations gives the continuity equations for species i.

In multiphase flow the volume occupied by phase i cannot be occupied by the remaining phases at the same position in space at the same time. This distinction introduced the concept of the volume fraction of phase i. It is denoted by both α_i and ε_i in two-phase flow literature: the ε_i notation is preferred by chemical engineers who have long used this concept. Consistent with common usage, the volume of phase i, V_i in a system of volume V is (Gidaspow, 1977)

$$V_i = \iiint_{V(t)} \varepsilon_i \, dV \quad \text{where} \quad \sum_{i=1}^{n} \varepsilon_i = 1. \tag{1.2}$$

Note that if the fluid is incompressible and there are no phase changes, V_i remains a constant. Then, application of the Reynolds transport theorem given by Eq. (1.1) yields the incompressible continuity equation in multiphase flow found in Eq. (1.3):

$$\frac{\partial \varepsilon_i}{\partial t} + \nabla \cdot \varepsilon_i \mathbf{v}_i = 0, \tag{1.3}$$

$$i = 1, 2, \ldots, n \text{ phases.}$$

1.2 Mass Balances

The mass of the fluid i can be given in terms of its density ρ_i in a multiphase system of volume V as

$$m_i = \iiint_{V(t)} \varepsilon_i \rho_i \, dV. \tag{1.4}$$

The Lagrangian mass balance on mass m_i moving with the velocity $\mathbf{v}_i$ is

$$\frac{d_i m_i}{dt} = \frac{d_i}{dt} \iiint \varepsilon_i \rho_i \, dV = \iiint m_i' \, dV, \tag{1.5}$$

where Eq. (1.5) defines the volumetric source of mass m_i'. An application of the Reynolds transport theorem and the usual contradiction argument applied to an arbitrary element of volume gives the well-known continuity equation for phase i:

$$\frac{\partial(\varepsilon_i \rho_i)}{\partial t} + \nabla \cdot (\varepsilon_i \rho_i \mathbf{v}_i) = m_i'. \tag{1.6}$$

In Eq. (1.6) the source term m_i' includes the volume fraction ε_i. It is better (Gidaspow, 1972) to define it in terms of a source per unit of volume of phase i, so that $m_i' = \varepsilon_i m_{is}'$, where m_{is}' is the specific rate of production of phase i by, say, chemical reaction of the solid with the fluid. For solid–gas reaction applications see Gidaspow (1972), and for phase transformations in storage batteries see Gidaspow and Baker (1973). Conservation of mass requires that the sum of m_i' over i phases be zero, that is,

$$\sum_{i=1}^{n} m_i' = 0. \tag{1.7}$$

Furthermore, m_i' must be invariant under a change of reference frames, known as Galileo's relativity principle or material objectivity in rational mechanics (for example, see Slattery, 1972). This implies that for two phase flow m_i' and a similar momentum source term, such as interface friction, must be a function of the relative phase velocity only, rather than a function of individual phase velocities.

1.3 Momentum Balances

The derivation of the momentum balances follows the approach of Bowen (1976) for a multicomponent mixture. The rate of change of momentum of the "particle" or system moving with the velocity v_i equals the sum of the forces acting on the system. In single phase flow (Slattery, 1972), external and contact forces are identified. In multiphase flow, similar to the multicomponent flow, there exists the force of interaction of phase i with the other phases. In rational mechanics this force is known as the momentum supply $\mathbf{p}_i$ (Bowen, 1976). There is also a force due to the addition of mass to a phase i. Similar to the multicomponent approach, it is assumed that the force is added to phase i at its velocity.

Mathematically these statements can be written as follows. The momentum balance for phase i is

$$\frac{d}{dt}\iiint_{V(t)} \rho_i \mathbf{v}_i \varepsilon_i \, dV = \mathbf{f}_i,$$

rate of change of momentum of phase i = forces acting on phase i
for $i = 1, 2, 3, \ldots n$ phases, where the force $\mathbf{f}_i$ is given by

$$\mathbf{f}_i = \underbrace{\oiint_{S(t)} \mathbf{T}_i \, da}_{\text{surface forces}} + \underbrace{\iiint_{V(t)} \rho_i \mathbf{f}_i \varepsilon_i \, dV}_{\text{external forces}} + \underbrace{\iiint \mathbf{p}_i \, dV}_{\text{forces of interaction between phases}} + \underbrace{\iiint m_i' \mathbf{v}_i \, dV}_{\text{momentum change due to phase changes}}. \tag{1.8}$$

In the (x, y, z) coordinate system the stress tensor $\mathbf{T}_i$ for phase i is given by the matrix

$$\mathbf{T}_i = \begin{pmatrix} T_{ixx} & T_{ixy} & T_{ixz} \\ T_{iyx} & T_{iyy} & T_{iyz} \\ T_{izx} & T_{izy} & T_{izz} \end{pmatrix}, \tag{1.9}$$

where its typical element T_{iyx} is the ith force in the x direction per unit area of the yth face:

$$T_{iyx} = \frac{\partial F_{ix}}{\partial A_y}. \tag{1.10}$$

See Fig. 1.1.

In thermodynamics of elasticity (Callen, 1960), such a stress component for a single phase is taken to be the gradient of internal energy U per unit volume V_o with respect to displacement:

$$T_{yx} = \frac{1}{V_o}\frac{\partial U}{\partial e_{yx}}, \tag{1.11}$$

where e_{yx} is the unit shearing strain in the x–y plane. An analogous interpretation has to exist for each phase i, since there exists an energy function for each

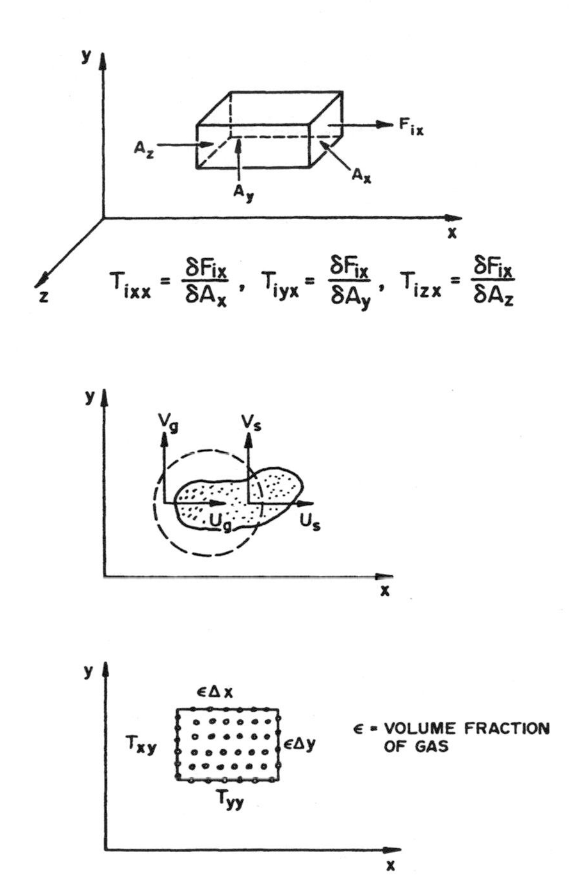

Fig. 1.1 Stresses, velocity components and porosity in multiphase balances.

phase i. In Eq. (1.8) the void fraction is not included separately in the partial stress $\mathbf{T}_i$. The vector surface element $d\mathbf{a} = \mathbf{n}\, dS$, where $\mathbf{n}$ is the outward drawn normal to the surface S.

An application of the divergence theorem gives

$$\oiint_a \mathbf{T}_i \, d\mathbf{a} = \iiint_V \nabla \cdot \mathbf{T}_i \, dV. \tag{1.12}$$

An application of the Reynolds transport theorem to the right side of Eq. (1.8) and the use of Eq. (1.12) in (1.8), followed by a standard contradiction argument, gives the three momentum balances for each phase i as shown in Eq. (1.13):

$$\frac{\partial(\rho_i\varepsilon_i\mathbf{v}_i)}{\partial t}+\nabla\cdot(\rho_i\varepsilon_i\mathbf{v}_i\otimes\mathbf{v}_i)=\nabla\cdot\mathbf{T}_i+\rho_i\varepsilon_i\mathbf{f}_i+\mathbf{p}_i+m_i'\mathbf{v}_i, \tag{1.13}$$

where ⊗ denotes a dyadic multiplication of the (3×1) velocity vector.by the (1×3) velocity vector $\mathbf{v}_i$ to give the 3×3 matrix. Operate with the divergence $\nabla\cdot=(\partial/\partial x,\partial/\partial y,\partial/\partial z)$ upon it to give the three-dimensional vector.

By differentiating the products on the left side of (1.13), the result can be written as

$$\rho_i\varepsilon_i\frac{\partial\mathbf{v}_i}{\partial t}+\rho_i\varepsilon_i\mathbf{v}_i\cdot\nabla\mathbf{v}_i+\mathbf{v}_i\left(\frac{\partial(\rho_i\varepsilon_i)}{\partial t}+\nabla\cdot\rho_i\varepsilon_i\mathbf{v}_i-m_i'\right)=\nabla\cdot\mathbf{T}_i+\rho_i\varepsilon_i\mathbf{f}_i+\mathbf{p}_i. \tag{1.14}$$

If the continuity equation (1.6) is used, the term in the brackets is zero. The acceleration in Eq. (1.14) in the x direction can be written as

$$\frac{du_i}{dt^i}=\frac{\partial u_i}{\partial t}+\mathbf{v}_i\cdot\nabla u_i=\frac{\partial u_i}{\partial t}+u_i\frac{\partial u_i}{\partial x}+v_i\frac{\partial u_i}{\partial y}+w_i\frac{\partial u_i}{\partial z},$$

where (1.15)

$$\mathbf{v}_i=(u_i,v_i,w_i).$$

Therefore, the momentum balance for phase i becomes

$$\rho_i\varepsilon_i\frac{d\mathbf{v}_i}{dt^i}=\nabla\cdot\mathbf{T}_i+\rho_i\varepsilon_i\mathbf{f}_i+\mathbf{p}_i \tag{1.16}$$

mass/vol.(acceleration of phase i) = momentum in-flow due to surface forces + body forces + interaction forces

Division of Eq. (1.16) by the void fraction ε_i shows it to be essentially identical to the momentum balance for the multicomponent mixture given by Bowen's (1976) Eq. (1.4.4).

To obtain meaningful balances, one needs to specify the interaction forces $\mathbf{p}_i$ and constitutive expressions for the stresses $\mathbf{T}_i$. The interaction forces consist of drag between the phases, of added mass forces that depend upon velocity gradients and of other interaction forces, such as collisional forces. The sum of the interaction forces $\mathbf{p}_i$ is clearly zero, that is,

$$\sum_{i=1}^{n\text{ phases}}\mathbf{p}_i=0. \tag{1.17}$$

The simplest expression for the stress, analogous to the single phase potential flow theory, is through the definition of a phase pressure p_i via the identity matrix **I**:

$$\mathbf{T}_i = -P_i\mathbf{I}. \tag{1.18}$$

Equation (1.18) is similar to the expression for partial stress given by Bowen's (1976) Eq. (2.1.1). For a mixture of fluids with a linear viscosity, he derived an expression for the stress involving a matrix of viscosities of individual components. An expression similar to his expression (2.1.3) should hold for a multiphase mixture and thus serve as a generalization of the Navier–Stokes equation for a multiphase mixture.

1.4 Energy Balances

In single phase fluid mechanics and in multicomponent mechanics (Bowen, 1976) it is customary to write an energy balance for the sum of internal and kinetic energies. Indeed, a balance similar to that given by Bowen's (1976) Eq. (1.5.1) can be written for the multiphase situation considered here. Such a balance would then be combined with the equation of motion dotted with the velocity vector for phase i. However, in multiphase flow some unexpected work terms arise (Gidaspow, 1977). In view of such complexities, a simpler approach is taken here, one presented previously by Gidaspow (1977) and used in elementary thermodynamics.

Consider an open system of mass m_i which gains mass and thus energy at a rate dm_i / dt^i. The energy balance moving with phase i then becomes

$$\frac{dU_i'}{dt^i} = \frac{dQ_i}{dt} - P_i\frac{dV_i}{dt^i} + D_{iss} + \left(U_i^n + P_i^n / \rho_i^n\right)\frac{dm_i}{dt^i},$$

where (1.19)

$$U_i' = \iiint\limits_{V(t)} \varepsilon_i \rho_i U_i \, dV,$$

and where the rate of heat transfer is related to the flux $\mathbf{q}_i$ by relations such as those used by Ishii (1975),

$$-\frac{dQ_i}{dt} = \oiint\limits_{A(t)} \mathbf{q}_i \varepsilon_i \, d\mathbf{a} = \iiint\limits_{V(t)} \left(\nabla \cdot \varepsilon_i \mathbf{q}_i\right) dV, \tag{1.20}$$

where $A(t)$ is the area enclosing the volume of the system at any instant of time. The differential element of surface area of system i was taken to be simply $\varepsilon_i\, da$, thus making no distinction between area and volume fractions. Heat transfer between phases may also be given in terms of heat transfer coefficients. These are often expressed per unit volume because of our lack of knowledge of instantaneous surface areas. Thus, surface areas need not be included in the formulation. For a one-dimensional system,

$$-\frac{dQ_i}{dt^i} = h_v\left(T_i - T_j\right)\Delta V, \tag{1.21}$$

where h_i is the volumetric heat transfer coefficient per unit volume ΔV, defined positive when the temperature of phase i is greater than the temperature of phase j.

In Eq. (1.19) only mechanical work done by phase i having its own pressure P_i was included. For instance, if the fluid does electrical work, an appropriate term for this work, such as voltage times current, must be included. The energy dissipation by means of friction is the dissipation term, D_{iss} in Eq. (1.19), which in the continuum is related to the energy supply by means of the redefinition

$$D_{iss} = \iiint\limits_{V(t)} u_i'\, dV. \tag{1.22}$$

The last term in Eq. (1.19) represents the gain of internal energy and work done by addition of mass to the system. The properties carry the superscript n to indicate possible non-equilibrium states. When Eq. (1.19) is applied to one-dimensional systems, it should also contain frictional dissipation of energy due to wall shear and external rate of heat transfer.

Now let's apply the Reynolds transport theorem, Eq. (1.1) , to the system bounded by a closed surface. Also assume existence of appropriate mean values and apply the balance, Eq. (1.19), to a typical element of a larger system. Observe that in transforming the pressure term $\psi = \varepsilon_i$ in the Reynolds transport theorem we obtain

$$\frac{\partial\left(\varepsilon_i \rho_i U_i\right)}{\partial t} + \nabla\cdot\left(\varepsilon_i \rho_i \mathbf{v}_i U_i\right) = -\nabla\cdot\varepsilon_i \mathbf{q}_i - P_i\frac{\partial \varepsilon_i}{\partial t} - P_i\,\nabla\cdot\varepsilon_i \mathbf{v}_i + h_i^n m_i' + u_i, \tag{1.23}$$

where h_i^n is defined as the enthalpy entering system i at possibly non-equilibrium conditions.

1.4.1 Enthalpy Representation

A strange term involving work of expansion of void fraction $P_i \partial\varepsilon_i/\partial t$ appeared in Eq. (1.23). If it is deleted, the energy equation will not satisfy the Clausius–Duhem inequality. However, before proceeding with the proof, let us rewrite the energy equation in terms of enthalpy to compare it to that given by Ishii (1975) and Boure *et al.* (1973). Since $h_i = U_i + P_i/\rho_i$,

$$\frac{dh_i}{dt^i} = \frac{dU_i}{dt^i} + P_i \frac{d\rho_i^{-1}}{dt^i} + \rho_i^{-1} \frac{dP_i}{dt^i}. \tag{1.24}$$

Using the phase i continuity equation in the form

$$\rho_i \frac{d\rho_i^{-1}}{dt^i} = \nabla \cdot \mathbf{v}_i + \frac{1}{\varepsilon_i}\frac{d\varepsilon_i}{dt^i} - m_i'/\rho_i\varepsilon_i, \tag{1.25}$$

we multiply both sides of Eq. (1.24) by $\varepsilon_i\rho_i$ and substitute Eq. (1.25) to give

$$\varepsilon_i\rho_i \frac{dh_i}{dt^i} = \varepsilon_i\rho_i \frac{dU_i}{dt^i} + P_i \nabla \cdot \varepsilon_i \mathbf{v}_i + P_i \frac{\partial\varepsilon_i}{\partial t} - m_i' P_i / \rho_i + \varepsilon_i \frac{dP_i}{dt^i}. \tag{1.26}$$

The first three terms on the right-hand side of Eq. (1.26) involve the energy and the work terms in the convective form of the internal energy equation. The energy equation (1.23) is now written in convective form. The phase i continuity equation is used. Equation (1.26) is used to express the enthalpy in terms of internal energy. The result is an energy equation in terms of enthalpy,

$$\varepsilon_i\rho_i \frac{dh_i}{dt^i} = -\nabla \cdot \varepsilon_i \mathbf{q}_i + \varepsilon_i \frac{dP_i}{dt^i} + u_i + m_i'\left(h_i^n - h_i\right). \tag{1.27}$$

Equation (1.27) is essentially Eq. IX–1.30, found on page 151 in Ishii (1975). His equation contains a contribution due to a difference of interphase pressure and phase i pressure and viscous dissipations which are included in U_i'. Equation (1.27) is similar to the one-component single-phase energy balance. Thus, working backwards one sees that the term $P_i \partial\varepsilon_i/\partial t$ had to appear in Eq. (1.23).

1.4.2 Entropy Representation

The entropy form of the energy equation can be obtained by using the fact that the internal energy of phase i depends upon the entropy of phase i and upon the specific volume of phase i for a one component system, that is,

$$U_i = U_i\left(S_i, \rho_i^{-1}\right). \tag{1.28}$$

Using the definitions of temperature and pressure, convective differentiation of Eq. (1.28) gives

$$\frac{dU_i}{dt^i} = T_i \frac{dS_i}{dt^i} - P\frac{d\rho_i^{-1}}{dt^i}. \tag{1.29}$$

Substitution of relation (1.29) multiplied by $\varepsilon_i\rho_i$ into the convective form of the internal energy equation (1.23) using (1.25) gives an energy equation in terms of entropy,

$$\varepsilon_i\rho_i T_i \frac{dS_i}{dt^i} = -\nabla\cdot\varepsilon_i\mathbf{q}_i + \left(h_i^n - h_i\right)m_i' + u_i. \tag{1.30}$$

The terms on the right-hand side of Eq. (1.30) are all dissipations for an adiabatic process, $q_i = 0$. Equation (1.30) shows that for a reversible adiabatic process the phase i entropy stays constant when moved with phase velocity $\mathbf{v}_i$. The energy equation in the form (1.30) was used by Gidaspow *et al.* (1973) and Lyczkowski *et al.* (1978). It is very similar to the energy equation given by Kalinin (1970).

The entropy equation can be written in conservative form using the continuity equation as

$$T_i\frac{\partial\left(\varepsilon_i\rho_i S_i\right)}{\partial t} + T_i\nabla\cdot\left(\varepsilon_i\rho_i\mathbf{v}_i S_i\right) = -\nabla\cdot\varepsilon_i\mathbf{q}_i + \left(h_i^n - h_i - T_i S_i\right)m_i' + u_i. \tag{1.31}$$

The only change is the contribution due to mass transfer on the right-hand side of Eq. (1.31).

1.5 Entropy Balance

Existence of the entropy S_i for phase i implies that a balance can be made for it. We choose again the system of mass m_i moving with velocity $\mathbf{v}_i$. If σ_i is the rate of production of entropy in phase i per unit volume of the system of volume V and $\mathbf{q}_i$ is the net rate of heat out flow per unit area out of phase i, then the balance becomes

$$\frac{d}{dt}\iiint_{V(t)} \varepsilon_i \rho_i S_i \, dV + \oiint_{A(t)} \frac{\mathbf{q}_i \varepsilon_i \, d\mathbf{a}}{T_i} - \iiint m_i' S_i \, dV = \iiint \sigma_i \, dV. \tag{1.32}$$

Accumulation of entropy in the system moving with velocity $\mathbf{v}_i$	Rate of entropy outflow due to flow of heat across area A	Rate of entropy flow at equilibrium due to phase change	Rate of entropy production in phase i

An application of the Reynolds transport theorem to the first integral in Eq. (1.32), the divergence theorem to the second integral, and the usual contradiction argument lead to the entropy balance given in Eq. (1.33).

$$\frac{\partial(\varepsilon_i \rho_i S_i)}{\partial t} + \nabla \cdot \varepsilon_i \rho_i S_i \mathbf{v}_i + \nabla \cdot \frac{\mathbf{q}_i \varepsilon_i}{T_i} - m_i' S_i = \sigma_i \tag{1.33}$$

For reversible processes (Prigogine, 1955) σ_i is zero. For irreversible processes σ_i is greater than zero.

Assuming the phase change will occur at equilibrium, that is, neglecting the difference between h_i^n and h_i in Eq. (1.31), a comparison of the entropy balance given by Eq. (1.33) with the energy balance given by Eq. (1.31) shows that σ_i is

$$\sigma_i = -\frac{1}{T_i}\nabla \cdot \varepsilon_i \mathbf{q}_i + \nabla \cdot \frac{\mathbf{q}_i \varepsilon_i}{T_i} + \frac{u_i}{T_i} = -\frac{\mathbf{q}_i \varepsilon_i \cdot \nabla T_i}{T_i^2} + \frac{u_i}{T_i} \geq 0. \tag{1.34}$$

If we use the Fourier's law for heat conduction for phase i given by

$$\mathbf{q}_i = -k_i \nabla T_i, \tag{1.35}$$

the local rate of entropy production becomes

$$\sigma_i = \frac{\varepsilon_i k_i (\nabla T_i)^2}{T_i^2} + \frac{u_i}{T_i}, \tag{1.36}$$

which restricts the sign of the thermal conductivity for phase i to be positive. Correlations for such a conductivity are given in the literature.

1.6 Mixture Equations

The mixture equations for mass, momentum, energy, and entropy are simply the sum over the phases of Eqs. (1.6), (1.13), (1.23), and (1.33), respectively.

Analogous to the method employed by Bowen (1976) for multicomponent mixtures, these equations must reduce themselves to the form of the single-phase equations through the vanishing of phase interactions and natural definitions.

Before proceeding, it is useful to review some mixture properties from thermodynamics. Extensive properties, such as volume, energy, and entropy are additive. Thus, the mixture volume V_t is the sum of the individual phase volumes:

$$V_t = \sum_{i=1}^{n} V_{t,i} . \tag{1.37}$$

In terms of the specific volumes V_i and masses, m_i this relation reads

$$mV = \sum_{i=1}^{n} m_i V_i . \tag{1.38}$$

Let W_i be the mass fraction of phase i,

$$W_i = m_i / m. \tag{1.39}$$

Then Eqs. (1.38) and (1.39) show that

$$V = \sum_{i=1}^{n} W_i V_i . \tag{1.40}$$

Similarly, the mixture energy and entropy are

$$U = \sum_{i=1}^{n} W_i U_i , \tag{1.41}$$

$$S = \sum_{i=1}^{n} W_i S_i . \tag{1.42}$$

If, for each phase i,

$$U_i = U_i (S_i , V_i), \tag{1.43}$$

the definitions of temperature and pressure for phase i give

$$dU_i = T_i \, dS_i - P_i \, dV_i . \tag{1.44}$$

Multiplication of Eq. (1.44) by W_i and summation gives the mixture equation for equal phase temperatures and pressures,

$$dU = T\,dS - P\,dV. \tag{1.45}$$

Thus, for isentropic process of the mixture,

$$dU = -P\,dV,$$

where V is now the mixture specific volume. A mean specific heat at a constant volume can be defined by considering Eq. (1.41) at a constant composition and volume:

$$dU = \sum_{i=1}^{n} W_i \left(\frac{\partial U_i}{\partial T}\right)_{V_i} dT = \sum_{i=1}^{n} C_{V_i} W_i\,dT. \tag{1.46}$$

We let the mean specific heat C_v be

$$C_v = \sum_{i=1}^{n} C_{V_i} W_i\,. \tag{1.47}$$

Similarly, a specific heat at a constant pressure is defined. As a useful example, consider isentropic expansion or compression of a gas–solid mixture. The internal energy of the solid is assumed to be a function of temperature only and the gas is assumed to be ideal. Then the entropy change is

$$dS = W_g\left[\left(\frac{\partial S_g}{\partial T}\right)_P dT + \left(\frac{\partial S_g}{\partial P}\right) dP\right] + \left(1 - W_g\right)\frac{\partial S_s}{\partial T} dT = 0, \tag{1.48}$$

where the subscripts g and s refer to the solid and gas, respectively. Using a Maxwell relation $\partial Sg/\partial P = -\partial V/\partial T$ and the ideal gas $PV = RT/M$, where M is the molecular weight of the gas, Eq. (1.48) becomes

$$\left[W_g C_{pg} + \left(1 - W_g\right)C_{ps}\right]\frac{dT}{T} - \frac{W_g R}{M}\frac{dP}{P} = 0. \tag{1.49}$$

Use of the mean specific heat C_p,

$$C_p = W_g C_{pg} + \left(1 - W_g\right)C_{ps}, \tag{1.50}$$

and integration gives

$$\frac{T}{T_o} = \left(\frac{P}{P_o}\right)^{\frac{W_g R}{C_p M}} . \tag{1.51}$$

Equation (1.51) shows that the equation reduces itself to that of the isentropic expansion of gas only, for W_g of one, while it gives a near zero increase in temperature upon expansion of a concentrated mixture of a solid. This equation is useful for estimating compression work and critical flow of mixtures.

With this review of thermodynamics, we can proceed to the mixture equations.

1.6.1 Mixture Conservation of Mass

The sum of Eq. (1.6) gives

$$\sum_{i=1}^{n} \frac{\partial(\varepsilon_i \rho_i)}{\partial t} + \sum_{i=1}^{n} \nabla \cdot (\varepsilon_i \rho_i \mathbf{v}_i) = \sum_{i=1}^{n} m_i' . \tag{1.52}$$

For this equation to reduce to the single phase form, it is natural to define the mixture density ρ in the usual way in two phase flow, as

$$\rho = \sum_{i=1}^{n \text{ phases}} \varepsilon_i \rho_i . \tag{1.53}$$

Note that the weight fraction W_i is given by

$$W_i = (\varepsilon_i \rho_i)/\rho . \tag{1.54}$$

The second term in Eq. (1.52) shows that it is natural to define the weighted average velocity $\mathbf{v}$ as

$$\rho \mathbf{v} = \sum_{i=1}^{n} \varepsilon_i \rho_i \mathbf{v}_i . \tag{1.55}$$

Equation (1.54) shows that this velocity is weighted averaged with the weight fractions.

Since $\sum_{i=1}^{n} m_i' = 0$ by conservation of mass, the mixture equation looks like the single phase equation

$$\frac{\partial \rho}{\partial t} + \nabla \cdot \rho \mathbf{v} = 0. \tag{1.56}$$

1.6.2 Mixture Identities

To relate the momentum equations and the energy equations to those for a single phase, it is necessary to derive some identities analogous to those given by Bowen (1976) for a multicomponent mixture. Define the relative velocity with respect to the weighted average velocity as

$$\mathbf{v}_{i,\mathrm{rel}} = \mathbf{v}_i - \mathbf{v}. \tag{1.57}$$

Then the sum of Eq. (1.57), multiplied by $\varepsilon_i \rho_i$ and the definition of $\mathbf{v}$, Eq. (1.55), shows that

$$\sum_{i=1}^{n} \varepsilon_i \rho_i \mathbf{v}_{i,\mathrm{rel}} = 0. \tag{1.58}$$

Now consider any differentiable function Γ. Motion with phase i gives us

$$\frac{d\Gamma}{dt^i} = \frac{d\Gamma(t, \mathbf{x}_i(t))}{dt} = \frac{d\Gamma}{dt} + \nabla\Gamma \cdot \frac{d\mathbf{x}_i}{dt} = \frac{\partial \Gamma}{\partial t} + \nabla\Gamma \cdot \mathbf{v}_i \tag{1.59}$$

by chain rule and definition of $\mathbf{v}_i$, while motion with the mixture gives

$$\frac{d\Gamma}{dt} = \frac{d\Gamma(t, \mathbf{x}_i(t))}{dt} = \frac{d\Gamma}{dt} + \nabla\Gamma \cdot \frac{d\mathbf{x}}{dt} = \frac{\partial \Gamma}{\partial t} + \nabla\Gamma \cdot \mathbf{v}. \tag{1.60}$$

We saw previously that properties such as mixture entropy and energy, are defined by

$$\Gamma \rho = \sum_{i=1}^{n} \varepsilon_i \rho_i \Gamma_i, \tag{1.61}$$

where $\varepsilon_i \rho_i / \rho$ was the weight fraction of i. We like to relate these properties moving with phase i to those of the mixture moving with the weighted average velocity $\mathbf{v}$.

The difference between Eqs. (1.59) and (1.60) gives

$$\frac{d\Gamma}{dt^i} - \frac{d\Gamma}{dt} = \nabla\Gamma \cdot \mathbf{v}_{i,\mathrm{rel}}. \tag{1.62}$$

Differentiation of Eq. (1.61) moving with $\mathbf{v}$ and use of Eq. (1.62) gives

$$\rho\frac{d\Gamma}{dt} + \Gamma\frac{d\rho}{dt} = \sum_{i=1}^{n} \varepsilon_i\rho_i\left(\frac{d\Gamma_i}{dt^i} - \nabla\Gamma_i \cdot \mathbf{v}_{i,\mathrm{rel}}\right) + \Gamma_i\left(\frac{d(\varepsilon_i\rho_i)}{dt^i} - \nabla\varepsilon_i\rho_i \cdot \mathbf{v}_{i,\mathrm{rel}}\right), \tag{1.63}$$

where first let Γ be Γ_i and then $\varepsilon_i\rho_i$. But the continuity equation (1.56) can be written as

$$\frac{1}{\rho}\frac{d\rho}{dt} = -\nabla \cdot \mathbf{v}, \tag{1.64}$$

while the conservation of mass of phase i can be rearranged to read in terms of a partial density $\varepsilon_i\rho_i$ as

$$\frac{1}{\varepsilon_i\rho_i}\frac{d(\varepsilon_i\rho_i)}{dt^i} = -\nabla \cdot \mathbf{v}_i + \frac{m_i'}{\varepsilon_i\rho_i}. \tag{1.65}$$

Substitution of (1.64) and (1.65) into (1.63) and rearrangement gives the mixture identity

$$\rho\frac{d\Gamma}{dt} = \sum_{i=1}^{n} \varepsilon_i\rho_i\frac{d\Gamma_i}{dt^i} - \nabla \cdot \varepsilon_i\rho_i\Gamma_i\mathbf{v}_{i,\mathrm{rel}} + \Gamma_i m_i', \tag{1.66}$$

analogous to Bowen's Eq. (1.2.17).

1.6.3 Mixture Momentum

The sum of momentum balances given by Eq. (1.16) over n phases gives

$$\sum_{i=1}^{n} \rho_i\varepsilon_i\frac{d\mathbf{v}_i}{dt^i} = \sum_{i=1}^{n} \nabla \cdot \mathbf{T}_i + \sum_{i=1}^{n} \rho_i\varepsilon_i\mathbf{f}_i + \sum_{i=1}^{n} \mathbf{p}_i. \tag{1.67}$$

To relate the motion along the phases to those moving with the mixture, let Γ be $\mathbf{v}$ in the identity (1.66) and use Eq. (1.57) to obtain

$$\rho\frac{d\mathbf{v}}{dt} = \sum_{i=1}^{n} \varepsilon_i\rho_i \frac{d\mathbf{v}_i}{dt^i} - \nabla\varepsilon_i\rho_i\mathbf{v}_{i,\mathrm{rel}} \otimes \mathbf{v}_{i,\mathrm{rel}} + \mathbf{v}_i m_i' . \tag{1.68}$$

Use of the identity (1.68) in Eq. (1.67) and the natural definitions similar to those used in multicomponent theory (Bowen, 1976) gives the mixture momentum balance in convective form, expressed by Eq. (1.69):

$$\rho\frac{d\mathbf{v}}{dt} + \sum_{i=1}^{n} \nabla\varepsilon_i\rho_i\mathbf{v}_{i,\mathrm{rel}} \otimes \mathbf{v}_{i,\mathrm{rel}} = \nabla\cdot\mathbf{T} + \rho\mathbf{f}, \tag{1.69}$$

where $\mathbf{f}$ is defined by

$$\rho\mathbf{f} = \sum_{i=1}^{n} \varepsilon_i\rho_i\mathbf{f}_i . \tag{1.70}$$

The mixture stress tensor $\mathbf{T}$, corresponding to the inner part of the stress in multicomponent systems, was defined by

$$\mathbf{T} = \sum_{i=1}^{n} \mathbf{T}_i . \tag{1.71}$$

The interactions between the phases had to disappear as shown in Eq. (1.72):

$$\sum_{i=1}^{n} \mathbf{p}_i + m_i'\mathbf{v}_i = 0. \tag{1.72}$$

The mixture equation (1.69) shows that the momentum balance moving along the weighted average velocity contains contributions due to the relative phase velocities given by the second term on the left-hand side of Eq. (1.69). In multicomponent theory (Bowen, 1976) these are grouped into a mixture stress tensor, a procedure that serves no purpose in this case.

The case of hydrostatics is of importance. Defining the mixture pressure P by

$$\mathbf{T} = -P\mathbf{I}, \tag{1.73}$$

we obtain the very useful manometer formula for mixtures from Eq. (1.69):

$$\nabla P = \mathbf{g}\sum_{i=1}^{n} \varepsilon_i \rho_i , \tag{1.74}$$

where the body force $\mathbf{f}_i$ was replaced by the gravitational acceleration g.

1.6.4 Mixture Entropy Restriction

The mixture energy equation can be obtained as the sum of Eqs. (1.23) and written in various ways using identity (1.66). It can also be expressed in terms of enthalpy and entropy. Since these operations are straightforward, only the restriction provided by the entropy inequality will be dealt with.

To accomplish this, the continuity equation (1.6) is first rewritten in a convenient form. Differentiation of the product in Eq. (1.6) and use of the definition of the motion along phase i, represented by d/dt^i, shows that

$$\rho_i \frac{\partial \varepsilon_i}{\partial t} + \rho_i \nabla \cdot \varepsilon_i \mathbf{v}_i + \varepsilon_i \frac{\partial \rho_i}{\partial t} + \varepsilon_i \mathbf{v}_i \nabla \rho_i = m_i' \tag{1.75}$$

can be written as

$$\frac{\partial \varepsilon_i}{\partial t} + \nabla \cdot \varepsilon_i \mathbf{v}_i = -\frac{\varepsilon_i}{\rho_i} \frac{d\rho_i}{dt^i} + \frac{m_i'}{\rho_i} = \varepsilon_i \rho_i \frac{d\rho_i^{-1}}{dt^i} + \frac{m_i'}{\rho_i}, \tag{1.76}$$

where the left-hand side of Eq. (1.76) is the continuity equation for an incompressible mixture and the right-hand side has been expressed in terms of the specific volume, ρ^{-1}.

For simplicity of the argument consider the energy equation (1.23) with zero $\mathbf{q}_i$, U'_i and m_i'. Then, using the continuity equation (1.6) in its conservative form, Eq. (1.23) can be rewritten in convective form as follows:

$$\varepsilon_i \rho_i \frac{dU_i}{dt^i} = -P_i \frac{\partial \varepsilon_i}{\partial t} - P_i \nabla \cdot \varepsilon_i \mathbf{v}_i . \tag{1.77}$$

For a single component phase i, its energy is a function of its entropy and its specific volume:

$$U_i = U_i\left(S_i, \rho_i^{-1}\right). \tag{1.78}$$

Convective differentiation of Eq. (1.78) moving with phase i gives the usual thermodynamic relation shown by Eq. (1.79), where definitions of temperature and pressure were used:

$$\frac{dU_i}{dt^i} = T_i \frac{dS_i}{dt^i} - P_i \frac{d\rho_i^{-1}}{dt^i}. \tag{1.79}$$

Multiplication of Eq. (1.79) by $\varepsilon_i \rho_i$ and substitution into the energy balance (1.77) gives the expression for the rate of change of entropy:

$$\varepsilon_i \rho_i T_i \frac{dS_i}{dt^i} = \varepsilon_i \rho_i P_i \frac{d\rho_i^{-1}}{dt^i} - P_i \frac{\partial \varepsilon_i}{\partial t} - P_i \nabla \cdot \varepsilon_i \mathbf{v}_i. \tag{1.80}$$

Substitution for $d\rho^{-1} / dt^i$ from the continuity equation (1.76) for zero m_i' gives the expression

$$\varepsilon_i \rho_i T_i \frac{dS_i}{dt^i} = P_i \frac{\partial \varepsilon_i}{\partial t} + P_i \nabla \cdot \varepsilon_i \mathbf{v}_i - P_i \frac{\partial \varepsilon_i}{\partial t} - P_i \nabla \cdot \varepsilon_i \mathbf{v}_i, \tag{1.81}$$

where all the terms on the right-hand side cancel. Thus, in the absence of heat transfer and energy dissipation, the entropy of the ith phase stays a constant along its own stream line.

Now, suppose the work term $P_i(\partial \varepsilon_i / \partial t)$ was not present in the energy balance. Then the entropy balance for phase i, as seen from Eq. (1.81), would read as follows:

$$\varepsilon_i \rho_i T_i \frac{dS_i}{dt^i} = -P_i \frac{\partial \varepsilon_i}{\partial t} \qquad \text{(wrong form).} \tag{1.82}$$

Then the entropy production σ for two fluids for which $\varepsilon_i + \varepsilon_j = 1$ is the sum of Eq. (1.82) written in conservative form to obtain the entropy production σ:

$$\sigma = \left(\frac{P_i}{T_i} - \frac{P_j}{T_j} \right) \frac{\partial \varepsilon_i}{\partial t} \geq 0. \tag{1.83}$$

Now consider expansion of volume fraction of phase i, $\partial \varepsilon_i / \partial t > 0$. Then inequality (1.83) states that $P_i / T_i - P_j / T_j > 0$. For equal phase pressures the temperature of phase j must be larger than of phase i. This is correct if temperature T_j begins larger than T_i. It is clearly false if T_j is sufficiently lower than T_i. Thus, the second law dictates to some extent the proper form of the energy equation.

This discussion is presented here in view of earlier violations of the second law of thermodynamics by some of the energy equations in the literature (Lyczkowski *et al.*, 1982).

1.7 Multicomponent Multiphase Flow

The design of fluidized bed reactors and other contacting devices in which chemical reactions occur involves the solution of multicomponent equations of change with reaction. To apply the present hydrodynamic theory to such situations, multicomponent, multiphase equations must be derived. In principle, one can write conservation of species, momentum, and energy for each component in each phase. Since all components in a given phase are usually at the same temperature and since diffusion can be neglected as a first approximation, we only need to consider species balances in each phase. If we neglect diffusion, then each species in a given phase moves with the phase velocity.

The mass of species i in phase g, say the gas phase, is given by

$$m_{ig} = \iiint_{V(t)} \varepsilon_g \rho_g Y_i \, dV, \tag{1.84}$$

where Y_i is the weight fraction of species i in phase g, ε_g the volume fraction of phase g and ρ_g is its density. This notation and the neglect of molecular diffusion avoids the use of double subscripts in the principal variables. If the species react in phase g according to a molar rate of reaction r_i per unit volume, then the mass balance reads

$$\frac{d}{dt} \iiint_{V(t)} \varepsilon_g \rho_g Y_i \, dV = M_i \iiint r_i \, dV, \tag{1.85}$$

where M_i is the molecular weight of species i and r_i is its molar rate of production per total phase volume. Application of the Reynolds transport theorem and the usual contradiction argument leads to the gas phase conservation of species balance,

$$\frac{\partial}{\partial t}\left(\varepsilon_g \rho_g Y_i\right)+\nabla \cdot\left(\varepsilon_g \rho_g Y_i \mathbf{v}_g\right)=r_i M_i . \tag{1.86}$$

If diffusion needs to be taken into account, $\mathbf{v}_g$ in Eq. (1.86) must be replaced by $\mathbf{v}_{ig}$, that is, diffusion velocity of i in phase g.

If there are changes of species in the solid phase, for example, and if the solid is moving with a velocity $\mathbf{v}_s$, then clearly a balance similar to that given by Eq. (1.86) exists for, say, component C in the solid phase:

$$\frac{\partial}{\partial t}\left(\varepsilon_s \rho_s Y_c\right)+\nabla\left(\varepsilon_s \rho_s Y_c \mathbf{v}_s\right)=r_c M_c , \tag{1.87}$$

where Y_C is the weight fraction of C in phase S.

Exercises

1. *Concept of Volume Averaging*

The meaning of volume fraction can be illustrated by considering stratified flow, as show in Fig. 1.2. The volume fraction of the gas phase is ε_g and that of the oil is ε_o. The figure shows that for a duct of width R, or in general, for any region in space of width R,

$$\varepsilon_1=\frac{R_1}{R}, \quad \varepsilon_2=\frac{R_2}{R}, \quad \text{where } R=R_1+R_2 .$$

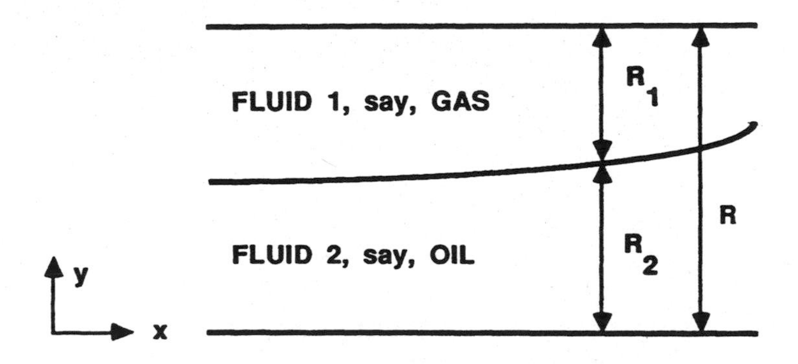

Fig. 1.2 Stratified flow.

Averaging operators over the width R are now defined (Stewart and Wendroff, 1984). Then the average velocity in the x direction for phase 1, u', becomes

$$u_1' = \frac{1}{R_1}\int_o^{R_1} u_1 dy.$$

Derive an unsteady state, one-dimensional continuity equation for each phase by integrating the single phase continuity equation

$$\frac{\partial \rho_i}{\partial t} + \frac{\partial(\rho_i u_i)}{\partial x} + \frac{\partial(\rho_i v_i)}{\partial y} = 0, \quad i = l, g,$$

over the width y. The result is Eq. (1.6) with zero phase change.

Hints:

i. Existence of an interface gives the condition $v_l = \frac{\partial R_1}{\partial t} + u_l \frac{\partial R_1}{\partial x}$. (Show this using chain rules.)

ii. In the derivation use identities of the type

$$\frac{\partial}{\partial x}\int_o^{R_1} u dy = \int_o^{R_1} \frac{\partial u}{\partial x} dy + u\frac{\partial R_1}{\partial x} \qquad \text{(why is this valid?)},$$

which gives

$$\frac{\partial(u' R_1)}{\partial x} = \int_0^{R_1} \frac{\partial u}{\partial x} dy + u\frac{\partial R_1}{\partial x}.$$

2. *Conservation of Total Energy for Phase i*

The balance for internal energy plus kinetic energy can be written similarly to that for component i (Bowen, 1976, p. 21, and Woods, 1985, p. 206). With the system moving at phase velocity $\mathbf{v}_i$, we have

$$\frac{d}{dt_i}\iiint \varepsilon_i \rho_i \left(U_i + \tfrac{1}{2} v_i^2\right) dV = \oiint_{A(t)} \left(\mathbf{T}_i \mathbf{v}_i - \varepsilon_i \mathbf{q}_i\right) d\mathbf{a} +$$

Change of internal energy + Kinetic energy — Power due to traction + Conduction

$$\iiint_{V(t)} \left(\varepsilon_i \rho_i \mathbf{f}_i \mathbf{v}_i + \mathbf{p}_i \mathbf{v}_i + \dot{q}_i\right) dV + \iiint m_i' \left(U_i + \frac{P_i}{\rho_i} + \tfrac{1}{2} v_i^2\right) dV.$$

Power due to body forces — Drag — Internal generation of heat by friction forces — Open system effects

(a) From the balance above, find the total energy partial differential equation.
(b) Use this equation to find the internal energy (U_i) equation.
(c) Express the internal energy equation in convective form, moving with velocity $\mathbf{v}_i$.
(d) Find an expression for the local rate of entropy production. Identify some of the irreversibilities.

3. *Energy Equation Using Bulk Density State Equation*

In the treatment below, consider the case of no phase change, $m_i = 0$. Dobran (1984) postulates an equation of state for fluid i to be

$$U_i = U_i\left(S_i, \rho_{B_i}{}^{-1}\right),$$

where ρ_{B_i} is the bulk or partial density $\rho_{B_i} = \varepsilon_i \rho_i$; then, the bulk pressure

$$\pi_i = -\left(\partial U_i / \partial \rho^{-1}{}_{B_i}\right)_{S_i} \text{ and } T_i = \left(\partial U_i / \partial S_i\right)_{\rho_{B_i}}.$$

(a) Express the equation of state in convective form (as an energy "balance"), dU_i/dt^i ?
(b) Compare the equation in part (a) to that in Exercise 2; note that $\nabla \cdot \mathbf{v}_i = (-1/\rho_{B_i})(d\rho_{B_i}/dt^i)$ (Continuity Equation). Find the equation for dS_i/dt^i.
(c) For inviscid fluids, $\mathbf{T}_i = -\pi_i \mathbf{I}$. Find dS_i/dt^i.
(d) Let $\mathbf{T}_i = -\pi_i \mathbf{I} - \tau_i$ and find the entropy production (Woods, 1985).

4. *Population Balance for a Fluidized Bed**

Consider a well-mixed fluidized bed reactor, such as shown in Fig. 1.3 from Levenspiel *et al.* (1969). In this figure, F_i are the mass flow rates (kg/s) and P_i are the size distributions (cm^{-1}) of the streams.

A population balance for the size distribution P can be written as

$$W \frac{d}{dt} \int_0^{R(t)} P dR = \text{Inflow} - \text{Outflow} + \text{Rate of Particle Growth},$$

* For Exercises 4-7, see Appendix D.2.

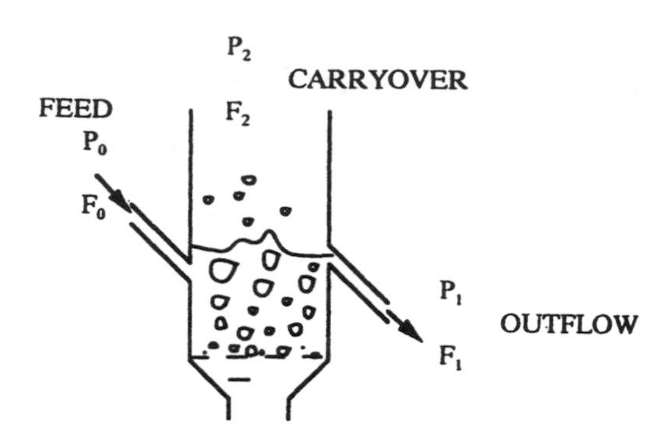

Fig. 1.3 Bubbling fluidized bed.

where W is the weight of the solids in the bed, t is time and R is the particle size. The rate of particle growth for spherical particles is expressed as $\frac{3}{R}\dot{R}PW$ in the above equation, where $\dot{R}$ is given by $\dot{R} = dR/dt$.

Derive the partial differential equation for P as a function of time and particle size.

5. *Population Balance for a Fluidized Bed with Changing Particle Size and Density*

Reklaitis and Overturf (1983) have generalized Levenspiel's population balance for a fluidized bed to the case in which the particles change in size and density.

Consider the fluidized-bed reactor shown in Fig. 1.4 in which P_0, P_1, P_2 and P_b are, respectively, the feed, overflow, fines, and bed–particle distribution functions, and W is the total weight of bed solids. Suppose all distribution functions are expressed in terms of size x_1 and density x_2.

Making the same assumptions as made by Levenspiel *et al.* (see Exercise 4), derive the first order partial differential equation for P_1. The assumptions are that the reactor is well mixed, that it achieved a steady state and that an elutriation constant can be defined to relate P_b to P_2.

Hints:

i. The balance is

$$W\frac{d}{dt}\iint P_b(x_1,x_2)\,dx_1 dx_2 = \text{Inflow} - \text{Outflow} + \text{Rate of Particle Growth.}$$

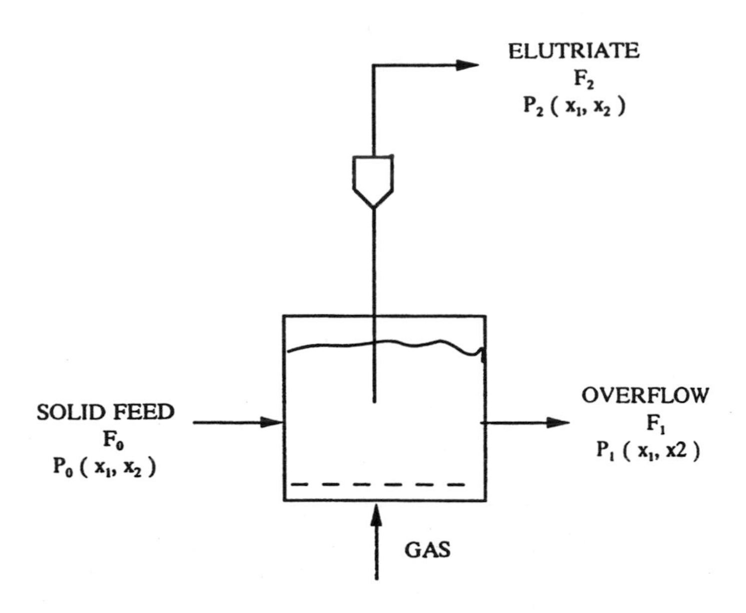

Fig. 1.4 Solid flows for determination of fluidized-bed reactor particle distributions.

Consider the rate of particle growth due to changing size, and changing density.
ii. Let $R = dx_1/dt$ and $R^1 = dx_2/dt$.

6. *Population Balance for a Riser*

Levenspiel's population balance (see Exercise 4) for a fluidized bed can be applied to a section Δx of a vertical pipe (shown in Fig. 1.5) carrying particles which react with a gas. The particle size distribution P' is assumed to be a function of time t, position x, and particle radius R. The balance then becomes

$$\frac{d}{dt}\int_{R(t)}^{R+\Delta R}\int_{x(t)}^{x+\Delta x} P'(t,x,R)\,dx\,dR = \int_{R(t)}^{R+\Delta R}\int_{x(t)}^{x+\Delta x} m_s'\,dx\,dR,$$

where $P' = \varepsilon_s \rho_s P$ and m_s' is the mass generation rate per unit volume per particle size. The volume fraction of solid is ε_s, and its density is ρ_s;

$$m_s' = m_{sp}'\varepsilon_s\rho_s P \quad \text{and} \quad m_{SP}' = \frac{3}{R}\dot{R},$$

where $\dot{R} = dR/dt$.

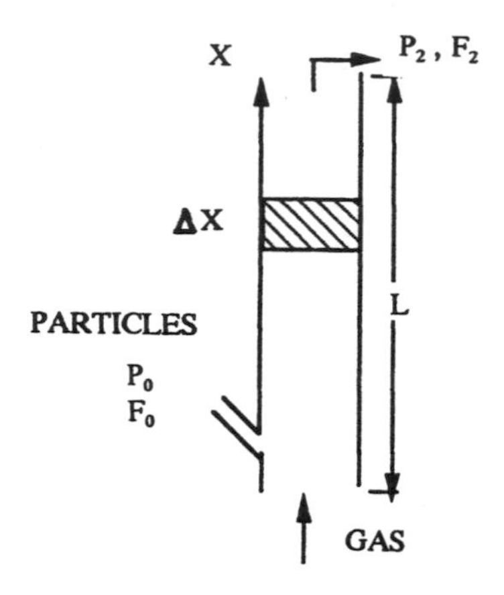

Fig. 1.5 Differential element in a riser.

(a) Obtain an equation for P (partial differential equation).
(b) Integrate it over L and reduce it to Levenspiel's case.

7. *Local Population Balance for a Fluidized Bed*

In multiphase flow applications such as sprays, crystallizers, and fluidized beds, the volume fraction of solids (or liquids, in the case of sprays), ε_s is a function of the particle size R. A mass balance in three dimensions can be written as

$$\frac{d}{dt}\int_{R(t)}^{R+\Delta R}\int_{x(t)}^{x+\Delta x}\int_{y(t)}^{y+\Delta y}\int_{z(t)}^{z+\Delta z} \varepsilon_s \rho_s (t,x,y,z,R)\,dx\,dy\,dz\,dR$$

$$= \int_{R(t)}^{R+\Delta R}\int_{x(t)}^{x+\Delta x}\int_{y(t)}^{y+\Delta y}\int_{z(t)}^{z+\Delta z} m_s'\,dx\,dy\,dz\,dR\,,$$

where ρ_s is the particle density and m_s' is the mass generation rate per unit volume per particle size.

(a) Express the balance in terms of a partial differential equation.
(b) Briefly indicate how to write a momentum balance for the dispersed phase for the situation above.
(c) Show how to reduce the population balance equation derived in part (a) to Levenspiel's well-mixed fluidized bed shown in Fig. 1.3 (Aldis and Gidaspow, 1989).

8. *Balances in Cylindrical and Spherical Coordinates*

For oil production through a pipe from an oil-saturated stratum of thickness h, the following cylindrical balance using the Eulerian balance of Appendix B can be made on the element shown in Fig. 1.6:

$$\frac{d}{dt}\int_r^{r+\Delta r} \rho_0 \varepsilon_0 2\pi r h\, dr + \rho_0 \varepsilon_0 v_r 2\pi r h]_r^{r+\Delta r} = 0,$$

where ρ_o is the oil density, ε_o is the volume fraction of oil in the stratum, h is its height, and v_r is the velocity.

(a) Use the mean value theorems of integral and differential calculus to obtain the partial differential equation.
(b) For a steady oil production of m', kg/s, find the velocity of oil in the stratum as a function of distance from the production well.
(c) Using Darcy's law,

$$v_r = -\frac{k}{\mu}\frac{dp}{dr},$$

where k is the permeability, m^2 and μ is the fluid viscosity, find the steady state pressure distribution for constant ε_o, k, and μ. Note that ε_o may in practice change as the stratum collapses due to depletion of oil.

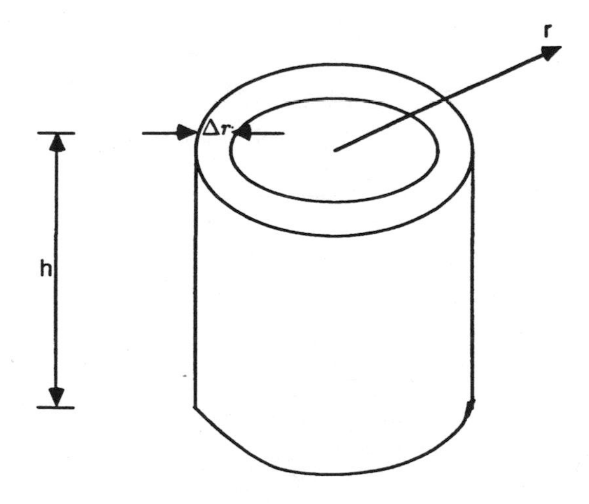

Fig. 1.6 Cylindrical coordinate element.

(d) Derive the conservation of mass equation for the oil without making an assumption of cylindrical symmetry. The system is the element $\Delta r \times r\Delta\Theta \times \Delta Z$.
(e) Derive the equation in spherical coordinates.

Literature Cited

Aldis, D. F., and D. Gidaspow (1989). "Combustion of a Polydispersed Solid Using a Particle Population Balance," *Powder Technology* **57**, 281–294.

Aris, R. (1962). *Vectors, Tensors, and the Basic Equations of Fluid Mechanics.* Englewood Cliffs, New Jersey: Prentice-Hall.

Boure, J. A., A. E. Bergles, and L. S. Tong (1973). "Review of Two-Phase Flow Instability," *Nuclear Engineering and Design* **25**, 165–192.

Bowen, R. M. (1976). "Theory of Mixtures" in A. C. Erigen, ed. *Continuum Physics,* Vol. III. New York: Academic Press.

Callen, H. B. (1960). *Thermodynamics.* New York: John Wiley & Sons.

Dobran, F. (1984). "Constitutive Equations for Multiphase Mixtures of Fluids," *Int'l J. Multiphase Flow* **10**, 273–305.

Gidaspow, D. (1972). "Separation of Gaseous Mixtures by Regenerative Sorption on Porous Solids, Part I, A Fluid Porous Solid Reaction Model with Structural Changes," pp. 59–70 in N. Li, ed. *Recent Developments in Separation Science,* Vol. II. Cleveland: Chemical Rubber Co.

Gidaspow, D., and B. S. Baker (1973). "A Model for Discharge of Storage Batteries," *J. Electrochem. Soc.* **120**, 1005-1010.

Gidaspow, D. (1977). *Fluid-Particle Systems,* presented as an invited lecture at the NATO Advanced Study Institute on Two Phase Flows and Heat Transfer, 1976. Pp. 115–128 in S. Kakac and F. Mayinger, eds. *Two-Phase Flows and Heat Transfer.* Hemisphere Publishing Co.

Gidaspow, D., R. W. Lyczkowski, C. W. Solbrig, E. D. Hughes, and G. A. Mortensen (1973). "Characteristics of Unsteady One-Dimensional Two-Phase Flow," *Trans. Am. Nuclear Soc.* **17**, 249–250.

Ishii, M. (1975). *Thermo-Fluid Dynamic Theory of Two-Phase Flow.* France: Eyrolles.

Kalinin, A. V. (1970). "Derivation of Fluid-Mechanics Equations for a Two-Phase Medium with Phase Changes," *Heat Transfer–Soviet Research* **2**(3), 83–96.

Levenspiel, O., D. Kunii, and T. Fitzgerald (1960). "The Processing of Solids of Changing Size in Bubbling Fluidized Beds," *Powder Technology* **2**, 87–96.

Lyczkowski, R. W., D. Gidaspow, C. W. Solbrig, and E. D. Hughes (1978). "Characteristics and Stability Analysis of Transient One-Dimensional Two-Phase Flow Equations and Their Finite Difference Approximations," *Nuclear Science and Engineering* **66**, 378–396.

Lyczkowski, R. W., D. Gidaspow, and C. W. Solbrig (1982). "Multiphase Flow-Models for Nuclear, Fossil and Biomass Energy Conversion," pp. 198–351 in A. S. Majumdar and R. A. Mashelkar, eds. *Advances in Transport Processes*. New York: Wiley-Eastern Publishers.

Prigogine, I. (1955). *Introduction to Thermodynamics of Irreversible Processes*. Springfield, Illinois: Charles Thomas.

Reklaitis G. V., and B. W. Overturf (1983). "Fluidized-Bed Reactor Model with Generalized Particle Balances," *AIChE J.* **29**, 813.

Slattery, J. C. (1972). *Momentum, Energy, and Mass Transfer in Continua*. New York: McGraw-Hill Book Co.

Stewart, H.B., and B. Wendroff (1984). *J. Comp. Phys.* **56**, 363–409.

Syamial, M., and D. Gidaspow (1985). "Hydrodynamics of Fluidization: Prediction of Wall to Bed Heat Transfer Coefficients," presented at the 21st ASME/AIChE Heat Transfer Conference, 1983. Published in *AIChE J.* **31**, 127–135.

Woods, L. C. (1985). *The Thermodynamics of Fluid Systems*. New York: Oxford University Press.

2

ONE-DIMENSIONAL STEADY GAS–SOLID FLOW

2.1 One-Dimensional, Steady Mixture Momentum Balance 31
2.2 One-Dimensional, Steady Gas Momentum Balance: Pressure Drop in Both Phases — Model A 33
2.3 Fluid Particle Drag 35
2.4 Buoyancy 37
2.5 Drag for Models B and C 39
2.6 Entrance Length in Pneumatic Transport 40
2.7 Pressure Drop 45
2.8 Pressure Drop Correlation in a Dilute Lift Line 51
Exercises 52
1. Unequal Velocity Slurry Flow through Venturi Meters 52
2. Differential Equations for Homogeneous One-Dimensional Gas–Liquid Two-Phase Flow 53
3. Desulfurization in a Transport Reactor 54
4. Transport Reactor Using a Shrinking Core Model 55
5. Direct Contact Heating of Solids in a Vertical Pipe 55
6. Simultaneous Heat and Mass Transfer in a Riser 56
7. Circulating Fluidized Bed Combustor 57
Literature Cited 58

2.1 One-Dimensional, Steady Mixture Momentum Balance

To illustrate the application of multiphase flow balances, the well-studied case of steady vertical gas–solid flow in a pipe is presented in detail. Although one-dimensional equations of change can be derived by the integration of the multidimensional equations presented in Chapter 1, it is much easier and more meaningful to rederive the momentum balances for the special case of one-dimensional flow. The equations are derived to be useful for the general case of particle–fluid flow. Fluid particle friction factors are also derived for both the dilute and the dense phase flow. These relations will be useful in all other chapters.

A steady-state momentum balance for a gas–solid mixture of gas volume fraction ε made on the element $a\Delta x$ fixed in space shown in Fig. 2.1 is as follows:

$$\rho_g v_g \varepsilon v_g a\Big]_x^{x+\Delta x} \quad + \quad \rho_s v_s \varepsilon_s v_s a\Big]_x^{x+\Delta x} \tag{2.1}$$

Net rate of momentum outflow $\rho_g v_g$ with a flow of $\varepsilon_g v_g a$ — Net rate of momentum outflow $\rho_s v_s$ with a flow of $\varepsilon_s v_s a$

$$+ Pa\Big]_x^{x+\Delta x} \quad + \quad \sigma a\Big]_x^{x+\Delta x} \quad - \quad \left(\bar{P} + \bar{\sigma}\right)\Delta a$$

Net rate of momentum outflow due to fluid pressure acting on fluid and solid phases — Net rate of momentum outflow due to solids normal stress — Normal force due to side wall pressure and stress due to area change where $\bar{P} = P(x')$, with $x < x' < x + \Delta x$

$$= -g\left(\rho_g \varepsilon + \rho_s \varepsilon_s\right) a \Delta x \quad - \quad \left(\tau_{wg} + \tau_{ws}\right)\pi D_t \Delta x.$$

Gravitational force acting downwards on the mixture — Rate of momentum dissipation due to wall shear

An application of the mean value theorem of differential calculus to Eq. (2.1), cancellation of the arbitrary element Δx yields the mixture momentum balance written for a constant cross-section a:

$$\frac{d}{dx}\left(\rho_g \varepsilon v_g{}^2 + \rho_s \varepsilon_s v_s{}^2\right) = -\frac{dP}{dx} - \frac{d\sigma}{dx} - g\left(\rho_g \varepsilon + \rho_s \varepsilon_s\right) - \frac{4\left(\tau_{wg} + \tau_{ws}\right)}{D_t}. \tag{2.2}$$

The wall shear for the solid and the gas is normally expressed by a generalization of the Fanning's equation commonly used in pipe flow, as shown below:

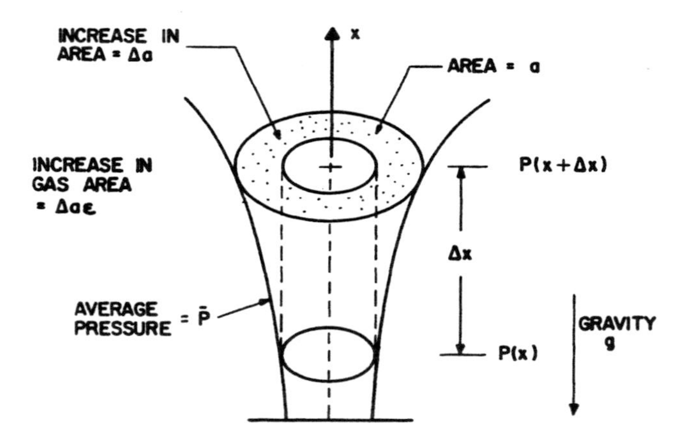

Fig. 2.1 Differential element for momentum balances.

$$\frac{4\left(\tau_{wg}+\tau_{ws}\right)}{D_t}=\frac{4}{D_t}\left(\tfrac{1}{2}\rho_g\varepsilon v_g{}^2 f_{gF}+\tfrac{1}{2}\rho_s\varepsilon_s v_s{}^2 f_{sF}\right). \tag{2.3}$$

In the definition of the Fanning's friction factors f_{gF} and f_{sF}, the corresponding volume fraction was used to enable each friction to vanish as ε and ε_s approach zero. As an alternative, corresponding volume fractions could have been added in Eq. (2.1) and (2.2) as, for example, $\tau_{ws}=\varepsilon_s\tau'_{ws}$. For the pressure P and the normal solids stress σ a constitutive equation must be supplied.

The mixture momentum Eq. (2.2) is almost identical to the one used by Jones and Leung (1985) for modeling standpipe flow. They have deleted the negligibly small contribution of the gas momentum appearing in the more general Eq. (2.2) and have taken their coordinate z in the opposite direction to the upward coordinate x used here.

For the dense downflow of solids, Jones and Leung (1985) further suggest that the wall shear τ_{ws} be related to the normal solids stress σ through the usual granular flow theory (Savage, 1983) where, neglecting velocity dependent components,

$$\tau_{ws}=\mu_w\sigma_r \quad \text{and} \quad \sigma_r=\sigma(1-\sin\delta)/(1+\sin\delta), \tag{2.3}$$

where σ is the area-averaged vertical normal stress in the solid, μ_w the coefficient of friction between solid and wall surface, σ_r the normal stress acting on the tube wall in the radial direction and δ the internal angle of friction.

Equation (2.3) provides an alternate way to correlate the wall shear due to solids, particularly for the case of dense flow treated by Jones and Leung (1985).

In the absence of wall shear, solids stress, and for developed flow, that is, when $\partial v_s/\partial x$ and $\partial v_g/\partial x$ are zero, Eq. (2.2) states that pressure drop is balanced by the weight of the bed. It is the so-called manometer formula:

$$-\frac{dP}{dx}=g\left(\rho_g\varepsilon+\rho_s\varepsilon_s\right). \tag{2.4}$$

2.2 One-Dimensional, Steady Gas Momentum Balance: Pressure Drop in Both Phases — Model A

A steady state momentum balance for the gas made on the element $a\Delta x$ fixed in space shown in Fig. 2.1 is

$$\rho_g v_g \varepsilon v_g a\Big]_x^{x+\Delta x} = -\rho_g g \varepsilon a \Delta x \tag{2.5}$$

Net rate of momentum, $\rho_g v_g$ outflow with a flow $\varepsilon v_g a$ — Gravitational force acting downwards on the gas

$$-Pa\varepsilon\Big]_x^{x+\Delta x} + \bar{P}\Delta a\varepsilon$$

Rate of momentum inflow due to fluid pressure acting on the fluid — Side wall normal force with a pressure $\bar{P} = P(x');\ x < x' < x$ due to gas area changes

$$-\beta_A\left(v_g - v_s\right)a\Delta x - \tau_{wg}\pi D_t \Delta x$$

Rate of momentum dissipation due to relative motion — Rate of momentum dissipation due to wall shear

An application of the mean value theorem of differential calculus to Eq. (2.5), cancellation of the arbitrary element Δx yields the gas momentum balance shown in Eq. (2.8) written for a constant cross-sectional area a. For a variable area the pressure forces become

$$-\frac{d(Pa\varepsilon)}{dx} + P\frac{d(a\varepsilon)}{dx} = -a\varepsilon\frac{dP}{dx} - P\frac{d(a\varepsilon)}{dx} + P\frac{d(a\varepsilon)}{dx} = -a\varepsilon\frac{dP}{dx}. \tag{2.6}$$

An alternative way to treat pressure is to assume that it acts on both the gas and the solid phases. Then, expressions two and three on the right-hand side of Eq. (2.8) are replaced by

$$\lim_{\Delta x \to 0}\frac{1}{\Delta x}\left(-Pa\Big]_x^{x+\Delta x} + \bar{P}\Delta a\right) = \lim_{\Delta x \to 0}\frac{1}{\Delta x}\left[-\frac{d(Pa)}{dx}\cdot\Delta x + \bar{P}\frac{da}{dx}\cdot\Delta x\right] = -a\frac{dP}{dx}, \tag{2.7}$$

$$\bar{P} = P(x');\quad x < x' < x + \Delta x.$$

In such a case the total pressure drop, $-dP/dx$, is in the gas phase only. Subtraction of this term from the mixture momentum equation (2.2) leads to Model B, one with no pressure drop in the solid phase. In view of its well-posedness for the corresponding transient problem (Lyczkowski *et al.*, 1982) it is in this sense superior to the ill-posed basic Model A.

With these observations in mind the Model A gas phase balance for a constant cross sectional area a becomes

$$\frac{d\left(\rho_g \varepsilon v_g^{\ 2}\right)}{dx} = -\varepsilon \frac{dP}{dx} - \rho_g \varepsilon_g - \beta_A \left(v_g - v_s\right) - \frac{4\tau_{wg}}{D_t}. \tag{2.8}$$

A corresponding momentum balance for the solid can be obtained by subtracting Eq. (2.8) from the mixture momentum balance, Eq. (2.2). Alternatively, the solids momentum balance can be derived similarly to that for the gas, keeping in mind that a different treatment of the normal stress will result in a different form of the equation. Table 2.1 summarizes the equations for three models. Model C, the relative velocity model, was derived by Gidaspow (1978) using the principles of nonequilibrium thermodynamics. It is sufficient at this stage to point out that it differs from Model B by a replacement of the solids velocity in Model B with the relative velocity. In Gidaspow's theory (1978) it arose due to the assumption that the entropy production can be a function of relative velocity, but not of individual phase velocities, to be invariant under a change of frame of reference. The extra terms that arise in Model C that are not present in Models A or B can be interpreted to be the classical added mass forces (Birkhoff, 1960).

2.3 *Fluid Particle Drag*

The friction coefficients between the fluid and the solid are obtained from standard correlations with negligible acceleration. With no acceleration, wall friction, or gravity, the gas momentum balance, Eq. (2.8), is

$$-\varepsilon \frac{\partial P}{\partial x} - \beta_A \left(v_g - v_s\right) = 0. \tag{2.9}$$

This equation is a statement of Darcy's law where the reciprocal of β_A is the permeability divided by the fluid viscosity.

The friction coefficient β_A is now obtained by comparing Eq. (2.9) to the Ergun equation, for example as given in Kunii and Levenspiel (1969), or in Bird *et al.* (1960):

$$\frac{\Delta P}{\Delta x} = 150 \frac{\varepsilon_s^{\ 2}}{\varepsilon^3} \frac{\mu_g U_o}{\left(\phi_s d_p\right)^2} + 1.75 \frac{\varepsilon_s \rho_g U_o^{\ 2}}{\varepsilon^3 \phi_s d_p}, \tag{2.10}$$

where U_o is the superficial velocity $U_o = \varepsilon(v_g - v_s)$. A comparison of Eqs. (2.9) and (2.10) shows that

Table 2.1
ONE-DIMENSIONAL, STEADY HYDRODYNAMICS MODELS

	Common Equations	
Gas Continuity	$\frac{d}{dx}\left[\varepsilon\rho_g v_g\right]=0$	(T2.1)
Solid Continuity	$\frac{d}{dx}\left[(1-\varepsilon)\rho_s v_s\right]=0$	(T2.2)
Mixture Momentum	$(1-\varepsilon)\rho_s v_s\frac{dv_s}{dx}+\varepsilon\rho_g v_g\frac{dv_g}{dx}=$ $-\frac{dp}{dx}-\frac{d\sigma}{dx}-g(1-\varepsilon)\rho_s-g\varepsilon\rho_g-\frac{4(\tau_{wg}+\tau_{ws})}{D_t}$	(T2.3)
	Solids Momentum Balances	
Pressure Drop in Both Phases	$(1-\varepsilon)\rho_s v_s\frac{dv_s}{dx}=$ $-(1-\varepsilon)\frac{dp}{dx}-\frac{d\sigma}{dx}-g(1-\varepsilon)\rho_s-\frac{4\tau_{ws}}{D_t}-\beta_A\left(v_s-v_g\right)$	(T2.4)
Pressure Drop in Fluid Phase Only	$(1-\varepsilon)\rho_s v_s\frac{dv_s}{dx}=$ $-\frac{d\sigma}{dx}-g(1-\varepsilon)\left(\rho_s-\rho_g\right)-\frac{4\tau_{ws}}{D_t}-\beta_B\left(v_s-v_g\right)$	(T2.5)
Relative Velocity Model: A Constitutive Equation for the Mixture	$(1-\varepsilon)\rho_s\left(v_s-v_g\right)\frac{d\left(v_s-v_g\right)}{dx}=$ $-\frac{d\sigma}{dx}-g(1-\varepsilon)\left(\rho_s-\rho_g\right)-\frac{4\tau_{ws}}{D_t}-\beta_C\left(v_s-v_g\right)$	(T2.6)

$$\beta_A=150\frac{(1-\varepsilon)^2\mu_g}{\varepsilon\left(d_p\phi_s\right)^2}+1.75\frac{\rho_g\left|v_g-v_s\right|(1-\varepsilon)}{\phi_s d_p},\quad \text{for } \varepsilon<0.8. \tag{2.11}$$

Note that β divided by the viscosity is the reciprocal permeability. It is proportional to the square of the particle diameter for low flows.

Wen and Yu (1966) have extended the works of Richardson and Zaki (1954) to derive an expression for pressure drop prediction in a particulate bed. For porosities greater than 0.8, if $\partial P/\partial x$ is replaced by Wen and Yu's expression for pressure drop, the friction coefficient in this porosity range becomes

$$\beta_A = \tfrac{3}{4} C_D \frac{\varepsilon |v_g - v_s| \rho_g (1-\varepsilon)}{d_p} f(\varepsilon), \tag{2.12}$$

where C_D, the drag coefficient, is related to the Reynolds number by (Rowe, 1961)

$$C_D = \frac{24}{(\mathrm{Re}_s)} \left(1 + 0.15 (\mathrm{Re}_s)^{0.687}\right); \qquad (\mathrm{Re}_s) < 1000, \tag{2.13}$$

$$C_D = 0.44; \qquad (\mathrm{Re}_s) \geq 1000, \tag{2.14}$$

where

$$(\mathrm{Re}_s) = \frac{\varepsilon \rho_g (v_g - v_s) d_p}{\mu_g}. \tag{2.15}$$

In Eq. (2.12) $f(\varepsilon)$ shows the effect due to the presence of other particles in the fluid and acts as a correction to the usual Stokes law for free fall of a single particle. Gidaspow and Ettehadieh (1983) have used

$$f(\varepsilon) = \varepsilon^{-2.65}, \tag{2.16}$$

although this correction was usually small. Syamlal *et al.* (1987) have extended this approach to the more general Richardson and Zaki expression to avoid switching between the Ergun equation and the drag relation.

The standard drag curve for a sphere is shown in Fig. 2.2, together with experimental data for transport of 520 μm glass beads in the developed region of a 0.0762 m Plexiglas tube. The deviation of data for the vertical transport shown from the standard single sphere curve is typical of related drag coefficient determinations summarized by Soo (1983), except at the highest solids flux, W_t, where clustering of particles may have occurred (Gidaspow *et al.*, 1987).

2.4 Buoyancy

A generalization of the usual balance — buoyancy equals drag (Bird *et al.*, 1960) — can be obtained by subtracting the momentum balance for the solid from

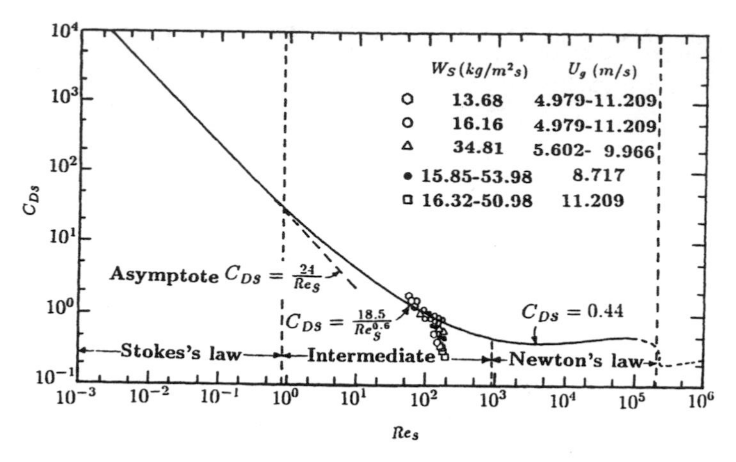

Fig. 2.2 Drag coefficient for spheres (Bird *et al.*, 1960) with data for vertical transport of 520 μm glass beads in a 0.0762 m Plexiglas tube (Luo, 1987).

the momentum balance for the gas. To eliminate the pressure in Model A, divide Eq. (2.8) by ε and Eq. (T2.4) by $(1-\varepsilon)$ and neglect acceleration. This procedure gives Eq. (2.17):

$$g(\rho_s-\rho_g) = \frac{\beta_A(v_g-v_s)}{\varepsilon\varepsilon_s} - \frac{4\tau_{ws}}{D_t\varepsilon_s} + \frac{4\tau_{wg}}{D_t\varepsilon_s} - \frac{1}{\varepsilon_s}\frac{d\sigma}{dx}, \quad (2.17)$$

Buoyant force = Drag - Solids wall friction + Gas wall friction - Solids pressure gradient

or, in terms of C_D using Eq. (2.12),

$$g(\rho_s-\rho_g)=\tfrac{3}{4}C_D\frac{\rho_g|v_g-v_s|(v_g-v_s)}{d_p}-\frac{4\tau_{ws}}{D_t\varepsilon_s}+\frac{4\tau_{wg}}{D_t\varepsilon}-\frac{1}{\varepsilon_s}\frac{d\sigma}{dx}. \quad (2.18)$$

In the absence of wall friction and with zero particle-to-particle contact, that is, zero solids stress, Eq. (2.18) is the simple balance of buoyancy due to gravity and drag force on the sphere, given in Eq. (2.19):

$$\tfrac{4}{3}\pi\left(\frac{d_p}{2}\right)^3 g(\rho_s-\rho_g)=\pi\left(\frac{d_p}{2}\right)^2\left(\tfrac{1}{2}\rho_g v_r^{\,2}C_D\right), \quad (2.19)$$

where v_r is the relative velocity. This relative velocity is known as the terminal settling velocity.

Integration of the creeping flow motion for a sphere in an infinite incompressible Newtonian fluid (Bird *et al.*, 1960) shows that in this regime C_D is as follows:

$$C_D = \frac{24}{\mathrm{Re}_s}. \tag{2.20}$$

Substitution of Stokes' drag coefficient into the balance of buoyancy equals drag gives the terminal velocity v_r:

$$v_r = \frac{d_p^2 \Delta\rho g}{18\mu_g}. \tag{2.21}$$

For the high Reynolds number flow in the Newton's regime, C_D is about 0.45 and Eq. (2.19) shows that the terminal velocity becomes

$$v_r = 1.72\left(\frac{d_p \Delta\rho g}{\rho_g}\right)^{1/2}. \tag{2.22}$$

In the limit of no motion, Eq. (2.17) gives

$$-\frac{d\sigma}{dx} = \varepsilon_s g\left(\rho_s - \rho_g\right), \tag{2.23}$$

which is the analogue of the manometer formula, Eq. (2.4). For a column filled with a fluid–solid mixture and open to the atmosphere or with zero stress at the top, the solids stress at the bottom becomes

$$\sigma = g\left(\rho_s - \rho_g\right)\int_0^x \varepsilon_s \, dx. \tag{2.24}$$

Solids pressure = Buoyancy

2.5 *Drag for Models B and C*

The dispersed phase equation (T2.5), was constructed to satisfy Archimedes' principle. This principle is immediately satisfied when considering the case of statics with no solids stress. For example, consider a dispersed gas–liquid system, with bubbles of density ρ_s and a fluid of density ρ_g. Then clearly the force per unit length acting on the bubbles is

$$\tau_{ws}\pi D_t = g\left(\rho_{\text{bubble}} - \rho_{\text{fluid}}\right)\frac{\pi D_t^2}{4}\cdot\varepsilon_{\text{bubble}}. \tag{2.25}$$

The fluid phase balance is obtained by subtracting the dispersed phase balance, Eq. (T2.5) from the mixture momentum balance, Eq. (T2.3). This yields

Model B Fluid Phase Momentum Balance (Generalized Darcy's Law)

$$\varepsilon\rho_g v_g \frac{dv_g}{dx} = -\frac{dp}{dx} - \beta_B\left(v_g - v_s\right) - g\rho_g - \frac{4\tau_{wg}}{D_t}. \tag{2.26}$$

The relation between β_B and β_A can be obtained by comparing Eq. (2.9) to Eq. (2.26) for the case of zero velocity gradient, zero gravity, and zero wall friction. The difference between these equations is the porosity that multiplies the gradient of pressure in the model in which the pressure drop is taken to be in the gas and solid phases. As an alternative, β_B can be obtained directly from the Ergun Equation (2.10). A similar derivation for Model C shows that the friction coefficients for Models B and C are equal.

Hence,

$$\beta_B = \beta_C = \beta_A/\varepsilon \tag{2.27}$$

Note that there is a lack of symmetry between the dispersed phase and the continuous phase momentum balances. Therefore, in applying these equations to describe boiling going from all liquid to all vapor, it is necessary to switch from liquid continuous phase to gas at some appropriate volume fraction of steam. Equations (T2.5) and (T2.6) can also be used to model flow of neutrally buoyant particles.

2.6 Entrance Length in Pneumatic Transport

Conveying of solids is a common unit operation. To understand the hydrodynamics of such systems, many experimental rigs (Leung, 1980) were constructed throughout the world. The Illinois Institute of Technology apparatus is shown in Fig. 2.3.

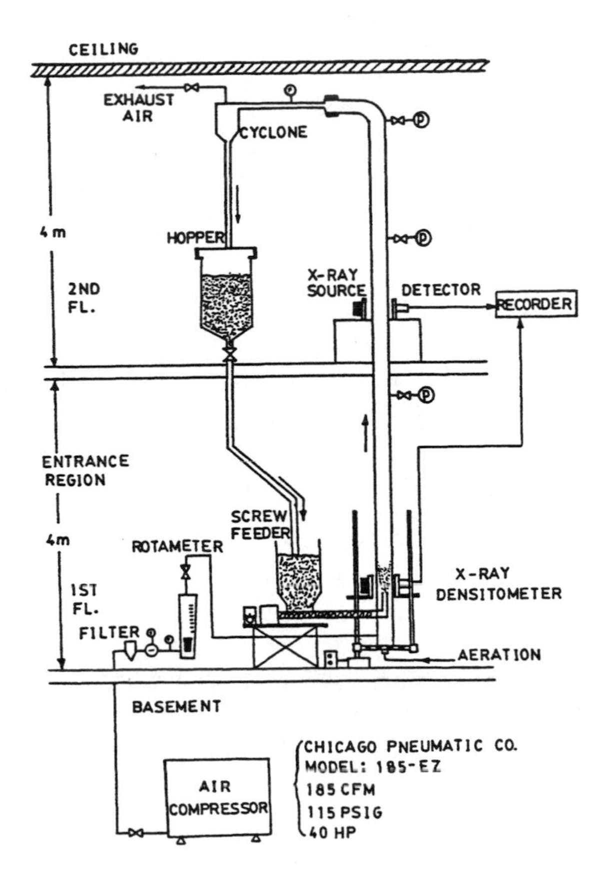

Fig. 2.3 Pneumatic conveying apparatus.

The transport loop extends through two floors with a compressor in the basement. The first floor has a specially constructed movable platform for an x-ray or a γ-ray densitometer, which is used to determine solids concentration in the entrance region. On the second floor densitometers and probes are used to determine flow properties in the developed region. The plastic pipe used for transport is grounded electrically to minimize interference due to triboelectric charging.

The behavior of particles in the entrance length can be obtained by solving the continuity and momentum equations shown in Table 2.1 for the variables v_g, v_s, ε and P (Arastoopour and Gidaspow, 1979a, b and c). The continuity equations for the steady-state one-dimensional flow show that

$$\varepsilon\rho_g v_g = U_o\rho_{go} = \text{inlet volumetric gas flux}$$
$$\text{where } U_o \text{ is the superficial gas velocity } \varepsilon v_g; \tag{2.28}$$

$$v_s(1-\varepsilon_s)\rho_s = W_s = \text{inlet volumetric solids flow per unit pipe area} \tag{2.29}$$

To solve the equations we need to specify either the inlet solids–volume fraction or the solids velocity at the inlet, and the pressure either at the inlet or at the outlet of the pipe. The three sets of equations can then be solved for the variables $(v_s, v_g, \varepsilon, P) = \mathbf{y}$ simultaneously as an initial value problem:

$$\frac{d\mathbf{y}}{dx} = f(\mathbf{y}), \quad \mathbf{y}(0) = \mathbf{y}_o, \tag{2.30}$$

by Runge–Kutta or other methods or as a boundary value problem if the pressure is given at the pipe outlet (Gidaspow *et al.*, 1972).

For an incompressible fluid, that is, for the case when the gas density does not change significantly with pipe length, both Models B and C can be reduced to a single differential equation for void fraction or velocity. Pressure can then be determined by quadrature of the mixture momentum equation. The simpler and more common case of Model B is illustrated in Eq. (2.31).

For dense phase flow, the normal stress is generally a strong function of the porosity of solids concentration (Gidaspow, 1986):

$$\sigma = \sigma(\varepsilon_s). \tag{2.31}$$

Then, by the chain rule,

$$\frac{d\sigma}{dx} = \frac{d\sigma}{d\varepsilon_s}\frac{d\varepsilon_s}{dx}. \tag{2.32}$$

For convenience, let the modulus of compressibility G be

$$G = d\sigma/d\varepsilon_s. \tag{2.33}$$

It has been defined here to have a sign opposite to that used by Gidaspow (1986) and others before him. Then, as indicated by Gidaspow (1986),

$$\text{Critical velocity in a vacuum} = \sqrt{G/\rho_s}, \tag{2.34}$$

where G is a strong function of porosity, vanishing for dilute flow, when there is no particle-to-particle contact.

With such a representation for the normal stress and the use of the solids continuity equation

$$\varepsilon_s \frac{dv_s}{dx} = -v_s \frac{d\varepsilon_s}{dx}, \tag{2.35}$$

the solids momentum balance (T2.5) can be written as

$$\left(-v_s^2 + \frac{G}{\rho_s}\right)\frac{d\varepsilon_s}{dx} = g\varepsilon_s - \frac{4\tau_{ws}}{\rho_s D_t} + \frac{\left(v_s - v_g\right)\beta_A}{\varepsilon \Delta\rho}. \tag{2.36}$$

Acceleration zone, = Buoyancy - Wall friction + Drag
zero in critical flow

The solids velocity and the relative velocity can be expressed by the algebraic equations

$$v_s^2 = \left(\frac{W_s}{\rho_s}\right)^2 \frac{1}{\varepsilon_s^2} \quad \text{and} \quad v_s - v_g = \left(\frac{W_s}{\rho_s \varepsilon_s} - \frac{U_o}{\varepsilon}\right), \tag{2.37}$$

and for the dilute regime, zero G, Eq. (2.36) can be integrated by quadrature.

A solids volumetric balance shows that the well-mixed solids volume fraction ε_{SM} is

$$\varepsilon_{SM} = \frac{W_s/\rho_s}{U_o + W_s/\rho_s}. \tag{2.38}$$

For transport of glass beads of density $\rho_s = 2620\ \mathrm{kg/m^3}$ and a solids flux of $W_s = 26.2\ \mathrm{kg/m^2\text{-}s}$, the solids superficial velocity W_s/ρ_s is 1 cm/s. For vertical transport of solids, the gas velocity exceeds the terminal velocity, which for 520 μm glass beads is about 4 m/s. Therefore,

$$U_o >> W_s/\rho_s. \tag{2.39}$$

Thus,

$$\varepsilon_{SM} = \frac{W_s/\rho_s}{U_o}, \tag{2.40}$$

and is the same value as for the case of zero relative velocity, as seen from Eq. (2.37). Note the for U_o of $5\ \mathrm{m/s}$, ε_{SM} is 0.002 when W_s/ρ_s is 1 cm/s.

After dimensionalizing Eq. (2.36), using ε_{SM} as the scale factor, it can be shown that the characteristic dimensionless distance $\bar{x}$ is

$$\bar{x} = \frac{gx\rho_s^2}{W_s^2} \cdot \varepsilon_{SM}^2 = \frac{gx}{U_o^2}. \tag{2.41}$$

Let us define the entrance length when $\bar{x}$ reaches a value of 0.3. Then

$$x_{\text{entrance}} = 0.3\frac{U_o^2}{g} \tag{2.42}$$

gives us a rough estimate of the length of pipe required to reach developed flow.

Figure 2.4 shows some typical data of the variation of the void fraction in the entrance region obtained by Luo (1987) using the x-ray densitometer shown in Fig. 2.5. The theoretical solution shown was obtained by integrating the equations of change, Eq. (2.36) for Model B with a solids friction factor and Set A in Table 2.1, using a standard gas friction factor. Since the term $(1-\varepsilon)dP/dx$ in Model A is nearly zero, both models give the same numerical values for the porosities.

For a dilute case, that is, zero G, Eq. (2.36) or directly Eq. (T2.5) can be written in terms of the drag coefficient C_D as follows in Eq. (2.43). For constant C_D Klinzig integrated this equation analytically,

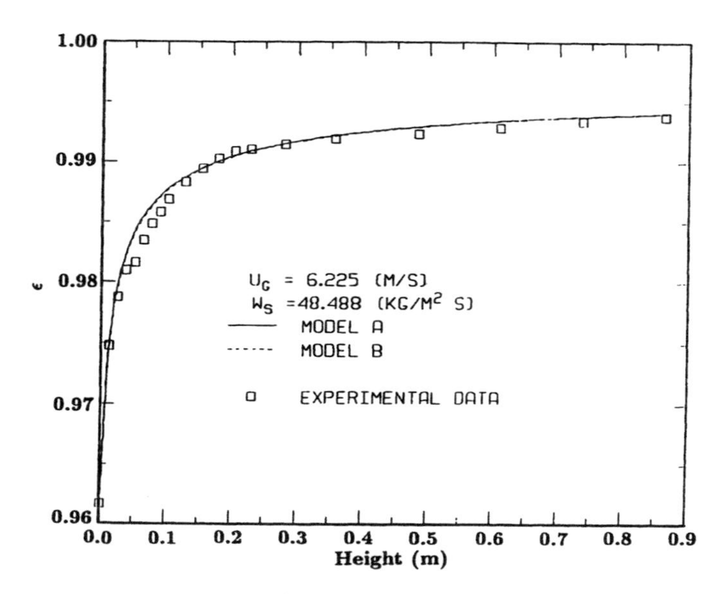

Fig. 2.4 Experimental and theoretically predicted porosities in the entrance region of a grounded Plexiglas pipe with 520 μm glass beads at U_g= 6.225 m/s and W_S = 48.488 kg/m^2-s (Luo, 1987).

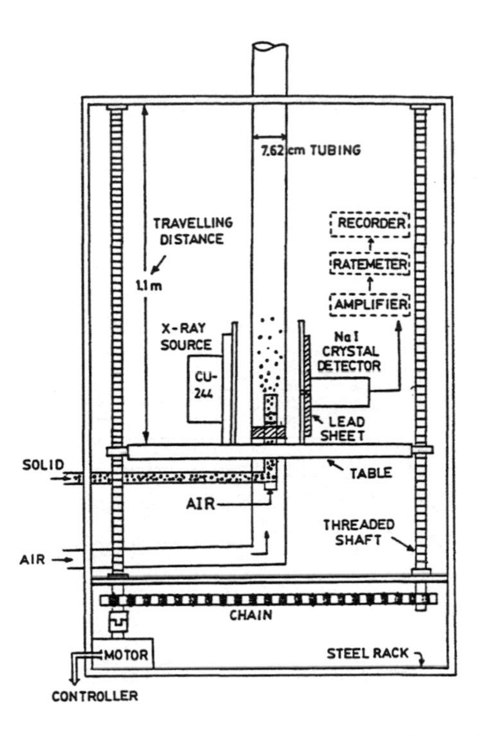

Fig. 2.5 A movable x-ray densitometer for measuring entrance region particle concentrations.

$$v_s \frac{dv_s}{dx} = -g - \left(\frac{4\tau_{ws}}{D_t \varepsilon_s \rho_s} \right) + \tfrac{3}{4} C_D \frac{\left(v_s - v_g\right)^2 \rho_g}{d_p \Delta\rho}, \tag{2.43}$$

$$x_{\text{acceleration}} = \int_{v_{s1}}^{v_{s2}} \frac{v_s \, dv_s}{\tfrac{3}{4} C_D \rho_g \left(v_s - v_g\right)^2 / d_p \Delta\rho - g - \left(4\tau_{ws} / D_t \varepsilon_s \rho_s\right)}, \tag{2.44}$$

where the velocity v_{s1} is computed knowing the inlet solids flux and v_{s2} is taken as some number that is sufficiently close to the developed velocity obtained from a balance of buoyancy equals friction.

2.7 Pressure Drop

The momentum balance for the mixture, Eq. (T2.3), shows that the total pressure drop, dP/dx, can be computed as the sum of the total contributions due to momentum of the gas and the solid. Furthermore, in dense flow or in dilute flow with a layer of solid at the wall, this pressure drop may have to be corrected for the effect of the normal solids stress transmitted by that contact of solid particles. Algebraically we obtain, with no solids stress,

$$\left(\frac{dP}{dx}\right)_{\text{total}} = \left(\frac{dP}{dx}\right)_{\text{momentum}} + \left(\frac{dP}{dx}\right)_{\text{friction}} + \left(\frac{dP}{dx}\right)_{\text{elevation}}. \tag{2.45}$$

The pressure drop due to the momentum, important in the entrance region, as given by Eq. (2.2) is

$$-\left(\frac{dP}{dx}\right)_{\text{momentum}} = \frac{d}{dx}\left(\varepsilon_s \rho_s v_s^2 + \varepsilon \rho_g v_g^2\right). \tag{2.46}$$

In this conservative form the momentum can be computed by integrating Eq. (2.46) and computing the momentum from the inlet to some length L, as shown below:

$$-\Delta P_{\text{momentum}} = \Delta\left(\varepsilon_s \rho_s v_s^2 + \varepsilon \rho_g v_g^2\right) L, \tag{2.47}$$

where Δ indicates the difference between the terms at position L minus position zero.

The pressure drop due to friction is

$$-\Delta P_{\text{friction}} = \frac{4}{D_t}\int_0^L \left(\tau_{wg} + \tau_{ws}\right) dx, \tag{2.48}$$

where τ_{wg} and often τ_{ws} are given by Fanning's equations (2.3). In the devel-oped region,

$$-\Delta P_{\text{friction due to gas}} = \frac{2 fg \varepsilon \rho_g v_g^2 L}{D_t}, \tag{2.49}$$

where for dilute flow it is reasonable to assume that the friction factor is that for the gas alone and is thus given by the Poiseulle expression in the laminar region:

$$fg = 16/\text{Re}_g \quad \text{for } \text{Re}_g \le 2100; \tag{2.50}$$

by the Blasius formula in the intermediate range,

$$fg = 0.0791/\text{Re}_g^{0.25}, \quad 2100 \le \text{Re}_g \le 100{,}000; \tag{2.51}$$

and by the transmission factor formula,

$$\frac{1}{\sqrt{fg}} = 2 \log\left(\mathrm{Re}_g \sqrt{f_g}\right) - 0.8, \tag{2.52}$$

for Reynolds numbers greater than 100,000 where

$$\mathrm{Re}_g = \frac{D_t v_g \rho_g \varepsilon}{\mu_g}. \tag{2.53}$$

As already discussed there exists no unique way of treating the friction due to the solids. One way is to relate the solids friction to the normal stress and to the internal angle of friction. A more common way in dilute phase flow is to use Fanning's equation for the solid phase. For coal the IGT-DOE data book (1982) recommends the use of the Fanning's equation with the friction factor given by the Konno–Saito correlation (1969). Using such relations in developed flow one can express the solids friction factor as shown in Eq. (2.54).

In Fanning's equation for the solids, Eq. (2.3), substitute the mass flux $W_s = \varepsilon_s \rho_s v_s$ and use the loading ratio θ:

$$\theta = W_s / U_g \rho_g, \tag{2.54}$$

where U_g is the superficial gas velocity εv_g. This gives the pressure drop for the solids in terms of the friction factor for the solids:

$$\Delta P_{\text{friction due to solids}} = \frac{2}{D_t} f_{sF}\, \theta U_g \rho_g v_s L. \tag{2.55}$$

But Konno and Saito (1969) in their experiments find that

$$2 f_{SF} v_s = 5.7 \times 10^{-2} \sqrt{g D_t}. \tag{2.56}$$

Thus, the contribution due to solids friction alone is

$$-\Delta P_{\text{friction due to solids}} = 5.7 \times 10^{-2} \frac{U_g \rho_g \theta L}{\sqrt{g D_t}}. \tag{2.57}$$

This correlation above cannot be widely applicable, since both Nakamura and Capes (1973) and Luo (1987) found that sometimes the solids friction is actually negative, as will be discussed later. It is presented here as the most reliable correlation, as evaluated by the IGT-DOE team.

The pressure drop due to the weight of the solids and the gas is given simply by the manometer formula, Eq. (2.4),

$$-\Delta P_{\text{elevation}} = g\int_0^L \left(\rho_g \varepsilon + \rho_s \varepsilon_s\right) dx. \tag{2.58}$$

Across the developed region of flow,

$$-\Delta P_{\text{elevation, developed}} = gL\left(\rho_g \varepsilon + \rho_s \varepsilon_s\right). \tag{2.59}$$

The decomposition of the total pressure drop in the developed region into the individual contributions due to weight of the bed, gas and solid wall friction is shown for the case of Luo's (1987) data in Fig. 2.6. The volume fraction of solids was measured by means of the calibrated x-ray densitometer shown in Fig. 2.5. The solids concentration was increased by decreasing the gas velocity as shown on the right-hand side of Fig. 2.6 at a constant solids flux. The pressure drop due to the weight of particles is given by Eq. (2.59) and is clearly linear with ε_s, since the gas density is negligibly small compared with the density of the solids. The total

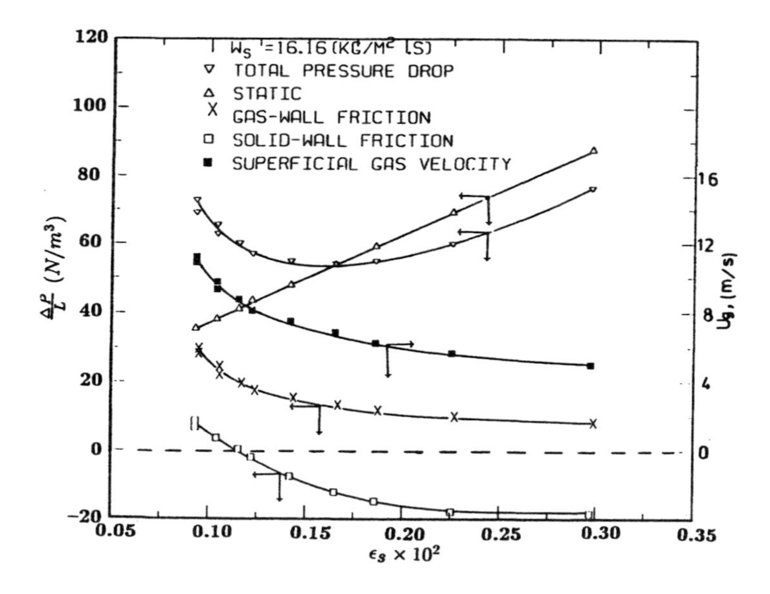

Fig. 2.6 Total pressure drop for transport of 520 μm glass beads in a 0.0762 m Plexiglas tube at a solids flux of 16.2 kg/m^{-2}-s decomposed into weight of particles and wall friction (Luo, 1987).

pressure drop was measured by a manometer and has a typical minimum. This minimum occurs due to the fact that as the velocity increases and the solids volume fraction decreases, the pressure drop due to wall friction, as given by the Fanning equations (2.49), increases with velocity of the gas. Such a minimum is a characteristic feature of many other vertical multiphase flow systems. The difference between the total pressure drop and that due to weight of solids should be friction. However, in Fig. 2.6 the weight of the solids exceeds the total pressure drop at high solids concentrations. This is due to the fact that for conditions of Fig. 2.6 at high solids concentrations, there is a layer of solids near the wall. The existence of such a layer of solids was confirmed by high-speed photography and by measuring radial solids distributions. A typical radial distribution is shown in Fig. 2.7. The situation is similar to annualar gas–liquid flow with a liquid film at the wall. The wall supports part of the weight of solids, resulting in a decreased pressure drop. This results in negative pressure drop in Fig. 2.6.

The calculated pressure drop, the gas and solid velocities, and the volumetric concentrations were compared with each other using three different models for countercurrent and co-current flow of air and spherical rape seed mixture through a vertical tube (Arastoopour and Gidaspow, 1979). See Fig. 2.8. The vertical tube and the particles studied here are the same as those used by Zenz (1949). The

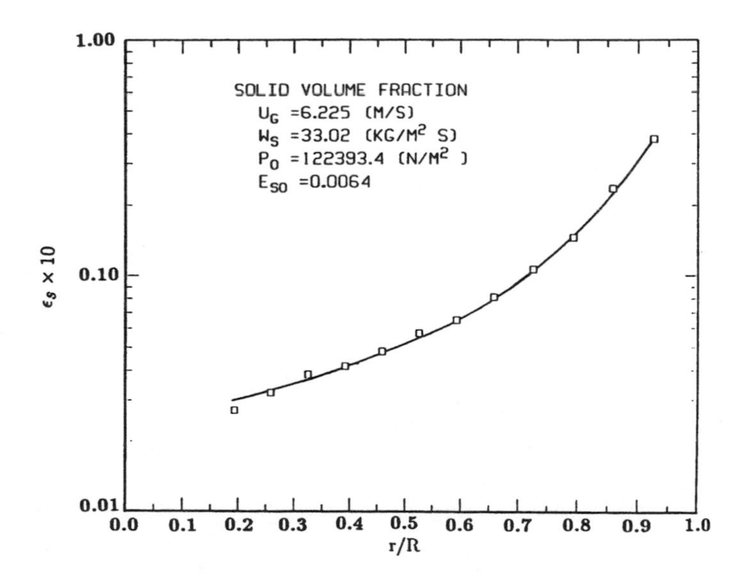

Fig. 2.7 Radial solids volume fraction with $U_g = 6.225$ m/s and $W_s = 33.02$ kg/m^2-s at a height of 3.4 m (Luo, 1987).

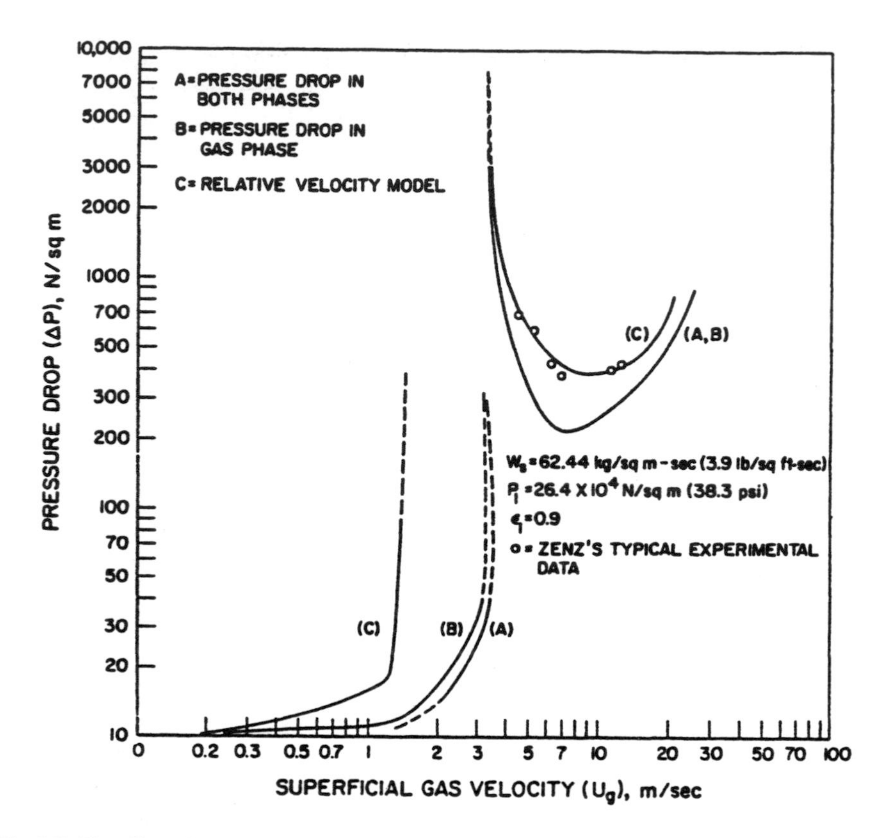

Fig. 2.8 The effect of superficial gas velocity on the pressure drop for countercurrent (bottom curves) and co-current (top curves) solid–gas flow through a vertical tube calculated using three unequal velocity models, and a comparison with Zenz's (1949; 1971) typical experimental data (Arastoopour and Gidaspow, 1979. Reproduced with permission from Pergamon Press Ltd., Headington Hill Hall, Oxford OX3 OBW, U.K.).

solids mass flow rate and the initial pressure are chosen as 62.44 kg/m^2-s and 26.4 $\times$ 10^4 N/m^2 with an assumed inlet void fraction, $\varepsilon_1 = 0.9$. Previous investigation did not measure the solids concentrations or velocities, although such work is now in progress in several laboratories.

The pressure drop in the gas phase model, Case B, predicts slightly larger values for the particles velocities than Case C, the relative velocity model.

Figure 2.8 also resembles Hewitt's (1977) well-known pressure drop curves for vertical gas-liquid flow in a tube. Also, Collier (1972) presents curves showing minima in pressure drops for constant heat fluxes in sub-cooled boiling. Collier computed void fractions using algebraic correlations.

2.8 Pressure Drop Correlation in a Dilute Lift Line

The IGT-DOE data book (1982) presents a correlation for pressure drop in dilute vertical pneumatic conveying, which is not an unreasonable approximation to reality in developed flow. The correlation was tested with data for coal, glass beads, and related materials. It can be justified as follows.

In developed flow the total pressure drop as described in Section 2.7 is the sum of the frictional loss for the gas by Eq. (2.49), frictional loss for the solids given by the Konno and Saito correlation, Eq. (2.57), and the pressure drop due to the weight of the bed, given by Eq. (2.59). For dilute flow the gas velocity is the same as the superficial velocity and the volume fraction of the solids can be expressed in terms of the mass flux of solids W_s. Then the pressure drop becomes

$$\Delta P_T = \frac{2 f_g \rho_g U_g{}^2 L}{D_t} + 0.057 \frac{U_g \rho_g \theta L}{\sqrt{g D_t}} + \left(\frac{w_s L}{v_s} + \rho_g L \right) g. \tag{2.60}$$

All the quantities in Eq. (2.60) are known in terms of the inlet gas and solid flow rates and the geometry, except for the velocity of the solid v_s. Klinzig (1981) has reviewed various correlations for estimating the solids velocity. Knowing the solids velocity is equivalent to knowing the solids volume fraction for steady flow. Thus void fraction correlations may be used for the same purpose. The IGT-DOE correlation assumes that

$$v_s = U_g - v_t, \tag{2.61}$$

where v_t is the terminal velocity. Computation of terminal velocity was already discussed in Section 2.4. To summarize, the terminal velocity can be computed using formulas (2.62) through (2.65):

$$v_t = \frac{d_p^2 (\rho_s - \rho_g) g}{18 \mu} \quad \text{for } \mathrm{Re}_p \le 2.0, \tag{2.62}$$

$$v_t = \frac{0.153 d_p^{1.14} g^{0.71} (\rho_s - \rho_s)^{0.71}}{\mu^{0.43} \rho_g^{0.29}} \quad \text{for } \quad 2.0 < \mathrm{Re}_p < 1000, \tag{2.63}$$

$$v_t = 1.74 \frac{\sqrt{g d_p (\rho_s - \rho_g)}}{\rho_g} \quad \text{for } 1000 < \mathrm{Re}_p < 250{,}000, \tag{2.64}$$

and

$$\mathrm{Re}_p = \frac{d_p U_t \rho_g}{\mu}. \tag{2.65}$$

Exercises

1. ***Unequal Velocity Slurry Flow through Venturi Meters***

Venturi meters are convenient devices for measuring flow rates of slurries. However, a surprising result of measurements and calculations (Shook and Masliyah, 1974; Arastoopour and Gidaspow, 1976) is that the discharge coefficient exceeds unity.

Consider a vertical venturi meter whose area, A, in terms of square meters is shown in Fig. 9 as a function of length, *x*, along with computed velocities for three models, A through C. The fluid, particle density and particle diameter, ρ_f, ρ_s and d are also shown. The inlet conditions are indicated by a subscript, where ε_1 is the fluid volume fraction entering the venturi.

(a) Formulate the conservation of mass and momentum for the fluid and solid phases for the venturi meter for Models A, B, and C.

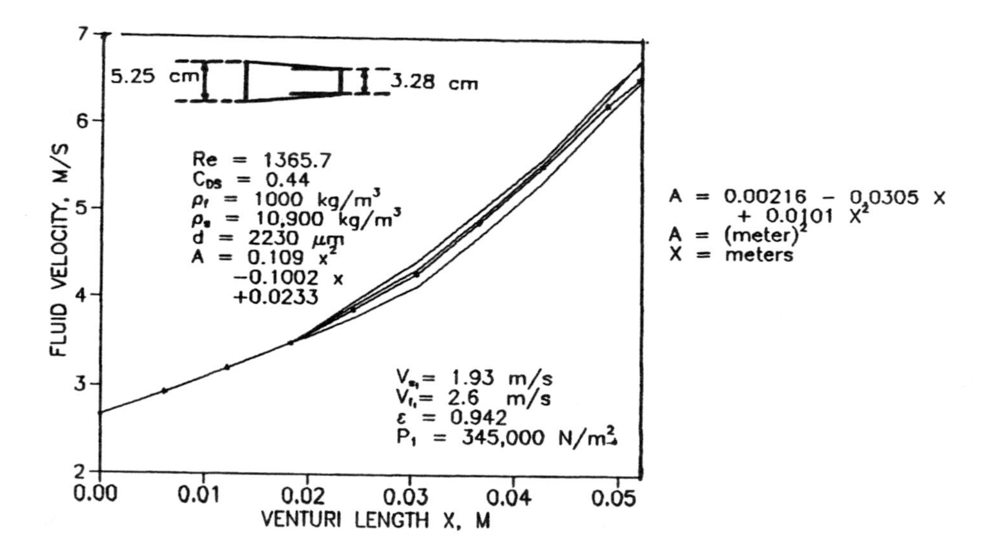

Fig. 2.9 Venturi with typical computed velocities for several models.

(b) Compute the fluid, solid velocities, pressure and solids fraction as a function of venturi height.
(c) Compute the discharge coefficient using the equation

$$C = \frac{\left[\varepsilon_1 v_{f_1} + (1-\varepsilon_1) v_{s_1}\right]}{n^2\left[\dfrac{2g_c(P_1 - P_2) + 2g_c\rho_D(z_1 - z_2)}{\rho_D(1-n^4)}\right]^{1/2}},$$

where

$$\rho_D = \frac{\rho_f \varepsilon v_f + \rho_s(1-\varepsilon)v_s}{\varepsilon v_f + (1-\varepsilon)v_s}$$

, $n = A_2/A_1$, and z = elevation above datum.

2. *Differential Equations for Homogeneous One-Dimensional Gas–Liquid Two-Phase Flow*

Consider equal velocity (homogeneous) equilibrium two-phase flow of a single-component fluid in a pipe of constant area, A, shown in Fig. 2.10. Heat is added to the fluid at a rate Q per unit volume; the friction acting on the wall per unit length and unit cross-sectional area is f_w. Pressure is related to the fluid density ρ and to the internal energy U by means of an equation of state:

$$P = P(\rho, U).$$

(a) Formulate the unsteady, one-dimensional equations of change for the four dependent variables: ρ, the mixture density; U, the mixture internal energy; v, the velocity, and P, the pressure.

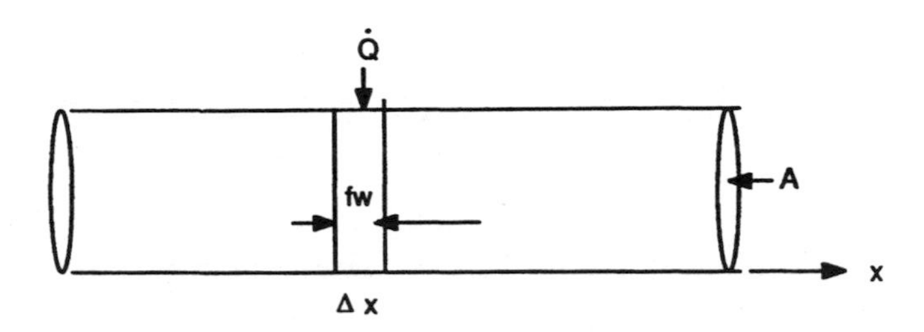

Fig. 2.10 Systems for one-dimensional flow with heat addition. Such a model was used in licensing nuclear reactors.

(b) Express the energy equation in terms of entropy and find the local rate of entropy production.

3. *Desulfurization in a Transport Reactor*

Combustion gases can be purified of SO_2 by treating them with calcined dolomite obtained by heating fine particles of the mineral dolomite, as shown in Fig. 2.11. The reaction taking place is

$$CaO + SO_2 + \tfrac{1}{2}O_2 \rightarrow CaSO_4.$$

The rate of reaction is first order in the concentration C_s of CaO equal to $\rho_s\varepsilon_s = C_s M_{CaO}$, where M_{CaO} is the molecular weight, ρ_s is the density of dolomite and ε_s its solid volume fraction. At 870°C the rate constant k is 2.3 min^{-1} (Borgwardt, 1970). Assume the initial porosity of the solid is sufficiently high so that the reaction occurs homogeneously throughout the particles.

One-millimeter diameter particles are injected into a pipe of cross-sectional area A, as shown in Fig. 2.11. The mass flow rate of solids is W_s and that of gases is W_g. The inlet SO_2 concentration is °C mol/m^3, the pressure is P_o and the temperature is T_o. Note that the solids are picked up by the gases from a zero velocity and accelerated up the reactor to a developed profile. Because of a higher solids hold-up at the bottom, there is more reaction in this region at the same CaO conversion.

Formulate the steady-state one-dimensional conservation equations for CaO and SO_2 concentrations, together with the other equations needed to determine the

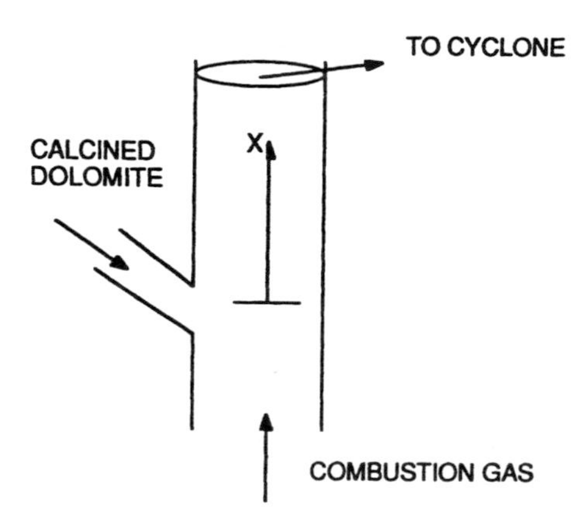

Fig. 2.11 Gas purification system using pneumatic transport.

desulfurization efficiency as a function of system parameters. Assume isothermal operation and neglect phase changes caused by the reaction.

4. *Transport Reactor Using a Shrinking Core Model*

Consider a steady-state one-dimensional transport reactor of diameter D. Catalyst particles of diameter d containing W_{co} carbon by weight are injected into the reactor at a rate G_C kg / m^2-s with air at a superficial velocity of U_{go}. The inlet temperature of air and particles is T_o, the inlet pressure is P_o, and the inlet volume fraction of particles is ε_{so}. The temperature of the transport line is maintained at the constant temperature T_v. The transport reactor is vertical. The objective is to determine the length of line needed to burn off the deposited carbon. At the high temperature of the reactor, the rate of reaction $C + O_2 \rightarrow CO_2$ can be expressed by the well-known shrinking core mechanisms (for a generalization of this model see Yoon *et al.*, 1978). This rate, r_i mol C/m^3 (gas + particle)-s, is given by

$$r = (1-\varepsilon)\frac{3DC_{O_2}}{R_o^2}\frac{W_c^{1/3}}{\left(1-W_c^{1/3}\right)},$$

where ε = void fraction in line, D = molecular diffusivity of O_2 through the catalyst pellet, R_o = $d/2$ of particle, W_c = weight fraction of carbon, and $C_{O_2} = O_2$ concentration, mol/m^3.

Set up the six ordinary differential equations for ε, P, V_g, V_s, W_c; the weight fraction of carbon; and Y_{O_2}, the weight fraction of oxygen needed to compute the profiles along the height of the reactor x.

5. *Direct Contact Heating of Solids in a Vertical Pipe*

Solids at a temperature T_{so} enter an adiabatic pipe of diameter D_t at a steady rate of W_s kg/m^2-s. Gas flowing at a superficial velocity U_{go} enters the pipe at a temperature T_{go} which is much higher than T_{so}. The heat transfer between the gas and the solid can be described by the following semi-empirical correlation:

$$Nu = \left(hd_p/k_g\right) = 2.0 + 0.65\,\mathrm{Re}_p^{0.5}\,\mathrm{Pr}^{1/3},$$

where

$$\mathrm{Re}_p = \frac{d_p\left|V_s - V_g\right|\rho_g}{\mu_g} \quad \text{and} \quad Pr = \frac{C_{p_g}\mu_g}{k_g}.$$

In these above equations h is the heat transfer coefficient per unit area, J/m^2-s of particles, d_p is the particle diameter, k_g is the thermal conductivity of the gas, μ_g is the gas viscosity, ρ_g is the gas density and C_{p_g} is the specific heat of the gas. Set up the six equations needed to compute the temperature profiles as a function of length x. The dependent variables in the six equations should be ε, P, V_g, V_s, T_g and T_s.

6. *Simultaneous Heat and Mass Transfer in a Riser*

Consider drying of alumina particles of diameter d_p in an adiabatic vertical pipe of diameter D_t, shown in Fig. 2.12. Solids at a temperature T_{so} and equilibrium weight fraction ω_o are fed into a pipe at a constant mass flow W_s kg/m^2-s. The hot gases enter the dryer at a temperature T_{go} and moisture weight fraction, Y_o at a constant superficial velocity of U_{go} m/s. The heat and mass transfer between the solid particles and the gas can be described by means of a heat transfer coefficient h, and a mass transfer coefficient k_y, as follows:

$$\text{Rate of heat transfer per unit particle area} = h\left(T_s - T_g\right),$$

$$\text{Rate of mass transfer per unit particle area} = k_y\left(Y_s - Y_g\right),$$

where T is the temperature and Y is the weight fraction, and the subscripts s and g refer to the solid and the gas, respectively. The equilibrium relationship for water in alumina is $\omega = \omega(T_s, Y_s)$, where ω is the weight fraction of water in the solid.

Set up the eight equations needed to compute the temperature and water profiles as a function of length x in the riser. The dependent variables should be ε, P, V_g, V_s, T_g, T_s, Y_g and ω.

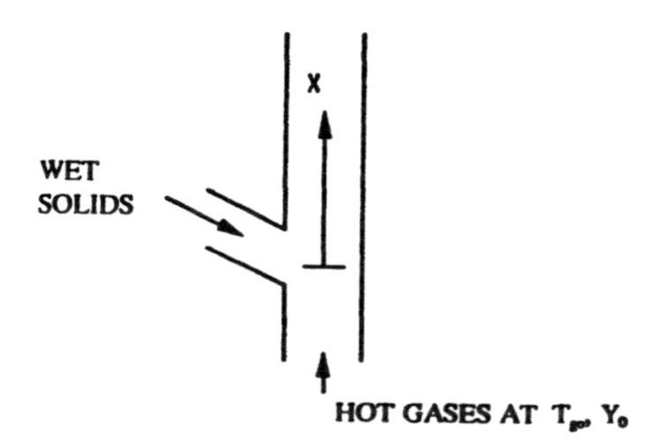

Fig. 2.12 Drying system using pneumatic transport.

7. *Circulating Fluidized Bed Combustor*

Consider combustion of a char particle in the circulating fluidized bed combustor in Fig. 2.13. For simplicity assume that the carbon in the char reacts with oxygen in the air to form carbon dioxide according to the following shrinking core model mechanism :

$$C + O_2 \rightarrow CO_2 ,$$

$$r_c = \frac{3DC_{O_2}}{R_o^2} \frac{\omega_c}{\left(1 - \omega_c^{1/3}\right)} ,$$

where r_c = molar rate of carbon consumption per unit volume of char (mol/s-m^3 char); C_{O_2}= molar concentration of oxygen (mol/m^3), D = diffusivity of oxygen through char (m^2/s), R_o = char particle radius, and ω_c =weight fraction of carbon in char.

Char is injected into the combustor at a temperature T_{so}, while the air and unreacted particles enter the combustion chamber at a temperature T_{go}. The rate of heat transfer between the particles and the gas is given by

$$q_{\text{wall-gas}} = U(T_w - T_g), \text{ J / m}^2\text{wall - s.}$$

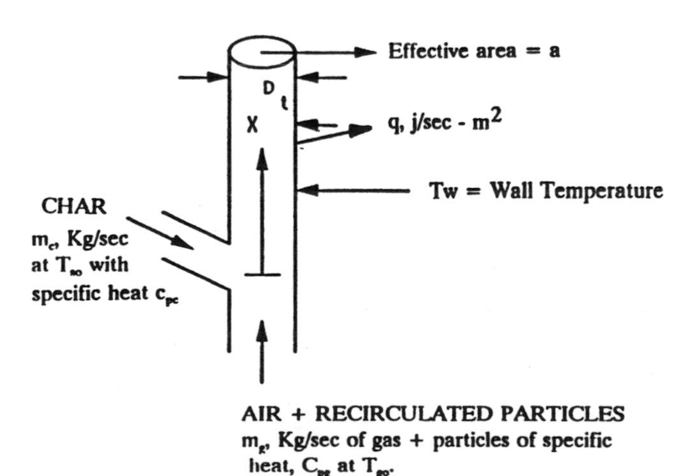

Fig. 2.13 Char combustion in pneumatic transport.

(a) Set up the eight ordinary differential equations for $\varepsilon, P, V_g, V_s, \omega_c, C_{O_2}, T_s$ and T_g needed to determine the profiles along the height of the reactor x and thus the combustion efficiency and the steam production.
(b) What simplification can you make in the hydrodynamically fully developed region?
(c) Since the surface area of the particles is large, the gas and the solid temperatures will become nearly equal. Set up the mixture energy equation for this case.
(d) In the developed region, find an analytical solution for oxygen concentration for the case of small carbon conversion.

For a more complete description of the kinetics in the system, see Breault (1987).

Literature Cited

Arastoopour, H., and D. Gidaspow (1976). "A Comparison of Four Unequal Velocity Models: Flow of a Slurry through a Venturi Meter," pp. 1333–1356 in T. N. Veziroglu and S. Kakac, eds. *Two-Phase Transport and Reactor Safety*, Vol. 1. Hemisphere Publishing Corporation.

Arastoopour, H., and D. Gidaspow (1979a). "Vertical Countercurrent Solids Gas Flow," *Chem. Eng. Sci.* **34**, 1033–1066.

Arastoopour, H., and D. Gidaspow (1979b). "Analysis of IGT Pneumatic Conveying Data and Fast Fluidization Using a Thermodynamics Model," *Powder Technol.* **22**, 77–87.

Arastoopour, H. and D. Gidaspow (1979c). "Vertical Pneumatic Conveying Using Four Hydrodynamic Models," *I&EC Fundamentals* **18**, 123–130.

Bird, R. B., W. E. Stewartt, and E. N. Lightfoot (1960). *Transport Phenomena.* New York: John Wiley & Sons.

Birkhoff, G. (1960). *Hydrodynamics: A Study in Logic, Fact and Similitude*, Rev. Ed. Princeton, New Jersey: Princeton University Press.

Borgwardt, R. H. (1970). "Kinetics of Reaction of SO_2 with Calcined Limestone," *Envir. Sci. and Tech.* **4**, 59–63.

Breault, R. W. (1987). "Theoretical Modeling of the Multi-solids Fluidized Bed Combustor: Hydrodynamics, Combustion and Desulfurization," pp. 770–775 in J. P. Mustonen, Ed. *1987 International Conference on Fluidized Bed Combustion*, Vol. II. New York: ASME.

Collier, J. G. (1972). Chapter 6 in *Convective Boiling and Condensation.* New York: McGraw-Hill.

Gidaspow, D. (1978). "Hyperbolic Compressible Two-Phase Flow Equations Based on Stationary Principles and the Fick's Law," pp. 283–298 in T. N. Veziroglu and S. Kakac, Eds. *Two-Phase Transport and Reactor Safety*, Vol. 1. Hemisphere Publishing Corp.

Gidaspow, D. (1986). "Hydrodynamics of Fluidization and Heat Transfer: Supercomputer Modeling," *Appl. Mech. Rev.* **39** (1), 1–23. Publication of AIChE Award Lecture: 1984 Donald Q. Kern award presented at the 1985 Heat Transfer Conference.

Gidaspow, D., and B. Ettehadieh (1983). "Fluidization in Two-Dimensional Beds with a Jet, Part II: Hydrodynamics Modeling," *I&EC Fundamentals* **22**, 193–201.

Gidaspow, D., W. Toczycki and C.W. Solbrig (1972). "Adiabatic Humidification in Boundary Layer Flow with a General Computer Program for Non-linear Boundary Value Problems," pp. 147–176 in R. Gordon, Ed. *Computer Programs for Chemical Engineering Education Vol. IV: Transport.* Houston, Texas: University of Houston; Austin, Texas: Aztec Publishing Co.

Gidaspow, D., K. M. Luo, and Y.P. Tsuo (1987). "Dilute, Dense-Phase and Maximum Solid–Gas Transport," *Final Technical Report to the U. S. Department of Energy, PETC Grant No. DE-FG22-84PC70057.* Available from the National Technical Information Service.

Hewitt, G. F. (1977). "Two-Phase Flow Patterns and Their Relationship to Two-Phase Heat Transfer," pp. 11–35 in S. Kakac and F. Mayinger, Eds., *Two Phase Flows and Heat Transfer,* Vol. I. Washington, D.C.: Hemisphere Publishing Co.

Institute of Gas Technology (1982). *Coal Conversion Systems Technical Data Book.* U.S. Department of Energy Report by IGT under Contract No. AC01-81FE05157. Available from National Technical Information Service.

Jones, P. J., and L.-S. Leung (1985). "Downflow of Solids through Pipes and Valves," in J. F. Davidson, R. Clift, and D. Harrison, Eds. *Fluidization.* New York: Academic Press.

Klinzig, G. (1981). *Gas–Solid Transport.* New York: McGraw-Hill.

Konno, H., and S. Saito (1969). "Pneumatic Conveying of Solid through Straight Pipes," *J. Chem. Eng. of Japan* **2** (2), 211–217.

Kunii, D. and O. Levenspiel (1969). *Fluidization Engineering.* New York: John Wiley & Sons, Inc.

Leung, L.-S. (1980). "The Ups and Downs of Gas–Solid Flow: A Review," in J. R. Grace and J. M. Matson, Eds. *Fluidization.* New York: Plenum.

Luo, K. M. (1987). "Experimental Gas–Solid Vertical Transport." Ph.D. Thesis. Chicago, Illinois: Illinois Institute of Technology.

Lyczkowski, R. W., D. Gidaspow, and C. W. Solbrig (1982). "Multiphase Flow Models for Nuclear, Fossil, and Biomass Energy Production," pp. 198–351 in A. S. Majumdar and R. A. Mashelkar, Eds. *Advances in Transport Processes,* Vol. II. New York: Wiley Eastern Publisher.

Nakamura, K., and C. E. Capes (1973). "Vertical Pneumatic Conveying: A Theoretical Study of Uniform and Annular Particle Flow Models," *Can. J. Chem. Eng.* **51**, 39–46.

Richardson, J. F., and Zaki, W. N. (1954). "Sedimentation and Fluidization: Part I," *Trans. Inst. Chem. Eng.* **32**, 35–53.

Rowe, P. N. (1961). "Drag Forces in a Hydraulic Model of a Fluidized Bed, Part II," *Trans. Inst. Chem. Engs.* **39**, 175–180.

Savage, S. B. (1983). "Granular Flows at High Shear Rates," in R. E. Meyer, Ed. *Theory of Dispersed Multiphase Flow*. New York: Academic Press.

Shook, C. A. and J. H. Masliyah (1974). "Flow of a Slurry Through a Venturi Meter," *Can. J. Chem. Eng.* **52**, 228–233.

Soo, S. L. (1983). *Fluid Dynamics of Multiphase Systems*. Urbana, Illinois: Soo & Associates.

Soo, S. L. (1989). *Particulates and Continuum*. New York: Hemisphere Publishing Co.

Syamlal, M., and T. J. O'Brien (1987). "A Generalized Drag Correlation for Multiparticle Systems," Morgantown Energy Technology Center DOE Report.

Teo, C. S., and L. S. Leung (1984). "Vertical Flow of Particulate Solids in Standpipes and Risers," in N. P. Cheremisinoff, Ed. *Hydrodynamics of Gas–Solids Fluidization*. Gulf Publishing.

Wen, C. Y., and Y. H. Yu (1966). "Mechanics of Fluidization," *Chem. Eng. Prog. Symp. Series.* **62**, 100.

Yoon, H., S. Wei, and M. M. Denn (1978). "A Model for Moving Bed Coal Gasification Reactors," *AIChE J.* **24**, 885-890.

Zenz, F. A. (1949). "Two Phase Fluid-Solid Flow," *Ind. Eng. Chem.* **41**, 2801–2806.

Zenz, F. A. (1971). "Regimes of Fluidized Behviour," Ch. 1 in J. F. Davidson and D. Harrison, Eds. *Fluidization*. New York: Academic Press.

3

DRIFT FLUX

3.1 Introduction 61
3.2 Hold-up in Homogeneous and Slip Flow 61
3.3 Single Particle Analysis 62
3.4 Ergun Equation Prediction 63
3.5 Two Regimes in a Bubble Column 66
3.6 Other Applications 68
Exercises 69
1. Prediction of Hold-up in Gas–Liquid–Solid Vertical Developed Flow 69
2. On Hold-up Correlations in the Literature 70
3. Application of Model A in Three-Phase Flow 71
Literature Cited 71

3.1 Introduction

Drift flux analysis is useful for estimating the volume fraction of the dispersed phase in multiphase flow systems. It was popularized by Wallis (1969) and Hewitt (1982) and extended to gas-particle systems by Leung (1980), Luo (1987), Matsen (1988) and others. It is based on a balance between buoyancy and drag for developed flow.

3.2 Hold-up in Homogeneous and Slip Flow

The volume fraction of the dispersed phase ε_s in a mass transfer device such as a bubble column or a three-phase fluidized bed is called the hold-up. This hold-up usually differs significantly from that computed by a single steady-state mass balance on the particular phase. Consider a vertical pipe into which solids from a screw-feeder are fed at a rate W_s kg/s and a gas at a rate W_g kg/s. The steady-state mass balance shows that the solids weight fraction discharged at the top, $Y_{s,\text{ out}}$, is simply

$$Y_{s,\,\text{out}} = \frac{\dot{W}_s}{\dot{W}_s + \dot{W}_g}. \tag{3.1}$$

However, because of slip or recirculation of the solids inside the vertical pipe, the weight fraction of the solids inside the pipe can be an order of magnitude or more greater than that given by the simple mass balance, Eq. (3.1). In terms of a slip ratio, S_L,

$$S_L = \frac{v_s}{v_g}. \tag{3.2}$$

Hewitt (1982) and others express these volume fractions in terms of respective volumetric flow rates $\dot{Q}_i$. For a steady state, continuity of phases for incompressible fluids shows that

$$\frac{\dot{Q}_i}{a} = \varepsilon_i v_i\,, \quad i = g, s. \tag{3.3}$$

Then an algebraic rearrangement of Eq. (3.2) shows that

$$\varepsilon_s = \frac{\dot{Q}_s}{\dot{Q}_s + S_L \dot{Q}_g}. \tag{3.4}$$

For the homogeneous model $v_g = v_s$, $S_L = 1$; the relation is similar to that obtained by the mass balance, Eq. (3.1). For nonhomogeneous flow, slip correlations exist for various flow regimes (Hewitt, 1982).

3.3 Single Particle Analysis

A balance between buoyancy and drag for developed flow with negligible wall effects,

$$\tfrac{4}{3}\pi\left(\frac{d_p}{2}\right)^3 g(\rho_s - \rho_g) = \pi\left(\frac{d_p}{2}\right)^2 \left(\tfrac{1}{2}\rho_g v_t^2 C_D\right), \tag{3.5}$$

gives the terminal velocity v_t in terms of the standard drag coefficient C_D as discussed in Chapter 2,

$$v_t = \left(\frac{4}{3} \frac{\Delta \rho g d_p}{C_D \rho_G} \right)^{1/2}. \tag{3.6}$$

It is now assumed that this single particle terminal velocity is the relative velocity between the particle and the gas phase in a vertical pipe:

$$v_t = v_g - v_s. \tag{3.7}$$

The gas and the solids velocities are expressed in terms of the constant gas and solids feed rates:

$$\varepsilon v_g = U_g, \tag{3.8}$$

$$\varepsilon_s v_s = W_s / \rho_s . \tag{3.9}$$

Solution of these equations for ε_s gives the hold-up relation in terms of gas and solids feed rates:

$$\varepsilon_s = \frac{W_s / \rho_s}{U_g / \varepsilon - \left(\frac{4}{3} \frac{\Delta \rho g d_p}{C_D \rho_g} \right)^{1/2}}, \tag{3.10}$$

$$\text{Hold-up} = \frac{\text{Solids volumetric flow}}{\text{Gas velocity - Terminal velocity}}.$$

For small particle sizes or a small terminal velocity, this equation reduces itself to the equations obtained from a mass balance, since W_s / ρ_s is normally very small. Equation (3.10) provides a rough estimate of the expected hold-up in dilute flow. It becomes invalid for circulating fluidized beds, since the assumptions made in the derivation are no longer true. It does show that there will be segregation in the vertical transport pipe by size and by density. Interaction between different sized particles has, of course, not been taken into account, leading to rough predictions only.

3.4 Ergun Equation Prediction

The buoyancy analysis presented in Chapter 2, Eq. (2.17), shows that, neglecting wall friction, a balance between the buoyant force and friction, expressed in terms of a friction coefficient β_A is as follows:

$$\varepsilon_s g(\rho_s - \rho_g) = \frac{\beta_A}{\varepsilon} v_r = \frac{\Delta P}{\Delta x}. \tag{3.11}$$

Buoyant force on the solid = Drag = Pressure drop given by the Ergun equation

The friction factor β_A was obtained from the Ergun equation and is given by Eq. (2.11) in terms of porosities. The relative velocity can be expressed in terms of the feed rates given by Eq. (3.8) and (3.9). Then Eq. (3.11) shows that

$$v_r = \frac{\varepsilon_s U_o - (W_s/\rho_s)\varepsilon}{\varepsilon\varepsilon_s} = \frac{\varepsilon\varepsilon_s g(\rho_s - \rho_g)}{\beta_A}. \tag{3.12}$$

For a given inlet superficial velocity U_o and mass flow rate per unit area W_s, Eq. (3.12) permits a computation of the solids volume fraction without any experiment. The accuracy of the prediction depends upon the validity of the Ergun approximation to the moving fluid-particle system. Clearly, wall effects and reverse flow, as occur in circulating fluidized beds, will invalidate the prediction.

Traditionally, the calculation of the hold-up is presented in terms of the drift flux j_{gs}, which is the flux of gas relative to the weighted average velocity as shown in Eq. (3.13):

$$j_{gs} = \varepsilon(v_g - \bar{v}), \tag{3.13}$$

where

$$\bar{v} = \varepsilon_s v_s + \varepsilon v_g. \tag{3.14}$$

Then, an algebraic manipulation of Eqs. (3.13) and (3.14) shows that

$$j_{gs} = \varepsilon\varepsilon_s v_r. \tag{3.15}$$

Use of the friction factor β_A from the Ergun equation given by Eq. (2.11) shows that Eq. (3.12) can be written in terms of the drift flux as follows:

$$j_{gs} = \frac{\varepsilon^3 (d_p \phi_s)^2 \Delta\rho g}{150\mu_g} + \frac{\varepsilon^3 \varepsilon_s^{\,2} \Delta\rho g d_p \phi}{1.75\rho_g j_{gs}}. \tag{3.16}$$

Solution of this quadratic for j_{gs} gives the expression

$$j_{gs} = \frac{2\varepsilon^3 \varepsilon_s{}^2 (\rho s - \rho g) g (d_p \phi_s)^2}{150 \varepsilon_s{}^2 \mu g + \sqrt{(150 \varepsilon_s{}^2 \mu g)^2 + 7\rho_g (\rho s - \rho g) g \varepsilon^3 \varepsilon_s{}^2 (d_p \phi_s)^3}}. \tag{3.17}$$

Figure 3.1 shows a plot of Eq. (3.17) as a function of porosity for transport of 520 micrometer round glass beads, with a sphericity of one. It also shows a plot of the operating line given by

$$j_{gs} = (1-\varepsilon) U_o - (W_S/\rho s)\varepsilon. \tag{3.18}$$

Equation (3.18), obtained from Eqs. (3.15), (3.8), and (3.9), is a straight line with two intersection points at the boundaries:

$$j_{gs} = -W_S/\rho_s \quad \text{at} \quad \varepsilon = 1, \tag{3.19}$$

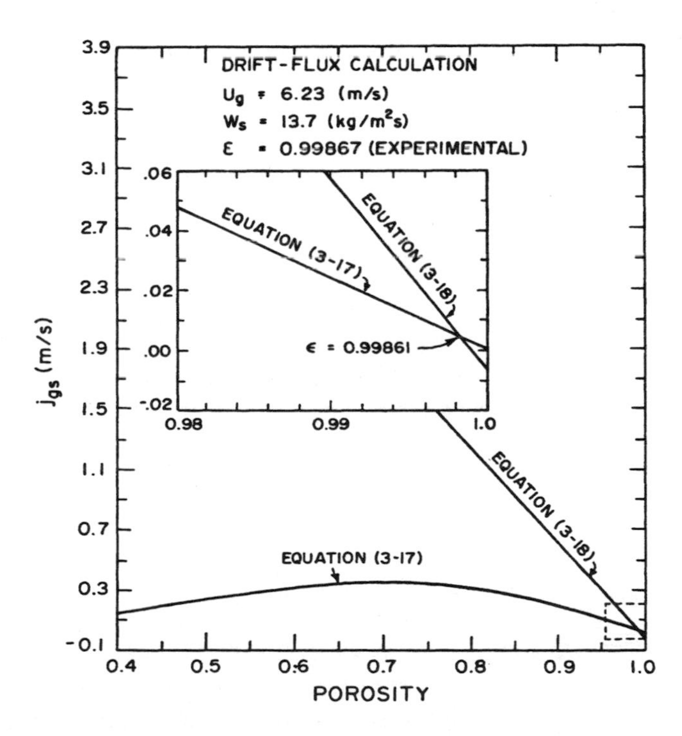

Fig. 3.1 A graphical method to predict porosity for a given solids flux and superficial gas velocity using a drift flux equation based on the Ergun equation (from data in Luo and Gidaspow, 1991).

$$j_{gs} = U_o \quad \text{at} \quad \varepsilon = 0. \tag{3.20}$$

For a given inlet solids feed rate W_S and inlet superficial gas velocity U_g, the intersection of the operating line and the hydrodynamic line gives the porosity in the developed region in the tube. Figure 3.1 illustrates this for typical inlet flow rates obtained in the pneumatic transport apparatus shown in Fig. 2.3. The computed value of the solids volume fraction of 1.39×10^{-2} is close to the measured value of 1.33×10^{-2} in the developed flow region in the 0.0762 m tube. Similar good agreement was obtained for other gas and solids feed rates in the dilute regime always obtained by means of a screw solids feeder (Luo, 1987; Gidaspow *et al.*, 1987). This illustrates how hold-up can be computed in developed dispersed flow.

3.5 Two Regimes in a Bubble Column

The quadratic nature of Eq. (3.15) for a constant relative velocity together with the operating line given by Eq. (3.18) suggests that two co-existent steady states are possible for some specified dispersed and continuous flows. Figure 3.2 shows the co-existence of loose and dense bubble phases in an Elgin type column reported by Lackme (1974). The volumetric concentrations of a dispersed phase in the two regions can be found by means of the drift flux analysis. Lackme (1974) and others (Hewitt, 1982; Wallis, 1969) related the relative velocity in Eq. (3.15) to the terminal velocity of the dispersed phase, corrected by means of an empirical packing effect:

$$v_r = v_t \varepsilon_s^{n-1}. \tag{3.21}$$

Then the drift flux, as given by Eq. (3.15) becomes

$$j_{gs} = v_t \varepsilon \varepsilon_s^n. \tag{3.22}$$

Hewitt (1982) gives various expressions for the terminal velocity of bubbles with their respective values of n for several regimes. For the Stokes flow regime, his value of n is two, in agreement with the example presented by Lackme (1974), to be described in the next paragraph. In gas–liquid systems α is usually taken to be the volume fraction of the dispersed phase that corresponds to ε_s.

Lackme (1974) found that the relative flux of nitrogen bubbles in a kerosene–heptane mixture divided by the volume fraction of the bubbles was proportional to the square of the volume fraction of the continuous phase. Thus,

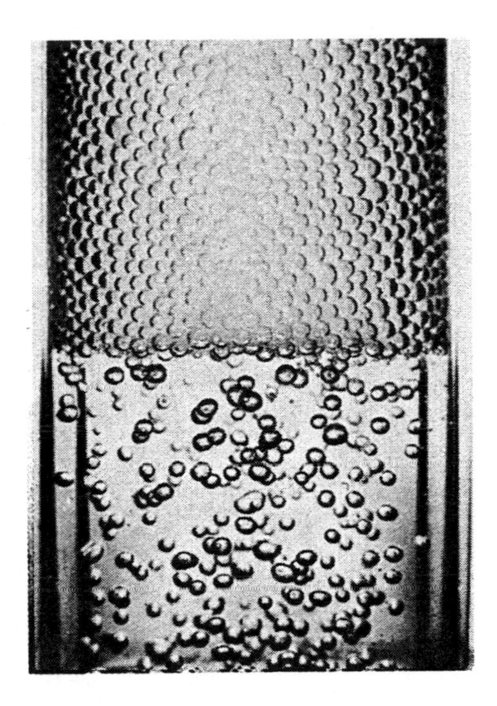

Fig. 3.2 Interface between loose and dense packing air–water system. From C. Lackme, "Two Regimes of a Spray Column in Countercurrent Flow," AIChE Symposium Series **70** (138), 57–63. Reproduced by permission of the American Institute of Chemical Engineers,

$$j_{12} = V_o \alpha (1-\alpha)^2, \tag{3.23}$$

where V_o is the terminal velocity of the single bubble. Figure 3.3 shows a plot of his data. The straight lines in the figure are the operating lines for various flow rates per unit tube area $\dot{Q}_c$ given by

$$j_{12} = (1-\alpha)\dot{Q}_D + \alpha \dot{Q}_c \tag{3.24}$$

for the countercurrent situation. The pairs of intersections of the operating lines and the hydrodynamic curve were found to agree well with data of volume fraction versus the volumetric rate of continuous phase flow for a fixed flow rate of the dispersed phase.

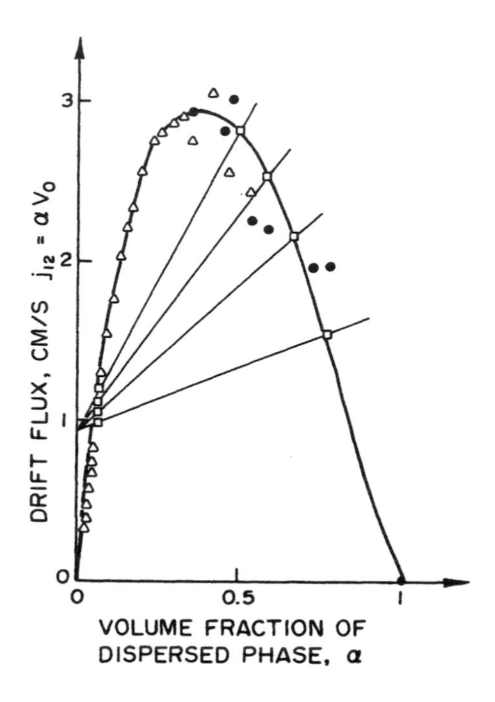

Fig. 3.3 Drift flux for nitrogen (kerosene–heptane), $\rho_c = 0.773$ g/cm. From C. Lackme, "Two Regimes of a Spray Column in Countercurrent Flow," AIChE Symposium Series **70** (138), 57–63. Reproduced by permission of the American Institute of Chemical Engineers, Copyright © 1974 AIChE. All rights reserved.

3.6 Other Applications

Matsen (1988) uses the type of semi-empirical analysis described by Eq. (3.22) to obtain a hydrodynamic curve that allows the intersection of an operating line for a gas–solid system in two places for co-current upward flow. He interprets these two states to correspond to the dense and dilute regions in circulating fluidized beds (see Fig. 3.4). To obtain the hydrodynamic curve, Matsen lets

$$v_r / v_t = 10.8\varepsilon_s^{0.293} \tag{3.25}$$

in Eq. (3.15) for the drift flux. Although such an approach violates the assumptions of uniform flow and no wall effects made in the drift flux theory, it is a very useful method for estimating the solids concentrations in risers and perhaps circulating fluidized bed combustors.

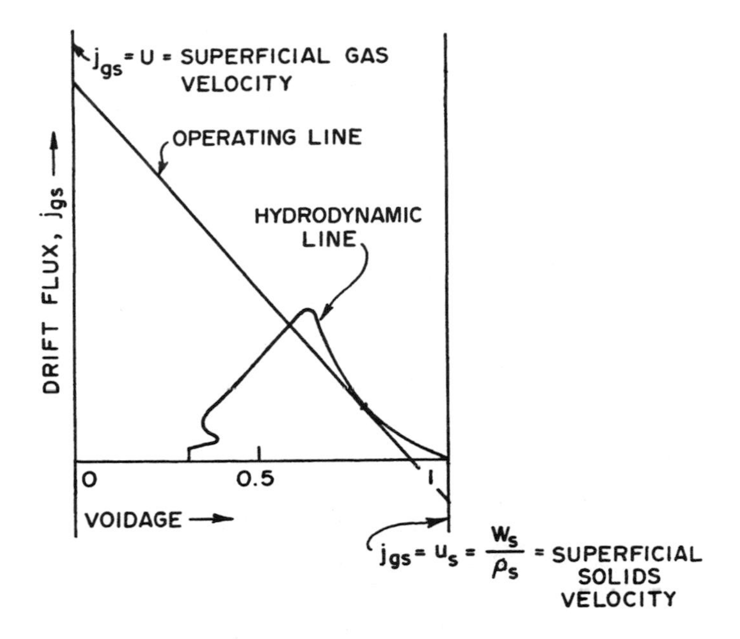

Fig. 3.4 Drift flux diagram for Type A powders. Operating line is set by gas and solids flux. System voidages are set by intersection with hydrodynamic line. From J. M. Matsen, "The Rise and Fall of Recurrent Particles," in *Circulating Fluidized Bed Technology II*. P. Basu and J. F. Large, eds. Reproduced by permission of the publisher. Copyright © 1988 Pergamon Press, PLC.

Extensions of the drift flux theory to estimate the hold-up in multiparticle systems, for gas-liquid-solid flow and for three-phase fluidization can be done using the separate phase momentum balances. As a minimum, such an approach should serve as a guide for elucidating the hold-up correlations normally obtained in these studies. As a bonus, it could provide a simple method of designing and thus costing a unit, such as a circulating fluidized bed combustor.

Exercises

1. *Prediction of Hold-Up in Gas–Liquid–Solid Vertical Developed Flow*

Consider steady state flow in a vertical vessel. A liquid slurry is pumped into the container at a rate of W_s kg/m^2-s of solids and W_l kg/m^2-s of liquid. Gas is injected through a distributor in the form of small bubbles of diameter d_B at a superficial velocity (volumetric flow rate divided by the total area of the vessel) of U_g. Due to a low gas velocity, bubble coalescence can be ignored.

Develop two algebraic relations to predict the volume fractions of the solid and the gas in the developed region of the vessel.

(a) Note that for incompressible flow the fluxes of the three phases are constant. Show this using the phase mass balances.
(b) A generalization of Eq. (3.11) to three phases is

$$\varepsilon_s g(\rho_s - \rho_l) = \beta_{ls}(v_s - v_l) + \beta_{gs}(v_s - v_g)$$

Buoyancy = Solid–Liquid drag + Liquid–Solid collisional interaction

Explain the assumptions in this momentum balance on the solid and make a similar balance on the gas.
(c) Formulate the equation for the pressure drop in the developed flow making assumptions that are consistent with the Model B momentum balances in part (b).
(d) Neglect the collisional gas–solid interaction and formulate the two algebraic equations to give the two hold-ups in the container in terms of the three inlet flow rates.
(e) Express the momentum balances in terms of the drift flux relations j_{gl}, j_{sl} and show that they are functions of the volume fractions of the solid and the gas, such as shown by Eq. (3.22) for the gas–solid system. Such relations may be useful for the dispersed bubble regime, but will likely break down for the case when the gas bubbles coalesce, that is, in the coalesced bubble regime.

2. *On Hold-up Correlations in the Literature*
L. S. Fan (1989) has summarized three-phase hold-up (volume fractions of liquid and gas) correlations reported in the literature. Two typical hold-up correlations for millimeter size glass beads suspended in water with air follow.

Correlation I. *Ostergaards and Michelsen's Correlation*

$$\text{Liquid volume fraction} = \varepsilon_l = 2.16 U_l^{0.397} 10^{-2.3 U_g},$$
$$2\times10^{-2}\,\text{m/s} \le U_l \le 11\times10^{-2}\,\text{m/s}; \quad 0 \le U_g \le 2.2\times10^{-2}\,\text{m/s}$$

where U_i is the superficial velocity of the fluid i, expressed in meters per second in the correlation.

Correlation II. *Razumer's Correlation for 30cm Column*

$$\varepsilon_s = 0.578 - 3.2U_l - 0.538U_g, \quad \varepsilon_l = 0.422 + 0.133U_l/dp - 1.82U_g.$$

(a) Explain Correlations I and II by referring to Eq. (3.10). Does the behavior with an increasing gas and liquid velocity make sense?
(b) Other correlations given by Fan (1989) are in terms of the Froude and Reynolds numbers, as defined by Eqs. (8.40) and (8.41). What are the natural dimensionless groups that arise from the momentum balances on the solid and on the gas?
(c) One of the correlations involves surface tension, expressed as

$$\varepsilon_g + \varepsilon_l = 1.4\, Fr_l^{0.17} \left(\frac{U_g \mu_l}{\sigma} \right)^{0.078},$$

where Fr_l is the Froude number for the liquid, μ_l is the liquid viscosity, and σ is the surface tension. Can such a correlation be deduced based on the momentum balances in part (a)?

Hint:

In gas–liquid flow a maximum bubble size can be estimated based on the critical Weber number (Fan, 1989):

$$We_{\text{critical}} = \frac{d_{B,\max} \rho_l \left(v_g - v_s \right)^2}{\sigma} = 1.24 .$$

3. *Application of Model A in Three-Phase Flow*
(a) Generalize the mixture momentum equation, Eq. (T2.3) in Chapter 2, to three-phase flow of dispersed solid and gas bubbles in a liquid.
(b) Generalize the pressure drop in all phases, Model A, Eq. (T2.4), to three-phase flow.
(c) Show that by neglecting wall friction, solids pressure due to particle-to-particle collisions and by assuming developed flow, the equations will reduce themselves to the buoyancy equaling drag relations previously obtained in Exercise 1.

Literature Cited

Fan, L. S. (1989). *Gas–Liquid–Solid Fluidization Engineering*. Boston: Butterworth.

Gidaspow, D. (Principal Investigator), K. M. Luo, and Y. P. Tsuo (1987). "Dilute, Dense-Phase, and Maximum Solid–Gas Transport," *Final Technical Report to U. S. Department of Energy, DOE Grant No. DE-FG-22-84 PC70057.*

Hewitt, G. F. (1982). "Void Fraction," Ch. 2.3 in G. Hetsroni, Ed. *Handbook of Multiphase Systems.*

New York: McGraw-Hill.

Lackme, C. (1974). "Two Regimes of a Spray Column in Countercurrent Flow," in D. Gidaspow, Ed. *AIChE Symposium Series* **138** (70), 57–63.

Leung, L. S. (1980). "The Ups and Downs of Gas–Solid Flow: A Review," in J. R. Grace and J. M. Matson, Eds. *Fluidization*. New York: Plenum Press.

Luo, K. M. (1987). *Experimental Gas–Solid Vertical Transport*. Ph.D. Thesis. Chicago, Illinois: Illinois Institute of Technology.

Luo, K. M. and D. Gidaspow (1991). "Computed Particle Hold-up in a Vertical Pneumatic Conveying Line," *Adv. Powder Technol.* **2** (4), 225–264.

Matsen, J. M. (1988). "The Rise and Fall of Recurrent Particles: Hydrodynamics of Circulation," Preprinted for the 2nd International Conference on Circulating Bed Technology, held in France.

Wallis, G. G. (1969). *One-Dimensional Two-Phase Flow*. New York: McGraw-Hill.

4

CRITICAL GRANULAR FLOW

4.1 One-Dimensional Granular Flow Momentum Balance 73
4.2 One-Dimensional Statics: Jannssen Equation 75
4.3 Incompressible, Frictionless Flow and Discharge 77
4.4 Thermodynamics of Powders: Compressibility 79
4.5 Critical Flow Theory for Granular Materials 82
4.6 Example of Critical Granular Flow 86
4.7 "Sound Speed" from Kinetic Theory 88
Exercises 89
1. Steady Gas–Solid Critical Flow 89
2. Maximum Homogeneous Discharge Rates 91
3. Critical Flow Using the Relative Velocity Model 93
4. Numerical Values of Critical Flow for an Air–Water System 94
Literature Cited 95

4.1 One-Dimensional Granular Flow Momentum Balance

Granular flow will be referred to as flow of a powder in a vacuum. Hence, there will be no fluid to support the particles. In this case the momentum balance on an element of a variable area a, as shown in Fig. 2.1, can be derived similarly to that of the derivation of the mixture equations (2.1). The Eulerian or shell balance is as follows:

$$\frac{d}{dt}\int_x^{x+\Delta x} \varepsilon_s \rho_s v_s a\, dx \quad + \quad a\varepsilon_s \rho_s v_s v_s \Big]_x^{x+\Delta x} +$$

Rate of accumulation of momentum in the volume $a\Delta x$ where the solid occupies a volume fraction ε_s

Net rate of momentum outflow across area $a\varepsilon_s$ with a velocity v_s

$$+\, a\sigma_s \Big]_x^{x+\Delta x} \quad - \quad \sigma_s \Delta a$$

Net rate of momentum outflow due to normal stress σ_s

Normal force due to curvature

$$= -\int_{x}^{x+\Delta x} g\varepsilon_s\rho_s a\,dx \quad + \quad \int_{x}^{x+\Delta x} \tau_{ws}\pi D_t\,dx.$$

Gravitational force acting downwards on the powder of true density ρ_s

Shearing force (friction) acting on the element of area $\pi D_t \Delta x$.

An application of the mean value theorem of differential calculus to the differences, evaluated at x and at $x + \Delta x$, of the mean value theorem of integral calculus to the integrals, followed by cancellation of the arbitrary element Δx, and application of the limit process to Δx, yields the one-dimensional momentum balance for granular flow:

Granular Momentum Balance

$$a\frac{\partial(\varepsilon_s\rho_s v_s)}{\partial t} + \frac{\partial(a\varepsilon_s\rho_s v_s v_s)}{\partial x} = -a\frac{\partial\sigma_s}{\partial x} - g\rho_s\varepsilon_s a + \tau_{ws}\pi D_t. \tag{4.1}$$

The momentum balance shown in Eq. (4.1) for the solid valid in a vacuum is complimented by the usual mass balance written in terms of the volume fraction ε_s and the "true" powder density ρ_s, which includes, however, the volume of the pores in the powder. The product of the true powder density and the solids volume fraction is almost always referred to as bulk density in the literature and is given the symbol of ρ or ρ_b. Hence, the solids mass balance is in a vacuum or in a mixture in the absence of phase formation:

Mass Balance

$$\frac{\partial(\varepsilon_s\rho_s)}{\partial t} + \frac{\partial(\varepsilon_s\rho_s v_s)}{\partial x} = 0. \tag{4.2}$$

It is an everyday experience that while the true powder density is approximately constant, the bulk density changes with, say, vibration. It is therefore convenient and in agreement with the convention used in the literature to define the new variable, ρ_b, as

$$\rho_b = \varepsilon_s\rho_s. \tag{4.3}$$

The following sections will explore the consequences of the mass and momentum balances derived.

4.2 *One-Dimensional Statics: Jannssen Equation*

For no motion, the granular momentum balance, Eq. (4.1), simplifies to the simple force balance equations

$$\frac{d\sigma_s}{dx} = -g\rho_b + \frac{\tau_{ws}\pi D_t}{a}. \tag{4.4}$$

For a fluid there is no shear in the absence of motion. Thus, the frictional force τ_{ws} would be zero. Then, for a height H, the stress at the bottom of the container would be

$$\sigma_{so} = g\rho_b H \qquad \text{(for a fluid)}. \tag{4.5}$$

Thus, the stress σ plays the same role as pressure. However, Eq. (4.5) is not valid for a powder, because a powder supports shear. The walls of the container support part of the weight of the powder.

The usual relation between the shear τ and the normal stress σ is given by the Coulomb failure condition (Sokolovsii, 1965; Abbott, 1966), which is the constitutive equation for a powder as represented in Fig. 4.1 and shown in Eq. (4.6):

$$\tau = k_c + \sigma \tan\theta. \tag{4.6}$$

Shear = Cohesion + Normal stress · Tangent of angle of internal friction

As represented in Fig. 4.1, this angle θ equals the angle of repose of a poured layer of particles in an idealized situation in which the mass of particles is homogeneous.

From statics it is known that a two-dimensional state of stress at any point, not necessarily in the failable state, can be represented by a circle in the σ–τ plane. This circle is usually called a Mohr circle.

A derivation clearly presented in McCabe and Smith's (1956) unit operations text for the case of non-cohesive solids shows that the ratio of applied stress to normal stress which they call pressure is given by the following ratio interpreted in terms of the radial stress σ_r to the applied stress σ_s:

$$\frac{\sigma_r}{\sigma_s} = \frac{1 - \sin\theta}{1 + \sin\theta}. \tag{4.7}$$

The wall shear is related to the radial stress in terms of the friction coefficient μ_w:

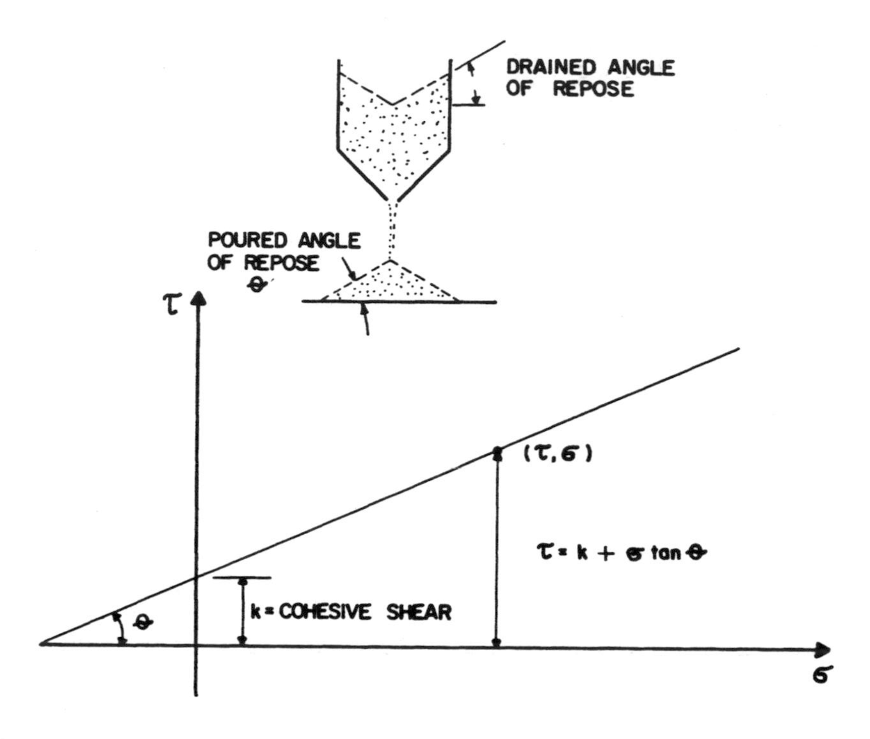

Fig. 4.1 Coulomb failure condition in the normal stress-shear plane. A constitutive equation for a powder.

$$\tau_{ws} = \mu_w \sigma_r . \tag{4.8}$$

McCabe and Smith (1956) give some typical values of the angle of repose for several materials. The angle of repose is low when the grains are smooth and rounded and high for sticky, very fine particles. For free-flowing granular materials the angle of repose is between 15 and 30 degrees and the ratio in Eq. (4.7) is between 0.35 and 0.6. A typical value of the friction coefficient in Eq. (4.8) is 0.5. Delaplaine (1956) presents typical values for smooth and roughened walls.

In terms of these definitions the static balance given by Eq. (4.4) becomes as follows for a circular tube:

$$\frac{d\sigma_s}{dx} = -g\rho_b + \left(\frac{4\mu_w}{D_t} \cdot \frac{1-\sin\theta}{1+\sin\theta} \right) \cdot \sigma_s . \tag{4.9}$$

Let σ_{so} be the stress at the bottom of the container at $x = 0$. If the container is filled with solids up to a height H, then at $x = H$, $\sigma_s = 0$. Then integration of Eq. (4.9) gives the Jannssen equation

$$\sigma_{so} = \frac{g\rho_s \varepsilon_s D_t}{K}\left(1 - e^{-Kx/D_t}\right), \tag{4.10}$$

where K is an effective friction coefficient

$$K = 4\mu_w \frac{1 - \sin\theta}{1 + \sin\theta}. \tag{4.11}$$

Expansion of the exponential in a Taylor series about zero shows that for small x / D_t, that is, for large vessel diameter, the stress at the bottom of the vessel is given by the hydrostatic formula for a fluid, Eq. (4.5). For a small vessel diameter, a large x, Eq. (4.10) gives the asymptotic value of the stress at the bottom of the vessel given by

$$\sigma_{so} = g\rho_b D_t / K. \tag{4.12}$$

Equation (4.12) shows that the stress at the bottom of the vessel is proportional to the vessel diameter and is independent of the height of powder in the container. The walls of the container support the weight of the powder and cause this reduced stress at the bottom of the vessel.

4.3 Incompressible, Frictionless Flow and Discharge

The steady-state granular momentum balance given by (4.1) can be written as

$$\varepsilon_s \rho_s v_s \frac{dv_s}{dx} = -\frac{d\sigma_s}{dx} - g\rho_s \varepsilon_s + \frac{4\tau_{ws}}{D_t}. \tag{4.13}$$

Although granular flow differs from the flow of fluids in that shear is a function of the normal stress, it is instructive to pursue the case of a frictionless granular flow. In such a case, Eq. (4.13) assumes the form of a steady Bernoulli's equation for fluids as shown in Eq. (4.14):

Bernoulli Analogue

$$v_S\, dv_S + \frac{1}{\rho_B}\, d\sigma_s + g\, dx = 0. \tag{4.14}$$

Clearly the solids stress plays the role of pressure. Then, for a constant solids volume fraction, Eq. (4.14) can be written as for fluids in the form below:

$$d\left(\frac{1}{2}v_s^2 + \frac{\sigma_s}{\rho_B} + gx\right) = 0. \tag{4.15}$$

Kinetic energy + Elastic energy + Potential energy

Equation (4.15) states that for frictionless flow the change of kinetic, elastic and potential energy is zero. If this equation is applied to the flow out of a vessel, shown in Fig. 4.2, it shows that

$$\frac{1}{2}v_{so}^2 = gL, \tag{4.16}$$

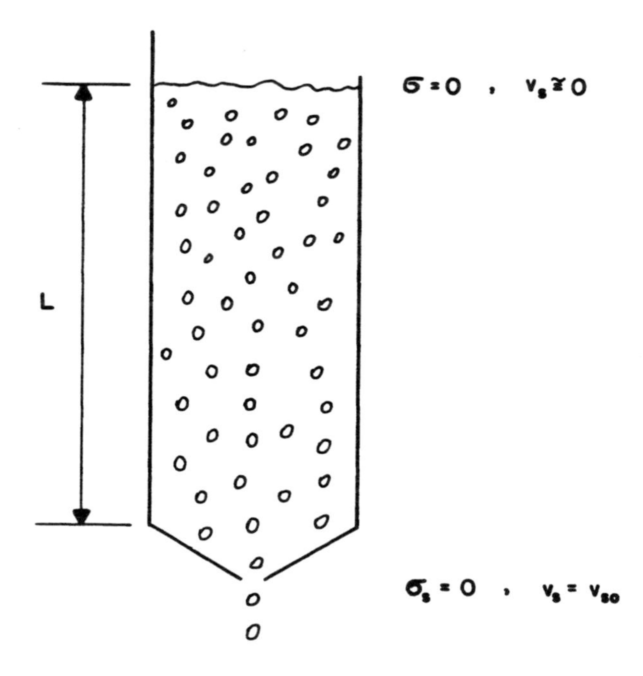

Fig. 4.2 Static stresses during hopper discharge.

since at $x = 0$ and at $x = L$, the stress σ_s is zero. Then, the velocity of discharge of granular materials should be given by, approximately,

$$v_{so} = \sqrt{2gL}. \tag{4.17}$$

However, it is a well-known experimental fact that the discharge rate of granular materials through orifices is independent of the height of the granular powder. (Nedderman *et al.*, 1982). The flow rate m' is usually expressed as

$$m' = \text{Constant} \cdot \rho_B D_o^2 \sqrt{gD_o}, \tag{4.18}$$

where D_o is the orifice diameter. Sometimes, the diameter under the square root sign is somewhat altered. In other words, the driving force, the potential energy due to the height, does not enter into the empirical equation for rates of discharge. The theoretical explanations in the literature (Nedderman *et al.*, 1982; Kaza and Jackson, 1982) involve the solution of the equations of motion for the granular flow in cylindrical or spherical geometries with the normal stress related to the shear as given by Eq. (4.6), but assume that the granular flow is incompressible. These solutions give the flow rate in the form of Eq. (4.18) with the addition of angles of repose similar to the value of K in Eq. (4.11). Kaza and Jackson (1982) state that the gap between the theory and the experiments is not closed. The thesis of this chapter is that this independence of flow of the material height in the container is due to the compressibility of the powder. The powder flow behaves more like a gas and reaches a critical flow. Bosley *et al.* (1969) and others (Gidaspow *et al.*, 1986) report a variation of the porosity of powders from the outlet orifice to the bulk, substantiating the claim of the effect of bulk density variation.

4.4 Thermodynamics of Powders: Compressibility

Thermodynamics of powders must start with the specification of the variables which define its state (Callen, 1960). In terms of the fundamental extensive variables per unit mass, these variables are the specific entropy, the bulk specific volume, the composition, and possibly some other variables such as the strains in a solid in more than one direction. Suppressing the latter, the internal energy of a powder U, is then

$$U = U(S, V_B), \tag{4.19}$$

where

$$V_B = 1/\rho_s \varepsilon_s. \tag{4.20}$$

Then the stress σ_s is defined like the pressure in a fluid by means of the equation

$$\sigma_s = -\frac{\partial U}{\partial V_B}. \tag{4.21}$$

In terms of a constant particle density of the solid ρ_s, at a constant entropy,

$$dU = \frac{\sigma_s}{\rho_s}\frac{d\,\varepsilon_s}{\varepsilon_s^{\,2}}, \tag{4.22}$$

which shows that expansion of the powder decreases its internal energy. Since the only new variable introduced here is the bulk powder specific volume, the temperature of the powder is still given by

$$T = \left(\frac{\partial U}{\partial S}\right)_{V_B}. \tag{4.23}$$

Thermodynamic stability (Callen, 1960)

$$d^2U > 0 \tag{4.24}$$

leads to the relation well known for fluids,

$$\left(\frac{\partial \sigma_s}{\partial V_B}\right)_{T \text{ or } S} < 0, \tag{4.25}$$

or to the more convenient expression

$$\left(\frac{\partial \sigma_s}{\partial \rho_B}\right)_{T \text{ or } S} > 0. \tag{4.26}$$

Since at constant temperature or entropy

$$\sigma_s = \sigma_s(\varepsilon_s), \quad \left(\frac{\partial \sigma_s}{\partial \rho_B}\right) = \frac{1}{\rho_s}\left(\frac{\partial \sigma_s}{\partial \varepsilon_s}\right). \tag{4.27}$$

The elastic bulk modulus for powders G, similar to the Young's modulus for solids, is defined to be

$$G = \left(\frac{\partial \sigma_s}{\partial \varepsilon_s} \right)_{T \text{ or } S} . \tag{4.28}$$

A subscript to denote isothermal or isentropic expansion is not included in the definition, since the differences may not be large, as in the case of solids. The sign of the modulus is also taken, the opposite of that previously used by Gidaspow (1986) and others, to avoid a negative sign in the empirical expression for G in terms of the void fraction. This thermodynamic stability shows that

$$\frac{\partial \sigma_s}{\partial \rho_B} = \frac{1}{\rho_s} \left(\frac{\partial \sigma_s}{\partial \varepsilon_s} \right) = \left(\frac{G}{\rho_s} \right) > 0 . \tag{4.29}$$

The square root of G divided by ρ_s is clearly the velocity of propagation through the powder as $\sqrt{\partial P / \partial \rho}$ is the critical velocity through a fluid. A solid supports both the compressional waves and shear waves, with the latter traveling slower (Tolstoy, 1973).

This development shows that the powder modulus G is a thermodynamic property of the powder. Therefore it should be measurable using thermodynamic, pressure–volume methods, in addition to critical flow or wave propagation methods. Some attempts at such measurements have been reported in the gun or ballistics community (Gough, 1979; Sandusky *et al.*, 1981). Gough (1979) has used the form

$$C_s = \begin{cases} C_{s_o} \varepsilon_o / \varepsilon & \varepsilon \le \varepsilon_o \\ C_{s_o} \exp\left[-\kappa (\varepsilon - \varepsilon_o) \right] & \varepsilon_{oo} < \varepsilon < \varepsilon_s \\ 0 & \varepsilon > \varepsilon_{oo} \end{cases} , \tag{4.30}$$

where C_s is the velocity defined by

$$C_s = \sqrt{G / \rho_s} , \tag{4.31}$$

ε_o is the settling porosity of the bed, κ is a stress attenuation factor, and ε_{oo} is some porosity above which C_s is zero. The value of C_{s_o} is 470 m/s for NACO propellant. Such a high value must correspond to a very high powder compaction.

Rietma and Mutsers (1973) used a tilted fluidized bed experiment to obtain the powder modulus for cracking catalysts, glass beads and polypropylene. Their data can be fitted by equations such as

$$G = 10^{-8.76\varepsilon+5.43}\,\mathrm{N/m^2}. \tag{4.32}$$

Similarly to Gough's expression (4.30), as the porosity ε approaches unity, the modulus becomes zero. Near minimum fluidizaiton velocity and below, the modulus becomes non-zero relative to other terms in the fluidization equations.

A convenient form for the modulus G was found to be (Bouillard *et al.*, 1989)

$$G = G_o \exp\left[-\kappa(\varepsilon - \varepsilon')\right], \tag{4.33}$$

with a typical value of $\kappa = 600$, $G_o = 1.0$ Pa, and $\varepsilon' = 0.376$.

Hopper-type critical flow experiments (Gidaspow *et al.*, 1986) produced a value of the velocity C_s near one meter per second at a porosity of about 0.4 for glass beads in the range of 1500 to 200 μm. Such values are typical of discharge velocities of granular materials from vessels.

To complete the thermodynamic treatment of powders one must define specific heats at constant porosity and at a constant stress:

$$C_\varepsilon = \left((\partial U)/(\partial T)\right)_\varepsilon , \tag{4.34}$$

$$C_\sigma = \left((\partial U)/(\partial T)\right)_\sigma . \tag{4.35}$$

Then thermodynamic stability requires that

$$C_\sigma > C_\varepsilon , \tag{4.36}$$

analogous to C_p being greater than C_v or constant stress heat capacity being greater than constant strain heat capacity. It would be of interest to measure such heat capacities of powders to determine whether they differ significantly from those of constituent solids.

4.5 Critical Flow Theory for Granular Materials

Critical flow is defined as the condition wherein the mass flow rate becomes independent of the downstream conditions. For a fluid, this is the condition at which a further reduction in pressure does not result in an increase of the mass flow rate. A limit occurs because signals no longer propagate upstream. This limit takes place when the fluid velocity just equals the propagation velocity, as illustrated in Fig. 4.3.

The theory of partial differential equations (Courant and Hilbert, 1952; Garabedian, 1964) shows that information flows along the characteristic directions $\lambda_i, i = 1, 2, 3...n$ for an nth order system. For multiphase flow the relevant partial differential equations are the conservation laws for mass and momentum for each phase.

In a vacuum, gas–particle flow reduces itself to granular flow. Compressible granular flow becomes almost completely analogous to compressible fluid flow, as treated in standard texts (von Mises, 1958). In one dimension, the momentum and mass balances are given by Eqs. (4.1) and (4.2). The constitutive equation for the stress σ_s can be expressed analogously to that of a fluid as

Constitutive Equation for the Stress

$$\sigma_s = \sigma_s(\varepsilon_s) + \sigma_s(\nabla \mathbf{v}_s, \varepsilon_s), \tag{4.37}$$

Thermodynamic "coulomb" stress + Rate dependent stress

where σ_s is a second order tensor. The second part of the right side of Eq. (4.37) gives rise to granular flow stresses (Savage, 1983) which are due to momentum

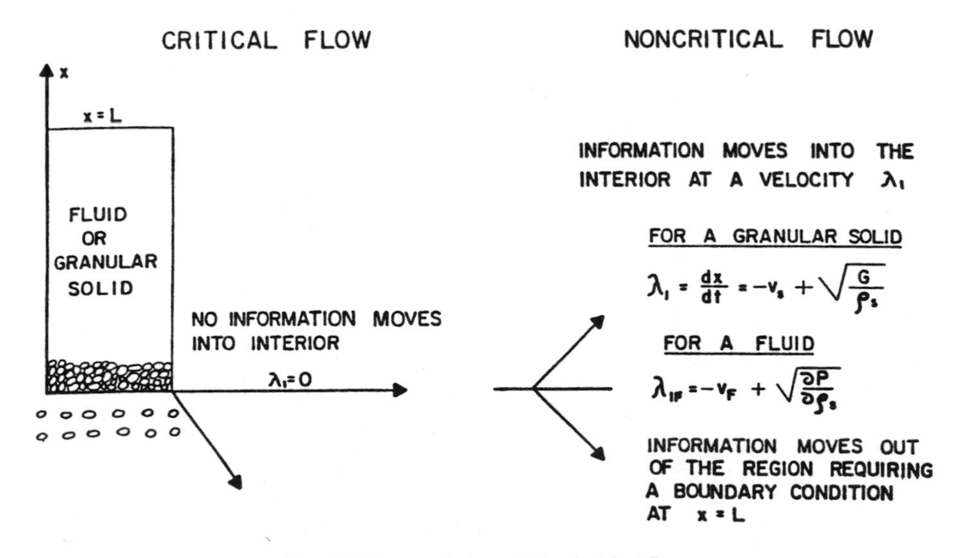

Fig. 4.3 Characteristic analysis of critical flow.

transfer during collisions between particles. As in compressible flow theory, these are neglected in the analysis of critical flow.

In matrix form the continuity and the momentum equations for granular flow, Eqs. (4.1) and (4.2) can be written as

$$\begin{pmatrix} \frac{\partial \rho_B}{\partial t} \\ \frac{\partial v_s}{t} \end{pmatrix} + \begin{pmatrix} v_s & \rho_B \\ \frac{G}{\rho_B \rho_S} & v_s \end{pmatrix} \begin{pmatrix} \frac{\partial \rho_B}{\partial x} \\ \frac{\partial v_s}{\partial x} \end{pmatrix} = \begin{pmatrix} 0 \\ -g + \frac{4\tau_{ws}}{D_t \rho_B} \end{pmatrix}. \tag{4.38}$$

The characteristic determinant is

$$\begin{vmatrix} v_s - \lambda & \rho_B \\ G/\rho_s \rho_B & v_s - \lambda \end{vmatrix} = 0. \tag{4.39}$$

The characteristic roots, λ_i, of the determinant are

$$\lambda_{1,2} = -v_s \pm \sqrt{G/\rho_s}\,. \tag{4.40}$$

Critical flow occurs when $\lambda_i = 0$, that is, when the downward flow is just balanced by $\sqrt{G/\rho_S}$, as shown in Fig. 4.3. Then for flow in a vacuum, the information that the stress fell to zero beyond the opening is not propagated into the container.

The compatibility conditions, that is, the equations along the characteristics given by Eq. (4.40) can be obtained by several methods (Abbott, 1966; von Mises, 1958; Gidaspow *et al.*, 1983). Since the result is the same as for gas dynamics or for homogeneous two-phase flow, it is merely quoted below for completeness. As used previously, let the critical velocity be

$$C_s = \sqrt{G/\rho_s}\,. \tag{4.41}$$

Then along the characteristic,

$$\frac{dx}{dt^{(1)}} = -v_s + C_s, \tag{4.42}$$

$$\frac{d\rho_B}{dt^{(1)}} + \frac{\rho_S}{C_s} \frac{dv_s}{dt^{(1)}} = \frac{\rho_S}{C_s}\left(-g + \frac{4\tau_{ws}}{D_t \rho_B}\right), \tag{4.43}$$

and along the second characteristic,

$$\frac{dx}{dt^{(2)}} = -v - C_s, \tag{4.44}$$

$$\frac{d\rho_B}{dt^{(2)}} + \frac{\rho_s}{C_s}\frac{dv_s}{dt^{(2)}} = \frac{\rho_s}{C_s}\left(-g + \frac{4\tau_{ws}}{D_{tB}}\right). \tag{4.45}$$

These equations show how information flows under non-critical conditions. At a critical condition Boure (1978) and Boure *et al.* (1976) have pointed out how the steady state equations can be used to obtain critical conditions quickly. Although Bilicki and Kestin (1983) have criticized Boure's procedure as mathematically inexact, it gives useful information once one is sure that we indeed have a critical state, as presented by the characteristic analysis. Boure's procedure involves looking at the steady state region of the conservation equations (4.38). Then solution for derivatives using determinants gives the following equation:

$$\frac{d\rho_B}{dx} = \frac{\begin{vmatrix} 0 & \rho_B \\ -g + \frac{4\tau_{ws}}{D_t} & v_s \end{vmatrix}}{\begin{vmatrix} v_s & \rho_B \\ \frac{G}{\rho_s \rho_B} & v_s \end{vmatrix}} = \frac{-g\rho_B + \frac{4\tau_{ws}}{D_t}}{v_s^2 - \frac{G}{\rho_s}} = \frac{0}{0}. \tag{4.46}$$

In Eq. (4.46) we assume that the flow is critical, given by

$$v_s = \pm\sqrt{G/\rho_s}\,. \tag{4.47}$$

Then, the numerator determinant must also be zero to have a finite derivative. Thus, when the flow is critical, the gravitational force is exactly balanced by friction, or

$$-g\varepsilon_s\rho_s + \frac{4\tau_{ws}}{D_t} = 0. \tag{4.48}$$

The maximum discharge rate can also be obtained from the steady-state horizontal momentum balance given by Eq. (4.1). For maximum flow, friction is zero. Under these conditions Eq. (4.1) assumes the simple form shown below:

$$\frac{d}{dx}\left(\rho_B v_s^2 + \sigma_s\right) = 0. \tag{4.49}$$

By Eq. (4.2) at a steady state, the flux, J_s, is constant:

$$J_s = \rho_B v_s. \tag{4.50}$$

In view of the constant flux, Eq. (4.49) can be rewritten as

$$J_s^2 \frac{d\rho_B^{-1}}{dx} + \frac{d\sigma_s}{dx} = 0. \tag{4.51}$$

Since $\sigma_s = \sigma_s(\rho_B)$,

$$J_s^2 = -\frac{d\sigma_s}{d\left(\rho_B^{-1}\right)} = \rho_B^2 \frac{d\sigma_s}{d\rho_B}. \tag{4.52}$$

With the definition of the powder modulus G given by Eq. (4.28), the maximum solids flux is

$$J_s = \rho_B \sqrt{G/\rho_s} = \rho_B C_s. \tag{4.53}$$

This completes the analogue between the compressible granular critical flow and gas dynamics and homogeneous two-phase flow. Other analogies, such as the observation that granular frictionless flow bulk density obeys the wave equation (Gidaspow, 1986), can be obtained.

4.6 Example of Critical Granular Flow

A crude attempt was made by Ding (Gidaspow *et al.*, 1986) to measure the powder modulus G from discharge rates of hoppers. Figure 4.4 shows the mass flow hopper used. The bin was made of Plexiglas® except for the curved section which was made of Teflon® to reduce friction between the particles and the wall. In a mass flow hopper, all particles are in motion. In a flat bottomed hopper of the type used by Bosley *et al.* (1969), the particles away from the outlet may not move. A dead zone forms in such hoppers. The hopper was filled with various sized glass beads and with sand, as shown in Table 4.1. The mass flow rates were determined by weighing the discharged particles on a balance. The values shown in Table 4.1 are an average of about 10 measurements. The outlet porosity was measured with a calibrated x-ray densitometer and a scintillation detector as shown in Figure 4.4. The solids velocity was obtained by divided the mass flow rate by the outlet area, which is 0.75×0.5 square inches, and by the product of the particle density and the measured porosity. The powder modulus G was then obtained from Eq. (4.47).

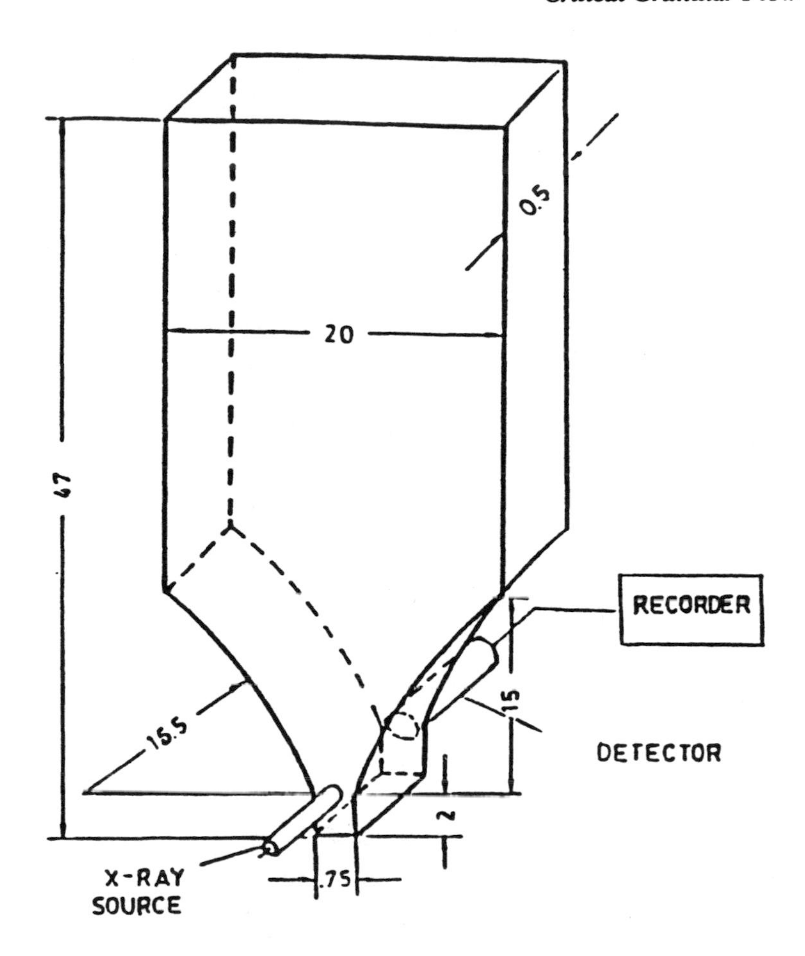

Fig. 4.4 Mass flow hopper for critical flow determination.

Table 4.2 shows that when the hopper was filled to the three heights shown, the measured mass flow rates were the same, within the experimental error, except for the possibly cohesive 86 μm particles. The independence of flow on height is one of the necessary conditions for the existence of critical flow. Various correlations of discharge of particles through orifices give flow rates that are somewhat smaller than those obtained in this specially constructed hopper. Verification of the measured modulus is necessary by an independent means. A measurement of the velocity of voidage propagation of particles in an evacuated tube seems like a more direct but more difficult way to obtain this void propagation velocity.

Table 4.1 MEASURED POROSITY, MASS FLOW RATE AND G

Material	d_p (μm)	ε_s	M (g/s)	v_s (cm/s)	G (dynes/cm^2)
Glass Bead	86	0.6170	203.7	56.39	7695
	200	0.6075	410.1	114.83	32042
	460	0.5275	374.6	120.30	35312
	500	0.5475	364.7	112.84	31068
	800	0.5700	351.1	103.92	26458
	1550	0.5750	386.8	113.49	31556
Sand	603	0.5650	375.3	105.19	28879

Table 4.2 INFLUENCE OF HEAD ON MASS FLOW RATE

Material	d_p (μm)	h = 116 cm	Mass Flow Rate (g/s) h = 88 cm	h = 58 cm
Glass Bead	86	204	191	177
	200	—	410	412
	460	374	372	364
	500	—	364	377
	800	351	354	357
	1550	387	390	373
Sand	603	375	369	363

4.7 "Sound Speed" from Kinetic Theory

In the 1980s, dense phase kinetic theory of gases (Chapman and Cowling, 1970) was applied to granular flow of solids by Savage and his colleagues (Jenkins and Savage, 1983; Lun *et al.*, 1984; Savage, 1988). This theory gives explicit expressions for the solids pressure or the normal stress in terms of the fluctuation velocity of the particles. Although the kinetic theory derivation is explained in a separate chapter, the resulting "equation of state" for a powder provides a useful interpretation of critical granular flow, and hence is included in this chapter for completeness.

Lun *et al.* (1984) have shown that the solids pressure consists of the kinetic pressure due to the motion of the particles and a collisional pressure due to inelastic collision of particles with a restitution coefficient of e. They find that

$$\sigma_s = \left[\varepsilon_s\rho_s + \varepsilon_s^2\rho_s(1+e)g_o\right]\Theta. \tag{4.54}$$

Solids pressure = Kinetic pressure + Collisional pressure

Note the analogy for a dense gas,

$$\text{Pressure} = [\text{Density} + \text{Repulsive Interaction}] \times \text{Temperature}. \quad (4.55)$$

In Eq. (4.54) the granular temperature Θ is given by

$$\Theta = \tfrac{1}{3} < C^2 > , \quad (4.56)$$

where C is the instantaneous minus the hydrodynamic velocity of the particle and the triangular parenthesis indicate averaging over the velocity space. In Eq. (4.54) g_o represents the repulsive radial distribution function similar to the repulsive function between molecules. Hence, the critical flow rate or the "sound speed" C_s is given by

$$C_s = \sqrt{G/\rho_s} = [1 + 4\varepsilon_s(1+e)g_o]^{1/2}\Theta^{1/2}. \quad (4.57)$$

For dilute flow, that is, small ε_s, the critical flow is

$$C_s = \sqrt{\Theta} = \sqrt{\tfrac{1}{3} < C >^2}. \quad (4.58)$$

Thus, discharge rates are proportional to the oscillation velocity of the particles. Typical values of this oscillation velocity computed for fluidized beds (Ding and Gidaspow, 1990) are of the order of one-tenth to one meter per second. This is very close to the discharge velocity for mass flow hoppers. Savage (1988) obtained such velocities in vibrated beds. A better comparison requires a numerical investigation of discharge of hoppers and a careful comparison to experiments of the type conducted by Bosley *et al.* (1969), who measured porosity distributions and discharge rates.

Exercises

1. *Steady Gas–Solid Critical Flow*

Find the critical flow conditions, similar to those given by Eq. (4.46), for steady state gas–solid dispersed flow using the three models listed in Table 2.1. A partial solution follows.

The conservation of mass, momentum, and energy equations for each phase can be written as

$$\mathbf{A}\frac{\partial \mathbf{u}}{\partial t}+\mathbf{B}\frac{\partial \mathbf{u}}{\partial x}=\mathbf{c} \tag{4.59}$$

where **u** is a vector of dependent variables such as void fraction, pressure, gas and liquid velocities and entropies of the phases, t is time and x the distance in one dimensional flow. **A**, **B** and **c** are functions of **u** and differ from model to model. The characteristic determinant for Eq. (4.59) is

$$\left|\mathbf{A}\lambda_0+\mathbf{B}\lambda_1\right|=0. \tag{4.60}$$

When we set one of the characteristic slopes in Eq. (4.60) to zero, some information, such as pressure, cannot propagate upstream. Then, physically, the system governed by Eq. (4.59) does not know that, say, the pressure in the reservoir was reduced below a critical pressure. Setting λ_0 to zero in the characteristic polynomial gives

$$\left|\mathbf{B}\right|=0. \tag{4.61}$$

The same condition is obtained from the steady-state version of the conservation equations:

$$\mathbf{B}\frac{\partial \mathbf{u}}{\partial x}=\mathbf{c}. \tag{4.62}$$

For Model A in Table 2.1, the equations in matrix form are given by Eq. (4.63):

$$\begin{bmatrix} \rho_g V_g & \varepsilon V_g/C_g^2 & \varepsilon\rho_g & 0 \\ -\rho_s V_s & 0 & 0 & (1-\varepsilon)\rho_s \\ -G & 1 & \varepsilon\rho_g V_g & (1-\varepsilon)\rho_s V_s \\ -G/1-\varepsilon & 1 & 0 & \rho_s V_s \end{bmatrix} \begin{bmatrix} d\varepsilon/dx \\ dp/dx \\ dV_g/dx \\ dV_s/dx \end{bmatrix} = \begin{bmatrix} 0 \\ 0 \\ -\left(\rho_s(1-\varepsilon)+\varepsilon\rho_g\right)g \\ \left(\beta\left(V_g-V_s\right)/1-\varepsilon\right)-\rho_{sg} \end{bmatrix}, \tag{4.63}$$

where G is the "elastic" modulus due to the particle to particle force.

The critical condition as given by Eq. (4.61) in this case is

$$V_s^2=\frac{G}{\rho_s}+\frac{(1-\varepsilon)\rho_g V_g^2}{\varepsilon\rho_s\left(\left(V_g^2/C_g^2\right)-1\right)}. \tag{4.64}$$

The two limiting cases are

i. $V_g \approx 0$ or flow in a vacuum $\rho_g=0$, giving

$$V_s=\sqrt{G/\rho_s}\,. \tag{4.65}$$

ii. When $\varepsilon = 1$, Eq. (4.64) correctly reduces to the condition

$$V_g = C_g, \tag{4.66}$$

that is, for a pure gas we get choking at the sonic flow rate if Eq. (4.64) gives the choking condition at the exit of a pipe. But for choking to take place within the pipe (or within the domain of Eq. (4.63)) a compatibility condition as given below should be satisfied:

$$\begin{vmatrix} 0 & \varepsilon V_g / C_g^2 & \varepsilon \rho_g & 0 \\ 0 & 0 & 0 & (1-\varepsilon)\rho_s \\ -\left(\rho_s(1-\varepsilon)+\left(\varepsilon\rho_g\right)g\right) & 1 & \varepsilon\rho_g V_g & (1-\varepsilon)\rho_s V_s \\ \left(\beta\left(V_g - V_s\right)/1-\varepsilon\right)-\rho_s g & 1 & 0 & \rho_s V_s \end{vmatrix} = 0. \tag{4.67}$$

Upon simplifying Eq. (4.67) we get

$$\beta\left(V_g - V_s\right)\left(1-\varepsilon\frac{V_g^2}{C_g^2}\right) = (1-\varepsilon)\varepsilon\rho_s g\left(1-\frac{V_g^2}{C_g^2}\right). \tag{4.68}$$

For $V_g << C_g$, Eq. (4.68) is similar to the minimum fluidization condition. In a situation such as horizontal transport where $g = 0$, the compatibility condition reduces to

$$V_g = V_s, \tag{4.69}$$

implying a homogeneous flow.

For homogeneous flow a separate analysis is required. Repeat the analysis for the relative velocity and for Model B in Table 2.1.

2. *Maximum Homogeneous Discharge Rates*

Finely dispersed powders tend to travel at fluid velocities. Making such an assumption, the steady-state one-dimensional energy balance for the mixture is the same as that for a single component with the properties being those of the mixture. It is

$$\frac{d}{dx}\left(h_m + \bar{v}^2/2 + gz\right) = \dot{Q} - \dot{W}, \tag{4.70}$$

where z is the spatial coordinate, h_m is the mixture enthalpy, $\bar{v}$ is the homogeneous velocity, g the gravity, $\dot{Q}$ is the rate of heat addition to the system and $\dot{W}$ is the rate of work done by the system. With no heat addition, work done by the mixture and constant elevation,

$$dh_m + d\left(\bar{v}^2/2\right) = 0. \tag{4.71}$$

In terms of inlet conditions, i, Eq. (4.71) becomes

$$\left(h_m - h_{m_i}\right) + \left(\frac{\bar{v}^2}{2} - \frac{\bar{v}_i^2}{2}\right) = 0. \tag{4.72}$$

If the inlet velocity is small, that is, $\bar{v}_i << \bar{v}$, Eq. (4.72) gives the velocity at any position,

$$\bar{v} = \sqrt{2\left(h_{m_i} - h_m\right)}. \tag{4.73}$$

For an ideal gas and an incompressible solid, the enthalpy is a function of temperature only. For a constant specific heat, Eq. (4.73) becomes

$$\bar{v} = \sqrt{2C_{pm}\left(T_i - T\right)}. \tag{4.74}$$

(a) Show how Eqs. (4.70) and (4.72) can be obtained from the equations in Appendix A6.
(b) For isentropic expansion, the temperature and the pressure are related by Eq. (1.51). What mechanisms of energy transfer and dissipation are neglected?
(c) Using Eq. (1.51), show that the mass flow rate per unit area $F = \rho_m v$, where ρ_m is the mixture density, can be expressed as follows for the case when

$$\rho_s >> \rho_g,$$

$$F = \frac{MP_i}{xRT_i}\left[2C_{pm}T_i\left(r^{-\frac{2xR}{C_{pm}M}+2} - r^{-\frac{xR}{C_{pm}M}+2}\right)\right]^{1/2}, \tag{4.75}$$

where r is the pressure ratio $r = P/P_i$.
(d) Differentiate the equation in part (c) with respect to r and set the derivative equal to zero to obtain the maximum flow rate F_c. Also, show that for $x < 1$ this expression simplifies to

$$F_c = \frac{P_i}{\sqrt{T_i}} \cdot \sqrt{\frac{2M}{xR}}. \tag{4.76}$$

The corresponding equation for critical flow of an ideal gas is

$$\underset{\text{ideal gas}}{F_c} = \frac{P_i}{\sqrt{T_i}} \cdot \sqrt{\frac{kM}{R}\left(\frac{2}{k+1}\right)^{\frac{k+1}{k-1}}}, \tag{4.77}$$

where k is the ratio of specific heats at constant pressure to constant volume.
(e) For pumping of a gas–solid mixture maintained near minimum fluidization with the void fraction of 0.5, show that the expression in part (d) simplifies to

$$\text{Maximum expulsion rate} = F_{\max} = \sqrt{2\rho_s P_i}. \tag{4.78}$$

This form agrees with the experimental data of Fricke, Berman, and Sobieniak (ASME 72-MH-19).
(f) Compute the maximum discharge rate for pumping fluidized solids of a specific gravity of 1.73 from a pressure of 1.82 MPa. The observed rate is 0.28 of this theoretical value.

3. *Critical Flow Using the Relative Velocity Model*

Show that the critical flow condition $|B| = 0$ for the relative velocity model in Table 2.1 gives the condition

$$\hat{v} = v_i v_j \bar{v} / C_m^2, \tag{4.79}$$

where $\hat{v} = \varepsilon_i v_j + \varepsilon_j v_i$ and $\bar{v} = (\varepsilon_i \rho_i v_i + \varepsilon_j \rho_j v_j) / \rho_m$.

For homogeneous flow, $v_i = v_j$ and $v_i = \bar{v} = \hat{v}$; thus, we obtain the well-known critical flow condition $v^2 = C_m^2$, where C_m is given by

$$\frac{1}{\rho_m C_m^2} = \frac{\varepsilon_i}{\rho_i C_i^2} + \frac{\varepsilon_j}{\rho_j C_j^2}. \tag{4.80}$$

In terms of the slip ratio, $s = v_j / v_i$, the critical flow conditions can be written as

$$v_i = C_m \sqrt{\frac{\varepsilon_i s + \varepsilon_j}{x_i + x_j s}}. \tag{4.81}$$

In this form we see that it differs from the homogeneous critical flow condition by the presence of the slip ratio, the void fraction ε_i, and the weight fraction of phase i, x_i (Rasouli *et al.*, 1983).

4. *Numerical Values of Critical Flow for an Air–Water System*

J. F. Muir and R. Eichorn (1963) studied compressible flow of a bubbly air–water mixture through a two-dimensional, converging-diverging nozzle. Figure 4.5 shows their data of throat velocity versus an estimated volume fraction of air extrapolated to high volume fractions. The discharge of the air-water

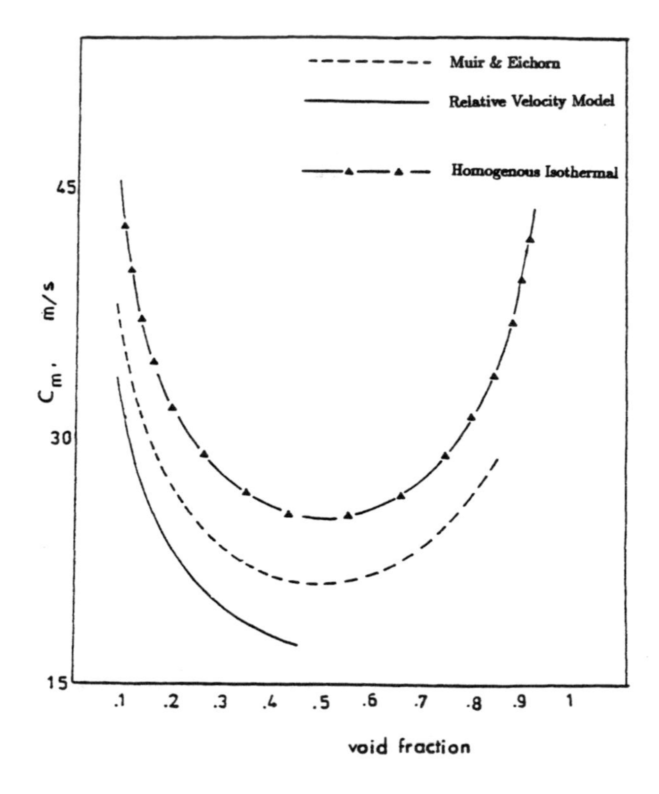

Fig. 4.5 Comparison of the sonic speed calculated from the critical flow condition to the experimental data.

mixture was to the atmosphere. The figure also shows the velocity computed using the relative velocity model, as in Exercise 4.3, and a plot of the homo-geneous iso-thermal model. Muir and Eichorn report a slip of about 8 m/s between the phases.

Verify the computations, when possible and plot the result for the homogeneous isentropic flow.

Literature Cited

Abbott, M. B. (1966). *An Introduction to the Method of Characteristics*. New York: Elsevier.

Bilicki, Z. and J. Kestin (1983). "Two-Phase Flow in a Vertical Pipe and the Phenomenon of Choking: Homogeneous Diffusion Model—I," *Int. J. Multiphase Flow* **9** (3), 269–288.

Bosley, J., C. Schofield, and C. A. Shook (1969). "An Experimental Study of Granular Discharge from Model Hoppers," *Trans. Inst. Chem. Eng.* **47**, T147–T153.

Bouillard, J. X., R. W. Lyczkowski, and D. Gidaspow (1989). "Porosity Distributions in a Fluidized Bed with an Immersed Obstacle," *AIChE J.* **35**, 908–922.

Boure, J. A. (1978). "Critical Two-Phase Flows," in *Two-Phase Flows and Heat Transfer with Application to Nuclear Reactor Design Problems*. Hemisphere Publishing Co.

Boure, J. A., A. A. Fritte, M. M. Giot, and M. L. Reocreus (1976). "Highlights of Two-Phase Critical Flow," *Int. J. Multiphase Flow* **3**, 1–22.

Callen, H. B. (1960). *Thermodynamics*. New York: John Wiley & Sons.

Chapman, S. and T. G. Cowling (1970). *The Mathematical Theory of Non-uniform Gases*, 3rd Ed. Cambridge, U.K: Cambridge University Press.

Courant, R. and D. Hilbert (1962). *Methods of Mathematical Physics, Vol. II: Partial Differential Equations*. New York: Wiley-Interscience.

Delaplaine, J. W. (1956). "Forces Acting in Flowing Beds of Solids," *AIChE J.* **2**, 127–138.

Ding, J. and D. Gidaspow (1990). "A Bubbling Fluidization Model Using Kinetic Theory of Granular Flow," *AIChE J.* **36**, 523–538.

Garabedian, P. R. (1964). *Partial Differential Equations*. New York: John Wiley & Sons.

Gidaspow, D. (1986). "Hydrodynamics of Fluidization and Heat Transfer: Supercomputer Modeling," *Appl. Mech. Rev.* **39** (1), 1–23.

Gidaspow, D., F. Rasouli, and Y. W. Shin (1983). "An Unequal Velocity Model for Transient Two-Phase Flow by the Method of Characteristics," *Nucl. Sci. and Eng.* **84**, 179–195.

Gidaspow, D., J. Ding, and K.-M. Luo (1986). "A Measurement of the Modulus of Elasticity Using Critical Flow Theory," *DOE Report, DE86011993*.

Gough, P. S. (1979). "Modeling of Two-Phase Flow in Guns," pp. 176–196 in H. Krier and N. Summerfeld, Eds. *Progress in Astronautics and Aeronautics, Vol. 66: Interior Ballistics of Guns.* AIAA.

Jenkins, J. T. and S. B. Savage (1983). "A Theory for the Rapid Flow of Identical, Smooth, Nearly Elastic Spherical Particles," *J. Fluid Mech.* **130**, 187–202.

Kaza, K. R. and R. Jackson (1982). "The Rate of Discharge of Course Granular Material from a Wedge-Shaped Mass Flow Hopper," *Powder Technol.* **33**, 2233–2237.

Lun, C. K. K., S. B. Savage, D. J. Jeffrey, and N. Chepamey (1984). "Kinetic Theories for Granular Flow: Inelastic Particles in Coette Flow and Slightly Inelastic Particles in a General Flowfield," *J. Fluid Mech.* **140**, 223–256.

McCabe, W. L. and J. C. Smith (1956). *Unit Operations of Chemical Engineering.* New York: McGraw-Hill.

Muir, J. F. and R. Eichorn (1963). "Compressible Flow of an Air–Water Mixture through a Vertical, Two-Dimensional, Converging–Diverging Nozzle," *Proceedings of Heat Transfer and Fluid Mechanics Institute.* Stanford, California: Stanford University.

Nedderman, R. M., U. Tuzin, S. B. Savage, and G. T. Houlsby (1982). "The Flow of Granular Materials, Part I: Discharge Rate From Hoppers," *Chem. Eng. Sci.* **37**, 1597–1609.

Rasouli, F., D. Gidaspow, and Y. W. Shin (1983). "Unequal Velocity, Two-Component, Two-Phase Flow by the Method of Characteristics and Critical Flow." *Argonne National Lab Report, ANL-83-98.*

Rietma, K. and S. M. P. Mutsers (1973). "The Effect of Interparticle Forces on the Expansion of a Homogeneous Gas-Fluidized Bed," *Proceedings of the International Symposium on Fluidization.* Toulouse, France.

Sandusky, H. W., W. L. Elban, K. Kim, R. R. Bernecher, S. B. Gross, and A. R. Clairmont, Jr. (1981). "Compution of Porous Solids of Inert Materials," *Proceedings of the 7th International Symposium on Detonation.* Annapolis, Maryland: Naval Surface Weapons Center.

Savage, S. B. (1983). "Granular Flows at High Shear Rates," pp. 339–358 in R. E. Meyer, Ed. *Theory of Dispersed Multiphase Flow.* New York: Academic Press.

Savage, S. B. (1988). "Streaming Motions in a Bed of Vibrationally Fluidized Dry Granular Material," *J. Fluid Mech.* **194**, 457–478.

Sokolovskii, V. V., translated from Russian by J. K. Luther (1965). *Statics of Granular Media.* Oxford, U.K.: Pergamon Press.

Tolstoy, I. (1973). *Wave Propagation.* New York: McGraw-Hill.

von Mises, R. (1958). *Mathematical Theory of Compressible Fluid Flow.* New York: Academic Press.

5

THE FLUIDIZED STATE

5.1 Introduction: Fluidization Regimes 97
5.2 Minimum Fluidization Velocity 98
5.3 Geldart's Classification of Powders 103
5.4 Kinetic Energy Dissipation Analysis 106
Exercises 109
1. Minimum Fluidization and Terminal Velocities 109
2. Minimum Fluidization Velocity and Segregation of a Binary Mixture with an Applied Electric Field 109
3. Energy Dissipation: Attrition, Erosion 112
Literature Cited 113

5.1 Introduction: Fluidization Regimes

The phenomenon of fluidization can be best visualized in terms of a simple experiment shown in Fig. 5.1. Particles such as sand are poured into a tube provided with a porous plate distributor, which for good quality fluidization has a pressure drop of 10 to 20 percent of the weight of the bed. Gas or liquid is then forced upward through the particle bed. This flow causes a pressure drop across the bed, which can be described by the Ergun equation depicted in Fig. 5.2. When this pressure drop is sufficient to support the weight of the particles, the bed is said to be at minimum fluidization. A further increase in the flow rate of gas often causes bubbling that can be seen through the walls of a transparent tube or at the top of the bed, as schematically illustrated in Figure 5.1. For narrow tubes and a sufficiently deep bed the bubbles coalesce and form a slug. These gas slugs then move up the bed in a fairly regular periodic motion. Large-size particles such as corn can be made to spout if the gas is brought in through a small central tube. A spout is a centrally located, upwardly moving particle region. Above such a spouted bed the particles drop down in a fountain and move down the tube in an annular region. As the gas velocity is increased substantially above the terminal velocity, the fluidized bed, the slugging bed, and the spouted bed are all blown out

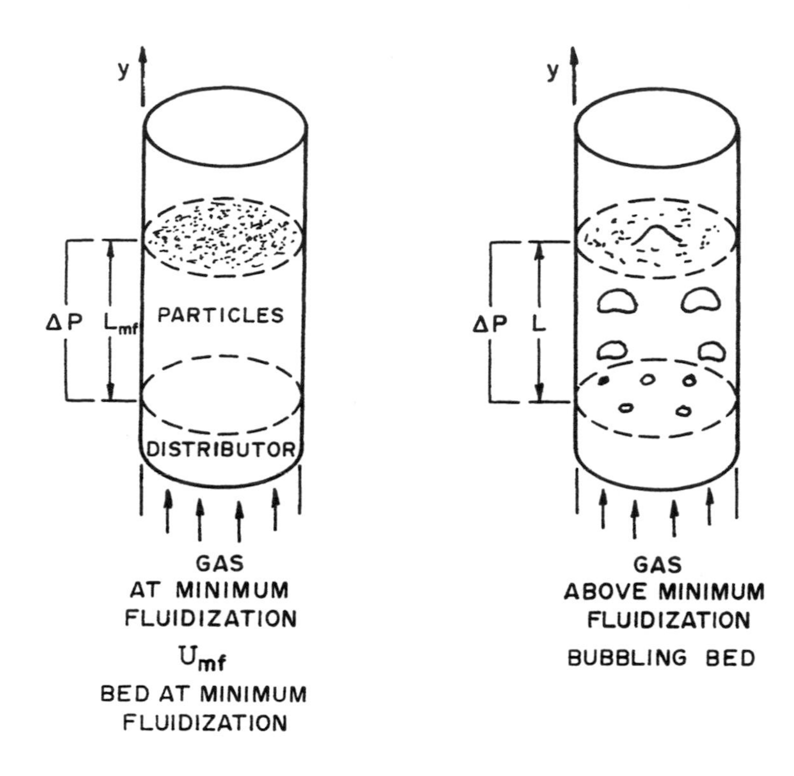

Fig. 5.1 Fluidization regimes.

of the tube. Continuous operation in the dense-phase regime can be continued by rapidly feeding solids into the bottom of the bed. The resulting flow regime is called recirculating fluidization. Dilute transfer of solids is known as pneumatic transport.

5.2 *Minimum Fluidization Velocity*

The minimum fluidization velocity is determined empirically by the intersection of the pressure drop versus the superficial velocity curve and the pressure drop equals the weight of the bed line, as illustrated in Fig. 5.2. At minimum fluidization the velocity of the solids is taken to be the velocity in the packed bed region, that is, zero. The porosity at minimum fluidization is determined from the height of the bed at minimum fluidization. For spherical particles of uniform size it is not

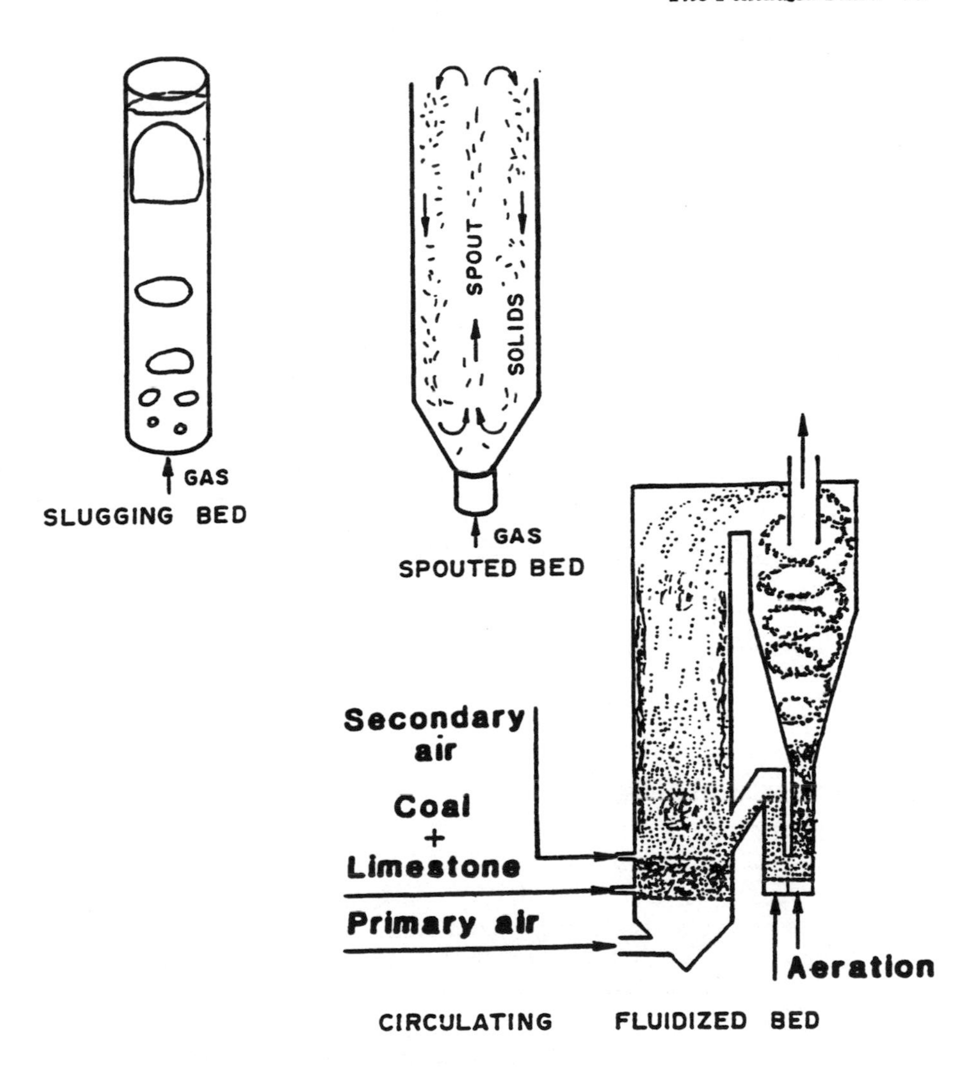

Fig. 5.1 (continued).

unreasonable to expect this porosity to be close to that of a bed packed with spheres in a cubic mode, with $\varepsilon_{mf} = 1 - \pi/6 = 0.476$. The actual porosity at minimum fluidization can differ considerably from this value, as shown by Wen and Yu (1966).

It is useful to see to what extent the conservation equations given in Chapter 2 can predict the minimum fluidization velocity. The mixture momentum balance

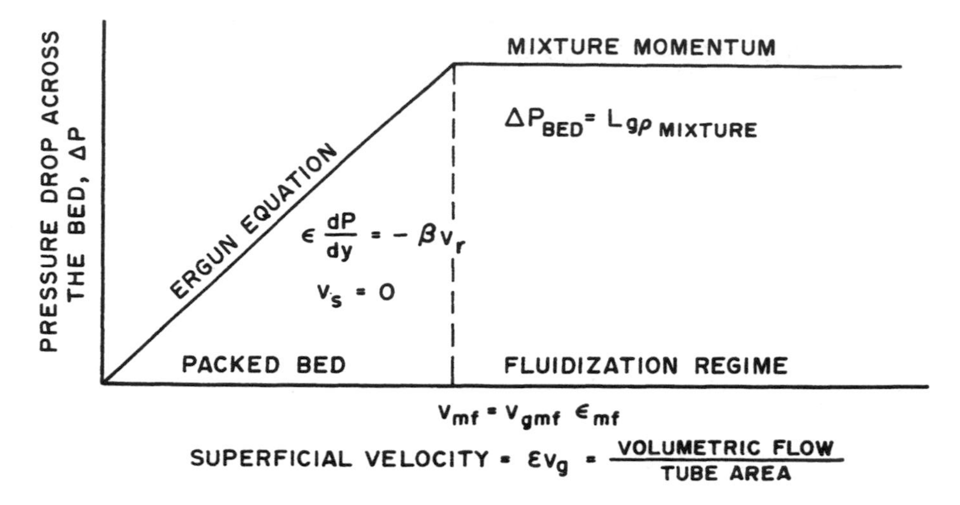

Fig. 5.2 Determination of minimum fluidization velocity.

with negligible acceleration and solids stress transmitted by the particles was shown to be

$$-\frac{dp}{dy} = g\rho_{mix}, \tag{5.1}$$

where the mixture density, ρ_{mix} is

$$\rho_{mix} = \rho_g \varepsilon + \rho_s (1 - \varepsilon). \tag{5.2}$$

With the assumptions made, these equations determine the porosity at minimum fluidization from a known pressure drop. In a design calculation the pressure drop is not known, and hence this equation is not useful for such purposes.

A momentum balance that is useful for an estimate of minimum fluidization is the buoyancy-equals-drag balance derived in Chapter 2. In the absence of gas–wall friction and solids stress transmitted by the particles, the balance can be written as follows:

$$(1-\varepsilon)(\rho_s - \rho_g)g = \frac{\beta_A}{\varepsilon}(v_g - v_s), \tag{5.3}$$

Buoyant force = Drag

where the friction coefficient β_A was obtained from the Ergun equation to be

$$\beta_A = 150\frac{(1-\varepsilon)^2 \mu_g}{\varepsilon\left(d_p \phi_s\right)^2} + 1.75\frac{\rho_g\left|v_g - v_s\right|(1-\varepsilon)}{\left(\phi_s d_p\right)}. \tag{5.4}$$

By definition, at minimum fluidization velocity v_s is zero. Thus, Eqs. (5.3) and (5.4) give a relation between ε_{mf} and $U_{mf} = \varepsilon_{mf} v_{gmf}$. These equations can be solved either for the solids volume fraction, $1-\varepsilon$,

$$(1-\varepsilon) = \tfrac{1}{2} \pm \sqrt{\tfrac{1}{4} - \frac{\beta_A\left(v_g - v_s\right)}{\Delta\rho g}}, \tag{5.5}$$

or more traditionally, for the minimum fluidizaiton velocity.

For small particles or for small Reynolds numbers, more precisely, for

$$\mathrm{Re}_{mf} = \frac{d_p U_{mf} \rho_f}{\mu} < 20, \tag{5.6}$$

substitution of the first term of the Ergun equation (5.4) into the buoyancy-equals-drag balance Eq. (5.3) yields the following expression for the minimum fluidization velocity U_{mf}:

$$U_{mf} = \varepsilon_{mf} v_{mf} = \frac{d_p^{\,2} \Delta\rho g}{150\mu} \cdot \frac{\phi_s^{\,2} \varepsilon_{mf}^{\,3}}{1-\varepsilon_{mf}}. \tag{5.7}$$

From the data of porosity at minimum fluidization versus the sphericity, Wen and Yu (1966) find the approximate relation

$$\frac{1-\varepsilon_{mf}}{\phi_s^{\,2} \varepsilon_{mf}^{\,3}} = \mathrm{A} \cong 11. \tag{5.8}$$

The use of Eq. (5.8) in Eq. (5.7) gives the commonly used estimate for the minimum fluidization velocity:

$$U_{mf} = \frac{d_p^{\,2}\left(\rho_s - \rho_f\right) g}{1650\mu}, \quad \mathrm{Re}_{mf} < 20. \tag{5.9}$$

It is instructive to compare this expression to Stokes' law

$$v_{\text{Stokes'}} = \frac{d_p^2(\rho_s - \rho_f)g}{18\mu}. \tag{5.10}$$

It is clear that the minimum fluidization velocity is $1650/18 \cong 90$ times smaller than the terminal velocity of a single particle in an infinitely large fluid. Thus, a particle blown out of the bed by bubble-bursting above the bed, just above minimum fluidization, returns to the bed, since its terminal velocity is too low for it to be carried out of the bed. This phenomenon is due to the fact that the friction coefficient β_A in Eq. (5.4) is higher for a packed bed than for a bed of spheres very far apart. Note that the higher the solids volume fraction, $(1-\varepsilon)$, the higher the friction. This change of resistance with solids volume fraction gives rise to the oscillatory behavior of the fluidized bed. It is this oscillatory behavior that is responsible for many of the useful features of fluidized beds, such as good heat and mass transfer. Although a quantitative explanation of bubbling is left for later chapters, the primitive explanation is to be found in this resistance behavior. As the gas flows through the bed above minimum fluidization, the bed expands to minimize resistance. As the resistance becomes too low, the particles cannot be supported by the gas any longer and drop down. This process is repeated over and over again.

For larger particle Reynolds numbers, the buoyancy-equals-drag equation (5.3) with the friction coefficient given by the Ergun equation (5.4) can be rewritten as follows in terms of an Archimedes number, A_r, that is, similar to the Grashof number used in free convection laminar heat transfer:

$$150\text{A}\,\text{Re}_{mf} + 1.75\text{B}\,\text{Re}_{mf}^2 = \text{A}_r, \tag{5.11}$$

where

$$\text{A}_r = \frac{d_p^3(\rho_s - \rho_f)\rho_f g}{\mu^2}, \tag{5.12}$$

$$\text{B} = 1/\varepsilon_{mf}^3 \phi_s, \tag{5.13}$$

and A was the porosity group given by Eq. (5.8).

The shape-voidage functions A and B have been approximated by Wen and Yu (1966). Their values for A and B are 11 and 14, respectively. Equation (5.11) can be solved for the Reynolds number at minimum fluidization given below in terms of constants C_1 and C_2:

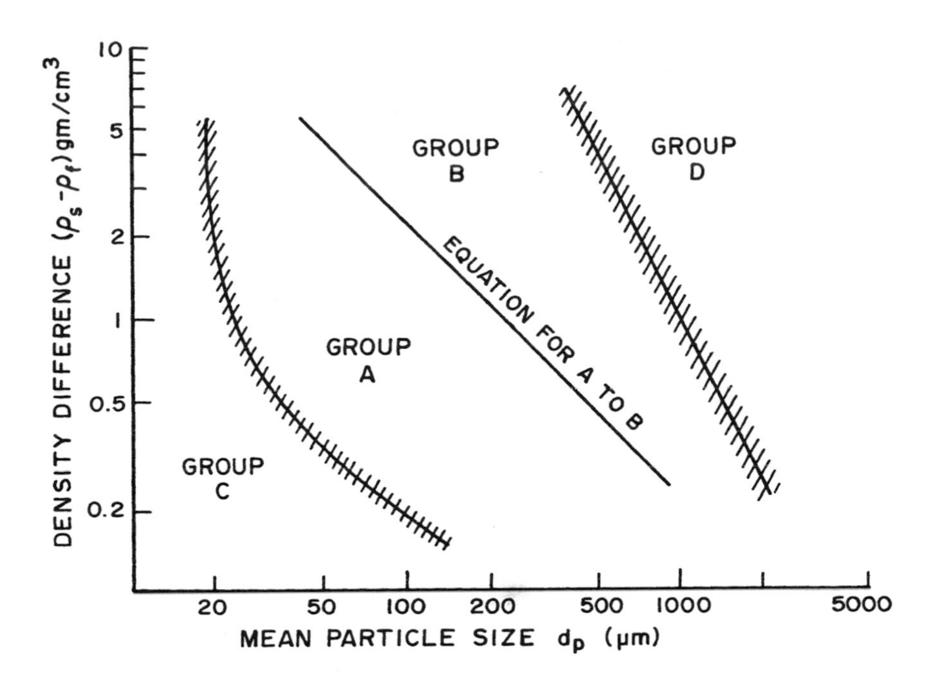

Fig. 5.3 Geldart classification of fluidized particles.

$$\mathrm{Re}_{mf} = \sqrt{C_1^2 + C_2 A_2} - C_1, \tag{5.14}$$

where

$$C_1 = 33.7 \text{ and } C_2 = 0.0408, \tag{5.15}$$

according to Wen and Yu (1966). Yang *et al.* (1985) tabulate such constants at elevated pressures. These constants differ slightly from those reported by Wen and Yu at atmospheric pressures.

For large particles, the second term in the Ergun equation is dominant and the minimum fluidization velocity can be expressed simply as

$$U_{mf}^2 = \frac{d_p(\rho_s - \rho_f)g}{24.5\rho_f}, \quad \mathrm{Re}_{mf} > 1000 \tag{5.16}$$

Equation (5.16) shows that at high pressures the minimum fluidization velocity is lower than at low pressures because of the higher fluid density at high pressures.

5.3 Geldart's Classification of Powders

The behavior of gas–solid fluidized systems depends upon the properties of the solid particles, size, density, fine content, cohesiveness, etc. Geldart (1973) suggested that uniformly sized powders can be classified into four types characterized by the density difference between the particle and the fluid and by the mean particle size. Figure 5.3 shows Geldart's diagram. It is valid for air at ambient conditions. The four groups considered by Geldart (1973) and accepted (Couderc, 1985) as a good beginning by the fluidization community follow.

Group A (Aerated)

Materials having a small mean size and a low density ($\rho_s < 1400\,\text{kg/m}^2$) are in this group. These powders expand considerably after the minimum fluidization velocity is reached and do not produce bubbles until some higher velocity, called the minimum bubbling velocity U_{mb}, is reached. The common cracking catalyst is an example of such a powder.

Fitzgerald (1985) argues that since the minimum fluidization velocity as given by Wen and Yu's expression is much smaller for these particles than the rise velocity of a single bubble U_B given by Davidson and Harrison's expression:

$$U_B = 0.711\left(gD_B\right)^{1/2}, \tag{5.17}$$

where D_B is the bubble diameter; therefore, any reasonable-sized bubbles will rise faster than the interstitial gas velocity, U_{mf}/ε_{mf}. Bubbles that rise faster than the interstitial velocity are called fast bubbles. Davidson (Davidson and Harrison, 1963) has shown that such fast bubbles possess the property that a spherical surface exists around the bubble that moves with the bubble velocity and through which there is a zero gas velocity. Therefore, any gas trapped in the bubble can leave the bubble only by the slow process of molecular diffusion. The implications for a catalytic reactor or for a reactor such as a coal gasifier is that the inlet gas, say oxygen, is trapped in the bubble. In the case of a gasifier, this trapped oxygen will then burn some of the gas already produced in the bed. This extra resistance due to molecular diffusion has given rise to a number of film models for the bubble phase to take this phenomenon into account. Although gasifiers were sometimes modeled using such a resistance concept, it should be added that plans for real gasifiers were made (Gidaspow *et al.*, 1985) using particle sizes that put them into the next category, Group B, in which fast bubbles are not the rule. Although fast bubbles do not occur except perhaps at shut-down in fluidized bed gasifiers using particle sizes of the order of 500 μm and fluidized bed combustors using millimeter-size particles, they could occur in catalytic reactors using fine particles, which are used to minimize the diffusional resistance into the catalyst particle. Hence it is important to understand the hydrodynamics of fluidization of this group of particles.

Group B (Bubbling)

The group contains materials in the approximate mean size and density range, $40\ \mu m < d_p < 500\ \mu m, 4000\ kg/m^3 < \rho_s < 1400\ kg/m^3$. Sand is such a typical powder. For these powders, bubbling starts at the minimum fluidization. At atmospheric pressure the bed expansion is not large. At equal bed heights and $U - U_{mf}$, the bubble sizes are almost independent of particle size.

Numerical simulation (Gidaspow *et al.*, 1986) has shown that bubbles produced by a continuous jet are slow bubbles in this size range. The bubble rises more slowly than the jet. Gas sweeps through the bubble offering little resistance. A fast bubble can be produced when the gas to the jet is shut off, maintaining slow flow of secondary gas. This fast bubble then traps the gas inside it. Such a simulation corresponds to the classical experiments conducted to show that the brown gas NO_X, injected into the bubble, travels with the bubble in a two-dimensional transparent bed with little diffusion.

Group C (Cohesive)

All cohesive powders that are difficult to fluidize belong to this group. In small-diameter tubes the powders lift as a plug. In larger or in two-dimensional beds with a jet the particles channel or rat-hole badly. This difficulty arises because the interparticle forces are greater than those which the fluid can exert on the particle. Geldart (1973) suggests the use of mechanical stirrers or vibrations to break up the stable channels or the use of sub-micron silica which coats the particles and increases flowability.

Group D

Powders in this group are made up of large or very dense particles. These powders can be made to spout easily. Chapter 6 shows that for such particles, when the particle Reynolds number exceeds about 1000, there will be a transition from the bubbling to the dispersion regime. No quantitative criterion exists today for the transition into this regime.

A quantitative criterion for distinguishing between Group A and B powders is based upon the experimentally determined minimum bubbling velocity. Geldart (1973) measured this velocity in a 5 cm diameter glass column having a filter-paper distributor. The minimum fluidization velocity was first measured as already discussed in the usual way, using the pressure drop–gas velocity curve. The air velocity was increased until the first clearly recognizable bubble, usually 0.5 cm diameter, was observed to break on the surface. This bed velocity was taken to be the minimum bubbling velocity. In the particle size range of 25 μm to about 250 μm, Geldart found the bubble velocity to be proportional to the particle size and independent of the particle density in the narrow range of 1 g/cm^3 to 1.5 g/cm^3. This equation for the minimum bubbling velocity is

$$U_{mb} = 100\, d_p, \tag{5.18}$$

where U_{mb} has the units of centimeters per second and d_p is in centimeters. A combination of this equation with the minimum fluidization velocity for small particle Reynolds numbers translates the criterion $U_{mb}/U_{mf} \geq 1$ into

$$\left(\rho_p - \rho_f\right) d_p' \leq 225, \tag{5.19}$$

where d_p' is the particle size in microns. This equation has been plotted in Fig. 5.3.

5.4 *Kinetic Energy Dissipation Analysis*

An energy dissipation analysis is useful for understanding some multiphase flow and fluidization phenomena. Consider gas and solid momentum balances, as derived in Chapters 1 and 2. For transient, one-dimensional flow using Model A, the gas and solid momentum balances are as follows:

Gas Momentum

$$\varepsilon\rho_g\left(\frac{\partial v_g}{\partial t} + v_g\frac{\partial v_g}{\partial x}\right) = -\varepsilon\frac{\partial p}{\partial x} - \varepsilon\rho_g g + \beta_A\left(v_s - v_g\right) - \frac{4\tau_{wg}}{D_t}, \tag{5.20}$$

Solid Momentum

$$\varepsilon_s\rho_s\left(\frac{\partial v_s}{\partial t} + v_s\frac{\partial v_s}{\partial x}\right) = -\varepsilon_s\frac{\partial p}{\partial x} - \frac{\partial \sigma}{\partial x} - \varepsilon_s\rho_s g + \beta_A\left(v_g - v_s\right) - \frac{4\tau_{ws}}{D_t}. \tag{5.21}$$

The corresponding kinetic energy equations are obtained in the same way as for single phase flow (Bird *et al.*, 1960). The gas momentum balance is divided by ε, multiplied by v_g, and written in convective (substantial) derivative form moving with the gas velocity. A similar operation is performed upon the solids momen-tum balance. Then the addition of the two kinetic energy balances yields the mixture kinetic energy balance in Eq. (5.22):

Mixture Kinetic Energy Balance

$$\sum_{i=g,s} \rho_i \frac{d\left(\frac{1}{2}v_i^2\right)}{dt^i} + v_i \frac{\partial P}{\partial x} + v_s \frac{\partial \sigma}{\partial x} + \rho_i v_i g =$$

Rate of change of kinetic energy + Rate of work due to pressure and normal stress + Power due to gravity =

$$-\left(v_g - v_s\right)^2 \frac{\beta_A}{\varepsilon\varepsilon_s} - \frac{4\tau_{wg} v_g}{D_t \varepsilon} - \frac{4\tau_{ws} v_s}{D_t \varepsilon_s} = -\dot{\mathcal{D}} \tag{5.22}$$

Rate of conversion of kinetic energy into interphase friction (drag) + Rate of conversion of K.E. into gas–wall friction + Rate of conversion of K.E. into solids–wall friction = Rate of dissipation

As discussed in Chapter 2, the relation between the friction coefficient for Model A and the standard drag coefficient C_D is

$$\beta_A = \tfrac{3}{4} C_D \frac{\varepsilon\left|v_g - v_s\right|\rho_g \varepsilon_s}{d_p}. \tag{5.23}$$

In terms of the standard drag coefficient, the dissipation of kinetic energy, $\dot{\mathcal{D}}$, is expressed as follows:

$$\dot{\mathcal{D}} = \tfrac{3}{4} C_D \frac{\rho_g \left(v_g - v_s\right)^2 \left|v_g - v_s\right|}{d_p} + \frac{4\tau_{wg} v_g}{D_t \varepsilon} + \frac{4\tau_{ws} v_s}{D_t \varepsilon_s}. \tag{5.24}$$

Equation (5.24) shows that the interphase frictional loss is proportional to the relative velocity between the phases, that it is large for small particle sizes, and that the loss due to the presence of the walls is a product of the respective shear forces and the velocities. The interphase dissipation term shows that for a given kinetic energy of the gas into a bed of particles, the bed will start moving to keep dissipation at a reasonably low value by reducing $(v_g - v_s)$. This will happen sooner for small particles than for large particles. For small particles, homogeneous flow, that is, equality of v_g and v_s, is favored. Other qualitative behavior can be seen by analyzing some special cases.

For packed bed or fluidized bed regimes for small Reynolds number β_A can be expressed in terms of the first term of the Ergun equation. This gives the expression

$$\dot{\mathcal{D}}_{\text{drag}} = \frac{150\mu(1-\varepsilon)\left(v_g - v_s\right)^2}{d_p{}^2\varepsilon^2}, \quad \text{Re}_p < 20. \tag{5.25}$$

In addition to the previous observations, Eq. (5.25) shows that for a given inlet v_g into a fluidized bed, in order to minimize $\dot{\mathcal{D}}_{\text{drag}}$, the fluidized bed will expand. Expansion should be more for smaller particles to keep $\dot{\mathcal{D}}$ bounded. For high Reynolds numbers the corresponding dissipation is

$$\dot{\mathcal{D}}_{\text{drag}} = \frac{1.75\left|v_g - v_s\right|\left(v_g - v_s\right)^2 \rho_g}{d_p\varepsilon}, \quad \text{Re}_p > 1000. \tag{5.26}$$

This expression shows that the dissipation is higher at higher pressures or higher gas densities. Thus, to minimize $\dot{\mathcal{D}}_{\text{drag}}$, the porosity increases as the pressure increases. This is the observed effect in fluidization. Note also that in this regime the dissipation is not a function of the fluid viscosity, while it was proportional to it in the low Reynolds number case, as expected from single-phase flow analysis.

With a negligible acceleration, the rate of energy dissipation $\dot{\mathcal{D}}$ can be rewritten in a different form using the relation between buoyancy and drag derived in Eq. (2.18). With zero wall friction this dissipation is of Onsager form

$$\dot{\mathcal{D}} = g \cdot \left(\rho_s - \rho_f\right)\left(v_f - v_s\right). \tag{5.27}$$

In the form of Eq. (5.27) it is clear that the kinetic energy dissipation is greater for a gas fluidized bed than it is for a liquid fluidized bed for the same relative velocity, since for a liquid fluidized bed $\rho_s - \rho_f$ is much smaller than for the gas fluidized bed.

With again negligible acceleration and wall friction, the dissipation $\dot{\mathcal{D}}$ as given by Eq. (5.22) can be rewritten in terms of the pressure drop using Darcy's type degenerate momentum balance given by Eq. (2.9):

$$\dot{\mathcal{D}} = \frac{\left(v_f - v_s\right)}{\varepsilon_s} \cdot \left(-\frac{\Delta p}{\Delta x}\right). \tag{5.28}$$

Dissipation = Flux Pressure Drop

In this form it is clear that the dissipation is due to the pressure drop across the bed. Thus, for a given sufficiently high fluid kinetic energy, the bed becomes fluidized to minimize the dissipation due to the pressure drop, which would otherwise continue increasing along the Ergun equation line as shown in Fig. 5.2.

Exercises

1. *Minimum Fluidization and Terminal Velocities*

(a) Compute the minimum fluidization velocity of *i.* 50 μm, *ii.* 500 μm, *iii.* 5 cm glass beads in air at 20°C and at atmospheric pressure and in water. The density of the beads is 2610 kg/m^3.

(b) What differences in velocities in part (a) do you expect between the measured values in large cylindrical vessels and in narrow two-dimensional beds?

(c) How would you obtain the values in part (a) for *i.* sand, *ii.* cylindrical pellets of length-to-diameter ratio 5.

(d) Estimate the terminal velocities for parts (a) to (c).

2. *Minimum Fluidization Velocity and Segregation of a Binary Mixture with an Applied Electric Field*

In an attempt to develop a process for separating pyrites from coal using the natural differences in surface charge of the highly negatively charged pyrite ($q_{pyrite} = 10^{-12}$ coulomb/particle), and the nearly neutrally charged coal, a direct electric field below the corona charge breakdown was applied between two horizontal walls of a two-dimensional fluidized bed with a jet (described in detail in Chapter 7) containing the mixture of coal and pyrite particles shown in Table 5.1. The strength of the electric field E was 500 volts/cm. The electric field was assumed to be uniform perpendicular to the direction of gas flow. The equations solved by Shih *et al.* (1987), typical computed bubbles and the calculated segregation are shown in Figures 5.4 to 5.7. The computations were done using Model A stabilized with the solids stress. There was a qualitative agreement between the observed and the computed pyrite segregation. In the experiment the pyrite that initially looks like gold could be seen to stick to the left metal wall that served as the positive electrode upon fluidization of an initially uniform mixture of particles and an application of a high-voltage electric field.

Table 5.1 PROPERTIES OF COAL AND PYRITE PARTICLES

	d_p, μm	ρ_p, g/cm^3	Est. U_{mf}, cm/s
Coal	120	1.27	5.1
Pyrite	150	4.6	6.6

Continuity Equations

$$\frac{\partial}{\partial t}\left(\rho_k \varepsilon_k\right) + \nabla\left(\varepsilon_k \rho_k \mathbf{v}_k\right) = 0 \tag{1}$$

Momentum Equations

$$\frac{\partial}{\partial t}\left(\rho_k \varepsilon_k \mathbf{v}_k\right) + \nabla \cdot \left(\varepsilon_k \rho_k \mathbf{v}_k \mathbf{v}_k\right) = -\varepsilon_k \nabla P + \rho_k \varepsilon_k \mathbf{g} + G_k \nabla \varepsilon_1 + \sum_{i=1}^{N} K_{k_i}\left(\mathbf{v}_1 - \mathbf{v}_k\right) + q_k \mathbf{E} \tag{2}$$

where

$$\sum_{k=1}^{N} \varepsilon_k = 1 \tag{3}$$

Equation of State

$$\rho_1 = \frac{P}{RT} \quad ; \quad \rho_k = \rho_{sk} \text{ for } k \geq 2 \tag{4}$$

$$K_{1k} = K_{k1} = \frac{150\left(1-\varepsilon_1\right)\varepsilon_k \mu_1}{\varepsilon_1\left(d_k \phi_k\right)^2} + \frac{1.75 \rho_1 \left|\mathbf{v}_1 - \mathbf{v}_k\right| \varepsilon_k}{\left(d_k \phi_k\right)} \quad 0.2 \leq \varepsilon_1 \leq 0.8$$
$$= \frac{3}{4} C_{Dk} \frac{\varepsilon_1 \left|\mathbf{v}_1 - \mathbf{v}_k\right| \rho_1 \varepsilon_k}{\left(d_k \phi_k\right)} f\left(\varepsilon_1\right) \quad 0.8 \leq \varepsilon_1 \leq 1.0 \tag{5}$$

where

$$f\left(\varepsilon_1\right) = \varepsilon_1^{-2.65} \tag{6}$$

$$C_{DK} = \frac{24}{\mathrm{Re}_k}\left(1 + 0.15 \mathrm{Re}_k^{0.687}\right) \quad \mathrm{Re}_k < 1.000$$
$$= 0.44 \quad \mathrm{Re}_k \geq 1.000 \tag{7}$$

and

$$\mathrm{Re}_k = \frac{d_k \left|\mathbf{v}_1 - \mathbf{v}_k\right| \rho_1 \varepsilon_1}{\mu_1} \tag{8}$$

$$K_{kl} = \frac{3}{2} \alpha (1+e) \frac{\rho_k \rho_1 \varepsilon_k \varepsilon_1 \left(d_k + d_1\right)^2}{\rho_k d_k^3 + \rho_l d_1^3} \left|\mathbf{v}_k - \mathbf{v}_1\right| \tag{9}$$

$$V_{s,\,\mathrm{crit}} = \left(-G/\rho_s\right)^{1/2} \tag{10}$$

$$G\left(\varepsilon_1\right) = -10^{-8.76\varepsilon_1 + 7.8} \; N/m^2 \tag{11}$$

$$G_k = 0 \text{ for } k = 1$$
$$= \varepsilon_k G\left(\varepsilon_1\right) \text{ for } k = 1, \ldots N \tag{12}$$

Fig. 5.4 Hydrodynamic Model A for electrostatic segregation of particles.

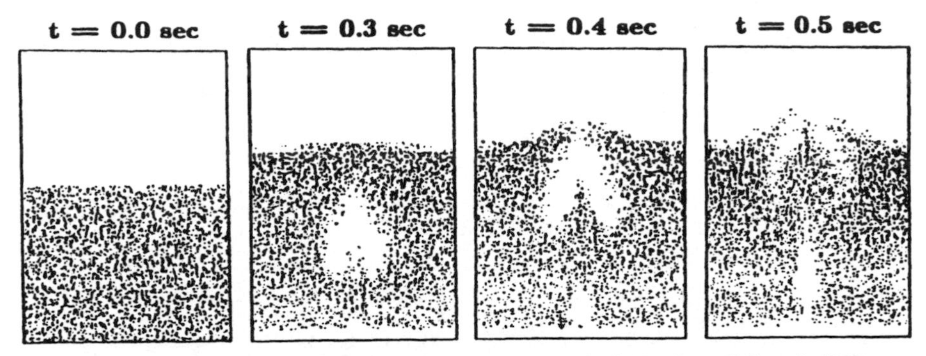

Fig. 5.5 Volume fractions of coal particles with no applied electric field. (From Shih *et al.*, 1987. Reproduced by permission of the American Institute of Chemical Engineers. Copyright © 1987 AIChE. All rights reserved.)

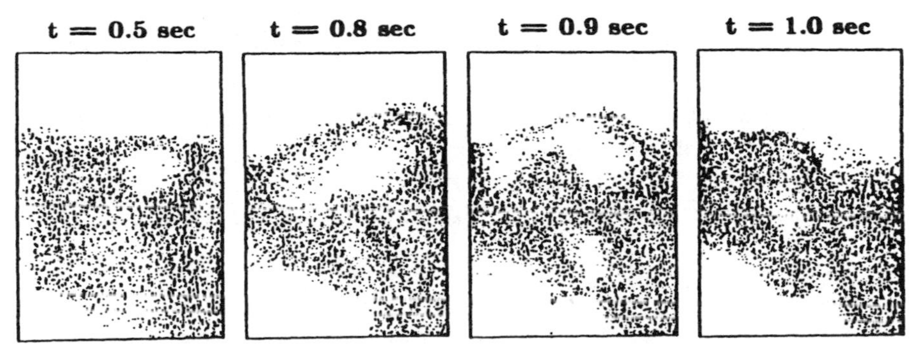

Fig. 5.6 Volume fractions of coal particles with 500 V/cm electric field strength. (From Shih *et al.*, 1987. Reproduced by permission of the American Institute of Chemical Engineers. Copyright © 1987 AIChE. All rights reserved.)

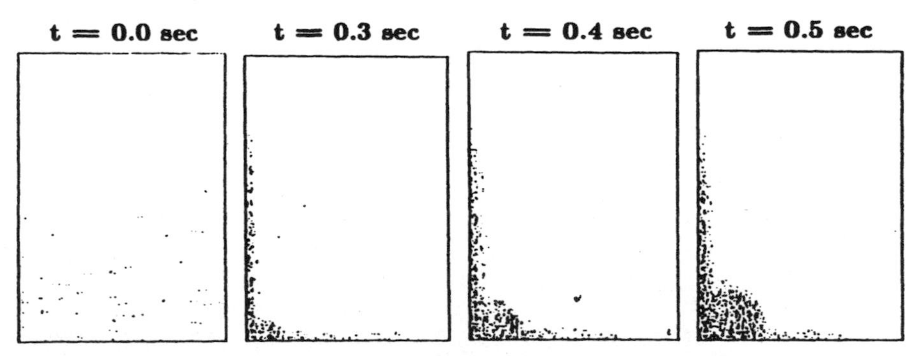

Fig. 5.7 Volume fractions of pyrite particles with 500 V/cm electric field strength. (From Shih *et al.*, 1987. Reproduced by permission of the American Institute of Chemical Engineers. Copyright © 1987 AIChE. All rights reserved.)

(a) Sum Eq. (2) over k phases to obtain the mixture momentum equation. Then obtain an expression for pressure drop to give

$$\textit{pressure drop} = \textit{weight of bed} + \textit{electrical force}$$

This equation generalizes Eq. (5.1).
(b) Show that when $E = -(dV/dy)$, where V is the applied voltage, the pressure drop across the bed increases linearly with the applied voltage and its slope can be used to determine the surface charge on the particles. Suggest a use of fluidized beds for space travel.
(c) Show that for zero acceleration, the elimination of pressure between the gas phase momentum balance and particle k balance gives the equation

$$\sum_{i=2}^{n} K_{gi}\left(\mathbf{v}_g - \mathbf{v}_i\right) = \varepsilon_g \sum_{k=2}^{n} \varepsilon_k \left(\rho_k - \rho_g\right)\mathbf{g} + \sum_{k=2}^{n} q_k \mathbf{E} \varepsilon_g .$$

This equation generalizes Eq. (5.3). Interpret the meaning of each of the terms.
(d) Show that the momentum balance in part (c) allows the following tentative definition of a minimum fluidization velocity U_{mf} for a mixture of particles of various sizes and densities by assuming that the particles do not move:

$$\sum_{i=2}^{n} K_{gi} U_{mf} = \varepsilon_g^{\,2} \sum_{k=2}^{n} \varepsilon_k \left(\rho_k - \rho_g\right) g .$$

(e) How many unknowns are there in the equation in part (d)? Is it possible to assign a zero velocity to all particles in a mixture? If not, how can the minimum fluidization of a mixture be obtained?

Hint: 200 and 500 μm particles segregate by size.

(f) Extend the definition of a minimum fluidization velocity to include the effect of the electric field. How does it vary with E?

3. *Energy Dissipation: Attrition, Erosion*
(a) Use the results of Exercise 2 of Chapter 1 or conventional single-phase analysis to show that the collisional energy dissipation in steady, laminar shear flow of particles is

$$\dot{\mathcal{D}} = \mu_s \left(\frac{\partial v_s}{\partial y}\right)^2 ,$$

where μ_s is the viscosity and $\partial v_s / \partial y$ is the solids velocity gradient between two walls.

(b) Relate this result to Eq. (5.24) and note which dissipation was neglected in part (a).

(c) Use the dense phase particulate viscosity given by Eq. (9.245) and the granular temperature given by Eq. (9.284) to show that the dissipation is

$$\dot{\mathcal{D}} = \frac{4}{5(15)^{1/2}} \varepsilon_s \rho_p d_p{}^2 g_0 \frac{(1+e)}{(1-e)^{1/2}} \left(\frac{\partial v_s}{\partial y} \right)^3, \quad e \neq 1.$$

(d) In particulate multiphase flow the energy dissipation will go into heat, into attrition — that is, into particle wear and break-up — and into erosion of surfaces. In the 1980s it was discovered that the erosion of horizontal heat exchanger tubes to be placed into bubbling bed coal combustors was so severe that the whole technology developed in the 1970s was threatened. Various fluidization consultants were called in by funding agencies to estimate the rate of tube wear and to suggest remedies. Carefully conducted experiments at Grimethorpe, England, confirmed the high rates of tube erosion.

The maximum rate of tube erosion can be estimated by equating the hardness of the metal tube times the rate of wear to the energy dissipation, as computed in part (c). The hardness of the metal is like a pressure. Hence $P\,dV$ is work of wear. Based on this analysis, estimate how the rate of erosion varies with

i. fluidization velocity
ii. particle diameter
iii. particle density
iv. hardness of the tube.

Where does the hardness of the particle come in?

For a review and other models see Bouillard and Lyczkowski (1991).

Literature Cited

Bird, R. B., W. E. Stewart, and E. N. Lightfoot (1960). *Transport Phenomena*. New York: John Wiley & Sons.

Bouillard, J. X., and R. W. Lyczkowski (1991). "On the Erosion of Heat Exchanger Tube Banks in Fluidized-Bed Combustors," *Powder Technol.* **68**, 37–51.

Couderc, J. P. (1985). "Incipient Fluidization and Particulate Systems," Ch. 1 in J. F. Davidson, R. Clift and D. Harrison, Eds. *Fluidization*. New York: Academic Press.

Davidson, J. F., and D. Harrison (1963). *Fluidized Particles*. Cambridge, U.K.: Cambridge University Press.

Fitzgerald, T. S. (1985). "Course Particle Systems," Ch. 12 in J. F. Davidson, R. Clift and D. Harrison, Eds. *Fluidization.* New York: Academic Press.

Geldart, D. (1973). "Types of Gas Fluidization," *Powder Technol.* 7, 285–292.

Gidaspow, D., B. Ettehadieh, and J. Bouillard (1985). "Hydrodynamics of Fluidization: Bubbles and Gas Compositions in the U-GAS Process," *AIChE Symposium Series No. 241* **80**, 57–64.

Gidaspow, D., M. Syamlal and Y. Seo (1986). "Hydrodynamics of Fluidization of Single and Binary Size Particles: Supercomputer Modeling," pp. 1–8 in *Fluidization V: Proceedings of the Fifth Engineering Foundation Conference on Fluidization.* New York: AIChE Engineering Foundation.

Shih, Y. T., D. Gidaspow, and D. T. Wasan (1987). "Hydrodynamics of Electrofluidization: Separation of Pyrites from Coal," *AIChE J.* **33**, 1322–1333.

Wen, C. Y., and Y. H. Yu (1966). "Mechanics of Fluidization," *Chem. Eng. Progress Symp. Series* **62**, 100–111.

Yang, W.-C., D. C. Chitester, E. M. Kornosky and D. L. Keairns (1985). "A Generalized Methodology for Estimating Minimum Fluidization Velocity at Elevated Pressure and Temperature," *AIChE J.* **31**, 1086–1092.

6

ON THE ORIGIN OF BUBBLES

6.1 Introduction **115**
6.2 Void Propagation in Incompressible Fluids **116**
6.3 Shocks and Dispersion with No Solids Stress **119**
6.4 Bubbling Criterion for Small Particles **121**
Acknowledgment **123**
Exercises **123**
1. Effect of Gas Compressibility **123**
2. Invariance to Models **124**
3. On the Prediction of Bubble Coalescence in Gas–Liquid–Solid Fluidization **124**
Literature Cited **125**

6.1 Introduction

Speculation concerning the origin of bubbles in fluidized beds has been going on for several decades. Jackson (1963) carried out a small perturbation analysis of his hydrodynamic equations and showed that a small perturbation in voidage will grow without bound. Davidson and Harrison (1963) analyzed the problem from the point of view of the existence of a maximum stable bubble size. Their stability condition was that the bubble velocity given by a semi-empirical expression in terms of bubble size had to be greater than the terminal velocity of the particles for the existence of a stable bubble. Rietema and Mutsers (1973) used a small perturbation analysis, reviewed in detail by Jackson (1985), to obtain in a criterion for bubbling. Rietema and Mutsers' work has been continued by Mutsers and Rietema (1977), Piepers *et al.* (1984) and Rietema (1984).

The realization that bubbles are shocks, that is, intersections of characteristic paths, in the mathematical sense, took some time to evolve. Verloop and Heertjes (1970) were apparently the first to state that shock waves will originate in fluidized systems when porosity waves rise faster than what they state to be the velocity of "an equilibrium disturbance." Then Verloop and Heertjes continued their square root of the elastic modulus of the bed divided by its density. Foscolo and Gibilaro (1984) found an expression for this velocity in terms of the gravitational constant and powder properties by making an analysis that neglected inertia of the particles. Rowe (1986) has extended Foscolo and Gibilaro's analysis with considerable success to give an explanation of the behavior of Geldart Type A and B powders.

The mathematical theory of shock formation in fluidized beds was first presented by Fanucci *et al.* (1979). They solved simultaneously the partial differential equations for porosity and velocity using the method of characteristics for the case when the elastic modulus of the powder divided by its density exceeded a reduced relative velocity. They assumed that the velocity of the particles at the grid varied sinusoidally with time. With such an input into the model they found that the characteristic paths intersected; hence, shocks formed. Since their method is numerical, no criteria for bubbling were found. This type of analysis was repeated by Rasouli (1981) with the relative velocity equations used here. Similar mild intersections of characteristics were found with the sinusoidal velocity input at the inlet of the bed. Pritchett *et al.* (1978), Klein and Scharff (1982) and Gidaspow and Ettehadieh (1983) solved the hydrodynamic equations numerically with realistic velocity input into the bed and found that bubbles will form in a natural way for Geldart Type B particles. These studies were fully reviewed by Gidaspow (1986).

To obtain an algebraic bubble criterion from the hydrodynamic equations of fluidization, approximations must be made. The objective here is to derive a bubbling criterion using shock theory of first-order partial differential equations. Such a theory is commonly used (Amundson and Aris, 1973) in chromatography.

6.2 Void Propagation in Incompressible Fluids

To obtain continuity shocks, it is necessary to derive a void propagation equation. The two incompressible continuity equations for the gas and the solid with no phase changes in one dimension can be written as follows:

$$\frac{\partial \varepsilon_i}{\partial t} + \frac{\partial(\varepsilon_i v_i)}{\partial x} = 0, \tag{6.1}$$

where $i = g, s$.

Equation (6.1) states that the volume of each phase i is constant as we move with velocity v_i. This follows directly from the Reynolds transport theorem. To obtain the desired void propagation equation, the two continuity equations are each solved for the gradient of velocity and the two gradients are subtracted to form the group

$$\frac{\partial(v_s - v_g)}{\partial x} = \frac{1}{\varepsilon\varepsilon_s}\left[\frac{\partial \varepsilon}{\partial t} + (\varepsilon v_s + \varepsilon_s v_g)\frac{\partial \varepsilon}{\partial x}\right]. \tag{6.2}$$

This algebraic manipulation produces an inversely weighted mean velocity

$$\hat{v} = \varepsilon v_s + \varepsilon_s v_g \tag{6.3}$$

and the relative velocity

$$v_r = v_s - v_g. \tag{6.4}$$

In terms of these new variables, the two continuity equations give a primitive form of the void propagation equation

$$\frac{\partial \varepsilon}{\partial t} + \hat{v}\frac{\partial \varepsilon}{\partial x} = \varepsilon\varepsilon_s \frac{\partial v_r}{\partial x}. \tag{6.5}$$

If the slip, v_r, is a constant, Eq. (6.5) shows that the void ε moves with the velocity $\hat{v}$. This is also true for very dilute flow, where ε_s is near zero. With such approximations, continuity shocks can be obtained, as will be discussed later.

It is most instructive to consider the relative velocity model (Gidaspow, 1978) before returning to the approximations mentioned previously. In one dimension, with normal stress σ and wall shear τ_{ws}, this model can be written as

$$\varepsilon_s \rho_s v_r \frac{\partial v_r}{\partial x} = -\frac{\partial \sigma}{\partial x} - g\varepsilon_s \Delta\rho - \beta_C v_r - \frac{4\tau_{ws}}{D_t}, \tag{6.6}$$

where the gravity g is taken in the opposite direction of the coordinate x, β_C is the interfacial drag and D_t is the actual or effective tube diameter. Solutions to various transport and fluidization problems using this equation were presented by Arastoopour and Gidaspow (1979), Liu and Gidaspow (1981) and Gidaspow *et al.* (1983). See Chapter 2, Eq. (T2.6).

Breault (1987) has applied this model to produce a computer code for recirculating fluidized bed combustors. As done previously (Gidaspow, 1986), a modulus of compressibility is introduced by noting that the normal stress is primarily a function of the porosity or volume fraction of the solid phase.

Thus,

$$\frac{\partial \sigma}{\partial \varepsilon_s} = \frac{\partial \sigma}{\partial \varepsilon_s} \cdot \frac{\partial \varepsilon_s}{\partial x}. \tag{6.7}$$

To have a positive modulus of powder compressibility G, let

$$G = \frac{\partial \sigma}{\partial \varepsilon_s} = -\frac{\partial \sigma}{\partial \varepsilon}. \tag{6.8}$$

The sign of this modulus is opposite to that used in the literature previously (Gidaspow, 1986). Using such a representation for the stress, substitution of the gradient of velocity from Eq. (6.6) into Eq. (6.5), gives the void propagation equation

$$\frac{\partial \varepsilon}{\partial t} + \left(\hat{v} - \frac{G\varepsilon}{\rho_s v_r} \right) \frac{\partial \varepsilon}{\partial x} = -\frac{\beta_C \varepsilon}{\Delta \rho} - \frac{g \varepsilon \varepsilon_s}{v_r} - \frac{4 \tau_{ws} \varepsilon}{D_t \Delta \rho v_r}. \tag{6.9}$$

Convection of voidage = Drag + Gravity + Wall friction

Equation (6.9) shows that the voidage moves with the velocity $\hat{v} - G\varepsilon / \rho_s v_r$. When the powder compressive modulus is zero, it moves with v, as already seen from the continuity equations above.

The ratio G / ρ_s is the square of the critical velocity. It has the same meaning as the velocity postulated by Verloop and Heertjes (1970) and Foscolo and Gibilaro (1984). This velocity square divided by the relative velocity opposes the main motion, as seen in Eq. (6.9). In developed flow the right-hand side of Eq. (6.9) is zero. Then buoyancy is balanced by drag and wall friction.

The interfacial drag β_C in the relative velocity equation is related to β used in the model of Gidaspow and Ettehadieh (1983) by the relation

$$\beta_C = \frac{\beta_A}{\varepsilon}. \tag{6.10}$$

The interfacial drag β was (Gidaspow, 1986) related to the drag coefficient for dilute flow or to the coefficients from the Ergun equation. The latter yield the expression for β_C as

$$\beta_C = 150 \frac{(1-\varepsilon)^2 \mu}{\varepsilon^2 \left(d_p \phi_s\right)^2} + 1.75 \frac{|v_r|(1-\varepsilon)\rho_g}{\left(\phi_s d_p\right)\varepsilon}. \tag{6.11}$$

To make further progress it is assumed as a first approximation that the velocities in Eq. (6.9) are equal to their fully developed or terminal velocities. They are then obtained by equating buoyancy to drag. Then in the absence of wall friction, the right-hand side of Eq. (6.9) shows that

$$-v_r = \frac{\Delta\rho\varepsilon_s g}{\beta_C} = \frac{\varepsilon\varepsilon_s \Delta\rho g}{\beta_A}. \tag{6.12}$$

This approximation has reduced the problem of void propagation to one first-order partial differential equation, Eq. (6.9).

6.3 Shocks and Dispersion with No Solids Stress

For developed flow, the combination of the two continuity equations as already given by Eq. (6.5) shows that the void moves with the velocity $\hat{v}$. The same result is given by Eq. (6.9) for zero stress modulus G. For convenience call this propagation velocity C.

Hence,

$$C = \hat{v} = \varepsilon v_s + \varepsilon_s v_g. \tag{6.13}$$

Next consider the case of void injection into a bed at minimum fluidization, where v_s is zero. Then using Eq. (6.12), the propagation velocity becomes as follows:

$$C = \varepsilon\varepsilon_s{}^2 \frac{\Delta\rho g}{\beta_A} = \varepsilon_s{}^2 \frac{\Delta\rho g}{\beta_C}. \tag{6.14}$$

This propagation velocity will have a different dependence on the void fraction depending upon the value of the friction coefficient β_A. If C were to be a constant, the void injected into a bed at minimum fluidization would move unchanged throughout the bed, as does a pulse of tracer gas injected into a chromatographic column with a linear isotherm and no diffusional resistance. The different behavior of the void injected into the bed is due to the different expressions for β_A in different regimes.

For low particle Reynolds number, the first term of the Ergun equation, as given by Eq. (6.11), describes the dependence of β_C on ε. Hence,

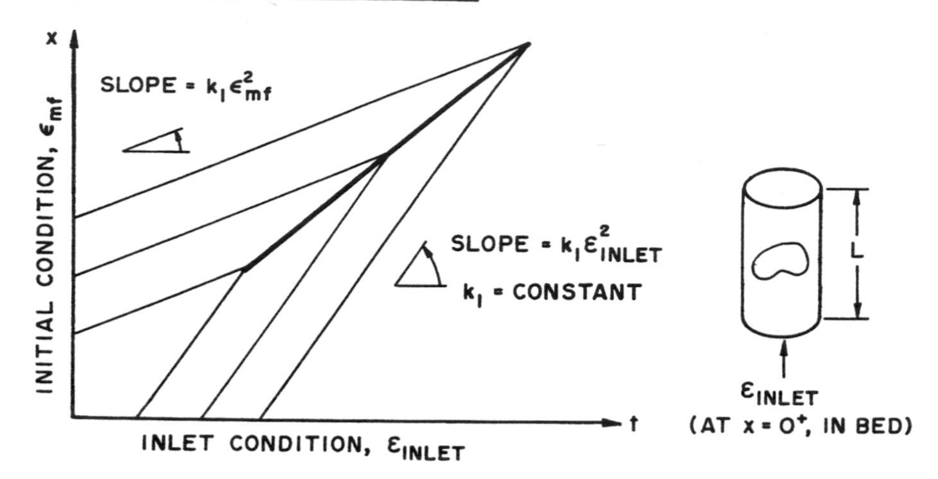

Fig. 6.1 Bubble formation in a fluidized bed initially at minimum fluidization. (From Bouillard and Gidaspow, 1991. Reproduced by permission of Elsevier Sequoia.)

$$C = \frac{\Delta\rho g\left(d_p\phi_s\right)^2}{150\mu}\varepsilon^2, \quad \mathrm{Re}_p < 1. \tag{6.15}$$

Equation (6.15) shows that the void injected at the bottom of the bed moves faster than the void at minimum fluidization velocity. Figure 6.1 shows that this results in the intersection of paths or characteristic directions and hence in the formation of a shock. The faster moving void injected at the bottom catches up with the slower moving void in the bed and forms a reinforced void, a shock.

Now consider the case of large particles or large particle Reynolds numbers. For such a situation the "turbulent" part of the Ergun equation is applicable. Hence, Eq. (6.11) gives

$$C = \left[\frac{\Delta\rho g\phi_s d_p}{1.75}\right]^{1/2}(1-\varepsilon)\varepsilon^{1/2}, \quad \mathrm{Re}_p > 1000. \tag{6.16}$$

For ε greater than $\frac{1}{3}$, this is a decreasing function of ε. Therefore, as seen in Fig. 6.2, the void injected into the bed moves more slowly than the void at minimum fluidization already in the bed. Hence, the void is dispersed, as in chromatogra-

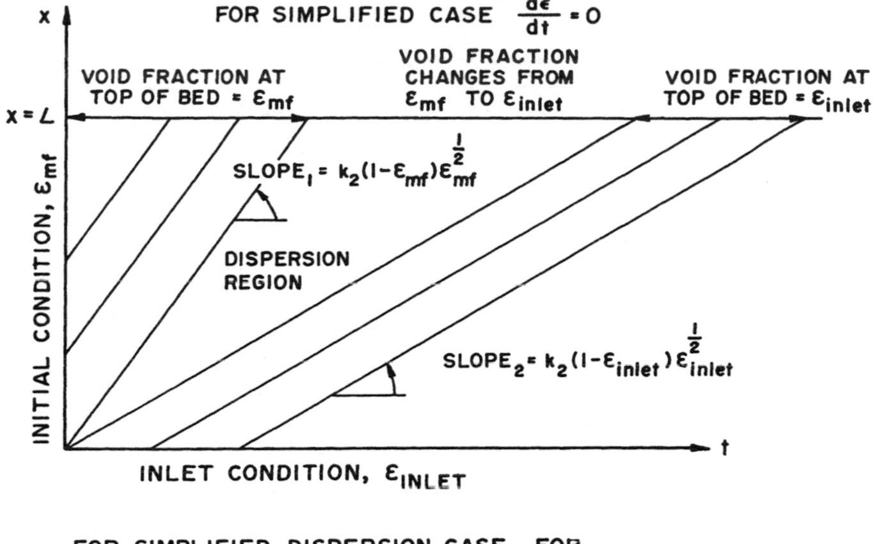

Fig. 6.2 Void dispersion in a fluidized bed initially at minimum fluidization. (From Bouillard and Gidaspow, 1991. Reproduced by permission of Elsevier Sequoia.)

phy with an unfavorable isotherm. Therefore, for sufficiently large particle sizes, bubbles will not form in the fluidized bed. This occurs for Geldart D type particles.

6.4 Bubbling Criterion for Small Particles

Equation (6.9) shows that the void propagates with the velocity C given by

$$C = \hat{v} - \frac{G\varepsilon}{\rho_s v_r}. \tag{6.17}$$

From the discussion given in the previous paragraph, it is clear that the criterion that must be satisfied for a shock formation is that

$$\frac{dC}{d\varepsilon} > 0 \text{ for bubble formation.} \tag{6.18}$$

For small particles C is the sum of Eq. (6.15) and the contribution due to non-zero G as given by Eq. (6.17). For convenience let the void independent part of the percolation velocity be

$$v_o = \frac{\Delta\rho g\left(d_p\phi_s\right)^2}{150\mu}. \tag{6.19}$$

Then the void propagation velocity can be written as follows for small particles:

$$C = v_o\varepsilon^2 + \frac{G}{\rho_s}\frac{(1-\varepsilon)}{v_o\varepsilon}. \tag{6.20}$$

A useful form for the powder modulus is

$$G = G_o \exp\left[-\left(\varepsilon - \varepsilon'\right)\mathrm{b}\right], \tag{6.21}$$

where G_o is the void independent part and ε' and b are empirical constants for the powder.

Equation (6.20) is the sum of an increasing function of ε and a decreasing function of ε. Hence, a maximum exists. For shock formation we must have

$$\frac{dC}{d\varepsilon} = 2v_o\varepsilon - \frac{1 + \mathrm{b}\varepsilon - \mathrm{b}\varepsilon^2}{\varepsilon^2 v_o}\frac{G}{\rho_s} \tag{6.22}$$

or

$$v_o > \left(\frac{1 + \mathrm{b}\varepsilon - \mathrm{b}\varepsilon^2}{\varepsilon^3}\right)^{1/2}\left(\frac{G}{\rho_s}\right)^{1/2}. \tag{6.23}$$

Percolation velocity > Constant × Critical flow velocity

Equation (6.23) shows that a bubble is formed only when the percolation velocity exceeds the critical flow velocity by some constant. A typical value of the critical velocity $(G/\rho_s)^{1/2}$ is one meter per second. Thus, bubbling will occur for sufficiently large particles but not for small particles according to the criterion given by Eq. (6.23), since v_o varies with the square of the particle size. It will occur in gas fluidized beds but not in liquid fluidized beds when $\Delta\rho$ is small. The bubbling criterion can even be stretched to the case when the particles are cemented together. In such a case G is the modulus of elasticity for the brick having a value of thousands of meters per second. No reasonable flow of gas can break such a cemented group of particles. Hence, no bubble forms.

The derived bubbling criterion can be regrouped to give the Geldart transition for Group A to Group B particles. For low particle Reynolds numbers, the velocity v_o used in Eq. (6.20) is related to the minimum fluidization velocity by means of the relation

$$v_o = 11\, U_{mf} \phi_s^2. \tag{6.24}$$

Then in terms of minimum fluidization velocity the bubbling criterion becomes

$$U_{mf} > \frac{1}{11\,\phi_s^2}\left(\frac{1 + b\varepsilon - b\varepsilon^2}{\varepsilon^3}\right)^{1/2}\left(\frac{G}{\rho_s}\right)^{1/2} \tag{6.25}$$

or

$$U_{mf} > U_{MB}, \tag{6.26}$$

where the bubbling velocity U_{MB} equals the expression on the right-hand side of inequality (6.25). Thus, the minimum bubbling velocity is related to how fast the solids are running down to fill the void at the critical velocity $(G/\rho_s)^{1/2}$, as well as to how the modulus G changes with the void. Although the order of magnitude of the minimum bubbling velocity computed here agrees with the values measured by Geldart, the high sensitivity of G with ε precludes a quantitative calculation of U_{MB} from G at the present time. However, when expression (6.26) is read as an equality, it clearly gives Geldart's boundary between A and B particles.

Acknowledgment

An extended version of this chapter was published while this book was in preparation. The reference is J. X. Bouillard and D. Gidaspow (1991). "On the Origin of Bubbles and Geldart's Classification," *Powder Technology* **68**, 13–22.

Exercises

1. *Effect of Gas Compressibility*
(a) Show that a generalization of the void propagation Eq. (6.2) to the case of a compressible gas is

$$\frac{\partial(v_s - v_g)}{\partial x} = \frac{1}{\varepsilon\varepsilon_s}\left[\frac{\partial\varepsilon}{\partial t} + \hat{v}\frac{\partial\varepsilon}{\partial t} + \left(\frac{\varepsilon^2\varepsilon_s}{\rho_g C_g^2}\right)\frac{dP}{dt^g}\right],$$

where dP/dt^g is the time derivative of the pressure in a frame moving with the gas velocity v_g, and C_g is the gas sonic velocity

$$C_g^2 = \left(\frac{\partial P}{\partial \rho}\right)_s .$$

(b) Using Eq. (6.9) show that the additional compressibility term is small and can be neglected. The effect of pressure on the transition to bubbling comes from the fact that at high pressure, as shown in Chapter 7, there is a greater bed expansion. The slope $dC/d\varepsilon$ decreases and eventually becomes negative as pressure increases, leading to no bubbling.

2. *Invariance to Models*

Show that the bubbling criteria derived are the same for Models A and B.

3. *On the Prediction of Bubble Coalescence in Gas–Liquid–Solid Fluidization*

There are two basic regimes in gas-liquid-solid fluidized beds, as discussed by Fan (1989). They are the dispersed bubble and the coalesced bubble regimes. They can be best understood by reference to Fig. 6.3. This figure shows a photograph of gas bubbles formed in a two-dimensional bed filled with 0.8 mm glass beads continuously fluidized by water with an approximately uniform injection of fine bubbles through a gas distributor mounted below a liquid porous distributor through which water was pumped at a uniform velocity. At much higher liquid velocities, that is, at higher bed expansions, visible bubbles tend to disappear. We are in the dispersed bubble regime. One may speculate that the particles act like Geldart Group A particles in gas fluidization or like particles in a gas at high pressure. Since the hydrodynamic three-phase computer code appears to be able to predict these observations, it should be possible to generalize the bubbling criterion developed in this chapter to predict bubble coalescence.

(a) Assume that the liquid–solid slurry is homogeneous, that is, that its velocities are equal, and find an approximate criterion for bubble formation.

Note that for gas bubble coalescence the slopes of the characteristics in Fig. 6.1 must intersect. The velocity of the fine bubbles at the injector must be greater than their velocities in the upper portion of the bed.

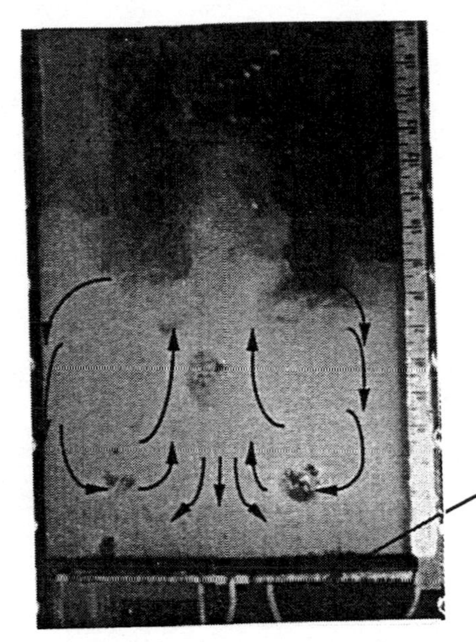

Fig. 6.3 Bubble coalescence in a gas–liquid–solid fluidized bed, with uniform liquid and gas injection.

Particles	0.8 mm leaded glass beads
Initial bed height of particles	17.8 cm
Superficial gas velocity	3.36 cm/s
Superficial liquid velocity	2.04 cm/s

(a) Note that the gas and liquid velocities are in the range of commercially useful flow rates, as reported by Tarmy and Coulaloglu (1992).
(b) Note that visible bubbles disappear when the liquid velocity is doubled (Bahary, 1994).

(b) Explain the location of the bubbles in Fig. 6.3. The arrows indicate the observed flow patterns, very roughly. See Chapter 10.

Literature Cited

Amundson, N. R., and R. Aris (1973). *Mathematical Methods in Chemical Engineering, Vol. 2: First Order Partial Differential Equations with Applications.* New York: Prentice Hall.

Arastoopour, H., and D. Gidaspow (1979). "Vertical Pneumatic Conveying Using Four Hydrodynamics Models," *I&EC Fundamentals* **18**, 123-130.

Bahary, M. (1994). Ph. D. thesis. Chicago, Illinois: Illinois Institute of Technology (in progress).

Bouillard, J. X. and D. Gidaspow (1991). "On the Origin of Bubbbles and Geldart's Classification," *Powder Technol.* **68**, 13–22.

Breault, R. W. (1987). "Theoretical Modeling of the Multi-solids Fluidized Bed Combustor: Hydrodynamics, Combustion, and Desulfurization," pp. 770–775 in *Proceedings of the 1987 International Conference on Fluidized Bed Combustion*, Vol. II. Boston: ASME.

Davidson, J. F., and D. Harrison (1963). *Fluidized Particles*. Cambridge, U.K.: Cambridge University Press.

Fan. L. S. (1989). *Gas–Liquid–Solid Fluidization*. Boston: Butterworth.

Fanucci, J. B., N. Ness, and R. H. Yen (1979). "On the Formation of Bubbles in Gas-Particulate Fluidized Beds," *J. Fluid Mech.* **94** (2), 353–367.

Foscolo, P. U. and L. G. Gibilaro (1984). "A Fully Predictive Criterion for the Transition between Particulate and Aggregate Fluidization," Chem. Eng. Sci. **39**, 1667–1675.

Gidaspow, D. (1978). Pp. 283–298 in T. N. Veziroglu and S. Kakac, Eds. *Two-Phase Transport and Reactor Safety, Vol. 1.* Washington: Hemisphere Publishing Co.

Gidaspow, D. (1986). "Hydrodynamics of Fluidization and Heat Transfer: Supercomputer Modeling," *Appl. Mech. Rev.* **39** (1), 1–23.

Gidaspow, D. and B. Ettehadieh (1983). "Fluidization in Two-Dimensional Beds with a Jet, Part II: Hydrodynamic Modeling," *I&EC Fundamentals* **22**, 193–201.

Gidaspow, D., Y. C. Seo, and B. Ettehadieh (1983). "Hydrodynamics of Fluidization: Experimental and Theoretical Bubble Sizes in a Two-Dimensional Bed with a Jet," *Chem. Eng. Comm.* **22**, 253–272.

Jackson, R. (1963). "The Mechanics of Fluidized Beds," *Trans. Inst. Chem. Eng.* **41**, 13–38.

Jackson, R. (1985). Chapter II (pp. 47–72) in J. F. Davidson, R. Clift, and D. Harrison, Eds. *Fluidization*, 2nd ed. London: Academic Press.

Klein, H. H. and M. F. Scharff (1982). "A Time Dependent Reactor Model of Fluidized Bed Chemical Reactors," *ASME Paper 82-FE-3*; also in H. H. Klein, D. E. Dietrich, S. R. Goldman, D. H. Laird, M. F. Scharff and B. Srinivas (1983). "Time Dependent Reactive Models of Fluidized Bed and Entrained Flow Chemical Reactors," Ch. 29 in N. Chereminisoff and R. Gupta, Eds. *Handbook of Fluids in Motion*. Ann Arbor Science Press.

Liu, Y. and D. Gidaspow (1981). "Solids Mixing in Fluidized Beds: A Hydrodynamic Approach," *Chem. Eng. Sci* **36** (3), 539–547.

Mutsers, S. M. P. and K. Rietema (1977). "The Effect of Interparticle Forces on the Expansion of a Homogeneous Gas Fluidized Bed," *Powder Technol.* **18**, 239–248.

Piepers, H. W., E. J. E. Cottar, A. H. M. Verkooijen and K. Rietema (1984). "Effects of Pressure and Type of Gas on Particle–Particle Interaction and the Consequences for Gas–Solid Fluidization Behavior," *Powder Technol* **37**, 55–70.

Pritchett, J. W., H. B. Levine, T. R. Blake and S. K. Garg (1978). "A Numerical Model of Gas Fluidized Beds," *AIChE Symposium Series* **176** (74) 134–148.

Rasouli, R. (1981). "One Dimensional Transient Unequal Velocity Two-Phase Flow by the Method of Characteristics," Ph.D. Thesis. Chicago, Illinois: Illinois Institute of Technology.

Rietema, K. (1984). "Powders: What Are They?" *Powder Technol.* **37**, 5–23.

Rietma, K., and S. M. P. Mutsers (1973). "The Effect of Interparticle Forces on Expansion of a Homogeneous Gas-Fluidized Bed," pp. 32–33 in *Proceedings of International Symposium on Fluidization*. Toulouse, France.

Rowe, P. H. (1986). "A Rational Explanation for the Behavior of Geldart Type A and B Powders When Fluidized," *AIChE Microfiche Preprint for the 1986 Miami Beach Annual AIChE Meeting.*

Tarmy, B. L., and C. A. Coulaloglu (1992). "Alpha-Omega and Beyond: Industrial View of Gas–Liquid–Solid Reactor Development," *Chem. Eng. Sci.* **47**, 3231–3246.

Verloop, J. and P. M. Heertjes (1970). "Shock Waves as a Criterion for the Transition from Homogeneous to Heterogeneous Fluidizaiton," *Chem. Eng. Sci.* **25**, 825-832.

7

INVISCID MULTIPHASE FLOWS: BUBBLING BEDS

7.1 Basic Equations 129
7.2 Compressible Granular Flow 131
7.3 Well-Posedness of Two-Phase Models 134
7.4 Homogeneous Flow and Pressure Propagation 139
7.5 Solids Vorticity and Void Propagation 141
7.6 Davidson Bubble Model 146
7.7 Computer Fluidization Models 149
7.8 Comparison of Computations to Observations 154
7.8.1 Bubbles with a Jet 154
7.8.2 Time-Averaged Porosities 164
7.8.3 Velocity Profiles 168
7.8.4 Pressure Effect 171
7.8.5 Fast Bubble 171
7.8.6 3-Meter KRW Fluid Bed 180
Exercises 183
1. Conical Moving-Bed Shale Retort Model 183
2. Desalting by Electrosorption: First Order PDE 185
3. Application of the Theory of Characteristic Mass Transfer in a Fluidized Bed 186
4 Counterflow Unsteady Heat Exchanger: Characteristics 187
5. The Method of Characteristics 188
6. Homogeneous Two Phase Flow 189
7. Homogeneous Equilibrium Boiling 190
8. Ill-Posedness for Model A 191
Literature Cited 192

7.1 Basic Equations

The conservation laws of mass and momentum for each phase k are

Continuity Equation for Phase k

$$\frac{\partial}{\partial t}(\varepsilon_k \rho_k) + \nabla \cdot (\varepsilon_k \rho_k \mathbf{v}_k) = m_k', \tag{7.1}$$

Momentum Equation for Phase k

$$\frac{\partial}{\partial t}(\varepsilon_k\rho_k\mathbf{v}_k)+\nabla\cdot(\varepsilon_k\rho_k\mathbf{v}_k\mathbf{v}_k)=\varepsilon_k\rho_k\mathbf{F}_k+\nabla\cdot\mathbf{T}_k+\sum_l\beta(\mathbf{v}_l-\mathbf{v}_k)+m_k'\mathbf{v}_k. \quad (7.2)$$

Acceleration of phase k = Body force + Stress + Drag force or particle–particle + Phase change momentum interaction

These equations were derived in Chapter 1. Equation (7.2) is Eq. (1.16) with the interaction force p_k' expressed in terms of the friction coefficient β fully discussed in Chapter 2 for solid–fluid interaction. Particle–particle interaction will be discussed further when multiphase effects are considered. The phase change momentum disappears when Eq. (7.2) is expressed as Eq. (1.16), moving with its phase velocity. As already alluded to in Chapter 1, a constitutive equation for the stress in the absence of viscous effects or cohesion can be chosen to be

Constitutive Equation for Stress with No Viscosity

$$\mathbf{T}_k=-P_k\mathbf{I}. \quad (7.3)$$

When Eq. (7.3) is substituted into the momentum balance, Eq. (7.2), the result is Model B, which was fully discussed in Chapter 2, and studied numerically for fluidization by Bouillard *et al.* (1989a). Each momentum equation contains its own pressure. Model A, favored by the gas–liquid two-phase flow investigators, involves a pressure gradient of the gas in the particulate phase momentum balance. Since the differences between the two models are small, Model B is the preferred model whenever instability problems arise. The mixture equation is the same for both models. In convective form, the momentum mixture equation is

$$\sum_{k=1}^{n\ \text{phases}}\left[\varepsilon_k\rho_k\frac{d\mathbf{v}_k}{dt^k}\right]=-\nabla\sum_{k=1}^{n}P_k-\rho_m\mathbf{g}. \quad (7.4)$$

Hence, in two-phase flow all differences between the models appear in the equation related to the difference between the phases, which involves the drag between the phases. Several other models have been used in multiphase literature for such a difference equation, such as the relative velocity equation given in Chapter 2, and the earlier slip correlations, which form the basis for the TRAC computer code (Vigil *et al.*, 1979) for studying transients in gas–liquid flow. The next several sections will explore the properties of the multiphase balances.

7.2 Compressible Granular Flow

Single-phase particulate flow in the absence of an interstitial fluid has been called granular flow by Savage (1983, 1988) and others (Jenkins and Savage, 1983). It has also been called bulk flow. For gas–particle flow, such a situation occurs in a vacuum and is also approximately true for flow of large grains, hence the name granular or bulk flow.

In the limit of zero fluid density, the conservation of mass and momentum equations for one-dimensional flow without phase changes and external forces are

Mass Balance

$$\frac{\partial \rho_b}{\partial t} + \frac{\partial (\rho_b U_s)}{\partial x} = 0, \tag{7.5}$$

Momentum Balance

$$\frac{\partial (\rho_b U_s)}{\partial t} + \frac{\partial (\rho_b U_s^2)}{\partial x} + \frac{\partial P_s}{\partial x} = 0, \tag{7.6}$$

where ρ_b is the bulk density equal to $\varepsilon_s \rho_s$. To proceed with the analysis, a constitutive equation for the solids pressure, P_s, is needed. Possible forms of constitutive equations were already discussed in the chapter on critical flow. The simplest form for particulate flow obtained from kinetic theory of solids is

$$P_s = \varepsilon_s \rho_s \Theta, \tag{7.7}$$

where Θ is the granular flow temperature, defined as one-third of the fluctuating energy. Whether particles oscillate or are simply compressed, the solids pressure is a function of the bulk density or volume fraction of solids for a constant particle density, as postulated in Eq. (7.8):

$$P_s = P_s(\varepsilon_s). \tag{7.8}$$

The modulus of elasticity G was defined as

$$G = \frac{\partial P_s}{\partial \varepsilon_s}. \tag{7.9}$$

Hence,

$$\frac{\partial P_s}{\partial x}=\frac{\partial P_s}{\partial \varepsilon_s}\frac{\partial \varepsilon_s}{\partial x}=G\frac{\partial \varepsilon_s}{\partial x}=\frac{G}{\rho_s}\frac{\partial \rho_b}{\partial x}=G'\frac{\partial \rho_b}{\partial x}, \tag{7.10}$$

where $G'=G/\rho_s$.

Further let the flux of powder flow be $F=\rho_b U_s$ and neglect the term involving the square of velocity in the momentum equation, for convenience and simplicity of analysis. Then the mass and momentum balance can be written as a set of two first-order partial differential equations for the two variables ρ_b and F:

Mass Balance

$$\frac{\partial \rho_b}{\partial t}+\frac{\partial F}{\partial x}=0, \tag{7.11a}$$

Momentum Balance

$$\frac{\partial F}{\partial t}+G'\frac{\partial \rho_b}{\partial x}=0. \tag{7.11b}$$

Such a set of equations is valid for any material with the approximation made. For a gas the modulus G' is the square root of the sonic velocity or the derivative of pressure with respect to density at a constant temperature or at a constant entropy.

Differentiation of (7.11a) with respect to x gives

$$\frac{\partial^2 \rho_b}{\partial x\,\partial t}=-\frac{\partial^2 F}{\partial x^2}, \tag{7.12}$$

and differentiation of (7.11b) with respect to t gives

$$\frac{\partial F}{\partial t^2}=-G'\frac{\partial^2 \rho_b}{\partial t\,\partial x}. \tag{7.13}$$

Elimination of the mixed partials of ρ_b in Eqs. (7.12) and (7.13) gives the same equation for the flux F:

$$G'\frac{\partial^2 F}{\partial x^2}=\frac{\partial^2 F}{\partial t^2}. \tag{7.14}$$

Repetition of the process by differentiating (7.11a) with respect to t and (7.11b) with respect to x gives a wave equation for ρ_b:

$$\frac{\partial \rho_b}{\partial t^2} = G' \frac{\partial^2 \rho_b}{\partial x^2}. \tag{7.15}$$

The characteristic directions for the wave equation (7.15) are

$$\frac{dx}{dt} = \pm\sqrt{G'} = \pm\sqrt{G/\rho_s}\,. \tag{7.16}$$

A measurement of the time required for a disturbance in bulk density to travel a distance x will give a value of G'. With specified initial and boundary conditions, as shown in Fig. 7.1, the solution to the wave equation can be obtained.

Let us suppose, however, that the sign of G in Eq. (7.9) is negative; that is, the pressure decreases as the solids volume fraction increases, then Eq. (7.15)

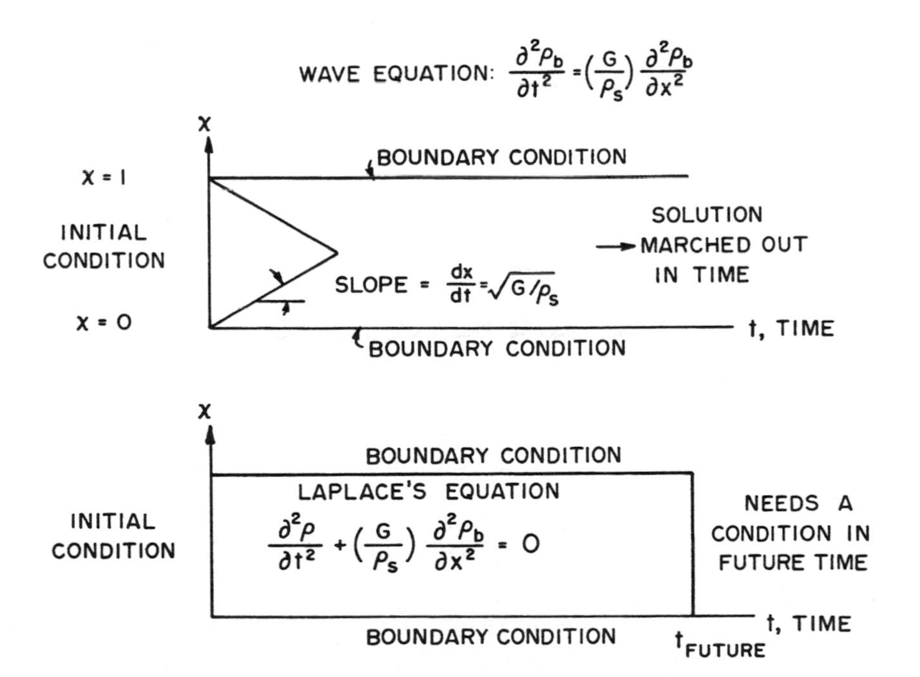

Fig. 7.1 Well-posed and ill-posed initial value problems for bulk density in granular flow.

becomes a Laplace's equation. As depicted in Fig. 7.1, a unique solution of a Laplace's equation requires the knowledge of some information at the boundary t_{future}. This means future events will influence the present. Such a violation of the principle of cause and effect does not make sense. This type of an occurrence is prevented by requiring the constitutive equation to satisfy the second law of thermodynamics. The stability condition (Callen, 1960) requires that

$$\left(\partial P_s / \partial \varepsilon_s\right) > 0. \tag{7.17}$$

This requirement ensures that the simultaneous solution of the conservation of mass and momentum will yield a well-posed or properly formulated initial value problem. Unfortunately, such violations are quite common (Gidaspow, 1974; Drew, 1983; Lyczkowski *et al.*, 1978, 1982) in two-phase flow, although they are not that easy to explain or to verify. In each case, they lead to numerical instabilities.

7.3 Well-Posedness of Two-Phase Models

The success of any numerical method depends on the requirement that the problem be well-posed. Otherwise numerical instabilities will produce non-physical results.

Conservation laws, such as those given by Eqs. (7.1) and (7.2), can be written as a system of quasilinear first-order partial differential equations:

$$\frac{\partial \mathbf{U}}{\partial t} + \sum_{j=1}^{3} \mathbf{A}_j \frac{\partial \mathbf{U}}{\partial x_j} = \mathbf{b} \tag{7.18a}$$

where the matrix multiplying the derivatives with respect to time has been reduced to a unit matrix by multiplying by its inverse. The initial value problem under consideration is to find a solution of system (7.18a) in some region

$$a_i \le x_i \le b_i \qquad (1 \le i \le 3), \qquad t \ge 0, \tag{7.18b}$$

subject to the initial condition

$$\mathbf{U}(0, \mathbf{x}) = G(\mathbf{x}) \tag{7.18c}$$

and the value of U prescribed on the boundaries.

The initial value problem (7.18) is called *well-posed* if it has a unique solution for all sufficiently differentiable initial values $G(\mathbf{x})$ (Lax, 1958; Courant and Hilbert, 1962). As a theorem (Lax, 1958; Pal, 1969) it can be proved that the initial value problem (7.18) is well-posed if and only if all linear combinations of $\sum A_j\mu_j$ the coefficient matrices A_j with real coefficients μ_j have only real eigenvalues.

The eigenvalues for the system of two first-order partial differential equations given by Eq. (7.11a) and (7.11b) are $\pm\sqrt{G'}$. The eigenvalues for the system with the sign G reversed are complex and equal to $\pm i\sqrt{G'}$.

The well-posed system is hyperbolic, while the ill-posed system is elliptic.

System (7.18) is hyperbolic, if all linear combinations $\sum A_j\mu_j$ of the coefficient matrices with real numbers μ_j have only real eigenvalues λ_1, λ_2, $\lambda_3, \ldots \lambda_n$, and n linearly independent eigenvectors, so that a nonsingular matrix $\mathbf{T}(\mu)$ exists such that

$$\mathbf{T}\left(\sum \mathbf{A}_j\mu_j\right)\mathbf{T}^{-1} = \Lambda = \begin{bmatrix} \lambda_1 & & & & 0 \\ & \lambda_2 & & & \\ & & \cdot & & \\ & & & \cdot & \\ 0 & & & & \lambda_n \end{bmatrix} \tag{7.19}$$

is symmetric and $\mathbf{T}$ depends smoothly on μ (Pal, 1969, adapted from Richtmyer and Morton, 1967). Sedney (1970) has stated that if the system is not completely hyperbolic, many possibilities of system classification exist and that not all types have been given names.

Thus, a parabolic system is well-posed. Characteristics analysis is an integral part of the computational aspects of multiphase flow. The major reasons for this are that the choice of numerical method and its stability when solved as an initial-value problem are intimately connected with the nature of the partial differential equations. The basic connection between the stability of finite difference schemes and the partial differential equations they approximate is given by the so-called Lax Equivalence Theorem (Lax and Richtmyer, 1956). This theorem states that if a finite difference approximation to a well-posed linear initial-value problem is consistent, then stability is both a necessary and a sufficient condition for convergence. If the partial differential equations posses complex characteristics, then they are not completely hyperbolic, hence ill-posed, and therefore no stable numerical scheme can be found at least for the linear system, since the Von

Neumann analysis predicts exponential growth. Lyczkowski *et al.* (1982) give a more complete discussion.

The characteristics for one-dimensional two-phase flow for the balances expressed by Eqs. (7.1) and (7.2) with the constitutive equation for each phase given by Eq. (7.3) called Model B in Chapter 2 can be obtained as follows. Table 2.1 gives these steady state equations. Call C_s the sonic velocity of the gas at a constant temperature or at a constant entropy. G is again the corresponding modulus of elasticity for the solid phase. Then conservation of mass and momentum equations for the gas and the solids particulate phase can be written in matrix form

$$\overline{A}\begin{pmatrix} \partial\varepsilon_g/\partial t \\ \partial P/\partial t \\ \partial v_g/\partial t \\ \partial v_s/\partial t \end{pmatrix} + \overline{B}\begin{pmatrix} \partial\varepsilon_g/\partial s \\ \partial P/\partial x \\ \partial v_g/\partial x \\ \partial v_s/\partial x \end{pmatrix} = \overline{C}\ , \tag{7.20}$$

where

$$\overline{A} = \begin{pmatrix} \rho_g & \varepsilon_g/C_s^2 & 0 & 0 \\ \rho_s & 0 & 0 & 0 \\ 0 & 0 & \rho_g\varepsilon_g & 0 \\ 0 & 0 & 0 & \rho_s\varepsilon_s \end{pmatrix}, \tag{7.21}$$

$$\overline{B} = \begin{pmatrix} \rho_g v_g & \varepsilon_g v_g/C_s^2 & \rho_g\varepsilon_g & 0 \\ \rho_s v_s & 0 & 0 & \rho_s\varepsilon_s \\ 0 & 1 & \varepsilon_g\rho_g v_g & 0 \\ -G & 0 & 0 & \varepsilon_s\rho_s v_s \end{pmatrix}, \tag{7.22}$$

and

$$\overline{C} = \begin{pmatrix} 0 \\ 0 \\ \beta_B\left(v_s - v_g\right) - 4\tau_{wg}/D_t - \rho_g g \\ -\beta_B\left(v_s - v_g\right) - 4\tau_{ws}/D_t - \varepsilon_s\Delta\rho_s g \end{pmatrix}, \tag{7.23}$$

where

$$C_s = \sqrt{\left(\partial P/\partial \rho_g\right)_T} . \tag{7.24}$$

The characteristic determinant can be shown to be (Tsuo and Gidaspow, 1990; Lyczkowski *et al.*, 1982)

$$\begin{vmatrix} v_s - \lambda & 0 & 0 & -\varepsilon_s \\ \frac{\rho_g C_s^2}{\varepsilon_g}\left(v_g - v_s\right) & v_g - \lambda & \rho_g C_s^2 & \frac{\varepsilon_s \rho_g C_s^2}{\varepsilon_g} \\ 0 & \frac{1}{\rho_g \varepsilon_g} & v_g - \lambda & 0 \\ -\frac{G}{\varepsilon_s \rho_s} & 0 & 0 & v_s - \lambda \end{vmatrix} . \tag{7.25}$$

The characteristic roots, λ_i, of this determinant are

$$\lambda_{1,2} = v_g \pm \sqrt{C_s^2/\varepsilon_g} \tag{7.26}$$

and

$$\lambda_{3,4} = v_s \pm \sqrt{G/\rho_s} . \tag{7.27}$$

Since $(C_s^2 / \varepsilon_g) > 0$, and $(G / \rho_g) > 0$, this equation set has real and distinct characteristics. Hence, the system is hyperbolic. The problem is well-posed as an initial value problem.

The characteristic directions also determine where the boundary conditions must be prescribed for a well-posed problem. Equation (7.26) shows that information about the gas must be prescribed at the inlet and at the exit, since the characteristic directions are positive and negative due to the large value of C_s. Although the form for the solid, as shown by Eq. (7.27), is similar to that for the gas, the value of G/ρ_s is small. Its square root is of the order of 1 m/s for dense flow for a volume fraction of solid of about 0.6. It is very low for dilute flow. Hence, the characteristic directions for the solid are normally both positive. For small values of G, the characteristics are nearly equal. The particles essentially move with their own velocity, with the wave effect negligible.

For the case of zero G and an incompressible solid, the characteristics are those of the Rudinger–Chang (1964) set, as generalized to the non-dilute case by Lyczkowski *et al.* (1982). The characteristics λ_3 and λ_4 can be obtained independently of the gas phase equations, as done in Chapter 4, since the system is coupled only through the non-derivative terms, matrix $\overline{C}$. Such a procedure permits a generalization of the classification to the multiphase set of Eqs. (7.1) to (7.3). For the dispersed phase k, the characteristics are

$$\lambda_k = v_k \pm \left[\left(\frac{\partial P_k}{\partial \varepsilon_k} \Big/ \rho_k \right) \right]^{1/2} \quad ; \quad k = 1, 2, 3, \ldots, m \text{ particulate phases.} \tag{7.28}$$

Hence, the multiphase system (7.1) to (7.3) is well posed as an initial value problem. For negligible particulate pressures, the characteristics are simply the trajectories or particle paths. They must clearly be real paths.

Because of the problems with buoyancy, as discussed in Chapter 2, the constitutive relation (7.3) was not used in earlier studies of dense phase flow and fluidization (Gidaspow, 1986). Instead, the continuous fluid phase pressure weighted by its volume fraction was used in the dispersed phase, in addition to the solids phase pressure. This is Model A, as described in Chapter 2. Fanucci *et al.* (1979) have obtained characteristics for this model for incompressible fluids.

Their paper shows one of their characteristics, C, to be

$$C = \left[\frac{1}{(\rho_g/\varepsilon) + (\rho_s/\varepsilon_s)} \right]^{1/2} \left[\frac{-\left(v_g - v_s\right)^2}{(\varepsilon/\rho_g) + (\varepsilon_s/\rho_s)} + \frac{G(\varepsilon)}{\varepsilon_s} \right]^{1/2} . \tag{7.29}$$

For zero G this model is ill-posed as an initial value problem, similar to the test case debated in 1974 at the International Heat Transfer Conference (Gidaspow, 1974) and described in greater detail by Lyczkowski *et al.* (1982). For a judicious choice of the stress modulus G based on some experimental data, such as

$$G(\varepsilon) = 10^{-8.76\varepsilon + 5.43} \, \mathrm{N/m^2}, \tag{7.30}$$

the characteristics are real and numerically stable calculations can be obtained.

Unfortunately, the stress modulus had to be adjusted (e.g., Bouillard *et al.*, 1989a) to prevent the void fractions from reaching impossibly low values and to satisfy the condition of real characteristics. A decade of experience with this conditionally stable model showed that unexpectedly non-physical results can be obtained, such as the appearance of particles in the free board in fluidization or disappearance of solid particles during detonation in impossible locations. In such cases, such a non-physical behavior can be traced to the numerical instability caused by imaginary characteristics. Hence, this model must be used with caution for transient flow.

7.4 Homogeneous Flow and Pressure Propagation

For homogeneous flow, that is, for the case when all phase velocities are equal, the mixture momentum equation (7.4) moving with the velocity v becomes as follows:

$$\rho_m \frac{dv}{dt} = -\frac{\partial P}{\partial x} - \rho_m g \tag{7.31}$$

when all particulate phase pressures are negligible with respect to the fluid pressure. For the continuous phase, say, the gas phase, the equation of state is

$$\rho_g = \rho_g(T, P). \tag{7.32}$$

The isothermal sound speed C_g is given by

$$C_g^2 = \left(\frac{\partial P}{\partial \rho_g}\right)_T. \tag{7.33}$$

A similar sound speed can be defined at a constant mixture entropy. The continuity equation for the gas phase with this sound speed with no phase change is

$$\frac{d\varepsilon}{dt} + \frac{\varepsilon}{\rho_g C_g^2}\frac{dP}{dt} + \varepsilon\frac{\partial v}{\partial x} = 0. \tag{7.34}$$

The continuity equation for the incompressible particulate phase is

$$-\frac{d\varepsilon}{dt} + \varepsilon_s \frac{\partial v}{\partial x} = 0. \tag{7.35}$$

Summation of the two continuity equations (7.34) and (7.35) gives

$$\frac{\varepsilon}{\rho_g C_g^2}\frac{dP}{dt} + \frac{\partial v}{\partial x} = 0. \tag{7.36}$$

Differentiation of Eq. (7.36) with respect to the convective time gives

$$\frac{d}{dt}\frac{\partial v}{\partial x}=-\frac{d}{dt}\left(\frac{\varepsilon}{\rho_g C_g{}^2}\frac{dP}{dt}\right). \tag{7.37}$$

Differentiation of the momentum equation (7.31) with no body forces with respect to x, and reversal of derivatives gives

$$\frac{d}{dt}\left(\frac{\partial v}{\partial x}\right)=-\frac{\partial}{\partial x}\left(\frac{1}{\rho_m}\frac{\partial P}{\partial x}\right). \tag{7.38}$$

Since the left-hand sides of Eqs. (7.37) and (7.38) are the same,

$$\frac{\partial}{\partial x}\left(\frac{1}{\rho_m}\frac{\partial P}{\partial x}\right)=\frac{d}{dt}\left(\frac{\varepsilon}{\rho_g C_g{}^2}\frac{dP}{dt}\right). \tag{7.39}$$

When the velocity v is small, the convective derivative

$$\frac{d}{dt}=\frac{\partial}{\partial t}+v\frac{\partial}{\partial x} \tag{7.40}$$

becomes a simple partial with respect to time, and Eq. (7.39) is simply the wave equation for the fluid pressure. Equation (7.39) shows that the pressure propagates with the mixture velocity C_m equal to

$$C_m{}^{-2}=\frac{\rho_m\varepsilon}{\rho_g C_g{}^2}, \tag{7.41}$$

where

$$\rho_m=\varepsilon\rho_g+(1-\varepsilon)\rho_s. \tag{7.42}$$

Under isothermal or isotropic conditions

$$C_g=\left(\frac{\gamma RT}{M}\right)^{1/2} \text{ or } \sqrt{\frac{RT}{M}}\text{, isothermally}, \tag{7.43a}$$

where γ is the ratio of specific heats of ideal gases.

For the case of not very dilute loading, $(1-\varepsilon)\rho_s > \varepsilon\rho_g$, Eq. (7.41) reduces itself to the simpler expression

$$\text{Mixture Sonic Velocity } = C_m = \sqrt{\frac{\rho_g}{\rho_s}}\frac{C_g}{\sqrt{\varepsilon(1-\varepsilon)}}. \tag{7.43b}$$

Equation (7.43b) shows that there is a minimum in C_m with respect to void fraction at $\varepsilon = 0.5$. It also shows that the sonic velocity of the mixture is much lower than it is for the gas; a typical value is 10 m/s. Pressure waves through fluidized beds and bubbly flow move at such velocities. Equation (7.43b) also shows that the mixture velocity vanishes as gas density goes to zero.

7.5 Solids Vorticity and Void Propagation

To understand multiphase flow phenomena, such as bubble formation near the grid for uniform flow, simpler models are frequently useful. In single-phase fluid mechanics, representation of the Navier–Stokes equation in terms of a vorticity transfer equation is often useful. Such an approach will be pursued here.

For inviscid two-phase flow with $\varepsilon\rho_g \ll \varepsilon_s\rho_s$ and $P_s \ll P_{gas}$, the mixture momentum equation (7.4) becomes

$$\varepsilon_s\rho_s\frac{d\mathbf{v}_s}{dt^s} = -\nabla P - \rho_m\mathbf{g}. \tag{7.44}$$

As done in free convection heat transfer (Batcheler, 1954; Poots, 1958; Churchill, 1966), consider static conditions defined by

$$\nabla P_0 = -\rho_{m0}\mathbf{g} \tag{7.45}$$

and define a dynamic pressure P_d as

$$P_d = P - P_0, \tag{7.46}$$

where the subscript "0" denotes the static state. Subtraction of Eq. (7.46) from (7.44) gives a momentum equation in terms of buoyancy,

$$\varepsilon_{s0}\rho_s\frac{d\mathbf{v}_s}{dt^s} = -\nabla P_d - \mathbf{g}\rho_s(\varepsilon_s - \varepsilon_{s0}), \tag{7.47}$$

where ε_s is assumed to stay at the static case except where it occurs as a difference.

With such an approximation normally made in free convection heat transfer, the stream function and the vorticity remain the same as in single-phase flow, since porosities only change where they occur as differences. Then for two-dimensional transient flow the continuity equation for the solid phase

$$\frac{\partial u_s}{\partial x}+\frac{\partial v_s}{\partial y}=0 \tag{7.48}$$

using the Green's theorem in the plane defines the solid stream function

$$u_s=\frac{\partial \psi_s}{\partial y} \quad \text{and} \quad v_s=-\frac{\partial \psi_s}{\partial x} \tag{7.49}$$

Let ς_s be twice the normal vorticity. Then

$$\varsigma_s=\frac{\partial v_s}{\partial x}-\frac{\partial u_s}{\partial y}=-\nabla^2\psi_s \tag{7.50}$$

The dynamic pressure in Eq. (7.47) is eliminated by differentiating the x component equation with respect to y and the y component equation with respect to x and then by subtracting the equations. The result is the vorticity transfer equation shown in Eq. (7.51) where gravity acts downward in the y direction,

$$\frac{\partial \varsigma_s}{\partial t}+u_s\frac{\partial \varsigma_s}{\partial x}+v_s\frac{\partial \varsigma_s}{\partial y}=\frac{g}{\varepsilon_{s0}}\frac{\partial \varepsilon_s}{\partial y}. \tag{7.51}$$

Equation (7.51) shows that in the absence of volume fraction gradients,

$$d\varsigma_s/dt^s=0. \tag{7.52}$$

Hence, if initially there is no rotation, the flow of solids remains irrotational. Equation (7.51) also shows that void fraction gradients set up circulation.

A useful void propagation equation can be obtained by combining two incompressible continuity equations, as done in Chapter 6. A generalization of Eq. (6.5) is

$$\frac{\partial \varepsilon}{\partial t}+\hat{\mathbf{v}}\nabla\varepsilon=\varepsilon\varepsilon_s\nabla\mathbf{v}_r, \tag{7.53}$$

where $\mathbf{v}_r$ is the relative velocity and $\hat{\mathbf{v}}$ is the inversely weighted mean velocity,

$$\hat{\mathbf{v}} = \varepsilon \mathbf{v}_s + \varepsilon_s \mathbf{v}_g. \tag{7.54}$$

The gradient of relative velocity can now be eliminated by various means. One approach is to use the relative velocity equation (Gidaspow, 1978) which gives a hyperbolic diffusion equation for porosity. Analytical solutions to such a linearized equation were used to describe solids tracer injection into fluidized beds (Liu and Gidaspow, 1981). A simpler but cruder approach is to use the concept of slip. Let us assume that in the direction of gravity, terminal velocity is reached and that in the perpendicular direction, there is homogeneous flow. Then using Eq. (6.12) we have

$$v_r = v_{\text{terminal}} = \frac{\varepsilon \varepsilon_s \Delta \rho g}{\beta_A}. \tag{7.55}$$

Then the first-order partial differential equation (7.53) shows that

$$\frac{d\varepsilon}{dt} = 0 \tag{7.56}$$

along the paths given by

$$\frac{dt}{1} = \frac{dx}{u} = \frac{dy}{\hat{v} - \varepsilon \varepsilon_s (\Delta \rho) g \frac{d}{d\varepsilon}\left(\varepsilon \varepsilon_s \beta_A{}^{-1}\right)}, \tag{7.57}$$

where u is either the gas or the solids velocity and the gradient of relative velocity contributed the non-linear term involving the derivative with respect to porosity. This latter effect gives rise to shocks or dispersion, as illustrated in Chapter 6 for the one-dimensional case.

Equation (7.57) is useful in analyzing regions of influence, such as that of a jet injected into a fluidized bed, as sketched in Fig. 7.2. Consider time average behavior. Then the slope of the constant porosity curves is given by

$$\frac{dy}{dx} = \frac{v_g - \varepsilon v_t - \varepsilon \varepsilon_s \frac{dv_t}{d\varepsilon}}{u_g}. \tag{7.58}$$

If the jet is moving straight up into the bed, μ_g is zero and the slope is infinite at point J. Curves starting in the grid region are at the constant porosity ε_{mf}. The

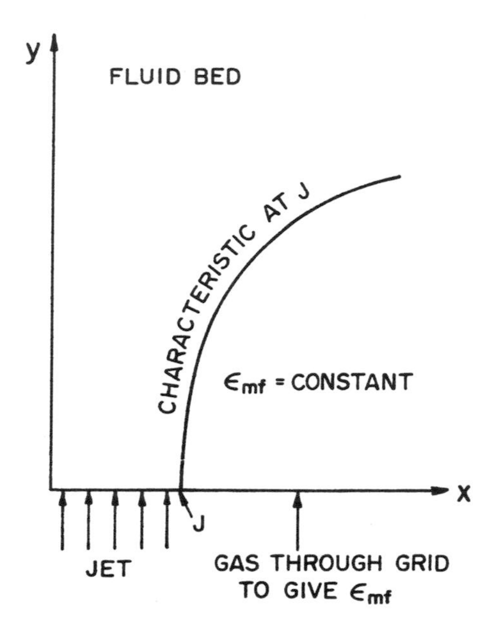

Fig. 7.2 Regions of influence for time average behavior of void fraction.

shape of the curve dividing the jet region and the grid region requires the solution of the complete hydrodynamic problem.

An analysis with more simplification can be used to show that a point source of gas injected into a two-dimensional fluidized bed will produce a circular bubble. Neglecting the non-linear effect of variation of relative velocity with porosity, Eq. (7.53) in cylindrical coordinates is

$$\frac{\partial \varepsilon}{\partial t} + \hat{v}_R \frac{\partial \varepsilon}{\partial R} = 0. \tag{7.59}$$

For a point source of width W and volumetric source strength $\dot{Q}$ m³/s,

$$\dot{Q} = W\pi R v_{Rg} = W\pi R v_{Rs}, \tag{7.60}$$

assuming the gas displaces the solid particles. Then porosity is one along the curve

$$\frac{dR}{dt} = \hat{v}_R = v_{SR} = \frac{\dot{Q}}{W\pi R}. \tag{7.61}$$

Integration of Eq. (7.61) gives

$$R^2 = x^2 + y^2 = 2\dot{Q}t/W\pi. \tag{7.62}$$

Equation (7.62) shows that constant time lines are circles or semi-circles for injection at the bottom boundary, as in Eq. (7.61). Similarly, we obtain spheres in three dimensions.

For the same level of approximation as used in deriving the vorticity equation (7.51), a Laplace's equation for pressure can be obtained. Neglecting all accelerations for the gas, the momentum balance (7.2) using the constitutive equation (7.3) becomes, in the absence of gravity and phase changes,

$$\nabla P = \beta(\mathbf{v}_g - \mathbf{v}_s). \tag{7.63}$$

Equation (7.63) is essentially Darcy's law for flow in porous media, which is usually written as

$$v_g - v_s = -\frac{k}{\mu}\nabla P, \tag{7.64}$$

where k is the permeability and μ is the fluid viscosity. It is useful to know this similarity, since there exists a large body of literature dealing with flow in porous media. Then for constant permeability k or constant friction coefficient β, the two continuity equations with negligible porosity variations,

$$\nabla \cdot \mathbf{v}_g = 0 \quad \text{and} \quad \nabla \cdot \mathbf{v}_s = 0, \tag{7.65}$$

give the Laplace's equation for pressure

$$\nabla^2 P = 0 \tag{7.66}$$

upon application of the divergence to Eq. (7.63) or (7.64).

The equations given in this section provide an approximate model for multiphase flow and for fluidization. They can be used to obtain $\mathbf{v}_s$, $\mathbf{v}_g$, ε and P. The solids velocity is obtained from the vorticity equations (7.51) in terms of porosity obtained from (7.57) and the integrated form of Eq. (7.50), expressed as

$$\psi_s = \iint G(\mathbf{r};\mathbf{r}')\varsigma_s \, dx\, dy, \tag{7.67}$$

where G is the Green's function for the region under consideration, which is a function of the observer coordinate $\mathbf{r}$ and the source coordinate $\mathbf{r}'$. Similarly,

Laplace's equation (7.66) can be solved by Green's functions (Lyczkowski and Gidaspow, 1971) to give the gas velocities relative to the solid velocities.

7.6 Davidson Bubble Model

Davidson (1961) analyzed bubble motion in a large fluidized bed. He predicted the existence of a spherical surface of zero velocity, called a cloud. Shortly after his prediction, the existence of a cloud was verified experimentally by inserting NO_x into a bubble injected into a fluidized bed near minimum fluidization. Since mixing across this cloud can only occur by diffusion, this cloud forms a resistance to gas diffusion. The concept of a cloud was considered to be so important that one denotes it to be a "phase" in the transfer coefficient models of Levenspiel and others (Wen, 1975; Babu *et al.*, 1976) for modeling reactions in fluidized beds. Although in many reactors, such as fluidized-bed gasifiers and combustors, the bubbles do not have clouds, the elegance and simplicity of Davidson's approach generates continued studies (Collins, 1982; Littman and Homolka, 1973).

Davidson assumed that the solids were in irrotational motion. Equation (7.51) shows that this is true in the absence of porosity gradients. Then for zero vorticity, in two dimensions, application of the Green's theorem in the plane shows that a potential function ϕ_s must exist for the solid, as show in Eq. (7.68):

$$0 = \iint \left(\frac{\partial v_s}{\partial x} - \frac{\partial u_s}{\partial y} \right) dx\, dy = \oint \left(u_s\, dx + v_s\, dy \right) = \oint d\phi_s = \int \left(\frac{\partial \phi_s}{\partial x} dx + \frac{\partial \phi_s}{\partial y} dy \right). \quad (7.68)$$

Hence,

$$u_s = \frac{\partial \phi_s}{\partial x} \quad \text{and} \quad v_s = \frac{\partial \phi_s}{\partial y}. \quad (7.69)$$

In general,

$$\mathbf{v}_s = \nabla \phi_s. \quad (7.70)$$

Substitution into the continuity equation with no porosity changes,

$$\nabla \cdot \mathbf{v}_s = 0, \quad (7.71)$$

gives the Laplace's equation for the solids potential

$$\nabla^2 \phi_s = 0. \quad (7.72)$$

It is possible to generalize this equation to transient flow and to the case when solids pressure and its derivative with respect to porosity, corresponding to sonic velocity, are significant, as done by Von Mises (1958) for compressible flow. However, usually it is the neglect of porosity gradients that will invalidate this simple analysis.

The remaining equations in Davidson's model are the continuity equations with no porosity changes, Eqs. (7.65) and the Laplace's equation for pressure, Eq. (7.66). The two continuity equations define into existence two stream functions, one for the gas and another one for the solid.

The solids potentials for a bubble in two and three dimensions can be obtained using classical potential flow theory. The solutions are given in Lamb's (1945) and in Davidson and Harrison's (1963) books. The boundary conditions to be satisfied by ϕ_s are

i. the velocity should be uniform and equal to the bubble velocity U_B at infinity, and

ii. the radial velocity should be zero at the bubble surface. Mathematically, these conditions are expressed as

$$BC1: \quad \text{at} \quad r=\infty, v_s = \frac{\partial \phi}{\partial y} = u_B \tag{7.73}$$

$$\text{or} \quad \phi_s = u_B y = u_B r \cos\theta,$$

$$BC2: \quad \text{at} \quad r = r_B, v_r = \frac{\partial \phi}{\partial r} = 0. \tag{7.74}$$

The solutions satisfying the boundary conditions consist of a superposition of dipoles and a uniform flow. For the cylinder it is

$$\phi_s = u_B \cos\theta \left(r + \frac{r_B^2}{r} \right), \tag{7.75}$$

and for the sphere the solution is

$$\phi_s = u_B \cos\theta \left(r + \frac{r_B^3}{2r^2} \right). \tag{7.76}$$

The boundary conditions for the Laplace's equation for pressure are

i. far away from the bubble the vertical pressure gradient equals the weight of the bed, as required by hydrostatics with negligible solids pressure, and
ii. the pressure of the bubble surface is a constant. These conditions are given by

$$BC1: \quad \text{at} \quad r = \infty, \quad \frac{\partial P}{\partial y} = -g\left(\varepsilon_{mf}\rho_g + \varepsilon_{sm}\rho_s\right),$$

$$\frac{\partial P}{\partial r} = g\cos\theta\,\rho_{bm}, \quad \text{where} \quad \rho_{bm} \equiv \varepsilon_{sm}\rho_s, \tag{7.77}$$

$$\text{and} \quad \frac{\partial P}{\partial r} = -\beta_0 u_0 \cos\theta \tag{7.78}$$

using Eq. (7.63).

The solution for the cylinder is

$$\frac{P - P_B}{\beta_0 u_0} = \frac{P - P_B}{g\rho_{bm}} = \cos\theta\left(r - \frac{r_B^2}{r}\right), \tag{7.79}$$

and for the sphere it is

$$\frac{P - P_B}{\beta_0 u_{go}} = \cos\theta\left(r - \frac{r_B^3}{r}\right). \tag{7.80}$$

Cloud formation will now be determined for the spherical bubble. The derivative of the pressure gradient, with pressure given by Eq. (7.80), gives the relative velocity between the solid and the gas as

$$-\beta_0\left(v_s - v_g\right) = -\frac{\partial P}{\partial r} = \beta_0 u_0 \cos\theta\left(1 + \frac{2r_B^3}{r^3}\right), \tag{7.81}$$

while the radial gradient of ϕ_s in Eq. (7.76) gives the solids velocity as

$$v_s = -u_B \cos\theta\left(1 - \frac{r_B^3}{r^3}\right). \tag{7.82}$$

From Eqs. (7.81) and (7.82) the gas velocity is as follows:

$$v_g = \cos\theta\left[\frac{r_B^3}{r^3}(u_B + 2u_0) - (u_B - u_0)\right]. \tag{7.83}$$

Zero gas velocity occurs for the radius r_c called the cloud radius, equal to

$$r_c = \left[\frac{u_B + 2u_0}{u_B - u_0}\right]^{1/3} r_B. \tag{7.84}$$

A cloud exists for the case u_B greater than the interstitial velocity u_o. Such a bubble is called a fast bubble. No cloud exists for u_B less than u_o, called the slow bubble. In two dimensions, Davidson found the result

$$\left(\frac{r_c}{r_B}\right) = \left[\frac{u_B + u_0}{u_B - u_0}\right]^{1/2}. \tag{7.85}$$

While u_o is the fluid velocity for which pressure drop equals the weight of the bed, u_B is not determined in this theory. It can be given by a correlation (Grace, 1982) such as

$$u_B = 0.711\sqrt{gD_B}, \tag{7.86}$$

where D_B is the bubble diameter.

7.7 Computer Fluidization Models

Table 7.1 gives the inviscid two phase flow hydrodynamic equations for Models A and B in Cartesian coordinates. In two dimensions there are the six variables corresponding to the equations as listed below.

Variable	Equation
Pressure, P	gas continuity
Porosity, ε	solids phase continuity
Gas velocity: u_g, v_g	x and y gas momentum
Solid velocity: u_s, v_s	x and y solids momentum

Table 7.1 HYDRODYNAMIC MODELS

Continuity Equations

Gas-Phase

$$\frac{\partial}{\partial t}\left(\rho_g \varepsilon\right)+\frac{\partial}{\partial x}\left(\rho_g \varepsilon u_g\right)+\frac{\partial}{\partial y}\left(\rho_g \varepsilon v_g\right)=\dot{m} \tag{T.1}$$

Solids-Phase

$$\frac{\partial}{\partial t}\left[\rho_s(1-\varepsilon)\right]+\frac{\partial}{\partial x}\left[\rho_s u_s(1-\varepsilon)\right]+\frac{\partial}{\partial y}\left[\rho_s v_s(1-\varepsilon)\right]=-\dot{m} \tag{T.2}$$

Momentum Equations: Model A

Gas-Phase Momentum in x-Direction

$$\frac{\partial}{\partial t}\left(\rho_g \varepsilon u_g\right)+\frac{\partial}{\partial x}\left(\rho_g \varepsilon u_g u_g\right)+\frac{\partial}{\partial y}\left(\rho_g \varepsilon v_g u_g\right)=-\varepsilon\frac{\partial P}{\partial x}+\beta_A\left(u_s-u_g\right)+\dot{m}u_g \tag{T.3}$$

Solids-Phase Momentum in x-Direction

$$\frac{\partial}{\partial t}\left[\rho_s(1-\varepsilon)u_s\right]+\frac{\partial}{\partial x}\left[\rho_s(1-\varepsilon)u_s u_s\right]+\frac{\partial}{\partial y}\left[\rho_s(1-\varepsilon)v_s u_s\right]$$
$$=-(1-\varepsilon)\frac{\partial P}{\partial x}+\beta_A\left(u_g-u_s\right)+G(\varepsilon)\frac{\partial \varepsilon}{\partial x}-\dot{m}u_s \tag{T.4}$$

Gas-Phase Momentum in y-Direction

$$\frac{\partial}{\partial t}\left(\rho_g \varepsilon v_g\right)+\frac{\partial}{\partial x}\left(\rho_g \varepsilon u_g v_g\right)+\frac{\partial}{\partial y}\left(\rho_g \varepsilon v_g v_g\right)=$$
$$-\varepsilon\frac{\partial P}{\partial y}+\beta_A\left(v_s-v_g\right)-\rho_g \varepsilon g+\dot{m}v_g \tag{T.5}$$

Solids-Phase Momentum in y-Direction

$$\frac{\partial}{\partial t}\left[\rho_s(1-\varepsilon)v_s\right]+\frac{\partial}{\partial x}\left[\rho_s(1-\varepsilon)u_s v_s\right]+\frac{\partial}{\partial y}\left[\rho_s(1-\varepsilon)v_s v_s\right]=$$
$$-(1-\varepsilon)\frac{\partial P}{\partial y}+\beta_A\left(v_g-v_s\right)-\rho_s(1-\varepsilon)g+G(\varepsilon)\frac{\partial \varepsilon}{\partial y}-\dot{m}v_s \tag{T.6}$$

Momentum Equations: Model B

Gas-Phase Momentum in x-Direction

$$\frac{\partial}{\partial t}\left(\rho_g \varepsilon u_g\right)+\frac{\partial}{\partial x}\left(\rho_g \varepsilon u_g u_g\right)+\frac{\partial}{\partial y}\left(\rho_g \varepsilon v_g u_g\right)$$
$$=-\frac{\partial P}{\partial x}+\beta_B\left(u_s-u_g\right)+\dot{m}u_g \tag{T.7}$$

Table 7.1 (CONTINUED)

Solids-Phase Momentum in x-Direction

$$\frac{\partial}{\partial t}\left[\rho_s(1-\varepsilon)u_s\right]+\frac{\partial}{\partial x}\left[\rho_s(1-\varepsilon)u_s u_s\right]+\frac{\partial}{\partial y}\left[\rho_s(1-\varepsilon)v_s u_s\right]=\beta_B\left(u_g-u_s\right)+G(\varepsilon)\frac{\partial \varepsilon}{\partial x}-\dot{m}u_s \tag{T.8}$$

Gas-Phase Momentum in y-Direction

$$\frac{\partial}{\partial t}\left(\rho_g \varepsilon v_g\right)+\frac{\partial}{\partial x}\left(\rho_g \varepsilon u_g v_g\right)+\frac{\partial}{\partial y}\left(\rho_g \varepsilon v_g v_g\right)=-\frac{\partial P}{\partial x}+\beta_B\left(v_s-v_g\right)-\rho_g g+\dot{m}v_g \tag{T.9}$$

Solids-Phase Momentum in y-Direction

$$\begin{aligned}&\frac{\partial}{\partial t}\left[\rho_s(1-\varepsilon)v_s\right]+\frac{\partial}{\partial x}\left[\rho_s(1-\varepsilon)u_s v_s\right]+\frac{\partial}{\partial y}\left[\rho_s(1-\varepsilon)v_s v_s\right]\\ &=\beta_B\left(v_g-v_s\right)-\left(\rho_s-\rho_g\right)(1-\varepsilon)g+G(\varepsilon)\frac{\partial \varepsilon}{\partial y}-\dot{m}v_s\end{aligned} \tag{T.10}$$

Constitutive Equations

Gas–Solid Drag Coefficients for Model A

Based on Ergun Equation, for $\varepsilon < 0.8$

$$\beta_A = 150\frac{\varepsilon_s^{\,2}\mu_g}{\varepsilon\left(d_p\phi_s\right)^2}+1.75\frac{\rho_g\varepsilon_s\left|v_g-v_s\right|}{d_p\phi_s} \tag{T.11}$$

Based on Single Sphere Drag, for $\varepsilon > 0.8$

$$\beta_A = \frac{3}{4}C_d\frac{\varepsilon\varepsilon_s\rho_g\left|v_g-v_s\right|}{\left(d_p\phi_s\right)}\varepsilon^{-2.65} \tag{T.12}$$

$$C_d=\begin{cases}\frac{24}{\mathrm{Re}_p}\left[1+0.15\left(\mathrm{Re}_p\right)^{0.687}\right], & \text{if } \mathrm{Re}_p<1000;\\ 0.44, & \text{if } \mathrm{Re}_p\ge 1000.\end{cases} \tag{T.13}$$

$$\mathrm{Re}_p=\left(\varepsilon\rho_g\left|v_g-v_s\right|d_p/\mu_g\right) \tag{T.14}$$

Gas–Solid Drag Coefficients for Model B

$$\beta_B=\beta_A/\varepsilon \tag{T.15}$$

In one dimension, the basic variables are the pressure, the porosity and the gas and the solids velocities. In homogeneous flow, as in compressible single phase pipe flow, the variables reduce themselves to just the velocity and the pressure, determined by their respective conservation of mass and momentum equations.

The solids stress determined by the modulus G in Eqs. (T.4) and (T.6), as already discussed, is needed to prevent the particles from compressing to unreasonably high volume fractions and for numerical stability in Model A. In the globally stable Model B, in Eqs. (T.8) and (T.10), G is needed only for predicting correct compression. It can be set to zero for reasonably dilute flow, where the solids pressure is negligible. In Model B, the pressure occurs only in the gas momentum equations. Thus, as discussed in Chapter 2, the force due to gravity must be modified to satisfy Archimedes' buoyancy principle. Also, the drag relation had to be modified due to the absence of the porosity multiplying the pressure gradient in Eq. (T.7) as compared to (T.3). The result was Eq. (T.15). The mixture momentum equations for both models are identical. Hence, the models can be expected to give a different value of the relative velocity only and not the average mixture quantities. It was nevertheless at first surprising to see the models produce nearly identical results for bubbling fluidization (Bouillard *et al.*, 1989a). In view of this and the sometimes unexpected numerical problems with Model A, it is recommended that only Model B be used. With zero G it is valid for dispersed gas–liquid flow where bubbles or droplets play the role of particles.

In view of continuing use of Model A by the gas–liquid two-phase flow investigators and the historical breakthrough of bubble computation in a fluidized bed, the hydrodynamic model used by the System-Science-Software group (Pritchett *et al.*, 1978; Blake *et al.*, 1979, 1980; Richner *et al.*, 1990) is presented in Table 7.2. Their computer code involved the same six variables, with the equations written in cylindrical coordinates with axial symmetry. The continuity equations, Eqs. (T.16) and (T.17), are identical to Eq. (T.1) and (T.2) in Cartesian coordinates with no phase change. The solids conservation of momentum equations (T.18) and (T.19) include an empirical solids viscosity of the type discussed in the next chapter. The authors also included an empirical solids pressure and a gas pressure into the solids momentum equations. Such an approach can be justified by looking at Eqs. (T.18) and (T.19) as mixture equations with a negligible gas momentum. Indeed, in the gas momentum Eqs. (T.20) and (T.21), the gas momentum has been neglected. The result of such an approximation is that simulation of high-speed jets became very inaccurate. The drag between the gas and the solid was modeled in an only slightly different way, since a similar friction data base was used. Because of this basic similarity, the System-Science-Software group produced bubbles in fluid beds.

Table 7.2 SYSTEMS-SCIENCE SOFTWARE (CHEMFLUB) HYDRODYNAMIC MODEL (in cylindrical coordinates; no phase change)

Continuity Equations

Gas-Phase

$$\frac{\partial}{\partial t}\left(\rho_g \varepsilon\right)+\frac{1}{r}\frac{\partial}{\partial r}\left(r\rho_g \varepsilon u_g\right)+\frac{\partial}{\partial y}\left(\rho_g \varepsilon v_g\right)=0 \tag{T.16}$$

Solids-Phase

$$\frac{\partial}{\partial t}\left(\rho_s \varepsilon_s\right)+\frac{1}{r}\frac{\partial}{\partial r}\left(r\rho_s u_s \varepsilon_s\right)+\frac{\partial}{\partial y}\left(\rho_s v_s \varepsilon_s\right)=0 \tag{T.17}$$

Momentum Equations

Radial Solids-Phase Momentum

$$\begin{aligned}
&\frac{\partial}{\partial t}\left(\varepsilon_s\rho_s u_s\right)+\frac{1}{r}\frac{\partial}{\partial r}\left(r\varepsilon_s\rho_s {u_s}^2\right)+\frac{\partial}{\partial y}\left(\varepsilon_s\rho_s u_s v_s\right)\\
&\quad=\frac{\partial}{\partial r}\left(P+P_s\right)+\frac{\partial}{\partial r}\left[\frac{1}{r}\left(\lambda_s-\frac{2}{3}\mu_s\right)\frac{\partial}{\partial r}\left(ru_s\right)\right]\\
&\quad+\frac{2}{r}\frac{\partial}{\partial r}\left(ru_s\frac{\partial u_s}{\partial r}\right)+\frac{\partial}{\partial r}\left[\left(\lambda_s-\frac{2}{3}\mu_s\right)\frac{\partial v_s}{\partial y}\right]\\
&\quad+\frac{\partial}{\partial y}\left[\mu_s\left(\frac{\partial u_s}{\partial y}+\frac{\partial v_s}{\partial r}\right)\right]+2\frac{\partial}{\partial y}\left(\mu_s\frac{\partial v_s}{\partial y}\right)
\end{aligned} \tag{T.18}$$

Axial Solids-Phase Momentum

$$\begin{aligned}
&\frac{\partial}{\partial t}\left(\varepsilon_s\rho_s v_s\right)+\frac{1}{r}\frac{\partial}{\partial r}\left(r\varepsilon_s\rho_s v_s\right)+\frac{\partial}{\partial y}\left(\varepsilon_s\rho_s {v_s}^2\right)=-\varepsilon_s\rho_g g-\frac{\partial}{\partial y}\left(P+P_s\right)\\
&+\frac{1}{r}\frac{\partial}{\partial r}\left[r\mu_s\left(\frac{\partial u_s}{\partial y}+\frac{\partial v_s}{\partial r}\right)\right]+\frac{\partial}{\partial y}\left[\left(\lambda_s-\frac{2}{3}\mu_s\right)\frac{\partial v_s}{\partial y}\right]+\frac{\partial}{\partial y}\left[\frac{1}{r}\left(\lambda_s-\frac{2}{3}\mu_s\right)\frac{\partial}{\partial r}\left(ru_s\right)\right]
\end{aligned} \tag{T.19}$$

Conservation of Gas-Phase Momentum

$$u_g=u_s-\frac{\varepsilon}{B(\varepsilon_s)}\frac{\partial P}{\partial r}\quad;\quad v_g=v_s-\frac{\varepsilon}{B(\varepsilon_s)}\frac{\partial P}{\partial y} \qquad \text{(T.20) and (T.21)}$$

Constitutive Equations

Solids-Phase Pressure

$$P_s(\varepsilon_s)=\begin{cases}\frac{1}{2}a^2{\rho_s}^2(\varepsilon_s-\varepsilon_{so})^2, \text{ if } \varepsilon_s>\varepsilon_{so}\\ 0, \text{ if } \varepsilon_s\le\varepsilon_{so}\end{cases} \tag{T.22}$$

where a is an empirical constant

Gas-Phase Pressure (T.23)

$$P=\rho RT$$

Table 7.2 (Continued)

Solids-Phase Shear Viscosity

$$\mu = \mu_{so}\left(\varepsilon_s/\varepsilon_{so}\right) \tag{T.24}$$

Solids-Phase Bulk Viscosity

$$\lambda_s = \lambda_{so} - \frac{2}{3}\mu_s \tag{T.25}$$

Drag Coefficient

$$B(\varepsilon_s) = \frac{\varepsilon^2\mu}{K}, \tag{T.26}$$

where the permeability K is defined as

$$K = 16 r_p{}^2 F_1(\varepsilon_s) F_2(N) \tag{T.28}$$

and where N is a Reynolds number

$$N = \frac{4\varepsilon\rho r_p\left(\mathbf{v}_g - \mathbf{v}_s\right)}{\mu}. \tag{T.29}$$

The function $F_1(\varepsilon_s)$ is given as

$$F_1(\varepsilon_s) = A\varepsilon^a \varepsilon_s{}^b, \tag{T.30}$$

where constants A, a and b are constants and the function $F_2(N)$ is given as

$$F_2(N) = \left(1 + B_1 N^c\right)^{-1}. \tag{T.31}$$

7.8 Comparison of Computations to Observations

Valid models must predict the observed physical phenomena. A key observation in gas fluidization was the photographing of bubbles (Rowe, 1971). The System Science Software code (Pritchett *et al.*, 1978) was the first computer model to predict the formation of bubbles in fluidized beds. Since no comparison to experiments was given, only the studies conducted at the Illinois Institute of Technology will be discussed. The origin of bubbles was discussed in Chapter 6. A correct dependence of the drag on the volume fraction of the solids is needed to compute bubble formation.

7.8.1 Bubbles with a Jet

The correct prediction of bubbles in fluidized beds is one of the key issues. Bubbles cause much of the stirring in gas fluidized beds, which give these contactors their unique properties. However, if in a hypothetical 10 m gasifier dreamed about at IGT, the bubble was to be a fast bubble containing oxygen, the potential for disaster would discourage the design of such a reactor. Indeed, as large-scale cold flow model data show, the bubbles in a 3 m bed are larger than one meter in diameter (Yang *et al.*, 1984). Following the work of Westinghouse cited above, a two-dimensional bed with a jet was constructed, instrumented and tested at IIT (Gidaspow *et al.*, 1983). The advantage of modeling fluidization with a jet is that in such a situation the jet establishes the flow pattern. Hence, from a modeling point-of-view, this problem is easier than fluidizaiton with a uniform inlet gas flow. Figure 7.3 shows a schematic of the IIT apparatus. The initial condition for bubble studies was that of minimum fluidization as depicted

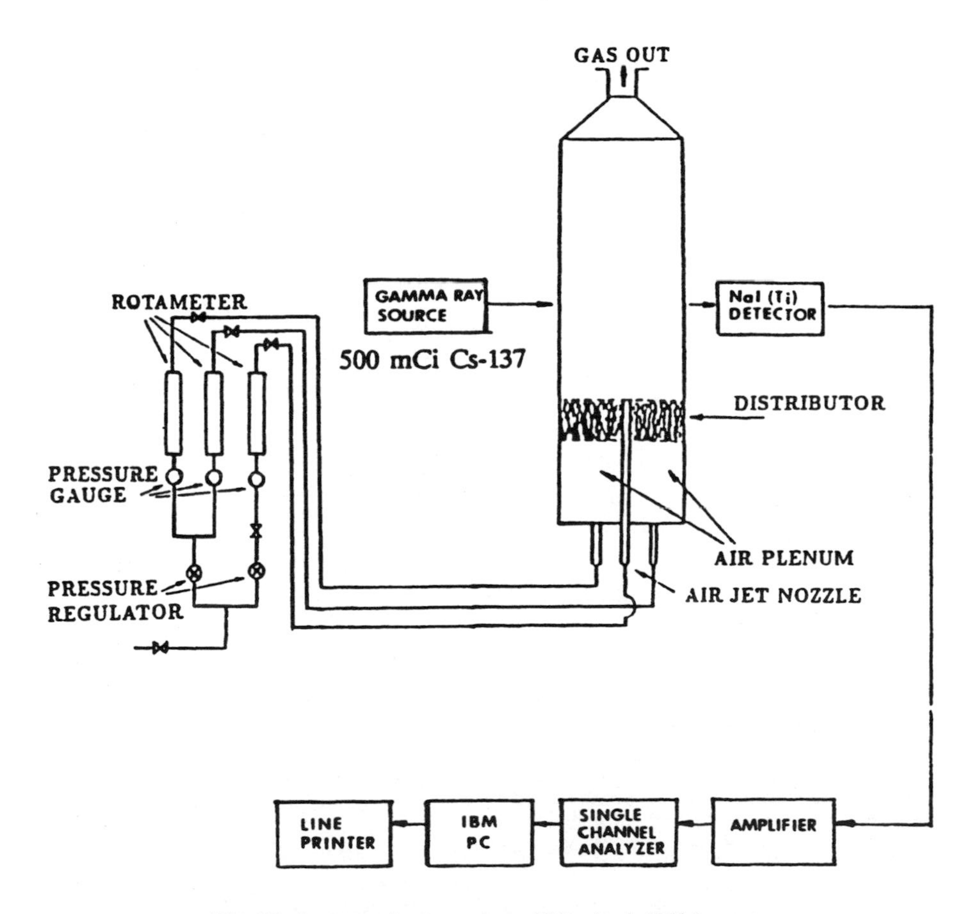

Fig. 7.3 Apparatus for measuring void fraction in IIT laboratory.

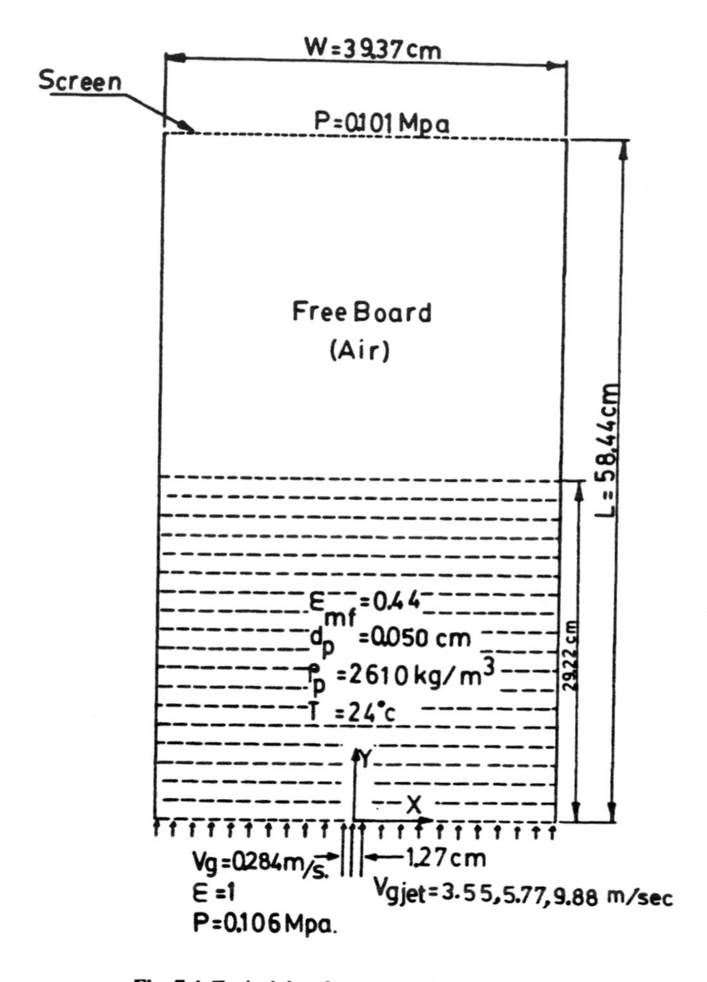

Fig. 7.4 Typical data for numerical computations.

in Fig. 7.4. The boundary conditions were those of no slip for the gas and for the solid at the left and the right walls.

In the experiment (Gidaspow *et al.*, 1983a) the bed was constructed from transparent Plexiglas sheets. The bed cross-section was 40 cm by 3.81 cm. The jet nozzle was a 1.27 × 3.81 cm rectangular slit. Both the distributor plates and the jet were covered with 80 mesh stainless steel wire mesh. In order to achieve a uniform fluidization, the distributor section was designed in such a way that the pressure drop through the distributor section was 10–20% of the total bed pressure drop (Kunii and Levenspiel, 1969).

Ottawa sand, which has an actual density of 2610 kg/m^3 and a settled bulk density of 1562 kg/m^3, was used as the bed material with the particle size of −20 + 40 mesh. The mean particle size was calculated to be 503 μm. The voidage was determined to be 0.402. The minimum fluidization superficial velocity was 28.2 cm/s. The minimum fluidization velocity was obtained by finding the intersection between pressure drop versus flow rate curves in the fixed bed and in the fluid bed region.

In a typical run, solids were first loaded into the rectangular column so as to give a static bed height of 28 cm. Then the bed was fluidized by introducing compressed air through the air plenum to give a fluidized bed height of 29.5 cm. Jet air was injected at a specific flow rate.

A photographic technique was used for measuring the bubble diameters using a 16 mm Fastax movie camera. Figures 7.5 to 7.7 show a comparison of the experimentally determined bubble shapes to the theoretically determined voids. In the computer calculations we assumed that a bubble is a region of void greater than 0.8. For all three velocities the theoretical bubble shown is the second bubble after start-up. The first bubble injected into a bed initially at minimum fluidization is not typical. The second, third and fourth theoretical bubbles appear to

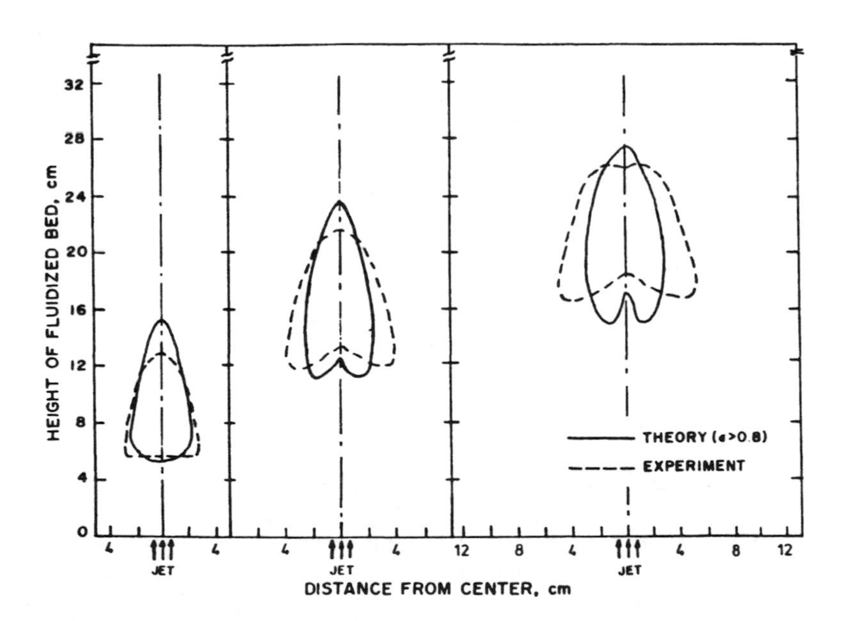

Fig. 7.5 A comparison of theoretically and experimentally determined bubbles in a fluidized bed with a jet. Jet velocity = 3.55 m/s (from Gidaspow *et al.*, 1983b).

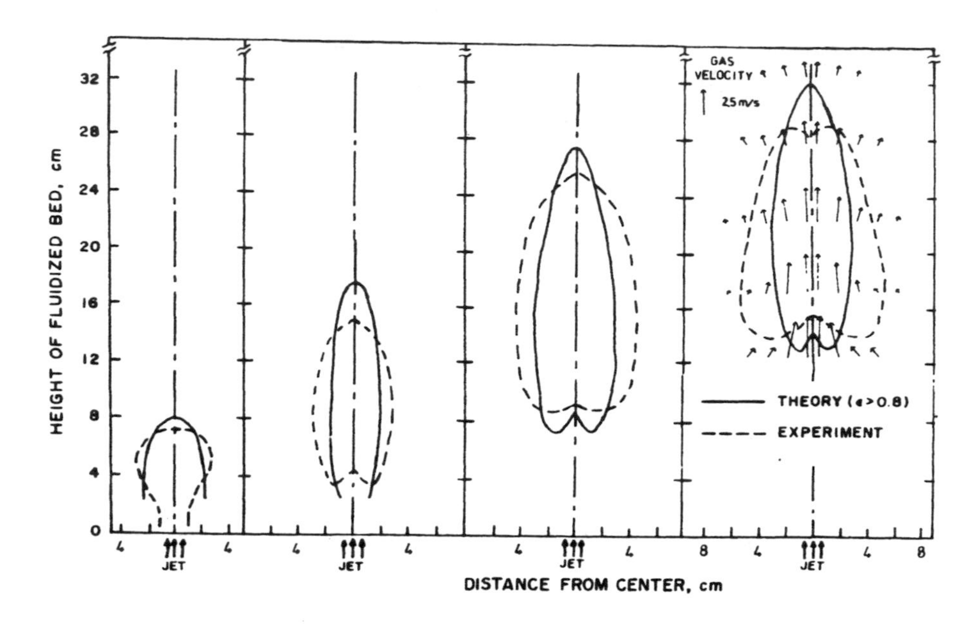

Fig. 7.6 A comparison of theoretically and experimentally determined bubbles in a fluidized bed with a jet. Jet velocity = 5.77 m/s. Arrows indicate computed gas velocities (from Gidaspow *et al.*, 1983b).

Table 7.3 FLUIDIZATION OF 800 μm GLASS BEADS

Width of bed	40 cm
Static bed height	29.2 cm
Jet opening	1.27 cm
Jet velocity	5.77 m/sec
Grid velocity	53.6 cm/sec
Particle diameter	800 μm
Particle density	2.42 g/cm^3
Radial grid size	0.635 cm
Axial grid size	1.08 cm
Time stop	1.0×10^{-4} sec

differ in a chaotic manner, as do the experimentally observed bubbles. For example, at a jet velocity of 5.77 m/s and 12 cm above the jet, the average diameters of the second through the fourth bubble are 8.32, 7.10 and 7.99 cm, respectively. The experimental bubble shown is a typical bubble seen at the indicated bed location. A sensitivity analysis to the variation of the solids stress was performed by Ettehadieh (1982). It was found that changing the solids pressure modulus does not appreciably alter the transient behavior of the bed. The effect is greater near minimum fluidization where this term in the equation becomes of appreciable magnitude.

Figures 7.5 to 7.7 show a good agreement between the theory and the experiment. After break-off from the jet inlet, the bubbles are elongated and all show a tendency to contain solids at their rear center. These solids may be brought into the bubble by the jet that moves much faster than the bubble. The jet drags the particles with it into the slow-moving large void. At higher jet velocities this may cause bubble splitting, as seen in Fig. 7.7. Although the agreement between the theory and the experiment is not perfect, as seen in Fig. 7.7, a better agreement is not to be expected because of the deletion of the viscosity in the computation. Figure 7.8 shows a good agreement of the measured, the computed and the effective bubble diameters obtained from literature correlations.

The computations and the experiment were repeated for fluidization of 800 micron glass beads, as summarized in Table 7.3 (Gidaspow *et al.*, 1986a). Figure

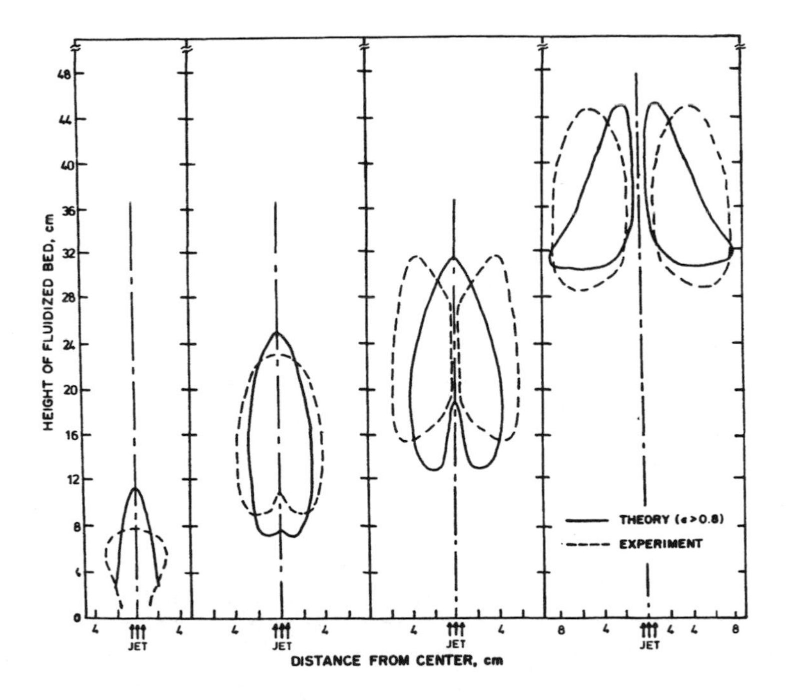

Fig. 7.7 A comparison of theoretically and experimentally determined bubbles in a fluidized bed with a jet. Jet velocity = 9.88 m/s (from Gidaspow *et al.*, 1983b).

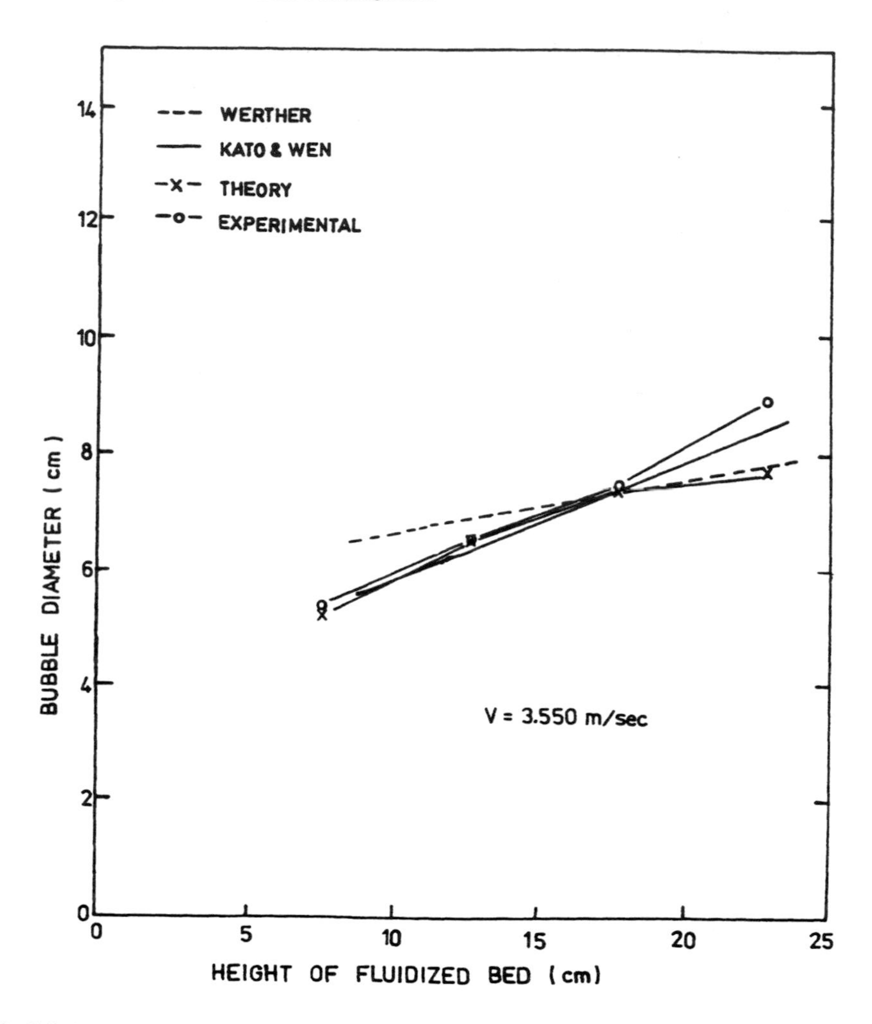

Fig. 7.8 A comparison of photographically determined bubble sizes to the theoretically computed sizes using a hydrodynamic model and to sizes computed from modified correlations at a jet velocity of 3.55 m/s (Gidaspow *et al.*, 1983b).

7.9 shows the computed formation, growth and bursting of the first bubble in the two-dimensional bed initially at minimum fluidization with a velocity of 0.536 m/s. At zero time a jet of a constant velocity of 5.77 m/s was turned on.

An experiment duplicating this computer run was also conducted using our two-dimensional bed with a jet. Figure 7.10 shows a comparison of the computed bubble (on the left) at 0.32 seconds after start-up to the experimental bubble (on the

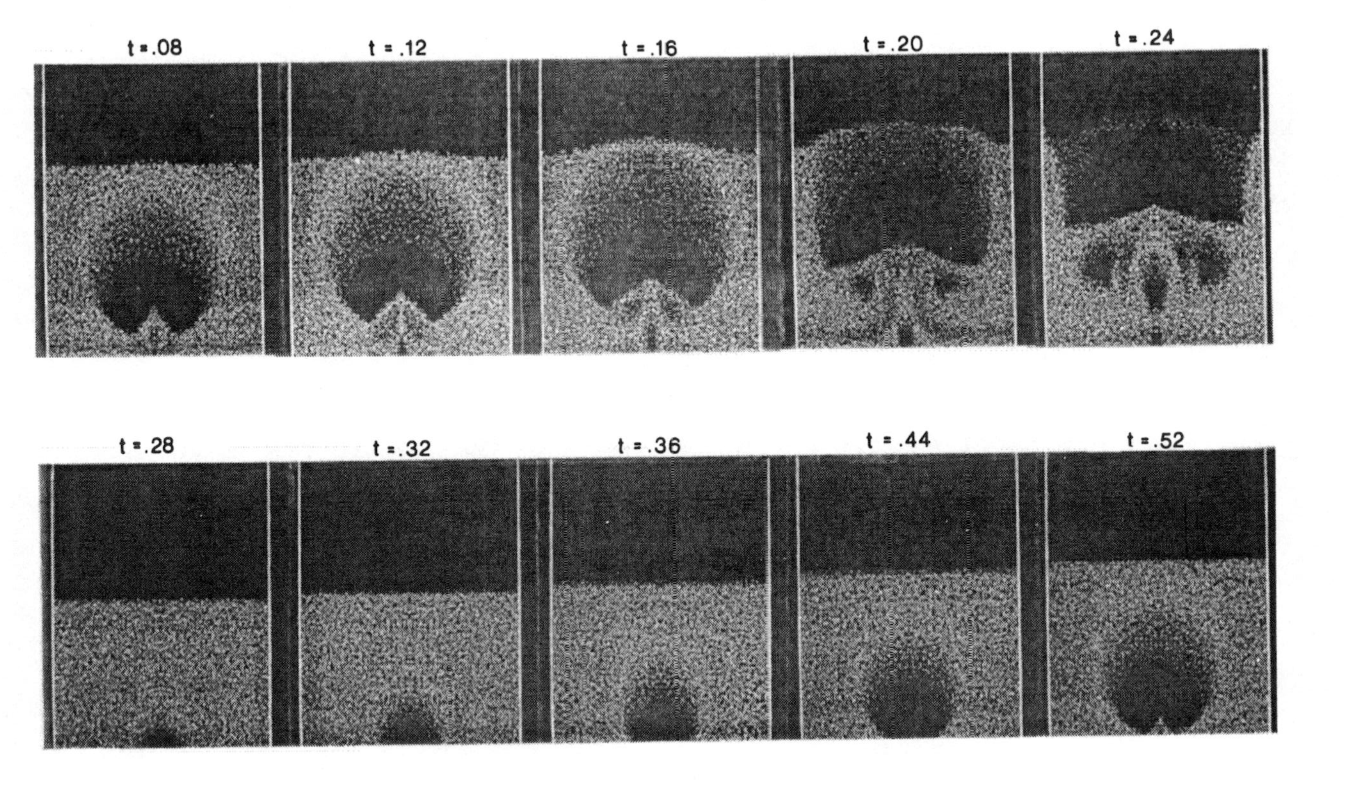

Fig. 7.9 Computed formation, growth and bursting of the first bubble in a two-dimensional fluidized bed with a jet (from Gidaspow *et al.*, 1986b).

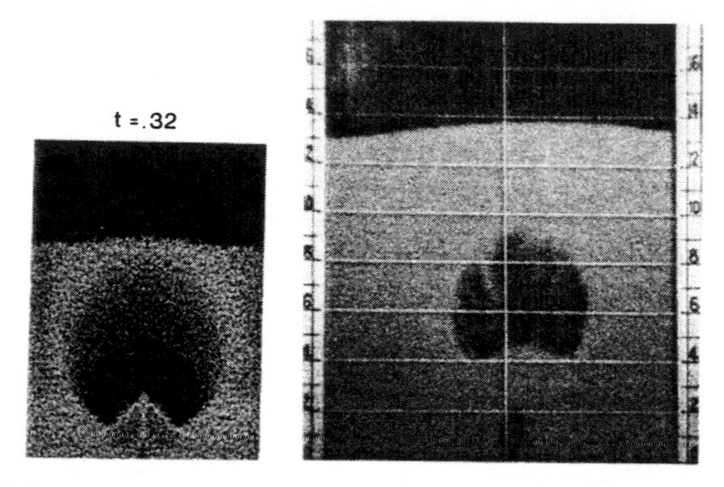

Fig. 7.10 Experimental (right) vs. computed (left) bubble at 0.32 s (from Gidäspow *et al.*, 1986a).

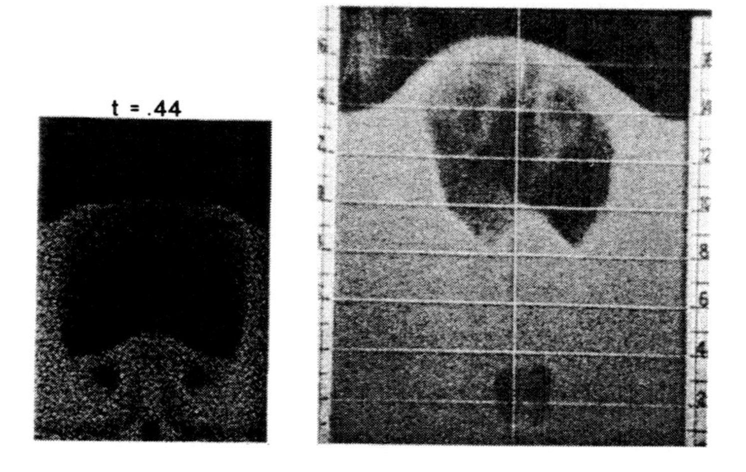

Fig. 7.11 Experimental (right) vs. computed (left) bubble at 0.44 s (from Gidaspow *et al.*, 1986a).

right) for the conditions summarized in Table 7.3. The computed bubble shape closely resembles the observed shape in the two-dimensional plastic bed. Figure 7.11 shows the comparison at 0.44 seconds after start-up from minimum fludization condition. The computed bubble is somewhat larger than the one observed in the narrow plastic bed. The difference may be due to the effect of the walls of the bed, which were not included in the computation.

Figure 7.12 shows the computed gas flux into the bubble at its point of detachment. This figure was constructed based on the computed void fractions and velocities around the bubble periphery of a void fraction of roughly 0.8. Since the bubble acts as a short-circuit for gas flow, the flow into the bubble is much larger than that supplied by the jet alone. The jet flow is 577 cm/s × 1.27 cm^2, compared to an inflow of 2426 versus an outflow of 1649.

In order to understand erosion in fluidized beds containing heat exchange tubes, such as fluidized bed combustors with rows of horizontal tubes, a few simple geometries with obstacles were studied. The simplest geometry was an obstacle placed above the jet in the two-dimensional bed already described. Figure 7.13

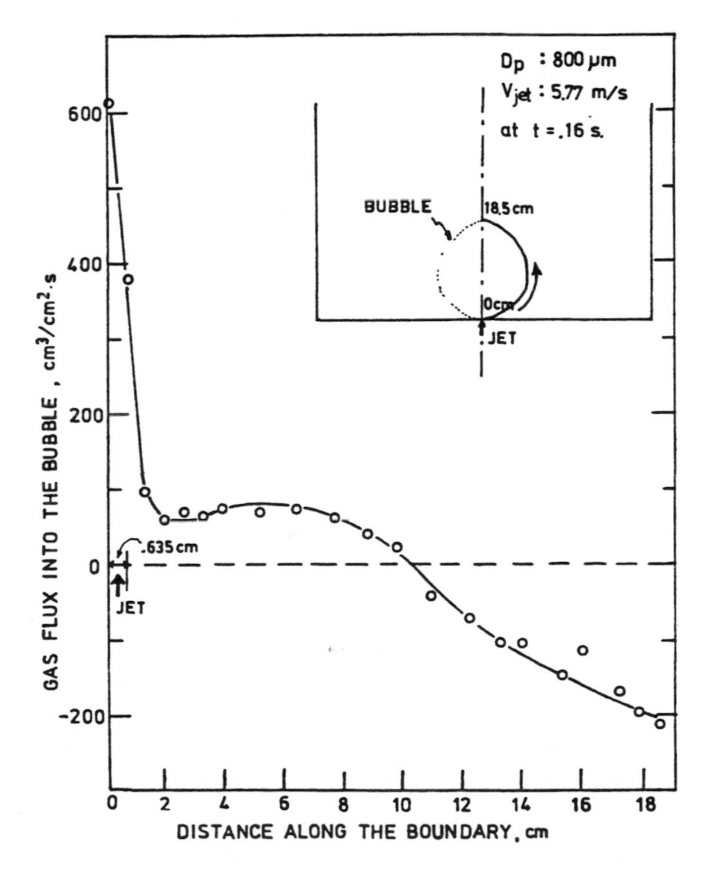

Fig 7.12 Computed gas flux into a bubble at its point of detachment corresponding to conditions of Fig. 7.9 (from Gidaspow *et al.*, 1986b).

shows the formation of the first bubble caused by suddenly increasing the jet flow to 5.78 m/s. Initially the bed was filled with 503 μm glass beads to a height of 29.2 cm with air supplied at a velocity of 26 cm/s to maintain the particles at minimum fluidization. The first bubble produced was the largest. Figure 7.14 shows the computed bubbles at approximately the same time as in the experiment. The principal difference between the inviscid theory and the experiment is the asymmetry. In the computation the calculations were done for half the bed only. It was not possible to obtain symmetry in the experiment despite an effort to have the same pressure drop on both sides of the obstacle. The asymmetry appears to be a natural phenomenon, perhaps related to Von Karman vortex shedding, as computed by Cook and Harlow (1986) for two phase bubbly flow around an obstacle. Computations with viscosity (Bouillard *et al.*, 1989b) do not change the results.

7.8.2 Time-Averaged Porosities

Porosities measurement is a sensitive test of the theory, since it is not prescribed at any of the boundaries in the simulation of the bubbling bed. Hence, as a test of the theory a two-dimensional bed with a jet roughly approximating the operation of a gasifier or a fluidized bed combustor was constructed (Gidaspow *et al.*, 1983a). Figure 7.3 shows a schematic of the apparatus used to determine time-averaged and instantaneous porosity distributions using a gamma-ray densitometer. Figure 7.15 shows a comparison of the experimentally determined porosity distributions for fluidization of Ottowa sand of mean diameter of 503 μm to computations using inviscid Model A. The sand was kept fluidized by supplying air at a velocity of 28.2 cm/s through a uniform distributor and by means of a jet at a velocity of 5.78 m/s. With a two-dimensional slit jet, the maximum porosity is always at the jet inlet, as expected. Figure 7.16 shows a comparison of the center-lane porosities for two jet velocities. However, with a circular jet the maximum porosity was not at the jet inlet. This maximum porosity occurred 4 to 6 cm away from the jet inlet. With increasing jet flow it moved further into the bed. This occurs because of a three-dimensional effect. Away from the jet, the time-averaged porosity increases slowly from the bottom to the top, and then increases rapidly near the free-board. However, as seen in Fig. 7.16, at the center line the porosity decreases to a minimum, associated with jet penetration, and then rises near the top of the bed. The inclined constant porosity contours in Fig. 7.15 can be explained by means of Eq. (7.58). Their slope is roughly the ratio of the gas velocity components. The differences between the theory and the experiment in Fig. 7.16 are due to the neglect of the front and back walls in the two-dimensional

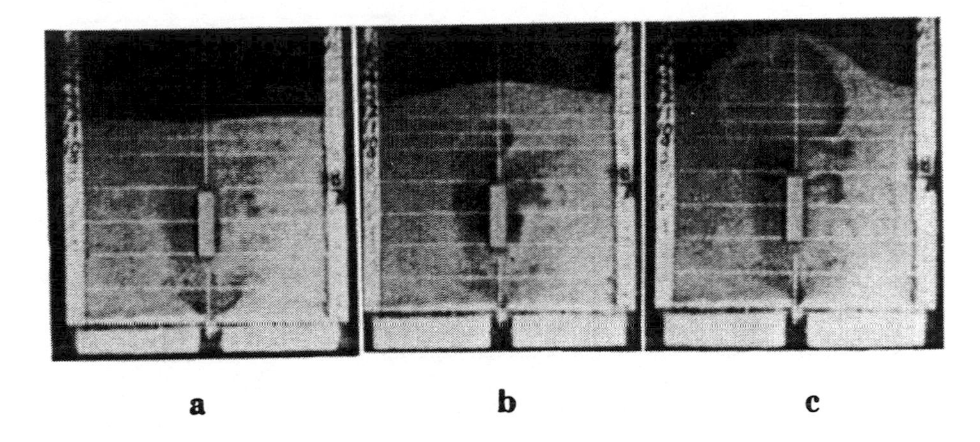

Fig. 7.13 Evolution of an experimental bubble in a two-dimensional rectangular fluidized bed with a central jet and an immersed obstacle (9.74 cm × 2.54 cm): a) bubble formation at the distributor, 0.105 s; b) bubble propagation around the obstacle, 0.236 s; c) bubble bursting at the top of the bed, 0.315 s. (From Bouillard *et al.*, 1989a. Reproduced by permission of the American Institute of Chemical Engineers. Copyright © 1989 AIChE. All rights reserved.)

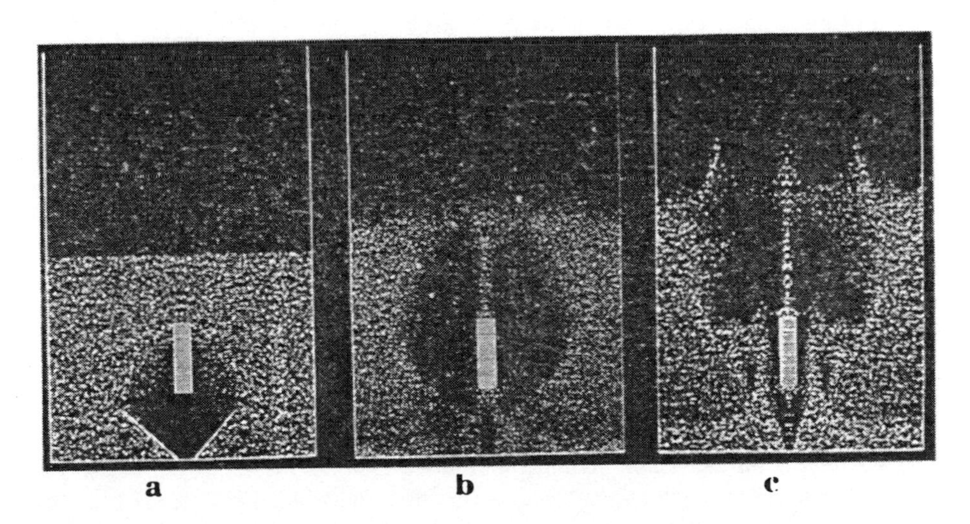

Fig. 7.14 Computer-generated porosity distribution for a two-dimensional fluidized bed with an obstacle, 31 × 48 nodes. (From Bouillard *et al.*, 1989a. Reproduced by permission of the American Institute of Chemical Engineers. Copyright © 1989 AIChE. All rights reserved.)

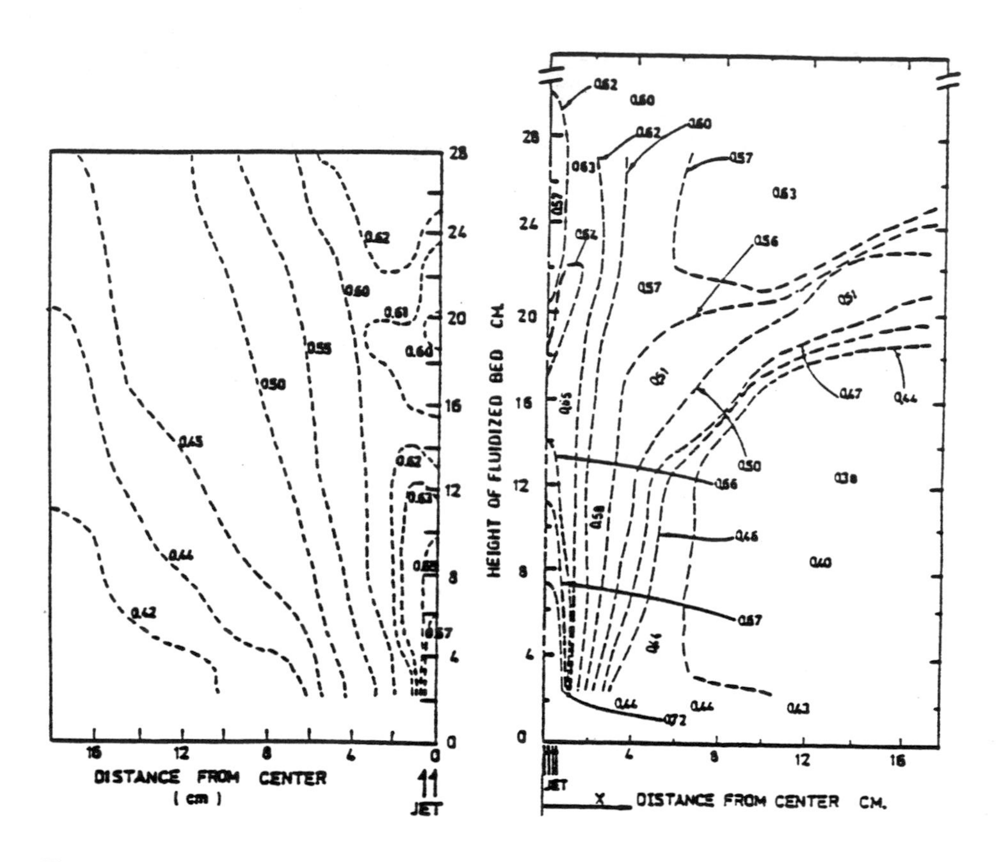

Fig. 7.15 *Experimental* (left) and *computed* (right) time-averaged porosity in a fluidized bed with a jet $V_{jet} = 5.78$ m/s, $V_{mf} = 0.234$ m/s, $d_p = 503$ μm (from Gidaspow *et al.*, 1983a and Gidaspow and Ettehadieh, 1983).

computation, due to not completely adequate time averaging and due to the neglect of the viscosity.

The solids stress played only a minor role in the computation. In the dilute region, that is, in the region significantly above minimum fluidization, the solids stress modulus G becomes zero. Below minimum fluidization it is needed to prevent the particles from overcompacting. In Model B in the dilute region it can be set to zero. In Model A it has to be retained and adjusted to prevent numerical instability. Hence, Model B contains no adjustable parameters.

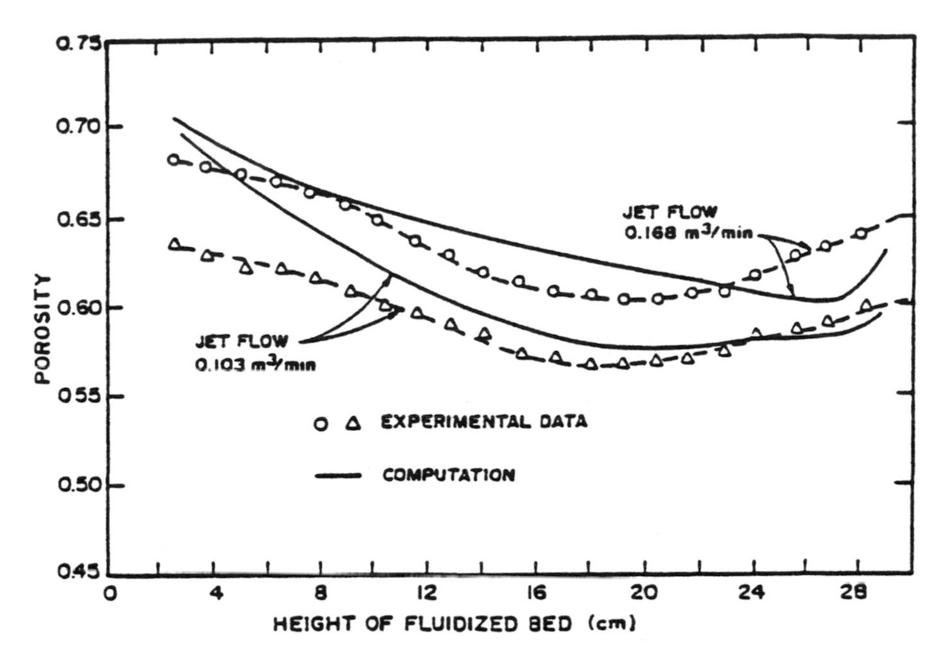

Fig. 7.16 Computed and experimental porosity profiles at center line of the fluidized bed with jet, d_p = 503 μm, a slit jet opening 1.27 cm wide (from Gidaspow *et al.*, 1983a and Gidaspow and Ettehadieh, 1983).

Figure 7.17 shows experimental time-averaged porosities in a two-dimensional bed with an obstacle. Porosity data were obtained only on one side of the bed. Symmetry was assumed in constructing Fig. 7.17. The corresponding time-averaged velocities are displayed in Fig. 7.18. Figure 7.19 shows a comparison of the time-averaged porosities computed using Models A and B to the experimental data away from the obstacle. As in the case without an obstacle, the porosity rises slowly from the bottom of the bed to near the top and then rises sharply in the free-board region. The agreement between the experiment and the theory (Bouillard *et al.*, 1989a) was good, except above the obstacle. At this location, the experiment showed a higher porosity caused by asymmetric motion of the bubble and partial sweeping of the region above the obstacle. In a perfectly symmetric experiment the agreement between the experiment and the model would be within a few percent, with the difference being due to neglect of the front and back walls and the solids viscosity. The most significant observation is, however, that Models A and B produce the same numerical results. This means that buoyancy can be handled either by including it as a pressure drop in the particulate phase, as done in Model A, or by including it directly as a difference in densities. See Table 7.1, Eq. (T.4)

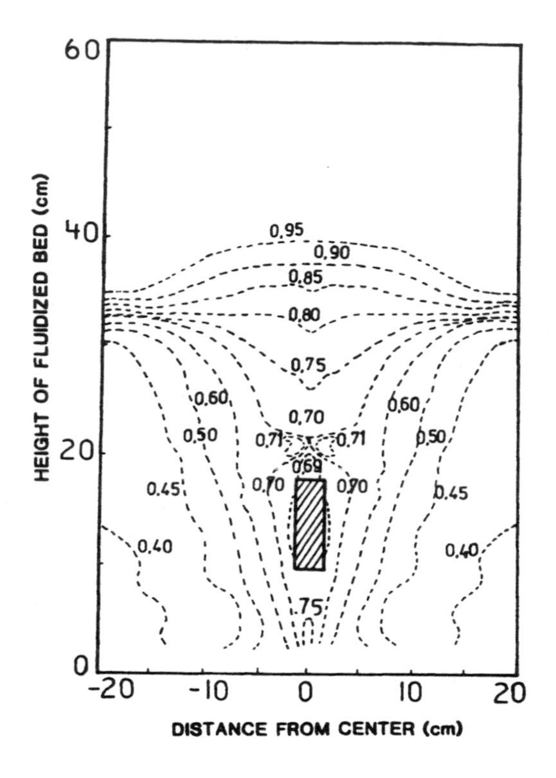

Fig. 7.17 Experimental time-averaged porosity distribution obtained from a two-dimensional fluidized bed with a slit jet and obstacle. V_{jet} = 5.77 m/s U_{mf} = 26 cm/s. (From Bouillard *et al.*, 1989a. Reproduced by permission of the American Institute of Chemical Engineers. Copyright © 1989 AIChE. All rights reserved.)

for Model A and Eq. (T.10) for Model B. Since Model B is well-posed as an initial value problem, it is recommended that Model A not be used in the future. Model A can give rise to unexpected instabilities. It needs the elastic modulus G in the dilute region purely for numerical reasons. The imaginary characteristics in Model A caused by a non-zero relative velocity are made real by a sufficiently large $G/(1-\varepsilon)$, as seen from Eq. (7.29).

7.8.3 Velocity Profiles

Complete verification of time average two-phase flow equations requires a comparison of computed to measured velocities, as well as porosities. Hence, a comparison of the theory to time averaged gas velocity profiles was made

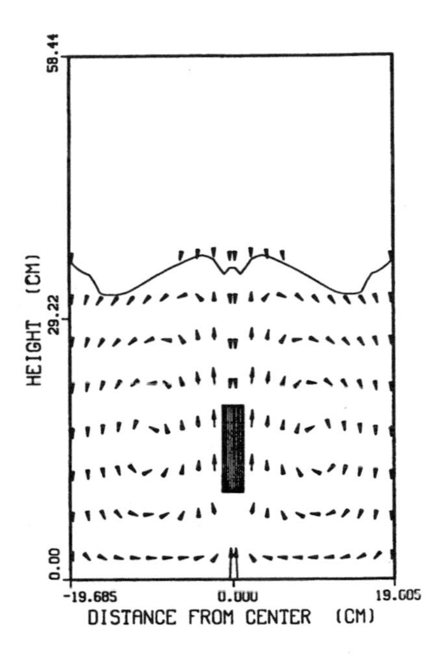

Fig. 7.18 Predicted solids velocity field, time-averaged over 2 s. Hydrodynamic Model B. (From Bouillard *et al.*, 1989a. Reproduced by permission of the American Institute of Chemical Engineers. Copyright © 1989 AIChE. All rights reserved.)

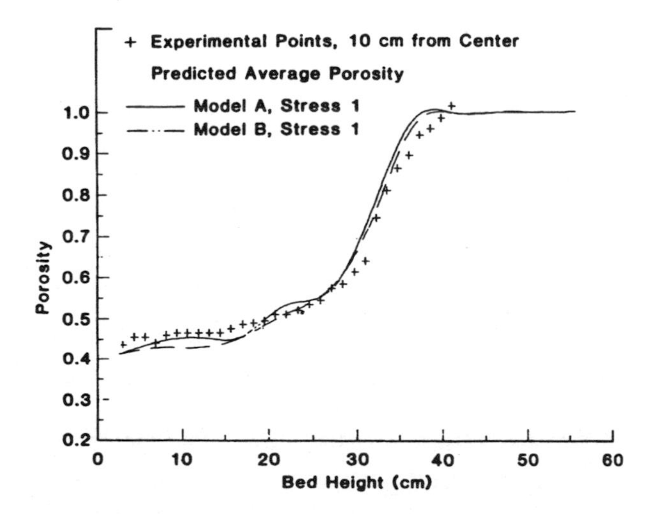

Fig. 7.19 A comparison of computed time-averaged porosities for Models A and B to experimental data shown in Fig. 7.17. (From Bouillard *et al.*, 1989a. Reproduced by permission of the American Institute of Chemical Engineers. Copyright © 1989 AIChE. All rights reserved.)

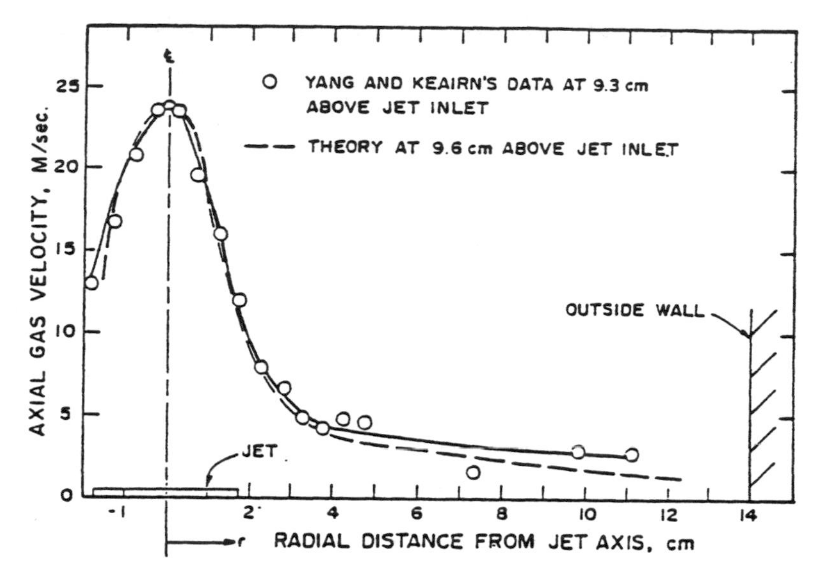

Fig. 7.20 A comparison of computed time-averaged axial gas velocities to Westinghouse data. From Ettehadieh *et al.*, 1984. Reproduced by permission from the American Institute of Chemical Engineers.

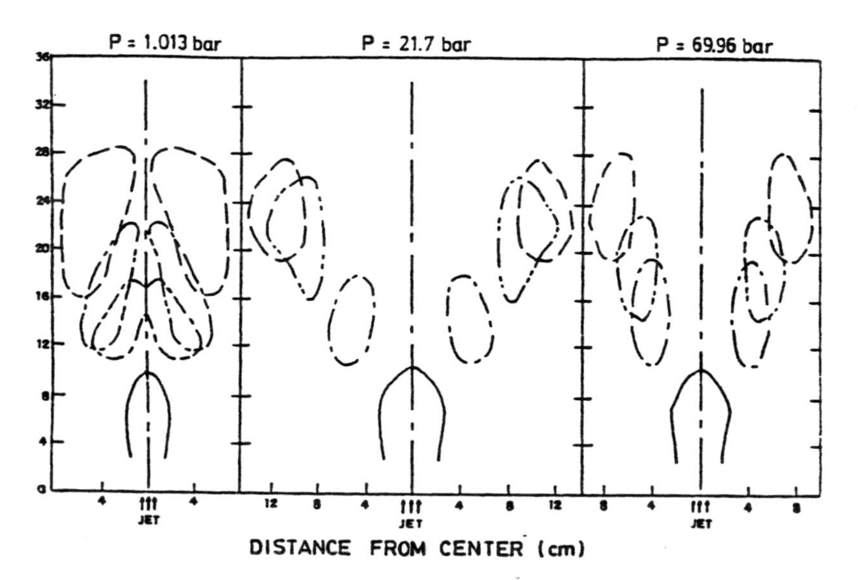

Fig. 7.21 Computed formation and splitting of the first bubble in our two-dimensional bed at various pressures, jet velocity = 3.55 m/s. (From Gidaspow *et al.*, 1983b; also Seo, 1985).

(Ettehadieh *et al.*, 1984). Figure 7.20 shows a comparison of the calculated time-averaged gas velocity profile to Westinghouse data obtained in a semi-circular fluidized bed. A semi-circular bed with a window to allow observation of the motion of the bubbles is a convenient device for understanding the dynamics of gas–solid motion. Yang and Keairns (1980) at Westinghouse have published numerous data on local pressure and gas velocities in such devices. They measured static and impact pressures at different locations in the bed by using U-tube and inclined manometers. The gas velocity was found by setting the kinetic energy of the gas equal to the difference between the two pressures. Donsi *et al.* (1980) have also measured axial solid and gas velocity profiles and interpreted their data using conventional single phase turbulence theory. Such a mixing length theory can indeed explain the shape of the curve in Fig. 7.20. Thus, note the amazing agreement between the Westinghouse data and the present hydrodynamic theory which uses no fitted parameters. Similar agreement was obtained for other Westinghouse data where solids were injected with the gas jet (Ettehadieh *et al.*, 1984).

7.8.4 Pressure Effect

To show that the model predicts the effects of pressure, bubbles were computed at elevated pressures corresponding roughly to the experiments conducted by T. Knowlton at IGT (Gidaspow *et al.*, 1986c; Seo, 1985). For the two-dimensional bed with a jet, Fig. 7.21 shows that bubble splitting increases with pressure or momentum of a jet. Knowlton's movie for a semicircular bed with an inlet central jet consistently shows bubble splitting in agreement with our computations. Figure 7.22 compares the effective spherical bubble diameters as a function of pressure. Both the experimental and the theoretical bubble sizes decrease with pressure. This gives rise to a decreased amplitude of oscillations (Seo, 1985) and hence, to so-called smoother fluidization at higher pressures.

It is well known in the fluidization community (Sobreiro and Monteiro, 1982; Rowe *et al.*, 1983; Piepers *et al.*, 1984) that under increasing pressure, Geldart's Type B powders undergo considerable expansion before bubbling. Figure 7.23 shows that with increasing pressure, the center-line porosity increases in the IIT two-dimensional bed with a jet (Gidaspow *et al.*, 1986c). This plot is similar to that displayed in Fig. 7.16 at atmospheric pressure and for higher jet velocities. There was good agreement between an experiment at 3.77 bars and the theory.

7.8.5 Fast Bubble

Davidson's fast bubble, i.e., a bubble with a velocity greater than the fluidizing velocity of the surrounding gas, was already shown to be contained in the

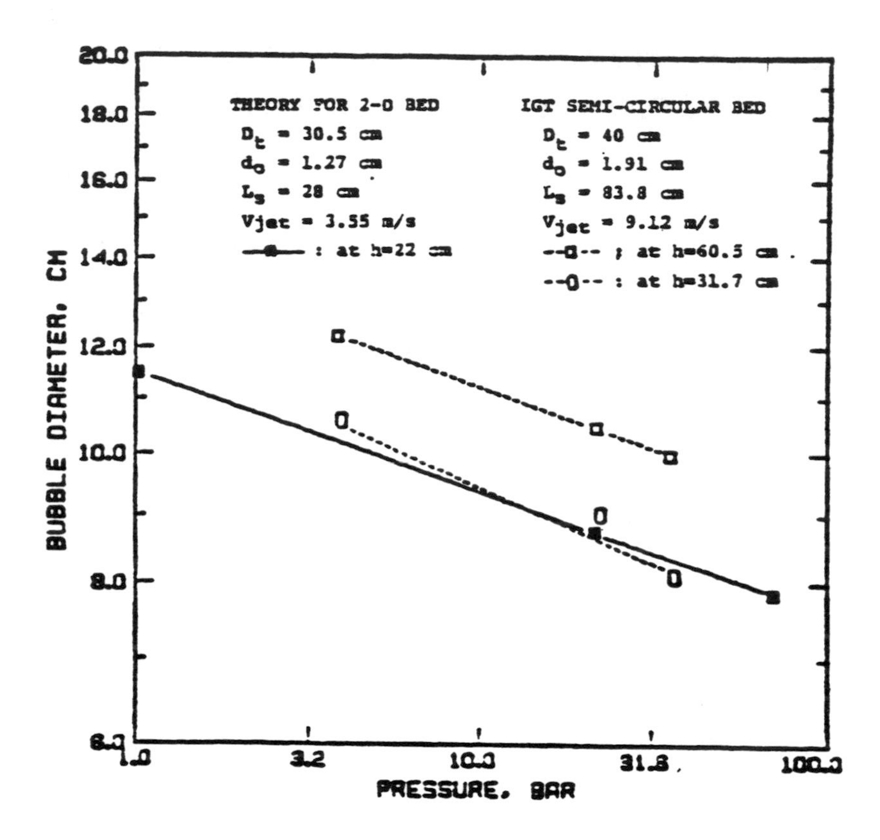

Fig. 7.22 Effect of pressure on bubble size (computed for a two-dimensional bed; experimental from Knowlton and Hirsan 1980 movie for a semicircular bed with an inlet section (from Gidaspow *et al.*, 1986c; also Seo, 1985).

hydrodynamic theory in paragraph 7.6. Hence, the computer model should give similar results. A bubble was computationally generated by starting with a two-dimensional bed filled with 500 μm particles at minimum fluidization, turning on a central jet to give a velocity of 5.7 m/s and injecting air until 0.15 s, then decreasing the jet flow back to the minimum fluidization velocity of 23.4 cm/s (Bouillard *et al.*, 1991). Such a computation corresponds to the experimental methods of generating a bubble. Figure 7.24 shows the formation, the propagation and the bursting of the fast bubble. The bubble volume remains approximately constant after its formation. After bubble eruption, a strong gas vortex is generated at the bed surface and a dome forms at the center of the bed. Figure 7.25 shows the computed bubble velocity as a function of time.

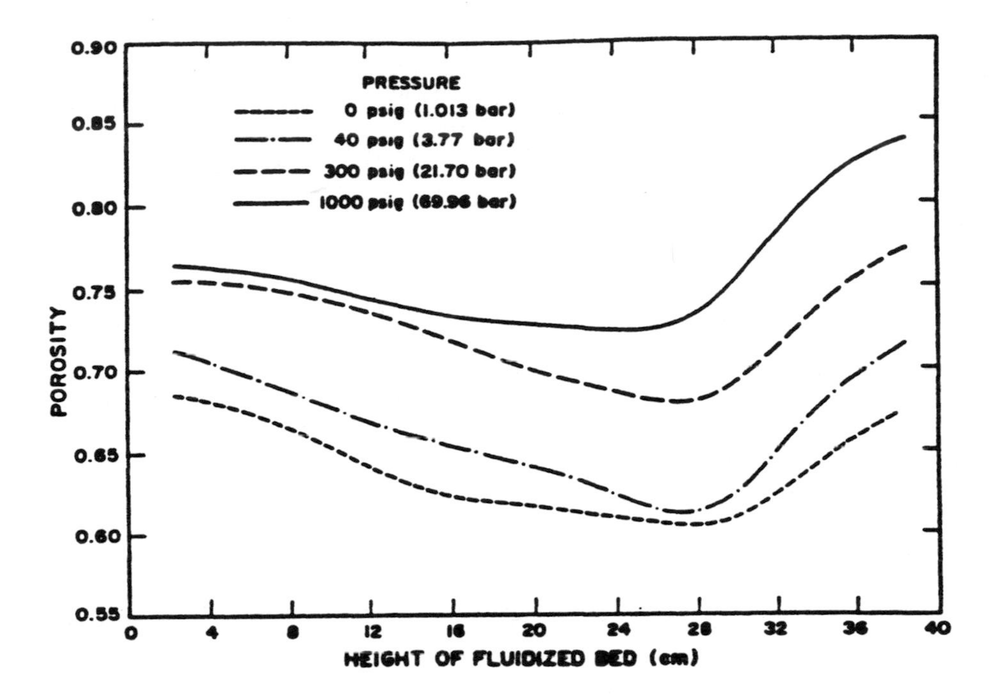

Fig. 7.23 Time averaged profiles along jet center in a two-dimensional fluidized bed, $d_p = 503$ μ, jet velocity = 3.55 m/s. (From Gidaspow *et al.*, 1986c).

The relationship between the bubble velocity u_b and the bubble radius r_B has been shown to be of the form (Grace, 1982)

$$u_b = C\sqrt{g r_B} \tag{7.87}$$

at minimum fluidization, and

$$u_b = u_g - u_{mf} + C\sqrt{g r_B} \tag{7.88}$$

at gas velocities in excess of minimum fluidization. The bubble-velocity coefficient C was derived by Davies and Taylor (1950) and found to be ⅔. At 0.25 s, our computed bubble velocity is somewhat larger than that predicted by Eqs. (7.87) and (7.88). Our value is closer to that computed for Eq. (7.86), a correlation based on the work of Davies and Taylor (1950) and Grace (1982).

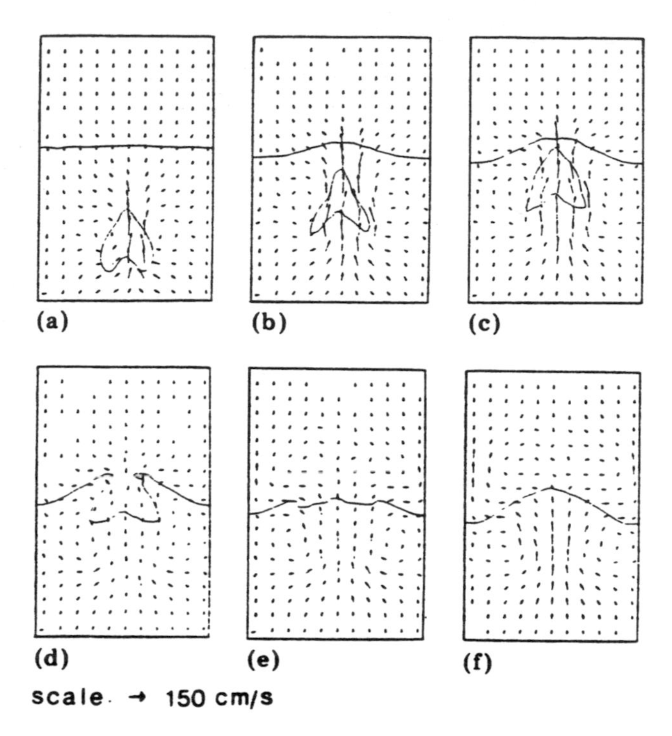

Fig. 7.24 Evolution and propagation of a two-dimensional fast bubble with superposed gas velocity vectors (a) 0.25 s, (b) 0.34 s, (c) 0.39 s, (d) 0.44 s, (e) 0.49 s. Slit jet is on for 0.15 s. Then, all flow is reduced to the minimum fluidization velocity. (From Bouillard *et al.*, 1991.)

From Fig. 7.24, at 0.25 s an equivalent bubble radius of r_B = 6.1 cm was estimated. Using this value, Eq. (7.86) predicts a bubble velocity of 78 cm/s, which is in fair agreement with the 83 cm/s predicted in our computations. This predicted value is the average bubble rise velocity over the period 0.2–0.3 s, at which time the bubble is in the middle of the bed and experiences minimum boundary influences. As can be seen for Figs. 7.24 and 7.25, the bubble never reaches a steady velocity, although its volume remains almost constant. Its shape, however, changes during its ascension in the bed and seems to be directly related to its changes in velocity, as if the bubble shape were a major aerodynamic factor of bubble velocity.

In view of Gidaspow's relative velocity model given in Chapter 2, one can derive a simplified expression for the bubble rise, which accounts for the shape of the bubble. Gidaspow's relative velocity equation can be written as

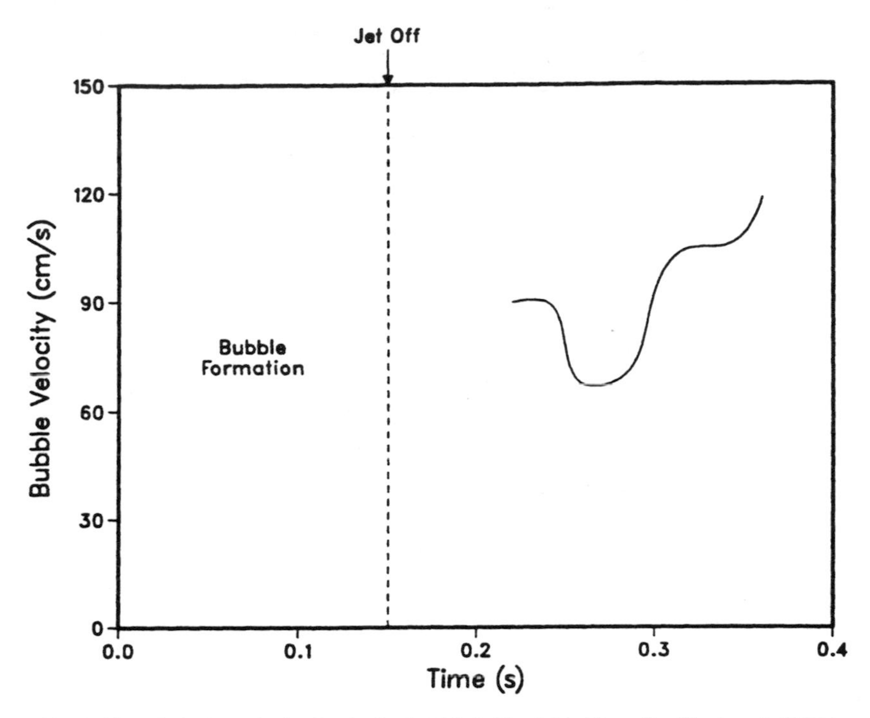

Fig. 7.25 Bubble rise velocity for the fast bubble in Fig. 7.24. (From Bouillard *et al.*, 1991.)

$$\tfrac{1}{2}\nabla \mathbf{v}_r^{\,2} = -\beta \mathbf{v}_r + \mathbf{g}, \tag{7.89}$$

where $\mathbf{v}_r = \mathbf{v}_g - \mathbf{v}_s$. Integrating Eq. (7.89) over the volume of the bubble V using Gauss's divergence theorem with the drag coefficient β equal to zero, because a porosity of $\varepsilon = 1$ within the bubble is assumed, one obtains

$$\iint\limits_{S(t)} \mathbf{v}_r^{\,2}\, dA = \iiint\limits_{V(t)} 2\mathbf{g}\, dV = 2\mathbf{g}V, \tag{7.90}$$

where $\mathbf{A}(t)$ is the surface area of the bubble.

If we assume $\mathbf{v}_g$ is approximately the bubble rise velocity u_b, and the solids phase is essentially motionless, we obtain

$$U_b = V_g = \left(\iint\limits_{A} \mathbf{v}_r^{\,2} \frac{dA}{A}\right)^{1/2} = \left(2g\frac{V}{A}\right)^{1/2}. \tag{7.91}$$

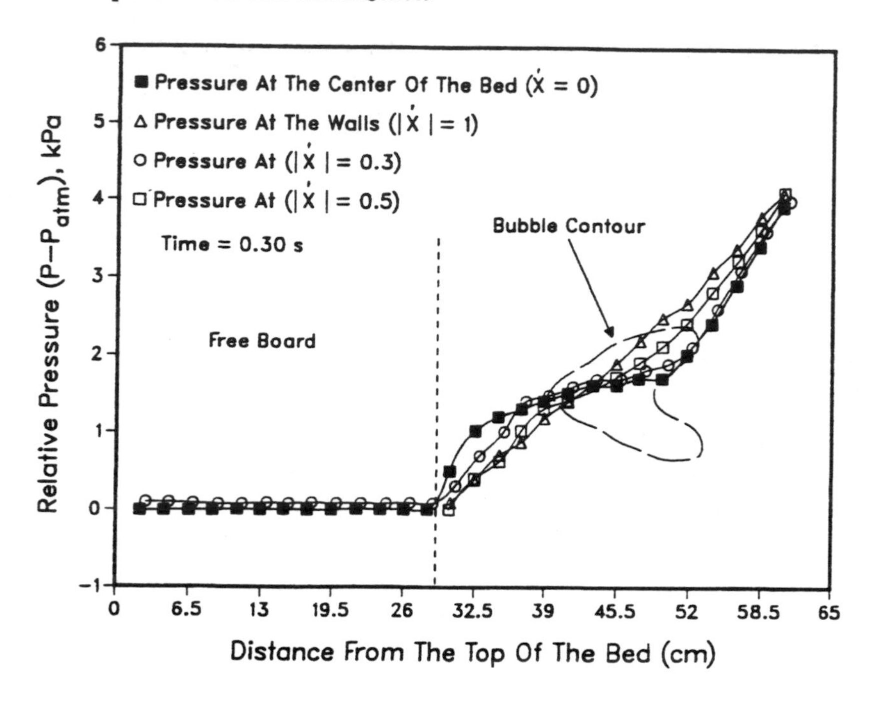

Fig. 7.26 Vertical pressure profiles at 0.30 s in two-dimensional fluidized bed with a fast bubble. (From Bouillard *et al.*, 1991.)

For a spherical bubble, Eq. (7.91) gives $u_b = 1\sqrt{3}(gD_b)^{1/2}$, predicting a bubble velocity that is bounded by Davies and Taylor's approximate formula ($u_b = 0.47(gD_b)^{1/2}$ and their experimental observations ($u_b = 0.711(gD_b)^{1/2}$. As can be seen from Eq. (7.91), the bubble rise velocity is a function of the ratio of the surface to the perimeter in two dimensions. We expect a maximum bubble rise velocity when the bubble is almost spherical or cylindrical. This behavior is indeed qualitatively predicted by our hydrodynamic model. Comparison of Fig. 7.24 with Fig. 7.25 shows that the bubble has a high velocity when its shape is almost spherical (0.25 s), slows down when the ratio of its surface to its length is at a minimum (0.3 s), and re-accelerates with a larger surface-to-length ratio.

Computed pressure profiles are analyzed at 0.30 s (in Fig. 7.26). Let x' represent the ratio of distance from the bed center to the bed half-width $x/(W/2)$. At the center, $x' = 0$; at the sidewalls, $|x'| = 1$. In Fig. 7.26 the pressure plotted is

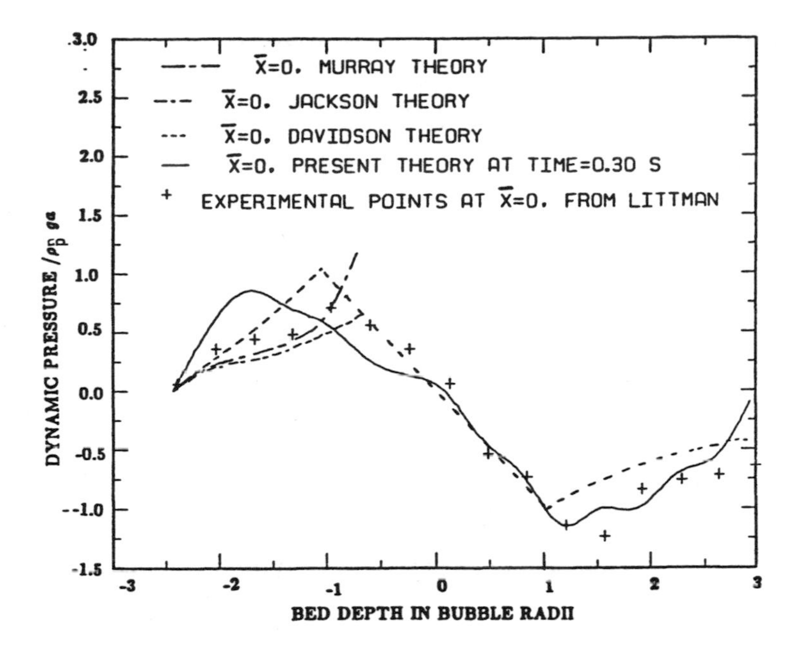

Fig. 7.27 Comparison of vertical dynamic pressure profiles at the center of a two-dimensional fluidized bed with fast bubble to Littman and Homolka's measurements (Bouillard *et al.*, 1991).

the relative pressure $(P - P_{\text{atm}})$, where $P_{\text{atm}} = 0.101$ MPa. The computed bubble has been superimposed in Fig. 7.26 (dashed lines) to show its position and shape at 0.30 s. The maximum bubble width is at $|x'| = 0.3$. The pressure is constant in the free-board region, and increases with the depth of the bed in the dense phase region because of the presence of solids.

The pressures at $|x'| = 0.5$ and $|x'| = 1.0$ increase almost linearly with the depth of the bed. On the other hand, the pressures at the center of the bed, at $|x'| = 0$, and at $|x'| = 0.3$ first increase in the region above the bubble, are virtually constant from the top to the bottom of the bubble and increase again in the wake of the bubble. In the top and bottom regions of the bubble, pressure gradients are large and characteristic of significant solids motion in these regions, as shown in Fig. 7.24.

Normalized dynamic pressure profiles, $P_d / (\rho_p g r_B)$, are shown at 0.30 s in Fig. 7.27. The parameter $\bar{x}$ represents the ratio of the lateral distance from the center of the bed to the bubble radius r_B. In these plots, four theories — those of

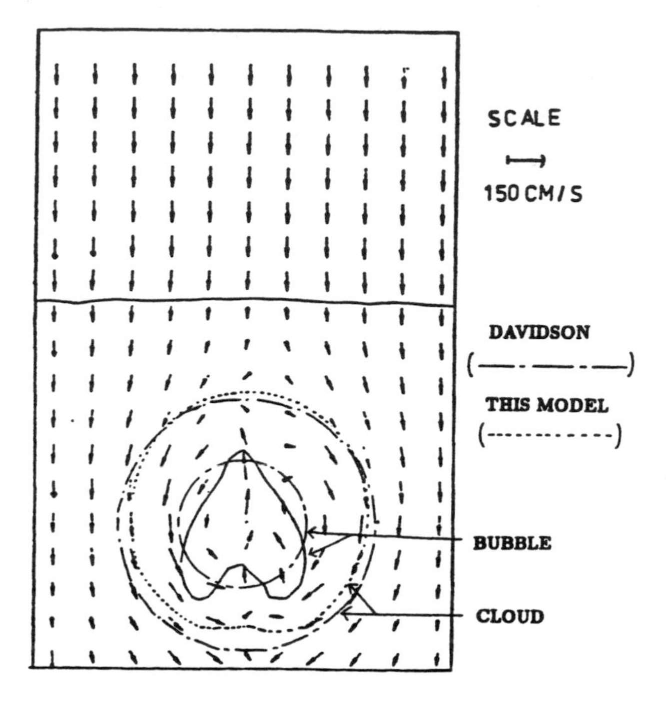

Fig. 7.28 A comparison of the predictions of Davidson's model and the inviscid hydrodynamic Model A of the cloud around a fast bubble in a two-dimensional bed at 0.25 s (also represented are the relative gas velocity vectors, $v_g - u_b$, and the equivalent cylindrical bubble).

Davidson (1961), Jackson (1963), Murray (1965) and Bouillard *et al.* (1991) — are compared with the experimental data of Littman and Homolka (1973).

Davidson's theory in Fig. 7.27 is presented as a plot of Eq. (7.79) at a dimensionless lateral distance from bubble center, $\bar{x}=0$. In Fig. 7.27 for $\bar{x}=0$, i.e., going through the bubble's center, Eq. (7.79) is valid only up to the bubble boundary. Here there is a discontinuity in the slope of the dynamic pressure. Inside the bubble where the pressure is assumed to be constant in the Davidson theory, the dimensionless dynamic pressure, from its definition given by

$$\frac{P_d}{\rho_m g r_B} = \frac{P - \rho_{mg}(H-y)}{\rho_m g r_B}, \tag{7.92}$$

varies linearly with y.

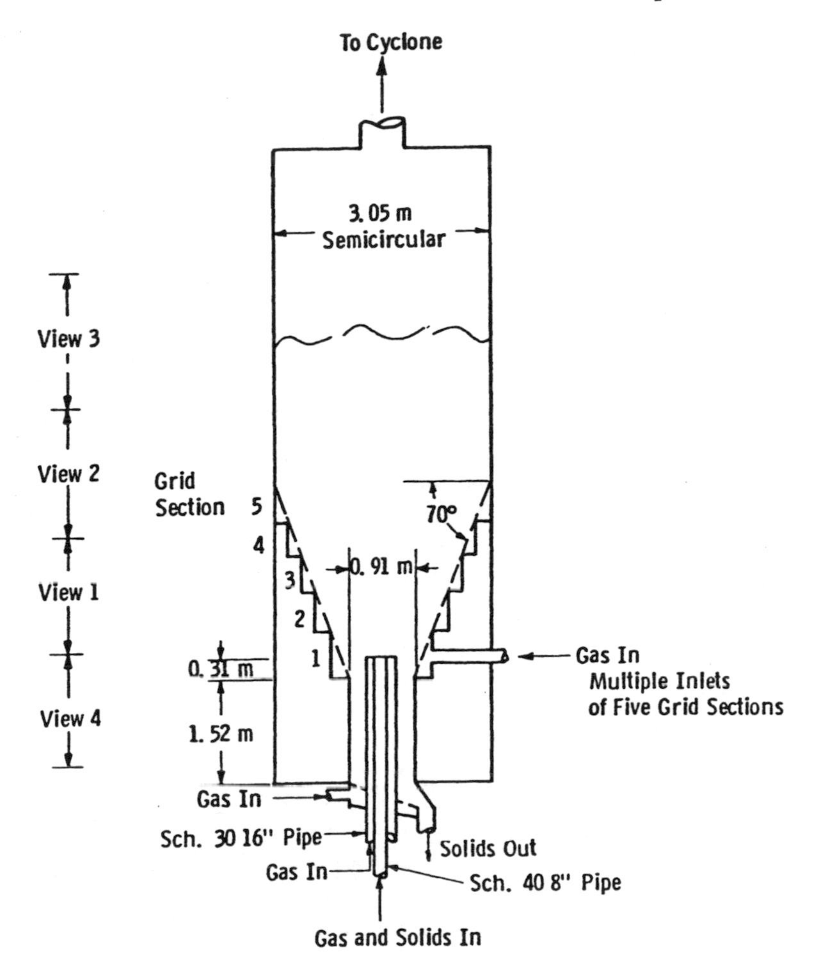

Fig. 7.29 Schematic of Westinghouse KRW 3-meter model. (From Yang *et al.*, 1984.)

The inviscid hydrodynamic model clearly predicts the measured trends of the dynamic pressure and compares, as well as can be expected, to the idealized circular bubble model of Davidson. Note that the computer model predicts the bubble velocity, while Davidson's theory requires it as an input.

Figure 7.28 shows the cloud formation predicted by the computer model. The cloud is defined as being a surface of zero normal velocity. In Fig. 7.28, the gas velocity relative to the bubble velocity is represented by vector plots. Our computer cloud radius is in good agreement with that predicted from Davidson's theory by Eq. (7.85). The computer cloud shape is almost circular, except in the lower region of the bubble, where it follows the bubble shape. The bubble shape predicted here is very similar to that used by Murray (1965).

7.8.6 3-Meter KRW Fluid Bed

A large-scale test facility in support of fluidized bed gasifier development was built and tested at Waltz Mill, Pennsylvania by Westinghouse Synthetic Fuel Division, which later became KRW (Yang *et al.*, 1981, 1984).

The vessel, as shown in Fig. 7.29, is semi-circular in cross-section, 3 m in diameter, with transparent Plexiglas plates at the front, and transparent windows at the circumferential side for flow visualization. Bed depths up to 6 m, gas velocities up to 4 m/sec, and particle sizes up to 0.6 cm can be used. Crushed acrylic plastic particles of a density and shape comparable to those of coal char particles were used. There are two semi-circular jet nozzles. The outer jet was designed to carry gas flow alone, while the inner jet was designed to simulate pneumatic feeding of coal. The conical grid is divided into five separate aeration sections.

Data on jet penetration, bubble diameter and bubble frequency were obtained through frame-by-frame analysis of high speed movies of either 24 or 64 frames/s. The data on bubble diameter were obtained by projecting movies frame by frame onto a screen with measuring grids.

The bubble frequency was obtained by counting the total number of bubbles and the total number of movie frames. From the known filming rate, the bubble frequency was determined. Table 7.4 shows some typical experimental conditions of the fluidized bed rig, while Fig. 7.30 presents Yang *et al.* (1984) correlations for the bubble volume. Following Davidson and Harrison (1963), Yang *et al.* (1984), assume that the bubble volume in a bed at minimum fluidization can be obtained by dividing the jet gas flow rate, G in Fig. 7.30, by n, the measured bubble frequency. Their correlations produce substantially lower bubble volumes. They account for the difference by postulating a leakage from the bubble to the remainder of the bed which may have been below minimum fluidization. A proper hydrodynamic model must give the correct bubble volume, frequency, jet penetration, etc.

A comparison of Yang *et al.*'s data to a simulation using the CHEMFLUB (Schneyer *et al.*, 1981) model was published by Richner *et al.* (1990). Table 7.2 gives the equations already described in paragraph 7.7. Since the solids viscosity used probably played only a minor role, the results are discussed in this inviscid chapter for completeness. Figure 7.31 shows the growth, the propagation and the bursting of the large bubble corresponding to the experimental conditions shown in Table 7.4. Note that the second to the last frame shows the bubble to be of toroidal shape, as computed by Ettahadieh *et al.* (1984) for the Westinghouse small semi-

Table 7.4 EXPERIMENTAL CONDITIONS IN THE KRW 300 cm DIAMETER COLD FLOW RIG TEST SERIES TP-M010-1 SET POINT 15

Particle diameter	1416 μm
Particle density	1.19 g/cm^3
Mass flow, air tube	68.2 g/s
Mass flow, annulus	11.1 g/s
Mass flow, shroud	4.4 g/s
Mass flow, conical grid	
Chamber 1	14.2 g/s
Chamber 2	28.6 g/s
Chamber 3	50.7 g/s
Mass flow, solids	9.3
transport	
Solids feed	0.0
Superficial velocity	82 cm/s
Reactor pressure	1.0 atm
Reactor temperature	293 K

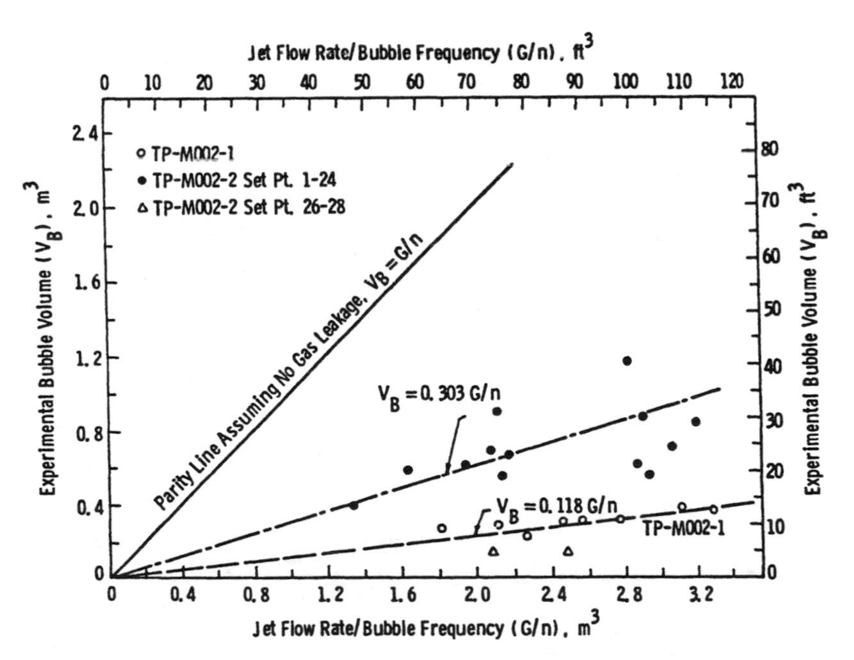

Fig. 7.30 Experimental data on gas leakage from the bubble to the emulsion phase. (From Yang *et al.*, 1984.)

circular bed with a jet. Richner *et al's.*, (1990) quantitative comparison of their calculations to Yang *et al.* data is shown in Table 7.5. The agreement is excellent except for the lower computed jet frequency, which may be due to the deletion of the gas momentum inertia in their model.

The good agreement between the experiments discussed in this section and two independently developed hydrodynamic models shows the potential utility of these models for scale-up and design of fluidized bed reactors.

Table 7.5 KRW 300 cm DIAMETER COLD FLOW RIG DATA AND CHEMFLUB CALCULATIONS

	CHEMFLUB	Exp.
Bubble radial dimension (cm)	96	94 ± 25
Bubble axial dimension (cm)	130	143 → 192*
Bubble velocity (cm/s)	310	185 ± 72
Bubble frequency (rate/min)	30	33
Jet penetration (cm)	180	160 ± 19
Jet frequency (rate/min)	90	117 → 135*

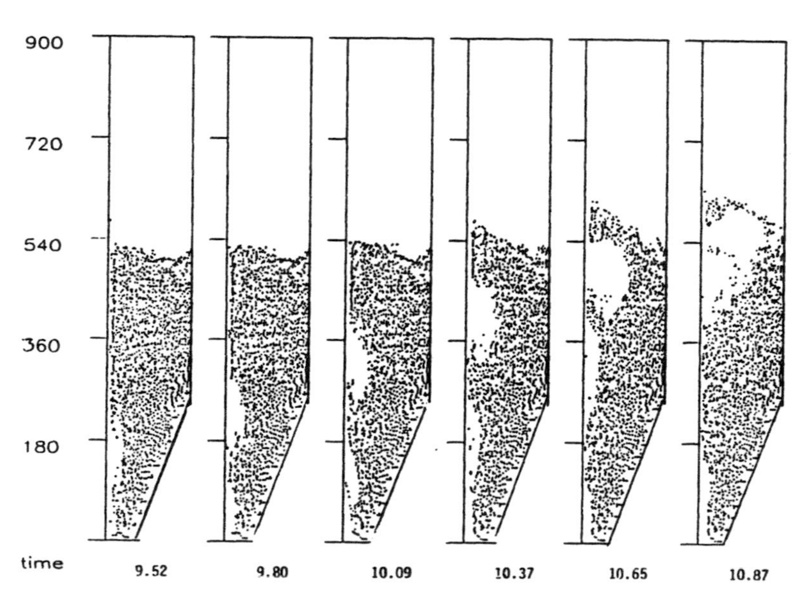

Fig. 7.31 CHEMFLUB prediction of bubbles in KRW 3-meter semi-circular fluid bed. (From Richner *et al.*, 1990. Reproduced by permission of the American Institute of Chemical Engineers. Copyright © 1990 AIChE. All rights reserved.)

Exercises

1. *Conical Moving-Bed Shale Retort Model*

Formulate two-dimensional steady state partial differential equations for the moving bed shale retort described below to give a description of the solid and gas flow, the solid and the gas temperatures and the kerogen conversion.

Model

The model, as described by Anderson *et al.* (1990), is as follows: The geometry of the conical retort shown in Fig. 7.32 reflects the basic features of the retort. The shell has a lower cone with walls slightly less steep than the upper cone. Shale is fed in the large-scale unit with a 3.05 m diameter piston. The pile of shale on top slopes with an assumed angle of repose of 38°. The boundaries of the present two-dimensional model (cylindrically symmetrical about the retort centerline) are shown in Fig. 7.33. Retort gas exit slots are located on the lower cone. An oil seal prevents gas flow through the base of the retort.

The model includes a description of gas and solids flow along with heat transfer and kerogen decomposition. Several simplifications and assumptions has been made in formulating the model. These include

- bed voidage and shale size are uniform,
- mass depletion of shale due to retorting (~10 wt. %) is neglected,
- additional gas flow from vaporized shale oil is neglected,
- condensed liquid oil flow is neglected,
- enthalpies of retorting, vaporization, or condensation are neglected in comparison to sensible heat transfer,
- heat conduction in each phase is negligible,
- intraparticle temperature gradients are neglected,
- constant gas and solid heat capacities,
- all physical properties are independent of variation in composition,
- raw shale has uniform kerogen content.

Boundary Conditions: Gas Flow

There is no flow across boundaries B_2, B_4, B_5, Fig. 7.33, which means they are streamlines with constant values of ψ. Because all gas enters at B_1 and exits across B_3, the streamline definition gives at steady state

$$\psi(B_2) = \psi(B_5) - \frac{m_g}{2\pi} \tag{7.93}$$

in cylindrical geometry, where m_g is the total gas mass flow rate. We set as the

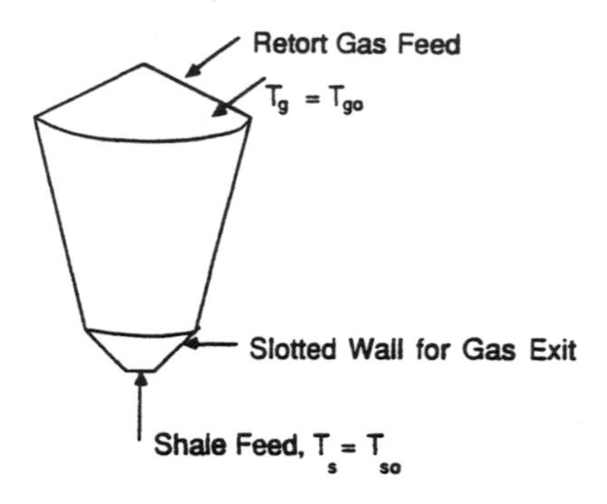

Fig. 7.32 Conical moving-bed shale retort. (From Anderson *et al.*, 1990. Reproduced by permission of the American Institute of Chemical Engineers. Copyright © 1990 AIChE. All rights reserved.)

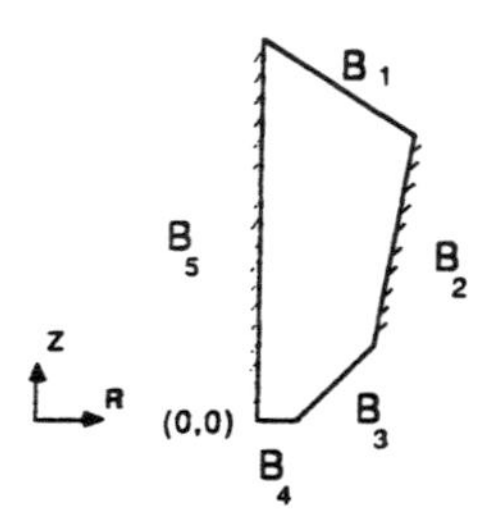

Fig. 7.33 Two-dimensional model boundaries. (From Anderson *et al.*, 1990. Reproduced by permission of the American Institute of Chemical Engineers. Copyright © 1990 AIChE. All rights reserved.)

$$\psi(B_4) = \psi(B_5) = 0 \tag{7.94}$$

reference streamline. Since pressure is uniform along boundaries B_1 and B_3, the gas velocity will be normal (perpendicular) to these faces. In terms of the stream function,

$$\begin{aligned} \sin\theta_1 \frac{\partial\psi}{\partial r} - \cos\theta_1 \frac{\partial\psi}{\partial z} = 0 \quad \text{on} \quad B_1, \\ \sin\theta_3 \frac{\partial\psi}{\partial r} - \cos\theta_3 \frac{\partial\psi}{\partial z} = 0 \quad \text{on} \quad B_3, \end{aligned} \tag{7.95}$$

where θ_i is the angle between the boundary and the horizontal direction.

Boundary Conditions: Solids Flow

On specifying a vertical entrance to the bed ($z = 0$),

$$\frac{\partial \psi_s}{\partial z} = 0, \quad z = 0. \tag{7.96}$$

The centerline ($r = 0$) is a streamline in cylindrical symmetry. It is specified as the reference streamline, that is,

$$\psi_s = 0, \quad r = 0. \tag{7.97}$$

As in gas flow,

$$\psi_s = \frac{m_s}{2\pi} \tag{7.98}$$

along the retort wall, where m_s is the total shale mass flow rate.

Shale velocity at the slanted face of the pile, B_1, is initially assumed to be normal to that face:

$$\sin\theta_1 \frac{\partial \psi_s}{\partial r} - \cos\theta_1 \frac{\partial \psi_s}{\partial z} = 0 \quad \text{on} \quad B_1. \tag{7.99}$$

(a) Formulate the two-dimensional partial differential equations to describe gas and solids flow in the retort. Solids flow is irrotational. What justifies this assumption?
(b) Formulate the equations for the gas temperature T_g and for the solids temperature T_s. The rate of heat transfer between the gas and the solid particles is given by $hA(T_g - T_s)$.
(c) Formulate the equation for the change of kerogen in the particle in terms of F, the fraction of initial kerogen remaining in the particle, where $1 - F$ is the conversion. The rate of kerogen decomposition into oil is given by

$$\text{Rate} = \rho_s \varepsilon_s k_o \exp[-E/RT_s] \cdot F. \tag{7.100}$$

(Reference: Anderson *et al.*, 1990.)

2. *Desalting by Electrosorption: First-Order PDE*

In desalination by electrosorption, a salt from, say, sea water is sorbed on a matrix when an electric potential is applied across the porous matrix. A. M. Johnson and John Newman (1971) have shown that the rate-controlling process is the unsteady flow of current due to capacitive and resistive effects in the matrix. During the sorption period they approximate the solution of the diffusion equation in the matrix by saying that the current decreases roughly as $1/\sqrt{t}$, where t is the time counted from the time the sorption began. With these approximations the material balance for the salt in a thin channel can be written as

$$\frac{\partial C}{\partial t}+v\frac{\partial C}{\partial x}=-\frac{\mathrm{K}}{\sqrt{t}}, \tag{7.101}$$

where v is the velocity in the channel, C is the salt concentration, x is the length coordinate, t is time, and K is a constant, proportional to potential difference and a function of the conductivities and the capacity of the matrix.

Sea water of concentration C_O has been flowing through the channel with no desalting. At time $t = 0$, an electric potential is applied that causes sorption according to the mechanism described above. The inlet concentration is still maintained equal to the initial concentration C_O.

(a) If the length of the channel is L, find the outlet concentration as a function of time.
(b) What is the outlet concentration for times much larger than the residence time L/v?

3. *Application of the Theory of Characteristic Mass Transfer in a Fluidized Bed*

According to Levenspiel and others, mass transfer in a fluidized bed may be treated as taking place between a bubble and an emulsion phase. Using some constant mass transfer coefficient k for transfer of a solute of concentration C_B in the bulk phase and C_E in the emulsion phase, the differential equations are

$$\frac{\partial C_B}{\partial t}+v_B\frac{\partial C_B}{\partial z}=k\left(C_E-C_B\right)$$
$$\text{and} \tag{7.102}$$
$$\frac{\partial C_E}{\partial t}+v_E\frac{\partial C_E}{\partial z}=k\left(C_B-C_E\right),$$

where v_B and v_E are the constant velocities of the bubble and the emulsion phases, respectively, and z is the bed height coordinate.

Suppose a tracer is injected at the bottom of the bed at $z = 0$ such that the scaled concentrations are both unity from time zero to a unit time. At $t = 1$ the tracer supply is cut off to zero sharply. Initially there is no tracer contained in either phase.

Determine the tracer distribution in both phases as a function of time and bed height.

(a) Set up the problem to be solved by the method of characteristics. Indicate the regions in which you know the solution without calculation.
(b) Find C_B and C_E for the bubble phase velocity twice the velocity of the emulsion phase. Let k and v_E both be one.
(c) Express the concentrations in closed form using Riemann's integration method.

Note: This problem is mathematically equivalent to the problem of unsteady state in heat exchangers (see Carslaw and Jaeger, 1959, p. 197).

4. *Counterflow Unsteady Heat Exchanger: Characteristics*

The unsteady operation of a double pipe type heat exchanger is governed by the following differential equations:

$$\frac{\partial T_c}{\partial t}+U_c\frac{\partial T_c}{\partial z}+a\left(T_c-T_h\right)=0,\quad \frac{\partial T_h}{\partial t}-U_h\frac{\partial T_h}{\partial z}-b\left(T_c-T_h\right)=0,\quad (7.103)$$

where

a, b = overall coefficient of heat transfer divided by the respective heat capacities of the fluids
T_c = temperature of the "cold" stream
T_h = temperature of the "hot" stream
t = time
U_c = velocity of the "cold" stream
U_h = velocity of the "hot" stream
z = length coordinate

Both streams are initially at zero temperature. At time zero, hot fluid enters the exchanger at $z = 1.0$. The entering temperature of this fluid is maintained at $T = 1.0$ until $t = 2.0$. Then the stream is switched back to the cold fluid at $T = 0$. In the meanwhile the "cold" stream was maintained at zero degrees.

Indicate how to find the temperature distribution of the two streams in the heat exchanger. Give the temperature in all the regions in which it is known without calculation.

5. *The Method of Characteristics*

Homogeneous Equilibrium Flow

The continuity and the momentum equations for one-dimensional transient one-component equilibrium flow are

$$\begin{pmatrix} \dfrac{\partial \rho_m}{\partial t} \\[2ex] \dfrac{\partial v}{\partial t} \end{pmatrix} + \begin{pmatrix} v & \rho_m \\[2ex] \dfrac{C^2}{\rho_m} & v \end{pmatrix} \begin{pmatrix} \dfrac{\partial \rho_m}{\partial x} \\[2ex] \dfrac{\partial v}{\partial x} \end{pmatrix} = \begin{pmatrix} 0 \\[2ex] g - \tau \end{pmatrix}, \tag{7.104}$$

where ρ_m is the mixture density, C is the two-phase equilibrium sound speed equal to $(\partial P / \partial \rho_m)_{Sm}$, g is gravity and τ is wall shear. In the above equation t is time, x the distance, and v the velocity.

Find the characteristics and the equation along the characteristics for the dependent variables ρ_m and v.

A Supersonic Nozzle: Characteristics

The governing equations of steady two-dimensional irrotational isentropic flow can be shown to reduce to the following pair of first-order equations for u and v:

$$\left(u^2 - c^2\right)\frac{\partial u}{\partial x} + uv\left(\frac{\partial u}{\partial y} + \frac{\partial v}{\partial x}\right) + \left(v^2 - c^2\right)\frac{\partial v}{\partial y} = 0, \quad \frac{\partial v}{\partial x} = \frac{\partial u}{\partial y}. \tag{7.105}$$

Obtain the characteristic directions. Note when the characteristics are real.

Isothermal Diffusion of Perfect Gases

For ideal gases the equations of motion relative to a fixed reference for constituent i are

$$\rho_i \mathbf{a}_i + \nabla P_i = \sum_{k=1}^{n} R_{ik} \rho_i \rho_k \left(\mathbf{v}_k - \mathbf{v}_i\right) + \rho_i \mathbf{f}_i . \tag{7.106}$$

This equation is Truesdell's (1962) Eq. (3). The momentum balance for the ith component and the corresponding mass balance for one-dimensional propagation can be written in matrix form as follows:

$$\begin{pmatrix} \dfrac{\partial v_i}{\partial t} \\ \\ \dfrac{\partial C_i}{\partial t} \end{pmatrix} + \begin{pmatrix} v_i & \dfrac{RT}{\rho_i} \\ \\ C_i & v_i \end{pmatrix} \begin{pmatrix} \dfrac{\partial v_i}{\partial x} \\ \\ \dfrac{\partial C_i}{\partial x} \end{pmatrix} = \begin{pmatrix} \sum_{k=1}^{N} R_{ik} J_k + f_i \\ \\ 0 \end{pmatrix}, \tag{7.107}$$

where v_i is the velocity of the ith species and C_i is the molar concentration related to the mass density through the molecular weight M_i by means of the relation

$$\rho_i = C_i M_i, \tag{7.108}$$

and to the partial pressure by means of the ideal gas law

$$P_i = C_i RT. \tag{7.109}$$

In Eq. (7.107), R_{ik} are the Onsager type friction factors and J_k are the kth fluxes relative to a barycentric average velocity. The external forces acting on species i are denoted by f_i. The independent variables are the time t and the distance x.

(a) Find the characteristic directions for the system of partial differential equations given by Eq. (7.107).
(b) Prescribe some initial and boundary data that will give a unique solution to some diffusion problem given by Eq. (7.107).
(c) Determine the equations along the characteristics for the case of small kinetic energy and constant temperature. Such a simplification is achieved by setting v_i to zero in the square matrix in Eq. (7.107) in the momentum equation (Eq. 7.107).

6. *Homogeneous Two Phase Flow*

The continuity equation for each phase with no mass transfer between phases for one-dimensional flow is

$$\frac{\partial(\rho_i \varepsilon_i)}{\partial t} + \frac{\partial(\rho_i \varepsilon_i v_i)}{\partial x} = 0, \quad i = 1, 2, \tag{7.110}$$

where the thermodynamic density of each phase ρ_i is a function of pressure P:

$$\rho_i = \rho_i(P). \tag{7.111}$$

In the homogeneous model it is assumed that the phases move at the same velocity v. The overall momentum equation in terms of the mixture density

$$\rho_m = \rho_i \varepsilon_i + \rho_j \varepsilon_j \tag{7.112}$$

is

$$\rho_m \frac{dv}{dt} + \frac{\partial P}{\partial x} = 0 \tag{7.113}$$

for frictionless flow. Using the definitions of sonic velocities C_i for each phase,

$$C_i^{-2} = \left(\frac{\partial \rho_i}{\partial P}\right) \tag{7.114}$$

and the relation $\varepsilon_i + \varepsilon_j = 1$.

Equations (7.110) and (7.113) can be written in matrix form as

$$\begin{pmatrix} \rho_i & \varepsilon_i/C_i^2 & 0 \\ -\rho_j & \varepsilon_j/C_j^2 & 0 \\ 0 & 0 & \rho_m \end{pmatrix} \begin{pmatrix} \dfrac{\partial \varepsilon_i}{\partial t} \\ \dfrac{\partial P}{\partial t} \\ \dfrac{\partial v}{\partial t} \end{pmatrix} + \begin{pmatrix} \rho_i v & \varepsilon_i v/C_i^2 & \varepsilon_i \rho_i \\ -\rho_j v & \varepsilon_j v/C_j^2 & \varepsilon_j \rho_j \\ 0 & 1 & \rho_m v \end{pmatrix} \begin{pmatrix} \dfrac{\partial \varepsilon_i}{\partial x} \\ \dfrac{\partial P}{\partial x} \\ \dfrac{\partial v}{\partial x} \end{pmatrix} = \begin{pmatrix} 0 \\ 0 \\ 0 \end{pmatrix}. \tag{7.115}$$

Find the characteristics.

7. ***Homogeneous Equilibrium Boiling***

In terms of the variables P, v, and s, transient homogeneous boiling in a pipe can be shown to be given by the equations below:

$$\begin{pmatrix} \dfrac{\partial P}{\partial t} \\ \dfrac{\partial v}{\partial t} \end{pmatrix} + \begin{pmatrix} v & \rho C^2 \\ \dfrac{1}{\rho} & v \end{pmatrix} \begin{pmatrix} \dfrac{\partial P}{\partial x} \\ \dfrac{\partial v}{\partial x} \end{pmatrix} = \begin{pmatrix} -\left(\dfrac{\partial \rho}{\partial s}\right)_P & \dfrac{C^2(\dot{q} + f_w v)}{\rho T} \\ f_w/\rho & -g \end{pmatrix} \tag{7.116}$$

and

$$\frac{ds}{dt} = \frac{\dot{q} + f_w v}{\rho T}, \text{ where } \frac{d}{dt} = \frac{\partial}{\partial t} + v \frac{\partial}{\partial x} \tag{7.117}$$

and $C^2 = (\frac{\partial P}{\partial \rho})_s$.

(a) Find the characteristic directions.
(b) Find the equations along the characteristics in terms of the variables P and v.
(c) For a pipe of length L, initially containing pressurized water at a pressure P_{in} and a sudden break at $x = L$, formulate the boundary conditions for flow into an atmosphere at a pressure P_a.
(d) Show how to solve the equations.

At a constant pressure with a negligible frictional dissipation, the energy equation in three dimensions can be written as

$$\frac{\partial T}{\partial t} + u\frac{\partial T}{\partial x} + v\frac{\partial T}{\partial y} + w\frac{\partial T}{\partial z} = \frac{\dot{q}}{\rho C_P}. \tag{7.118}$$

(e) Express Eq. (7.118) as a set of ordinary differential equations.

8. *Ill-Posedness for Model A*

For the 5th International Heat Transfer Conference in Tokyo, Japan (Gidaspow, 1974), the opinions of the world's experts on two-phase flow concerning the proper form of the equations were solicited. It was pointed out that the incompressible momentum balances with the pressure drop in each phase which were then favored by the majority of computer code developers had imaginary characteristics for unequal velocity flow. Show this to be true.

The equations considered were as follows, where α_i is the volume fraction of phase i, called ε_i in this book, following the conventional use of ε for porosity used by chemical engineers before the invention of two-phase flow theory:

$$\begin{pmatrix} 1 & 0 & 0 & 0 \\ -1 & 0 & 0 & 0 \\ 0 & 0 & \rho_i & 0 \\ 0 & 0 & 0 & \rho_f \end{pmatrix}\begin{pmatrix} \partial\alpha_i/\partial t \\ \partial P/\partial t \\ \partial v_i/\partial t \\ \partial v_j/\partial t \end{pmatrix} + \begin{pmatrix} v_i & 0 & \alpha_i & 0 \\ -v_j & 0 & 0 & \alpha_j \\ 0 & 1 & \rho_i v_i & 0 \\ 0 & 1 & 0 & \rho_j v_j \end{pmatrix}\begin{pmatrix} \partial\alpha_i/\partial x \\ \partial P/\partial x \\ \partial v_i/\partial x \\ \partial v_j/\partial x \end{pmatrix} = 0. \tag{7.119}$$

The step-by-step procedure is as follows.

The characteristic determinant for this system is

$$\begin{vmatrix} \lambda_0 + v_i\lambda_i & 0 & \alpha_i\lambda_i & 0 \\ -\lambda_0 + v_j\lambda_i & 0 & 0 & \alpha_j\lambda_i \\ 0 & \lambda_i & \rho_i\lambda_0 + \rho_i v_i\lambda_i & 0 \\ 0 & \lambda_i & 0 & \rho_j\lambda_0 + \rho_j v_j\lambda_i \end{vmatrix} = 0. \tag{7.120}$$

Evaluation of the characteristic determinant yields the expression

$$\begin{aligned}&(\lambda_o + v_i\lambda_1)(\alpha_j\lambda_1)(-\lambda_1)(\rho_i\lambda_o + \rho_i v_i\lambda_1)\\&\quad + (\alpha_i\lambda_1)(-\lambda_o - v_j)(\lambda_1)(\rho_j\lambda_o + \rho_j v_j\lambda_1) = 0.\end{aligned} \tag{7.121}$$

Let $\lambda = \lambda_1 / \lambda_o$. Two roots λ are then zero. The remaining two are given by

$$(\alpha_j\rho_i v_i^2 + \alpha_i\rho_j v_j^2)\lambda^2 + 2(\alpha_j\rho_i v_i + \alpha_i\rho_j v_j)\lambda + (\alpha_j\rho_i + \alpha_i\rho_j) = 0 \tag{7.122}$$

The dicriminant D for this quadratic is

$$\begin{aligned}D &= 4(\alpha_j\rho_i v_i + \alpha_i\rho_j v_j)^2\\&\quad - 4(\alpha_j\rho_i v_i^2 + \alpha_i\rho_j v_j^2)(\alpha_j\rho_i + \alpha_i\rho_j) - \alpha_i\alpha_j\rho_i\rho_j(v_i - v_j)^2.\end{aligned} \tag{7.123}$$

Literature Cited

Anderson, D. H., A. V. Sapre, C. R. Kennedy, and F. J. Krambeck (1990). "Conical Moving-Bed Shale Retort Model," *AIChE J.* 36(6), 809–816.

Babu, S. P. S., S. A. Weil and L. Anich (1976). "Modeling of Coal Gasification Fluidized-Bed Reactors for Scale-up and Commercial Design," *Institute of Gas Technology Annual Report to the American Gas Association.*

Batcheler, G. K. (1954). "Heat Transfer by Free Convection across a Closed Cavity between Vertical Boundaries at Different Temperature," *Q. Appl. Math* **12**, 209–233.

Blake, T. R., D. H. Brownell, Jr. and G. P. Schneyer (1979). "A Numerical Simulation Model for Entrained Flow Coal Gasification I: The Hydrodynamic Model," *The Alternate Energy Sources Conference*. Miami, Florida.

Blake, T. R., D. H. Brownell, Jr., P. J. Chen, J. C. Cook and G. P. Schneyer (1980). "Computer Modeling of Coal Gasification Reactors," *Annual Report to the Department of Energy, Morgantown Energy Tech. Center, Contract No. DE-AC21-76ET10242.*

Bouillard, J. S., R. W. Lyczkowski, and D. Gidaspow (1989a). "Porosity Distributions in a Fluidized Bed with an Immersed Obstacle," *AIChE J.* **35**, 908–972.

Bouillard, J. S., R. W. Lyczkowski, S. Folga, D. Gidaspow, and G. F. Berry (1989b). "Hydrodynamics of Erosion of Heat Exchanger Tubes in Fluidized Bed Combustors," *Can. J. Chem. Eng.* **67**, 218–229.

Bouillard, J. S., D. Gidaspow, and R. W. Lyczkowski (1991). "Hydrodynamics of Fluidization: Fast-Bubble Simulation in a Two-Dimensional Fluidized Bed," *Powder Technol.* **66**, 107–118.

Callen, H. B. (1960). *Thermodynamics.* New York: John Wiley & Sons.

Carslaw, H.S., and J. C. Jaeger (1959). *Conduction of Heat in Solids,* 2nd ed. London, U.K.: Oxford University Press.

Churchill, S. W. (1966). Invited paper at the Third International Heat Transfer Conference, "The Prediction of Natural Convection," VI, 15–30. New York: AIChE.

Collins, R. (1982). "Cloud Patterns around a Bubble Growing in a Gas-Fluidized Bed," *J. Fluid Mech.* **104**, 297–303.

Cook, K. T., and F. H. Harlow (1986). "Vortices in Bubble Two-Phase Flow," *Int. J. Multiphase Flow* **12**, 35–61.

Courant, R. and D. Hilbert (1962). *Methods of Mathematical Physics Vol. II: Partial Differential Equations.* New York: Wiley-Interscience.

Davidson, J. R. (1961). "Symposium on Fluidization—Discussion," *Trans. Inst. Chem. Eng.* **39**, 230–232.

Davidson, J. F., and D. Harrison (1963). *Fluidized Particles.* Cambridge, U.K.: Cambridge University Press.

Davies, R. M., and Taylor, G. I. (1950). "The Mechanics of Large Bubbles Rising through Extended Liquids and through Liquids in Tubes," *Proc. Roy. Soc. A* **200**, 375.

Donsi, G. M., L. Massimilla, and L. Colantuoni (1980). "The Dispersion of Axi-symmetric Jets in Fluidized Beds," pp. 297–304 in J. R. Grace and J. M., Eds. *Fluidization: Proceedings of the 1980 Fluidization Conference, Henniker, New Hampshire.* New York: Plenum Press.

Drew, D. A. (1983). "Mathematical Modeling of Two-Phase Flow," *Ann. Rev. Fluid Mech.* **15**, 261–291.

Ettehadieh, B. (1982). "Hydrodynamics Analysis of Gas–Solid Fluidized Beds," Ph.D. Thesis. Chicago, Illinois: Illinois Institute of Technology.

Ettehadieh, B., D. Gidaspow, and R. W. Lyczkowski (1984). "Hydrodynamics of Fluidization in a Semi-Circular Bed with a Jet," *AIChE J.* **30**, 529-536.

Fanucci, J. B., N. Ness, and R. H. Yen (1979). "On the Formation of Bubbles in Gas-Particulate Fluidized Beds," *J. Fluid Mech.* **94** (2), 3533–3567.

Gidaspow, D. (1974). "Round Table Discussion (RT-1-2): Modeling of Two-Phase Flow," from the 5th International Heat Transfer Conference, Tokyo, Japan, pp. 163–168 in *Heat Transfer 1974,* Vol. VII. Japan: JSME-SCEJ.

Gidaspow, D. (1978). "Hyperbolic Compressible Two-Phase Flow Equations Based on Stationary Principles and the Fick's Law" pp. 283–298 in T. N. Veziroglu and S. Kakac, Eds. *Two-Phase Transport and Reactor Safety I*. Hemisphere Publishing Company.

Gidaspow, D. (1986). 1984 Donald Q. Kern Award presented at the 1985 Heat Transfer Conference. Lecture: "Hydrodynamics of Fluidization and Heat Transfer: Supercomputer Modeling," published in *Appl. Mech. Rev.* **39** (1), 1–23.

Gidaspow, D., and B. Ettehadieh (1983). "Fluidization in Two-Dimensional Beds with a Jet, Part II: Hydrodynamic Modeling," *I&EC Fund.* **22**, 193–201.

Gidaspow, D., C. Lin, and Y. C. Seo (1983a). "Fluidization in Two-Dimensional Beds with a Jet, Part I: Experimental Porosity Distributions," *I&EC Fund.* **22**, 187–193.

Gidaspow, D., Y. C. Seo, and B. Ettehadieh (1983b). "Hydrodynamics of Fluidization: Experimental and Theoretical Bubble Sizes in a Two-Dimensional Bed with a Jet," *Chem. Eng. Comm.* **22**, 253–272.

Gidaspow, D., M. Syamlal, and Y. C. Seo (1986a). "Hydrodynamics of Fluidization: Supercomputer Generated vs. Experimental Bubbles," *J. Powder & Bulk Solids Tech.* **10** (3), 19–23.

Gidaspow, D., M. Syamlal, and Y. C. Seo (1986b). "Hydrodynamics of Fluidization of Single and Binary Size Particles: Supercomputer Modeling," pp. 1–8 in *Fluidization V: Proceedings of the 5th Engineering Foundation Conference on Fluidization*. New York: AIChE Engineering Foundation.

Gidaspow, D., Y. C. Seo, and B. Ettehadieh (1986c). "Hydrodynamics of Fluidization: Effect of Pressure," *Particulate Sci. & Tech.* **4**, 25–43.

Grace, J. R. (1982). "Fluidized Bed Hydrodynamics," pp. 8.5–8.83 in G. Hetsroni, Ed. *Handbook of Multiphase Systems*. Washington, D.C.: McGraw-Hill Hemisphere.

Jackson, R. (1963). "The Mechanics of Fluidized Beds," *Trans. Inst. Chem. Eng.* **41**, 13–28.

Jenkins, J. T. and S. B. Savage (1983). "A Theory for the Rapid Flow of Identical Smooth, Nearly Elastic, Spherical Particles," *J. Fluid Mech.* **130**, 197–202.

Johnson, A. M., and J. Newman (1971). "Desalting by Means of Porous Carbon Electrodes," *J. Electrochem. Soc.* **118**, 510–517.

Kunii, D. and O. Levenspiel (1969). *Fluidization Engineering*. New York: John Wiley & Sons.

Lamb, H. (1945). *Hydrodynamics*. Dover Publications, 1st Ed., 1897.

Lax, P. D. (1958). "Differential Equations, Difference Equations and Matrix Theory," *Comm. Pure and Appl. Math.* **11**, 175–194.

Lax, P. D. and R. D. Richtmyer (1956). "Survey of Stability of Linear Finite Difference Equations," *Comm. Pure Appl. Math.* **9**, 267.

Littman, H. and G. A. J. Homolka (1973). "The Pressure Field around a Two-Dimensional Gas Bubble in a Fluidized Bed," *Chem. Eng. Sci.* **28**, 2231–2243.

Liu, Y. and D. Gidaspow (1981). "Solids Mixing in Fluidized Beds: A Hydrodynamic Approach," *Chem. Eng. Sci.* **36** (3), 539–547.

Lyczkowski, R. W. and D. Gidaspow (1971). "Two-Dimensional Inert Accumulation in Space H_2–O_2 Fuel Cells," invited paper presented by Gidaspow at the 20th Meeting of CITCE in Strasbourg, Sept. 14–20, 1969. Published in *AIChE J.* **17**, 1131–1140.

Lyczkowski, R. W. and D. Gidaspow, C. W. Solbrig, and E. D. Hughes (1978). "Characteristics and Stability Analyses of Transient One-Dimensional Two-Phase Flow Equations and Their Finite Difference Approximations," *Nucl. Sci. & Eng.* **66**, 378–396.

Lyczkowski, R. W., D. Gidaspow, and C. W. Solbrig (1982). "Multiphase Flow-Models for Nuclear, Fossil and Biomass Energy Production," pp. 198–351 in A. S. Mujumdar and R. A. Mashelkar, Eds. *Advances in Transport Processes.* New York: Wiley-Eastern.

Murray, J. D. (1965). "On the Mathematics of Fluidization, Part I: Fundamental Equations and Waves Propagation," and "On the Mathematics of Fluidization, Part II: Steady Motion of Fully Developed Bubbles," *J. Fluid Mech.* **22**, 57–80.

Pal, S. K. (1969). "Numerical Solution of First Order Hyperbolic Systems of Partial Differential Equations," Ph.D. Thesis. Toronto, Canada: University of Toronto.

Piepers, H. W., E. S. E. Cottaar, A. H. M. Verkooijen, and K. Rietma (1984). "Effects of Pressure and Type of Gas on Particle–Particle Interaction and the Consequences for Gas–Solid Fluidization Behavior," *Powder Technol.* **37**, 55–70.

Poots, G. (1958). "Heat Transfer by Laminar Free Convection in Enclosed Plane Gas Layers," *Quarterly J. Mech. Appl. Math.* **11**, 257–273.

Pritchett, J. W., T. R. Blake, and S. K. Garg (1978). "A Numerical Model of Gas Fluidized Beds," presented at the AIChE 69th Annual Meeting, Chicago, Illinois, December 2, 1976. Published in *AIChE Symp. Series Vol. No. 176*, **74**, 134–148.

Richner, D. W., T. Minoura, J. W. Pritchett, and T. R. Blake (1990). "Computer Simulation of Isothermal Fluidization in Large-Scale Laboratory Rigs," *AIChE J.* **36**, 361–369.

Richtmyer, R. D. and K. W. Morton (1967). *Difference Methods for Initial-Value Programs.* New York: Interscience Publishers, Inc.

Rowe, P. H. (1971). "Experimental Properties of Bubbles," Ch. 4 in J. F. Davidson and D. Harrison, Eds. *Fluidization.* New York: Academic Press.

Rowe, P. N., P. U. Foscolo, A. C. Hoffmann, and J. G. Yates (1983). "X-ray Observation of Gas Fluidized Beds under Pressure," in D. Kunii and R. Joei, Eds. *Proc. 4th Int'l Conf. on Fluidization, Japan 1983.* New York: AIChE.

Rudinger, G., and A. Chang (1964). "Analysis of Nonsteady Two-Phase Flow," *Phys. of Fluids* **7**, 1747–1754.

Savage, S. B. (1983). "Granular Flow at High Shear Rates," pp. 339–358 in R. E. Meyer, Ed. *Theory of Dispersed Multiphase Flow.* New York: Academic Press.

Savage, S. B. (1988). "Streaming Motions in a Bed of Vibrationally Fluidized Dry Granular Material," *J. Fluid Mech.*, **19** (4), 457–478.

Schneyer, G. P. E., W. Peterson, P. J. Chen, D. H. Brownell, and T. R. Blake (1981). "Computer Modeling of Coal Gasification Reactors," Final Report for June 1975-1980 DOE/ET/10247 Systems, Science, Software.

Sedney, R. (1970). "The Method of Characteristics," pp. 159–225 in P. P. Wagener, Ed. *Non-equilibrium Flows, Part II*. New York: Marcel Dekker, Inc.

Seo, Y. (1985). "Fluidization of Single and Binary Size Particles," Ph.D. Thesis. Chicago, Illinois: Illinois Institute of Technology.

Sobreiro, L. E. L., and J. L. F. Monteiro (1982). "The Effect of Pressure on Fluidized Bed Behaviour," *Powder Technol.* **33**, 95–100.

Truesdell, C. J. (1962). "Mechanical Basis of Diffusion," *J. Chem. Phys.* **37**, 2336–2344.

Tsuo, Y. P., and D. Gidaspow (1990). "Computations of Flow Patterns in Circulating Fluidized Beds," presented at the 26th National Heat Transfer Conference, Philadelphia, August 1989. Published in *AIChE J.* **36**, 885–889.

Vigil, J. C. (1979). TRAC Code, Report NUREG/CR-1059, LA-8059-MS; for related references, see S. Fabic (1982). Ch. 6.6 in G. Hetroni, Ed. *Handbook of Multiphase Systems*. New York: McGraw-Hill.

Von Mises, R. (1958). *Mathematical Theory of Compressible Fluid Flow*. New York: Academic Press.

Wen, C. Y. (1975). "Optimization of Coal Gasification Processes," R&D Report No. 66 to the Office of Coal Research.

Yang, W. C. and D. L. Keairns (1980). "Momentum Dissipation of and Gas Entrainment into a Gas Solid Two-Phase Jet in a Fluidized Bed," pp. 305 in J. R. Grace and J. Matsen, Eds. *Fluidization*. New York: Plenum Press.

Yang, W. C., S. S. Kim, and J. A. Rylatt (1981). "A Large-Scale Cold Flow Scale-up Test Facility," in *Coal Proc. Technol. VII*. New York, AIChE.

Yang, W. C., D. Revay, R. G. Anderson, E. S. Chelan, D. L. Keairns and D. C. Cicero (1984). "Fluidization Phenomena in a Large-Scale, Cold-Flow Model," pp. 77–84 in D. Kunii and R. Toei, Eds. *Fluidization*. New York: AIChE.

8

VISCOUS FLOW AND CIRCULATING FLUIDIZED BEDS

8.1 Introduction 197
8.2 Multiphase Navier–Stokes Equation Model 198
8.2.1 Incompressible Flow 199
8.2.2 Compressible Flow 200
8.3 Dimensional Analysis: Scale Factors 201
8.4 CFB or Riser Flow: Experimental 207
8.5 Need for Clusters in One-D Modeling 214
8.6 Computation of Cluster Flow 216
8.7 Computation of Core–Annular Regime 224
8.8 Radial Profiles and Turbulence 228
Exercises 233
1. Scale Factors and Flow Regimes for Homogeneous Two Phase Flow 233
2. Boundary Layer on a Plate in a Two Phase Flow "Wind Tunnel" 234
3. Developed Time-Averaged Two Phase Flow 235
Literature Cited 235

8.1 Introduction

In Chapter 7, it was shown that inviscid models predicted the formation, the growth and the bursting of bubbles in a reasonable agreement with experiments when a flow pattern was established by means of a jet. In such a case, time-averaged porosities agreed well with experiments. Transient and time-average wall-to-bed heat transfer coefficients were also computed in rough agreement with data using empirical effective thermal conductivities (Syamlal and Gidaspow, 1985), since the inviscid model gives an approximately correct residence time for particles contacting a wall (Gidaspow, 1986). However, analogous to classical fluid dynamics, inviscid models do not predict the forces acting on the tubes or obstacles immersed in fluidized beds. This can be made clear by using a simple example. Consider steady incompressible laminar boundary layer flow over a flat plate at zero incidence with flow at a constant velocity U_∞ far away from the plate of width b and length L. For a fluid of viscosity μ and density ρ, Schlichting (1960) shows that the drag on both sides of the plate is as follows:

$$\text{Drag} = 1.328 b U_\infty^{3/2} (\mu \rho L)^{1/2}. \tag{8.1}$$

Drag, as well as the momentum boundary layer thickness δ,

$$\delta \approx \left(\frac{\mu L}{\rho U_\infty} \right)^{1/2}, \tag{8.2}$$

are both proportional to the square root of the fluid viscosity. Hence, a fluid of zero viscosity will exert no drag on a plate immersed in the fluid. A wall will not cause the development of a momentum boundary layer. Hence, steady flows influenced by walls cannot be described by inviscid models. Flows in circulating fluidized beds have significant wall effects. Particles in the form of clusters or layers can be seen to run down the walls. Hence, modeling of circulating fluidized beds (CFB) without a viscosity is not possible. However, in interpreting Eqs. (8.1) and (8.2) it must be kept in mind that CFB or most other two phase flows are never in a true steady state. Then the viscosity in Eqs. (8.1) and (8.2) may not be the true fluid viscosity to be discussed next, but an Eddy type viscosity caused by two phase flow oscillations usually referred to as turbulence. In view of the transient nature of two-phase flow, the drag and the boundary layer thickness may not be proportional to the square root of the intrinsic viscosity but depend upon it to a much smaller extent. As another example, in liquid–solid flow and settling of colloidal particles in a lamella electrosettler (Gidaspow *et al.*, 1989b; Jayaswal *et al.*, 1990) the settling process is only moderately affected by viscosity. Inviscid flow with settling is a good first approximation to this electric field driven process.

The physical meaning of the particulate phase viscosity is described in detail in the chapter on kinetic theory. Here the conventional derivation presented in single phase fluid mechanics is generalized to multiphase flow.

8.2 Multiphase Navier–Stokes Equation Model

A generalization of the Navier–Stokes equation to multiphase flow can be obtained in a very simple way. The conservation of mass and momentum for each phase were given by Eqs. (7.1) and (7.2). For inviscid flow, the traction $\mathbf{T}_k$ was simply the negative of the pressure of phase k. Now consider flow with frictional forces due to differences in velocity. For simplicity, let us first treat incompressible flow. In multiphase flow, this means the bulk density of phase k is constant.

8.2.1 Incompressible Flow

In single phase flow the traction $\mathbf{T}$ is a function of the symmetric gradient of the velocity. The driving force for the transfer of shear is the symmetric gradient of velocity rather than the ordinary gradient because of the need to satisfy invariance under a change of frame of reference under rotation, called objectivity in continuum mechanics or the Galileo relativity principle. Similarly, there exists a driving force of $\mathbf{T}_k$ due to its gradient of velocity. To meet the requirement of objectivity, let

$$\mathbf{T}_k = \mathbf{T}_k\left(\nabla^s \mathbf{v}_k\right). \tag{8.3}$$

Linearization of the $\mathbf{T}_k$ gives

$$\mathbf{T}_k = A_k \mathbf{I} + B_k \nabla^s \mathbf{v}_k. \tag{8.4}$$

For incompressible fluids, A_k is chosen to be the negative of the pressure of fluid k and the derivative of the traction with respect to the symmetric gradient is the viscosity of fluid k, as shown below:

$$A_k = -P_k, \tag{8.5}$$

$$B_k = 2\mu_k = \frac{\partial \mathbf{T}_k}{\partial \nabla^s \mathbf{v}_k}. \tag{8.6}$$

Then, using the tensor identity

$$\nabla\nabla^s \mathbf{v} = \tfrac{1}{2}\nabla^2 \mathbf{v} + \tfrac{1}{2}\nabla\nabla\cdot\mathbf{v} \tag{8.7}$$

and the assumption of constancy of bulk density,

$$\varepsilon_k \rho_k = \text{constant}, \tag{8.8}$$

$$\nabla\cdot\mathbf{v}_k = 0. \tag{8.9}$$

The traction for phase k becomes as follows:

$$\mathbf{T}_k = -P_k\mathbf{I} + 2\mu_k\nabla^s\mathbf{v}_k. \tag{8.10}$$

Then for a constant phase viscosity μ_k, the incompressible multiphase Navier–Stokes equation becomes

$$\frac{\partial(\varepsilon_k\rho_k\mathbf{v}_k)}{\partial t} + \nabla\cdot\varepsilon_k\rho_k\mathbf{v}_k\mathbf{v}_k = \varepsilon_k\rho_k\mathbf{f}_k + \sum_j \beta_j(\mathbf{v}_j - \mathbf{v}_k) + m_k'\mathbf{v}_k - \nabla P_k + \mu_k\nabla^2\mathbf{v}_k. \tag{8.11}$$

Equation (8.11) differs from the single-phase incompressible Navier–Stokes equation principally because of coupling of the momentum balances through drag.

8.2.2 Compressible Flow

For the more general case of compressible flow with negligible phase change, Eq. (8.9) is replaced by the rearranged form of Eq. (7.1) shown below:

$$\nabla\cdot\mathbf{v}_k = -\frac{1}{\varepsilon_k\rho_k}\frac{d(\varepsilon_k\rho_k)}{dt^{(k)}}. \tag{8.12}$$

Equation (8.12) states that the divergence of the velocity of phase k is the relative rate of change of bulk density of phase k moving with its own velocity.

Analogously to single phase flow, let

$$A_k = -P_k\mathbf{I} + \lambda_k\nabla\cdot\mathbf{v}_k \tag{8.13}$$

to take into account the additional variable. Then the traction becomes

$$\mathbf{T}_k = -P_k\mathbf{I} + 2\mu_k\nabla^s\mathbf{v}_k + \lambda_k\mathbf{I}\nabla\cdot\mathbf{v}_k. \tag{8.14}$$

Define the phase k pressure to be minus one-third of the trace of the traction in the three directions x, y and z, as shown in the following equation:

$$-3\overline{P}_k = T_{kxx} + T_{kyy} + T_{kzz}. \tag{8.15}$$

Using Cartesian tensor notation and Einstein summation convention, Eq. (8.14) can be expressed in terms of components of direction *lm* (Jeffreys, 1961) as

$$T_{klm} = -P_k \delta_{lm} + 2\mu_k \frac{1}{2}\left(\frac{\partial v_{kl}}{\partial x_m} + \frac{\partial v_{km}}{\partial x_l}\right) + \lambda_k \frac{\partial v_{kl}}{\partial x_l} \delta_{lm}. \tag{8.16}$$

Now perform a tensor contraction, i.e., let $l = m$, and sum over l:

$$T_{kxx} + T_{kyy} + T_{kzz} = -3P_k + 2\mu_k \frac{\partial v_{kl}}{\partial x_l} + 3\lambda_k \frac{\partial v_{kl}}{\partial x_l}. \tag{8.17}$$

If

$$\overline{P}_k = P_k, \tag{8.18}$$

then Eq. (8.17) shows that

$$2\mu_k + 3\lambda_k = 0, \tag{8.19}$$

or that the bulk viscosity λ_k is related to the shear viscosity by the simple relation

$$\lambda_k = -\tfrac{2}{3}\mu_k. \tag{8.20}$$

The compressible multiphase Navier–Stokes equation in convective form then becomes

$$\varepsilon_k \rho_k \frac{d\mathbf{v}_k}{dt^k} = \varepsilon_k \rho_k \mathbf{f}_k + \sum_j \beta_j (\mathbf{v}_j - \mathbf{v}_k) - \nabla P_k + \nabla\left(2\mu_k \nabla^s \mathbf{v}_k - \tfrac{2}{3}\mu_k \mathbf{I} \nabla \cdot \mathbf{v}_k\right). \tag{8.21}$$

Although the stress in Eq. (8.3) can be taken to depend upon other variables in addition to the velocity and the compressibility (Drew and Lahey, 1979), the kinetic theory justifies the phenomenological approach just presented.

8.3 Dimensional Analysis: Scale Factors

The multi-fluid model derived in this chapter can be used to establish the basic dimensionless groups that govern such flows, as is conventionally done for single phase flow (Schlichting, 1960; Bird *et al.*, 1960). These dimensionless groups, called scale factors, may be useful in design and scale-up and for correlation of experimental data. In fluidization, Fitzgerald (1985), Blake *et al.* (1990) and others have successfully used such an approach. The following analysis is an extension of the work of Glicksman (1984). He used the inviscid, incompressible form of the equations discussed below. Here, the assumption of incompressibility will be lifted and the solids pressure caused by particle collisions which gives rise to the modulus G will be retained. However, as is common in compressible single-phase flow, the compressibility effect in the multiphase Navier–Stokes equation will be considered to be negligible.

Then the well-posed Model B momentum balances for the fluid and for the solid are

Dispersed Solids Momentum Balance

$$\varepsilon_s\rho_s\left(\frac{\partial \mathbf{v}_s}{\partial t}+\mathbf{v}_s\cdot\nabla\mathbf{v}_s\right)=\beta_B\left(\mathbf{v}_g-\mathbf{v}_s\right)-G\nabla\varepsilon_s+\left(\rho_s-\rho_g\right)\varepsilon_s\mathbf{g}+\mu_s\nabla^2\mathbf{v}_s, \quad (8.22)$$

where $G=(\partial P_s/\partial\varepsilon_s)$; and

Continuous Fluid Momentum Balance

$$\varepsilon\rho_f\left(\frac{\partial \mathbf{v}_f}{\partial t}+\mathbf{v}_f\cdot\nabla\mathbf{v}_f\right)=\beta_B\left(\mathbf{v}_s-\mathbf{v}_f\right)-\nabla P+\rho_f\mathbf{g}+\mu_f\nabla^2\mathbf{v}_f. \quad (8.23)$$

Let us choose the same characteristic length and velocity used by Glicksman (1984). In fluidization the inlet velocity U_o is a good reference velocity. When wall effects are neglected, the particle diameter d_p is the only invariant reference for length. The reference pressure is P_o and the inlet gas density is ρ_{go}. Then, as in Bird *et al.* (1960), let

$$\overline{\mathbf{v}}_k=\frac{\mathbf{v}_k}{U_o},\ k=f,s;\quad \overline{P}=\frac{P-P_o}{\rho_{go}U_o^{\ 2}}; \quad (8.24;\ 8.25)$$

$$\overline{t}=\frac{tU_o}{d_p};\quad \overline{x}_i=\frac{x_i}{d_p},\ i=1,2,3; \quad (8.25;\ 8.26)$$

$$\overline{\nabla}=\frac{\nabla}{d_p};\quad \overline{\nabla}^2=\frac{1}{d_p^{\ 2}}\nabla^2. \quad (8.26;\ 8.27)$$

In terms of the dimensionless variables, the momentum balances become as follows:

Dimensionless Solids Momentum

$$\varepsilon_s \frac{d\bar{\mathbf{v}}_s}{d\bar{t}^s} = \left(\frac{\beta_B d_P}{\rho_s U_o}\right)(\bar{\mathbf{v}}_g - \bar{\mathbf{v}}_s) + \left(\frac{G}{\rho_s U_o^2}\right)\nabla\varepsilon$$

$$+\left(\frac{\Delta\rho\varepsilon_g g d_p}{\rho_s U_o^2}\right) + \frac{\mu_f}{\rho_f U_o d_P}\left(\frac{\rho_f}{\rho_s}\right)\left(\frac{\mu_s}{\mu_f}\right)\bar{\nabla}^2\bar{\mathbf{v}}_f. \tag{8.28}$$

In the dimensionless solids momentum balance, a group involving a particle Reynolds number was formed. The drag group is one used by Glicksman (1984):

Dimensionless Fluid Momentum

$$\varepsilon \frac{d\bar{\mathbf{v}}_f}{dt^f} = \left(\frac{\rho_s}{\rho_f}\right)\left(\frac{\beta_B d_P}{\rho_s U_o}\right)\left(\bar{\mathbf{v}}_s - \bar{\mathbf{v}}_g\right) - \bar{\nabla}\bar{P} + \left(\frac{g d_p}{U_o^2}\right) + \left(\frac{\mu_f}{\rho_f U_o d_p}\right)\bar{\nabla}^2\bar{\mathbf{v}}_f. \tag{8.29}$$

The fluid momentum balance differs from the single-phase momentum balance only because of the presence of a drag group and the porosity multiplying the fluid acceleration.

The common groups are the fluid Reynolds number and the Froude number.

In single-phase compressible fluid flow the Mach number is the dimensionless group that measures the extent of compressibility. With the pressure scaled as it is given by Eq. (8.25), the Mach number is obtained from the continuity equation which, in terms of the dimensionless flux $\bar{\mathbf{f}} = \rho v / \rho_{fo} U_o$, can be written as

$$\bar{\nabla}\bar{\mathbf{f}} + M^2 \frac{\partial \bar{P}}{\partial \bar{t}} = 0, \tag{8.30}$$

where

$$M = \left(U_o / C_s\right) = \text{Mach number}, \tag{8.31}$$

with the sonic velocity C_s defined as

$$C_s = \left(\frac{\partial P}{\partial \rho_g}\right)^{1/2}. \tag{8.32}$$

Equation (8.30) gives a measure of how fast the convective terms in the momentum equation change with time. It also provides a more natural scale for time than is given by Eq. (8.25). Equation (8.30) suggests that for a constant sonic velocity a more natural scale for time is obtained by incorporating M into time as follows:

$$\hat{t} = \frac{\bar{t}}{M^2} = \left(\frac{tC_s}{d_p}\right)\left(\frac{C_s}{U_o}\right). \tag{8.33}$$

The grouping (8.33) and the wave equation which is obtained by combining (8.30) with the inviscid fluid momentum balance, as done in Chapter 7, and an order of magnitude analysis

$$\text{Order}\left(tC_s/d\right) = 1 \tag{8.34}$$

strongly suggest the relation

$$C_s = \text{frequency} \cdot \text{length}, \tag{8.35}$$

where frequency is $1/t$. Equation (8.35) states that for a unit length, the frequency of oscillation is high for a high sonic velocity. Gas oscillations in turbulent flow are known to be of the order of 300 H and larger. A similar analysis performed for granular flow in Chapter 7 shows that the granular flux or the bulk density satisfy a wave equation with C_s equal to $(G/\rho_s)^{1/2}$. See Eq. (7.15) and (7.16). Since hopper flow experiments show that the critical flux or the granular "sound velocity" is of the order of one meter per second, for a length of one meter the characteristic frequency should be one hertz. Indeed most bubbling beds have a dominant frequency of oscillation of about one hertz.

This granular sonic group already appeared in Eq. (8.28), since the solids pressure in Eq. (8.22) was assumed to be a function of the solids volume fraction, as suggested by kinetic theory of granular flow and experiments. The fluid pressure in Eq. (8.23) was, however, treated as an independent variable. To bring the sonic velocity into the momentum balance, it could have been taken as a function of density and the Mach number then defined. The alternate way is to bring in the Mach number using the approach illustrated for single-phase flow in Eq. (8.30).

For two-phase flow with incompressible solid grains, the solids continuity equation provides no additional scaling factors. The gas continuity equation alone has a troublesome porosity variation with time. Only the sum of the two continuity equations gives the well-known mixture sonic velocity already discussed in Chapter

7. In terms of the mixture density ρ_m and the sum of the fluid and solid fluxes, the mixture continuity equation is

$$\frac{\partial \rho_m}{\partial t} + \nabla F_m = 0, \tag{8.36}$$

where

$$F_m = \varepsilon \rho_f v_f + \varepsilon_s \rho_s v_s. \tag{8.37}$$

In terms of the mixture sound velocity, C_m, defined by Eq. (7.41), the mixture continuity equation in dimensionless form becomes

$$\overline{\nabla} \overline{F}_m + \left(\frac{U_o}{C_m} \right)^2 \frac{\partial \overline{P}}{\partial \overline{t}} = 0. \tag{8.38}$$

Roy *et al.* (1990) measured C_m in fluidized beds.

The dimensionless groups that appeared in scaling the solid and the fluid momentum and mass balances are finally as follows:

$$\left(\frac{\beta_B d_p}{\rho_s U_o} \right) = \text{Dimensionless drag group}, \tag{8.39}$$

$$\left(g d_p / U_o^2 \right) = \text{Froude number}, \tag{8.40}$$

$$\frac{\rho_f U_o d_p}{\mu_f} = \text{Reynolds number}, \tag{8.41}$$

$$\left(\rho_f / \rho_s \right) = \text{Density ratio}, \tag{8.42}$$

$$\left(\mu_s / \mu_f \right) = \text{Relative viscosity}, \tag{8.43}$$

$$\left(\frac{G}{\rho_s} \right)^{1/2} \frac{1}{U_o} = \frac{\text{Granular "sonic" velocity}}{\text{Inlet fluid velocity}}, \tag{8.44}$$

$$\left(U_o / C_m \right) = \text{Mach number based on mixture sonic speed.} \tag{8.45}$$

The solids and the fluid momentum balances, Eqs. (8.28) and (8.29), show that for some typical values such as

$$\left(\frac{\rho_f U_o d_p}{\mu_f}\right) = 100, \quad \left(\frac{\rho_g}{\rho_s}\right) = 10^{-3}, \quad \frac{\mu_s}{\mu_g} = 10^2, \qquad (8.46\text{–}8.48)$$

the viscous effects vanish away from the surfaces. The momentum balances reduce themselves to the inviscid flow studied in Chapter 7 and used by Glicksman (1984).

The granular dimensionless sonic velocity defined by the scale factor, Eq. (8.44), vanishes for dilute flow. It is of the order of unity, however, near minimum fluidization. The Mach number based on the sonic velocity is small, except at high flows, of the order of 10 m/s. The deletion of these terms by Glickman (1984) in scaling fluidized beds will not generally lead to errors.

By using the Ergun equation for the friction coefficient, Glickman observed that the drag group, Eq. (8.39), could be expressed in terms of the particle Reynolds number, defined by Eq. (8.41) in the viscous limit. In this limit, his scale factors that are obtained from the momentum balances are then the Reynolds number, the Froude number, and the sphericity, which occurs in the Ergun equation. For high particle Reynolds numbers, the corresponding scale factors are the number, the Froude density ratio and the sphericity. As pointed out by Glicksman (1990) these groups help to explain Geldart's classification of particles. In Chapter 6 it has already been shown that the transition between the bubbling B group and the disperse D group depended upon the use of the first, viscous, or the second, inertial, term in the Ergun equation for A and D groups, respectively. The transition to the A group depends upon the minimum bubbling velocity. Hence, it involves the group given by Eq. (8.44), the granular sonic velocity group, not used by Glicksman. The transition to Geldart C category involves the cohesive force which was not used in the constitutive equation for traction. Electrical forces of importance for fine particles were also not included in the theory developed in this chapter. Application of the present theory to dispersed gas–liquid flow will involve consideration of the surface tension. This will lead to another fundamental group, called the Weber number (Lyczkowski *et al.*, 1982). Dimensionalization of the boundary, inlet, outlet and the initial conditions will lead to additional scale factors, most of which were recognized by Glicksman (1984). For no slip boundary conditions at the walls, the only dimensionless groups will be the geometric length ratios recognized by Glicksman (1984):

$$\left(L_i / d_p\right) = \text{Geometric scale factors, } i = 1, 2, 3. \qquad (8.49)$$

With particle slip at the wall, a slip ratio group may appear.

With a prescribed inlet fluid velocity U_o, a pressure P_{in}, and no entering solids — that is, for a batch system, such as a bubbling bed — there exists a pressure scale factor recognized by Glicksman (1984) and given below:

$$\left(P_{in}/\rho_{go}U_o^2\right) = \text{Pressure scale factor.} \tag{8.50}$$

The outlet pressure may be used as a reference pressure. For zero gradients at the outlet, there are no scale factors contributing to the preceding list.

For large times of practical interest in scaling, the influence of the initial conditions will disappear, except for the inventory of particles in a batch system. The inventory contributes an initial height of particles, L_{initial}, and the additional scale factor shown below:

$$\left(L_{\text{initial}}/d_p\right) = \text{Inventory height.} \tag{8.51}$$

The scale factor (8.51) is clearly an important group overlooked by Glicksman (1984) in scaling fluidized beds. It is well known that shallow beds and deep beds do not operate in the same flow regime. For example, deep beds will slug.

For a flow system, in place of the dimensionless inventory, an inlet solids flux or an inlet porosity must be prescribed to completely characterize the system (Arastoopour and Gidaspow, 1979). For example, pressure drop data for circulating fluidized beds are generally given as a function of gas flow rate and solids flux (Yerushalmi, 1986).

This will clearly lead to an additive dimensionless group. In such a flow system such a group will replace the scale factor (8.51).

Other geometrical factors and flow ratios will result from use of complex geometries with several inlet flows. The presence of so many scale factors that must be the same for complete similarity makes the traditional scale-up of fluidized beds quite difficult. An alternate approach is to simply numerically solve the equations of change.

8.4 CFB or Riser Flow: Experimental

Circulating fluidized beds (CFB) are common in the petroleum industry, in which the older bubbling fluidized bed reactors are being replaced by the circulating fluidized bed risers because of the recently developed highly active catalysts. The second major emerging application is to circulating fluidized bed combustors, which are being built to burn high-sulfur coals. These combustors offer the potential of burning coal and incinerating wastes with low SO_2 and NO_x emissions.

Figure 5.1 shows a schematic of flow in a CFB. The standpipe is filled with solids which may be fluidized to yield fast recirculation rates. The flow in the riser section discussed here lies in the range between the minimum fluidization volume fraction which may occur near the bottom of the riser and the dilute pneumatic transport discussed in Chapter 2. The pressure drop versus the superficial velocity behavior is to the left of the minimum shown in Fig. 2.8. Gas velocities are in the range of 5 to 15 m/s.

Although CFBs are widely used in the petroleum and chemical industries, the understanding of riser flow hydrodynamics came slowly. A systematic study began at the City College of New York. (Squires *et al.*, 1985; Yerushalmi, 1986). Pressure drop, solids flux and gas velocity data were correlated in terms of a slip, as shown below:

$$\text{Slip velocity} = v_g - G_s/\rho_s\varepsilon_s\,, \tag{8.52}$$

where G_s is the measured solids flux and where ε_s, the solids volume fraction, was determined by assuming that the pressure drop equals the weight of the solids. These slip velocities turned out to be many times the terminal solids velocity. The explanation given was that the solids formed clusters which behaved like large particles. Indeed, high-speed movies do show the formation of dense regions of particles, especially near the walls. However, later measurements at the City College of New York (Weinstein *et al.*, 1986), at Bradford in England (Geldart and Rhodes, 1986; Rhodes *et al.*, 1989), at IGT (Bader *et al.*, 1988), in France (Galtier *et al.*, 1989) and elsewhere showed that there exist large radial non-uniformities and that the flow can be characterized as a core-annular flow with particles descending down the walls either in the form of sheets or clusters. Figure 8.1 shows an idealized velocity profile constructed by Miller (1991). Solids are transported up the riser pipe in a dilute core. For them to flow up, the axial pressure drop must be larger than the weight of the dilute core in developed flow. At the walls there exists a thick dense annular region. The weight of this annulus exceeds the axial pressure drop. Hence, this annular region moves down slowly. The core is generally thinner at the bottom of the riser than near the top. There exists internal refluxing. Particles move from the annular region to the core near the bottom of the riser. They move from the core to the annulus near the top. The flow is never steady, but oscillates slowly.

Some typical data obtained by Miller (Miller and Gidaspow, 1992; Miller, 1991) are presented below to clarify the observed hydrodynamics. Time-averaged data were obtained for flow of 75 μm FCC catalyst particles in a 7.5 cm

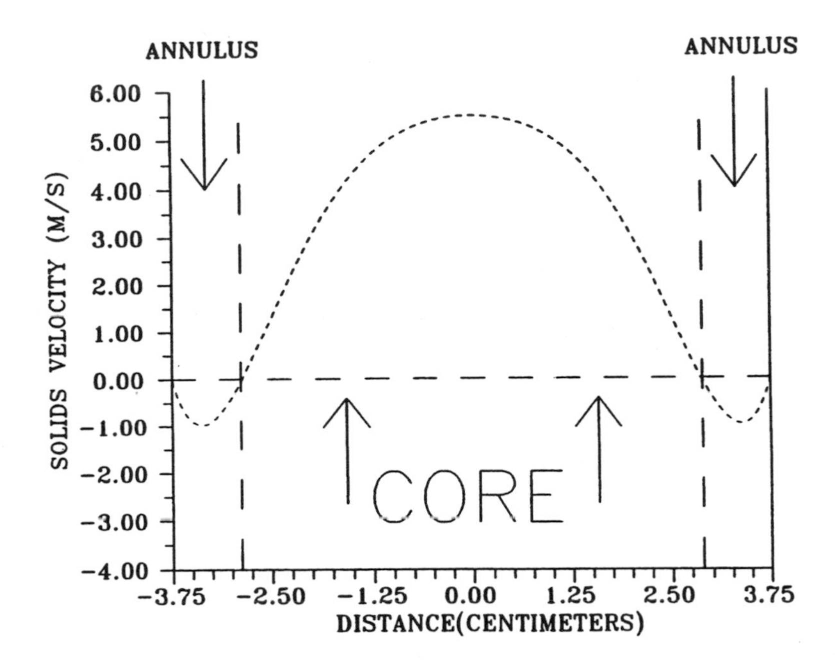

Fig. 8.1 Particle velocity profile for dense, vertical transport of 75 μm FCC catalyst particles in the CFB shown in Fig. 8.2. (From Miller and Gidaspow, 1992. Reproduced by permission of the American Institute of Chemical Engineers. Copyright © 1992 AIChE. All rights reserved.)

diameter clear acrylic riser shown in Fig. 8.2 for the central section as a function of gas and solids flow rates. Particle concentrations obtained using the probe in Fig. 8.3 are shown in Fig. 8.4. We see a dilute core and the central dense annulus. To obtain particle velocities, fluxes were measured using the extraction probe depicted in Fig. 8.5. Flux data are shown in Fig. 8.6 for three sections. The largest downflow at the wall is near the bottom. This flux exceeds the average feed by several times, illustrating the internal refluxing. A video of the experiment shows an oscillating downflow at the wall. A similar high-speed video of a more dilute flow of Geldart group B 500 μm particles showed the flow to be in the form of clusters (Gidaspow *et al.*, 1989a). They descended down the wall at a rate of about 1 m/s.

Flux and concentration data were used to construct radial particle velocity profiles. Figure 8.7 shows such typical curves for the three sections at one gas velocity. At the wall, the solids descend at velocities of 10 cm/s to 1.5 m/s.

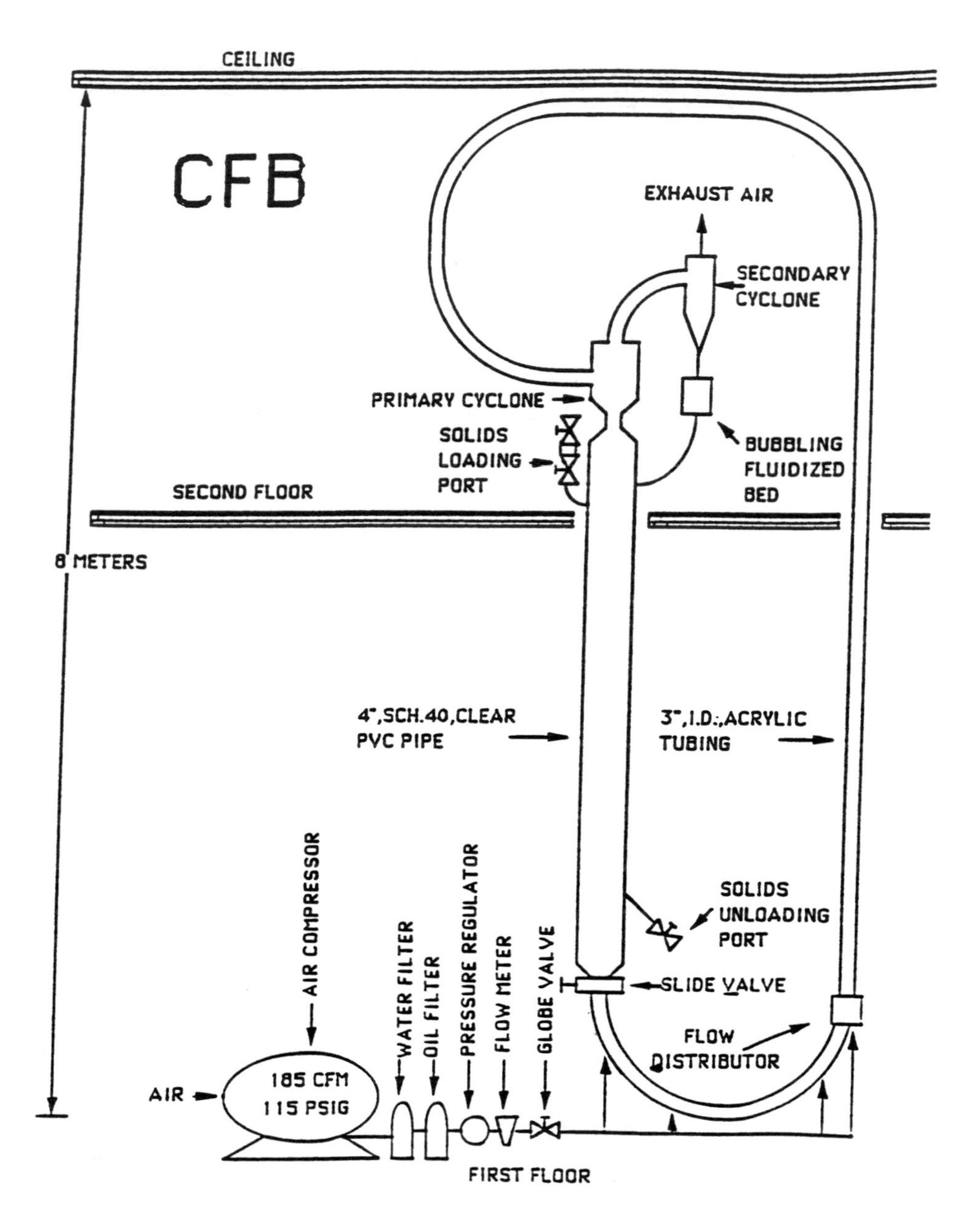

Fig. 8.2 Circulating fluidized bed (CFB) at IIT. (From Miller and Gidaspow, 1992. Reproduced by permission of the American Institute of Chemical Engineers. Copyright © 1992 AIChE. All rights reserved.)

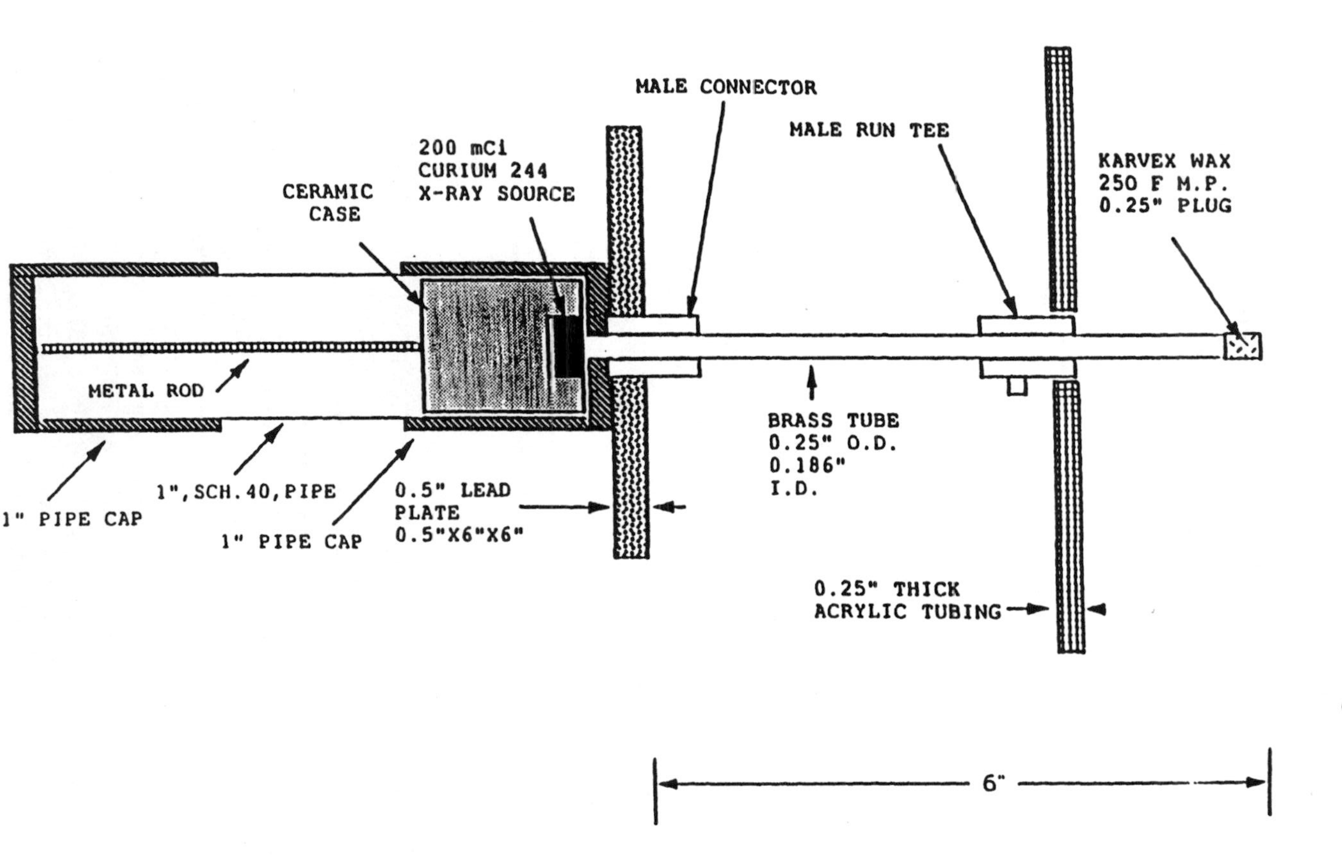

Fig. 8.3 X-ray probe constructed by Miller (1991) to measure particle concentrations. (From Miller and Gidaspow, 1992. Reproduced by permission of the American Institute of Chemical Engineers.)

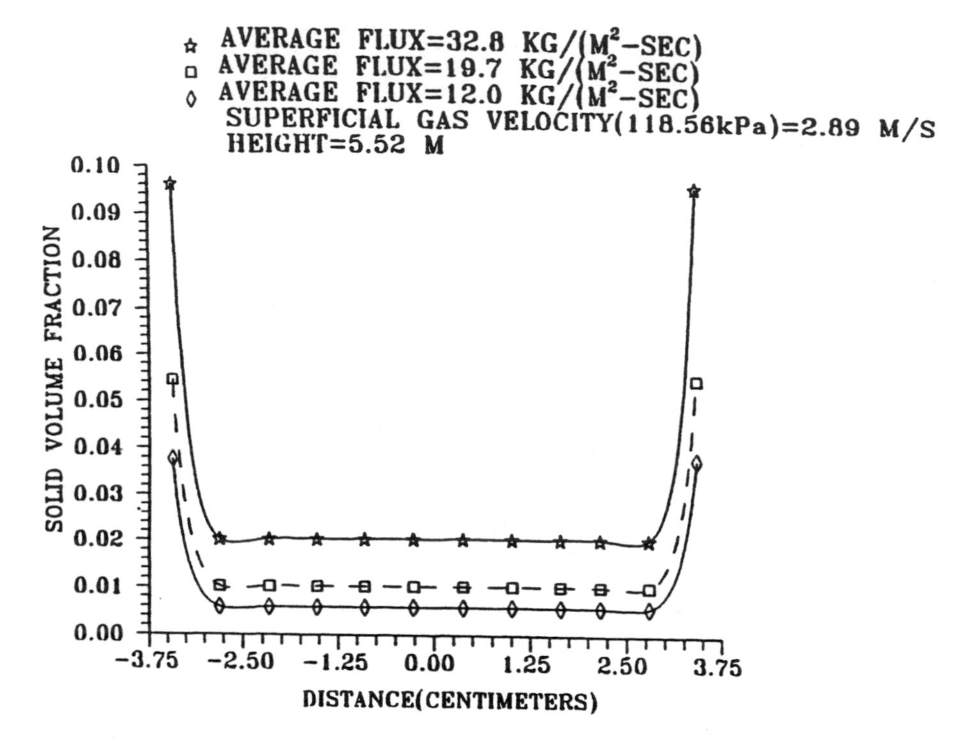

Fig. 8.4 Typical 75 μm FCC particle concentration: dense layer at the wall with a dilute core. (From Miller and Gidaspow, 1992. Reproduced by permission of the American Institute of Chemical Engineers. Copyright © 1992 AIChE. All rights reserved.)

A fully developed flow momentum balance in Fig. 8.8 shows how the velocity, concentration and pressure drop data can be used to obtain the mixture solids viscosities. The pressure drop minus the radially integrated weight of the bed over the core equals the shear at the core–annular interface. For developed flow, the solids shear equals the mixture viscosity times the radial velocity gradient that was computed from least square fits of data such as those shown in Fig. 8.7. Such computations produced the viscosities shown in Fig. 8.8. The dashed lines show the 95% confidence levels. When a correction is made for developing flow (Miller and Gidaspow, 1992; Miller, 1991) the values of the viscosities decrease a little, but not significantly. The viscosities shown in Fig. 8.8 extrapolate to the bubbling bed viscosities which are of the order of poises (Schügerl, 1971). Like the viscosities for liquid–solid systems reviewed by Jeffrey and Acrivos (1976), the viscosities measured for CFB flow correlate with the solids volume fraction.

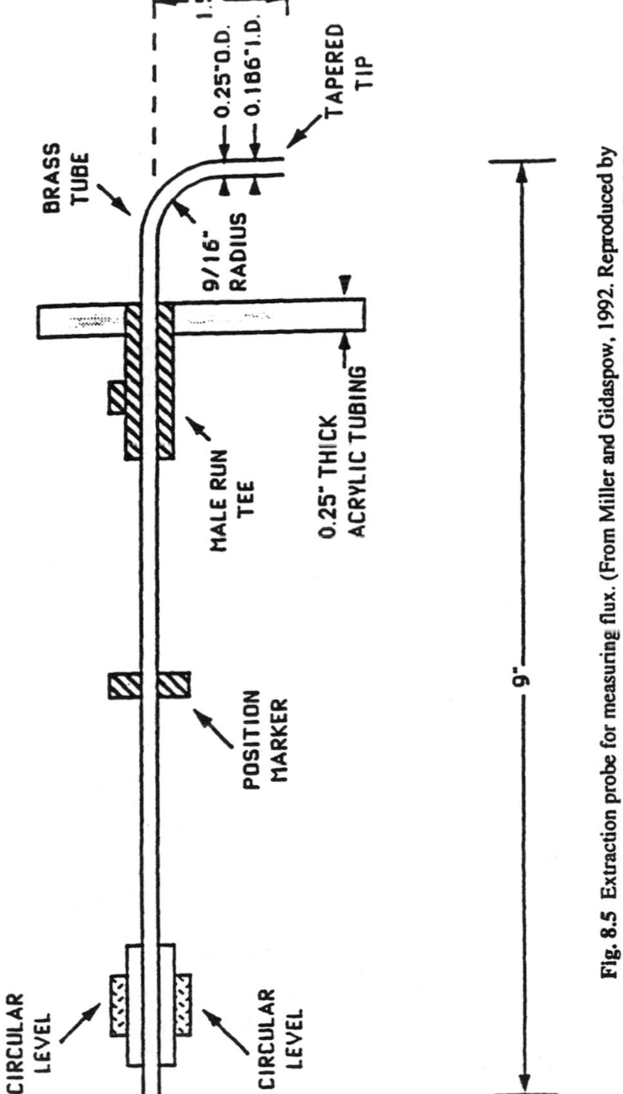

Fig. 8.5 Extraction probe for measuring flux. (From Miller and Gidaspow, 1992. Reproduced by permission of the American Institute of Chemical Engineers. Copyright © 1992 AIChE. All rights reserved.)

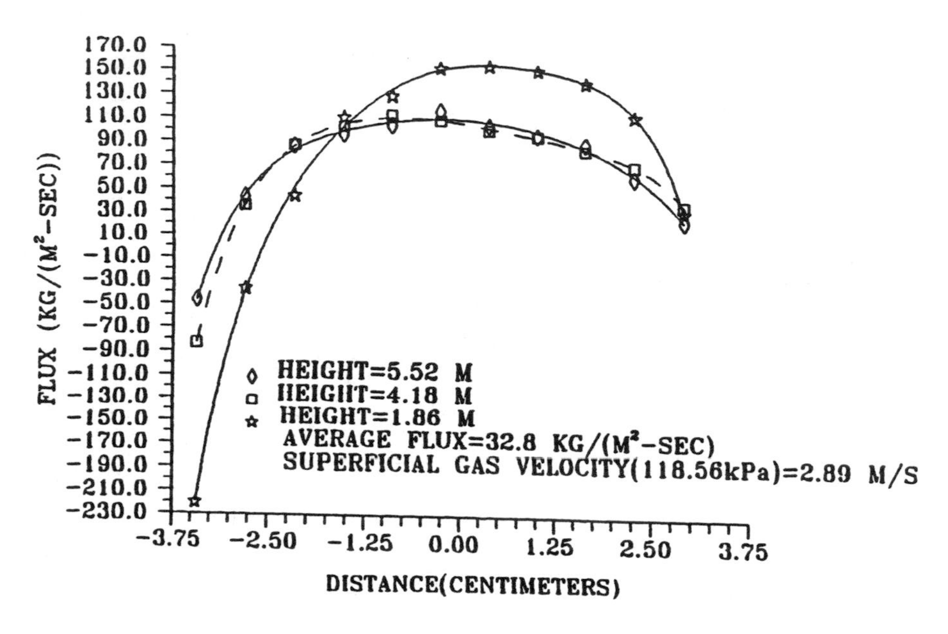

Fig. 8.6 Typical flux profile in IIT's CFB for 75 μm FCC particles. Note large downward flux near the wall. (From Miller and Gidaspow, 1992. Reproduced by permission of the American Institute of Chemical Engineers. Copyright © 1992 AIChE. All rights reserved.)

8.5 Need for Clusters in One-D Modeling

CFB or riser flow involves vertical transport of solids with downflow either in the form of cluster-like solid waves or in the form of a more continuous downward moving sheet of solids. One-dimensional modeling of the type described in Chapter 2 clearly cannot give two solids velocities without breaking up the pipe region into at least two parts: a downward-moving region and an upward-moving region. Insistence on the use of the convenient one-dimensional model led to the concept of a cluster. Slip was simply increased artificially by the use of a large cluster diameter in place of the actual particle size (Arastoopour and Gidaspow, 1979). This trick, which unfortunately can also be supported by visual observation of clusters, produced the correct large particle concentration in the pipe and the correspondingly large pressure drop. Other investigators had modified the drag correlations in different ways or simply presented the particle distribution as an empirical correlation. To show that one-dimensional modeling using standard drag relations produces a particle concentration in the riser that is too low, the

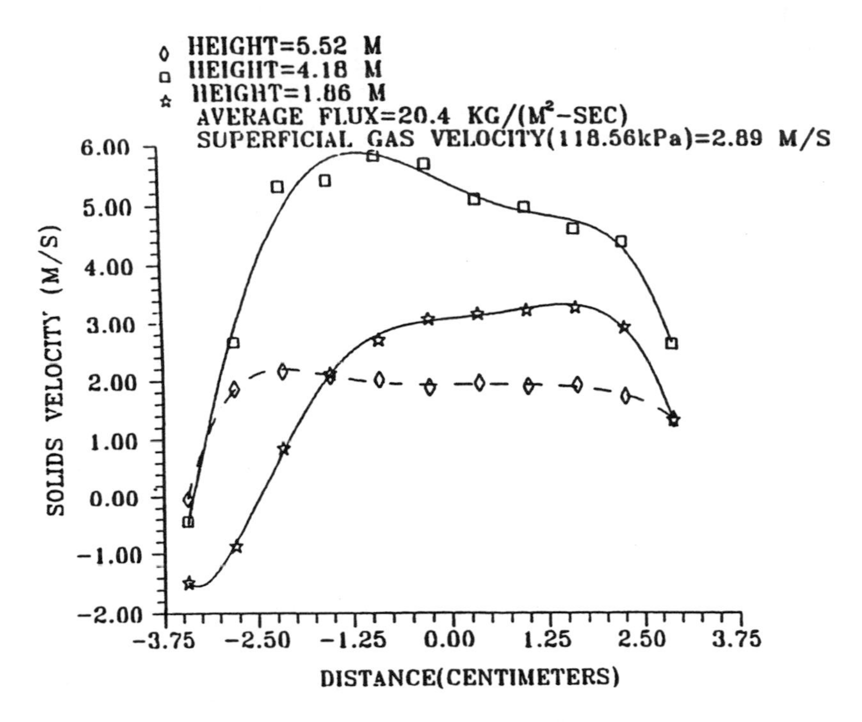

Fig. 8.7 Typical particle velocities in IIT's CFB for 75 μm FCC particles determined from flux and concentration measurements. Note acceleration and the deceleration due to top bend. (From Miller and Gidaspow, 1992. Reproduced by permission of the American Institute of Chemical Engineers. Copyright © 1992 AIChE. All rights reserved.)

following problem was solved by Tsuo and Gidaspow (1990). Glass beads 520 μm in diameter were conveyed vertically upward with a gas at a velocity of 5 m/s, with a solids flux of 25 kg/m^2 in a grounded 7.67 cm diameter acrylic plastic pipe in a set-up similar to that shown in Fig. 8.2. For this low gas velocity, a high speed motion picture showed that the solids were descending down the wall in the form of clusters. The clusters disappeared when the velocity was raised to 5.5 m/s. The axial particle concentration profile was similar to that shown in Fig. 2.4 for the pneumatic transport regime. In the cluster regime, at 5 m/s the fully developed particle volume fraction was 0.0115.

To obtain the computed axial concentration profile the transient version of the steady state one-dimensional equations fully discussed in Chapter 2 were solved. Table 8.1 shows these equations. As in the experiment, the pipe was initially free of solids. At zero time the solids were introduced into the pipe, as

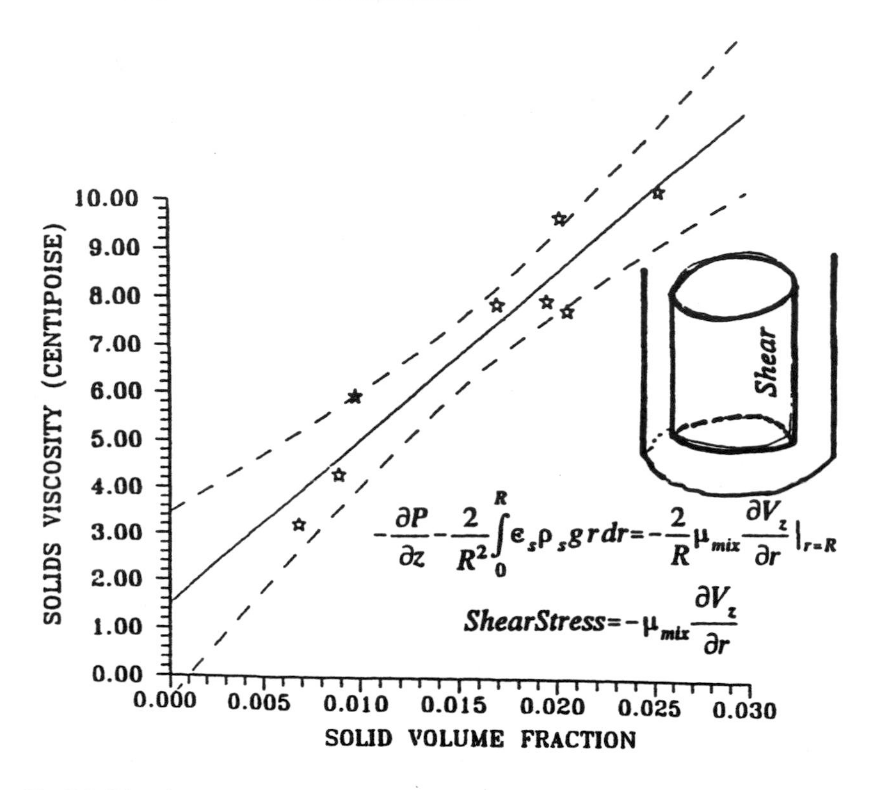

Fig. 8.8 Viscosity of 75 μm FCC particles without a correction for acceleration. (From Miller and Gidaspow, 1992. Reproduced by permission of the American Institute of Chemical Engineers. Copyright © 1992 AIChE. All rights reserved.)

shown in Fig. 8.9. This figure also shows the computed axial distribution as a function of time. The system reached a steady state after a period of five seconds. The computed solids volume fraction in the fully developed region was only 0.0065, or almost half the measured value. Two-dimensional analysis described in the next section produced much more realistic results.

8.6 Computation of Cluster Flow

Two-dimensional computations were done using the equations shown in Table 8.2 for the geometry shown in Fig. 8.10. As a crude approximation constant inlet conditions were used. Characteristics discussed in Chapter 7 demand that outlet conditions be prescribed. For naturally occurring downflow, there are no simple outlet conditions that can be assigned (Tsuo and Gidaspow, 1990). Hence, a bend

Table 8.1 ONE-DIMENSIONAL EQUATIONS OF CHANGE

Gas Continuity Equation

$$\frac{\partial(\rho_g \varepsilon_g)}{\partial t} + \frac{\partial(\rho_g \varepsilon_g u_g)}{\partial x} = 0 \qquad \text{(T.1)}$$

Solid Continuity Equation

$$\frac{\partial(\rho_s \varepsilon_s)}{\partial t} + \frac{\partial(\rho_s \varepsilon_s u_s)}{\partial x} = 0 \qquad \text{(T.2)}$$

Gas Momentum Equation

$$\frac{\partial(\rho_g \varepsilon_g u_g)}{\partial t} + \frac{\partial(\rho_g \varepsilon_g u_g u_g)}{\partial x} = -\frac{\partial P}{\partial x} + \beta(u_s - u_g) + \frac{2 f_g \varepsilon_g \rho_g u_g^2}{D} + \rho_g g \qquad \text{(T.3)}$$

Solids Momentun Equation

$$\frac{\partial(\rho_s \varepsilon_s u_s)}{\partial t} + \frac{\partial(\rho_s \varepsilon_s u_s u_s)}{\partial x} = \beta(u_g - u_s) - G\frac{\partial \varepsilon_s}{\partial x} + \frac{2 f_s \varepsilon_s \rho_s u_s^2}{D} + (\rho_s - \rho_g)\varepsilon_s g \qquad \text{(T.4)}$$

$$\varepsilon_g + \varepsilon_s = 1 \qquad \text{(T.5)}$$

The Ideal Gas Equation

$$\rho_g = (P/RT) \qquad \text{(T.6)}$$

Inter-phase Friction Coefficient β: Based on Ergun Equation

$$\varepsilon_g \le 0.8 \quad \beta = 150 \frac{\varepsilon_s^2 \mu_g}{(\varepsilon_g d_p \phi_s)^2} + 1.75 \frac{\rho_g |u_g - u_s| \varepsilon_s}{(\varepsilon_g d_p \phi_s)} \qquad \text{(T.7)}$$

Based on Single Sphere Drag

$$\varepsilon_g \ge 0.8 \quad \beta = \tfrac{3}{4} C_d \frac{|u_g - u_s| \rho_g \varepsilon_s}{d_p \phi_s} \varepsilon_g^{-2.65} \qquad \text{(T.8)}$$

with

$$C_d = \frac{24}{Re_s}\left(1 + 0.15 Re_s^{0.687}\right) \quad Re_s \le 1000 \qquad \text{(T.9)}$$

$$C_d = 0.44 \quad Re_s \ge 1000 \qquad \text{(T.10)}$$

Table 8.1 (CONTINUED)

$$Re_s = \frac{\left|u_g - u_s\right| d_p \rho_g \varepsilon_g}{\mu_g} \tag{T.11}$$

Solids Stress Modulus

$$G(e_g) = \frac{\partial P_s}{\partial \varepsilon_s} = 10^{-8.76\varepsilon_g + 5.43} \tag{T.12}$$

Friction Factors: Modified Hagen–Poiseuille Expression

$$f_g = (16/Re_g) \quad for \quad Re_g \leq 2100 \tag{T.13}$$

$$f_g = (0.0791/Re_g{}^{0.25}) \quad \text{for} \quad 2100 < Re_g \leq 100{,}000 \tag{T.14}$$

$$\frac{1}{\sqrt{f_g}} = 2 \log\left(Re_g \sqrt{f_g}\right) - 0.8 \quad \text{for} \quad Re_g > 100{,}000 \tag{T.15}$$

where

$$\text{Re}_g = \frac{D u_g \rho_g \varepsilon_g}{\mu_g} \tag{T.16}$$

Konno's Correlation

$$f_s = 0.0025 u_s{}^{-1} \tag{T.17}$$

---

in the pipe was included as part of the computational geometry. There continuous outflow was assigned, as shown in Fig. 8.9. At the walls a no slip boundary condition is a good approximation for both the gas and fine solids. A viscosity coefficient of 0.509 Pa-s was used in the computation.

The two-dimensional simulation was done for 18 s. It took a 4 s period to fully fill the empty pipe with particles. Tsuo (1989) shows the outlet solids mass flux as a function of time. The outlet solids mass flux reached the inlet solids mass flux of 25 kg/m^2-s at 11 s of real fluidization time. Outlet mass flux of solids oscillated at an average period of 5 s. This very low frequency of around 0.2 H is consistent with the observed and computed wall cluster descent speed of about 1 m/s. Similar speeds and low frequencies were reported by Weinstein *et al.* (1986) and Abed (1984) in their CFB systems.

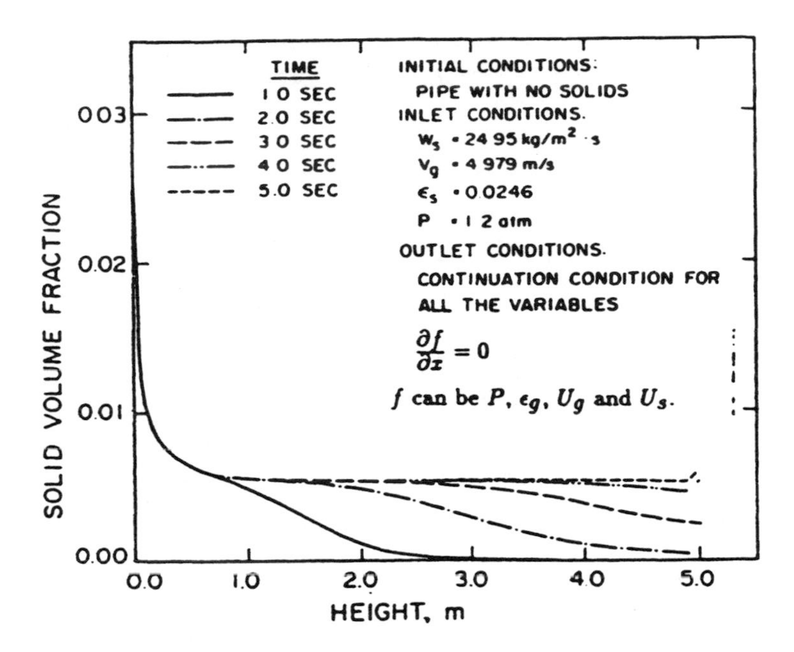

Fig. 8.9 One-dimensional simulation of transient vertical conveying. (Tsuo and Gidaspow, 1990. Reproduced by permission of the American Institute of Chemical Engineers. Copyright © 1992 AIChE. All rights reserved.)

Figure 8.10 shows the solids concentration profiles as a function of time. The first cluster of particles forms after 3.0 s in the center region (Tsuo, 1989). The cluster at first grows larger and denser, then it starts falling down. Later several clusters form near the wall and start to run down. A few clusters join together and create a significant downflow. The typical wall clusters observed descend at a velocity of 1.1 m/s. The average cluster size is about 2 to 3 cm. All these phenomena were observed in the high speed motion pictures of the experiment.

The cluster behaves as a hydrodynamic unit or as a particle of a large effective diameter, as arbitrarily assumed in the modeling study of Arastoopour and Gidaspow (1979). Figure 8.10 shows the negative velocities of the clusters.

Figure 8.11 shows typical computed time-averaged and measured radial profiles of solids concentration in a developed region. Both the model and the experiment show a much higher solids concentration at the wall of the tube due to the formation of clusters near the wall.

Figure 8.12 shows typical computed time-average and measured radial distribution of axial solids velocities at an axial location of 3.4 m. In general, a

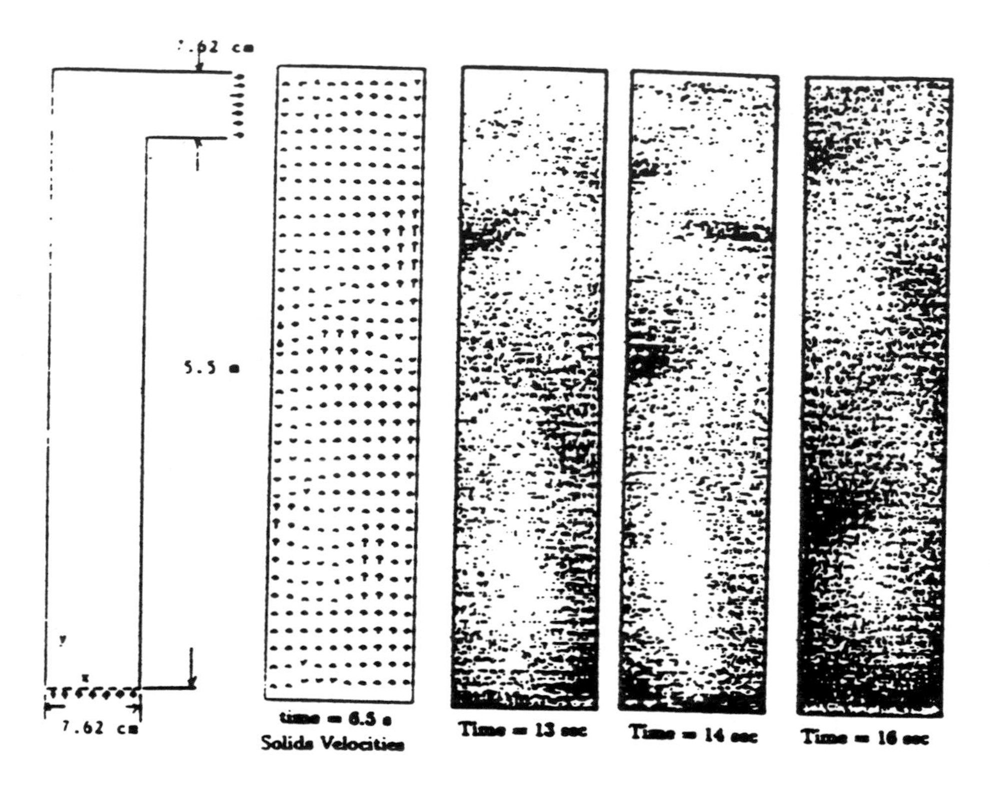

Fig. 8.10 IIT geometry and computed distributions of solids velocities and particle density for inlet and initial conditions in Fig. 8.9. $d_p = 520\mu m$; $\rho_p = 2.62 g/cm^3$. (Tsuo and Gidaspow, 1990. Reproduced by permission of the American Institute of Chemical Engineers. Copyright © 1992 AIChE. All rights reserved.)

comparison between the experimental and the computed radial distribution of axial solids velocities is good. There are some discrepancies in the center region. The solids volume fraction in the center region is around 0.005. The x-ray densitometer, which was used in Luo's (1987) experiment for measuring the solids concentration, is not considered to be accurate for such small values of solids concentration. Therefore, the experimental solids velocity, computed from the measured local flux using a ball probe and the measured local concentration, is also not accurate in the center region. Figure 8.13 shows typical computed time-average and measured radial distributions of axial gas velocity at an axial location of 3.4 m. A comparison between the experimental and the computed radial distributions of axial gas velocities is good.

Table 8.2 VISCOSITY COEFFICIENT HYDRODYNAMIC MODEL B

Hydrodynamic Model

Continuity Equations

Gas Phase

$$\frac{\partial(\rho_g \varepsilon_g)}{\partial t} + \nabla \cdot (\rho_g \varepsilon_g \mathbf{v}_g) = 0 \tag{T.1}$$

Solid Phase

$$\frac{\partial(\rho_s \varepsilon_s)}{\partial t} + \nabla \cdot (\rho_s \mathbf{v}_s \varepsilon_s) = 0 \tag{T.2}$$

Momentum Equations

Gas Momentum

$$\frac{\partial(\rho_g \varepsilon_g \mathbf{v}_g)}{\partial t} + \nabla \cdot (\rho_g \varepsilon_g \mathbf{v}_g \mathbf{v}_g) = -\nabla P + \beta(\mathbf{v}_s - \mathbf{v}_g) + \nabla \cdot \varepsilon_g \mathbf{T}_g + \rho_g \mathbf{g} \tag{T.3}$$

where

$$\mathbf{T}_g = 2\mu_{gc} \nabla^s \mathbf{v}_g - 2/3\mu_{gc} \nabla \cdot \mathbf{v}_g \mathbf{I} \tag{T.4}$$

$$\nabla^s \mathbf{v}_g = \frac{1}{2}\left[\nabla \mathbf{v}_g + \nabla \mathbf{v}_g{}^T\right] \tag{T.5}$$

Solids Momentum

$$\frac{\partial(\rho_s \varepsilon_s \mathbf{v}_s)}{\partial t} + \nabla \cdot (\rho_s \varepsilon_s \mathbf{v}_s \mathbf{v}_s) = \beta(\mathbf{v}_s - \mathbf{v}_g) \tag{T.6}$$

$$-G\nabla \varepsilon_s + \nabla \cdot \varepsilon_s \mathbf{T}_s + (\rho_s - \rho_g)\varepsilon_s \mathbf{g} \tag{T.7}$$

where

$$\mathbf{T}_s = 2\mu_{sc} \nabla^2 \mathbf{v}_s - 2/3\mu_{sc} \nabla \cdot \mathbf{v}_s \mathbf{I} \tag{T.8}$$

$$\nabla^s \mathbf{v}_s = \frac{1}{2}\left[\nabla \mathbf{v}_s + \nabla \mathbf{v}_s{}^T\right] \tag{T.9}$$

$$\varepsilon_g + \varepsilon_s = 1 \tag{T.10}$$

Table 8.2 (CONTINUED)

Constitutive Equations

Ideal Gas Equation

$$\rho_g = P/RT \tag{T.11}$$

Interphase Friction Coefficient β

$$\varepsilon_g \leq 0.8 \quad \beta = 150\frac{\varepsilon_s{}^2\mu_g}{\left(\varepsilon_g d_p \phi_s\right)^2} + 1.75\frac{\rho_g\left|\mathbf{v}_g - \mathbf{v}_s\right|\varepsilon_s}{\left(\varepsilon_g d_p \phi_s\right)} \tag{T.12}$$

Based on single sphere drag

$$\varepsilon_g \geq 0.8 \quad \beta = \frac{3}{4}\frac{\left|\mathbf{v}_g - \mathbf{v}_s\right|\rho_g\varepsilon_s}{d_p\phi_s}\varepsilon_g{}^{-2.65} \tag{T.13}$$

with

$$C_d = \frac{24}{Re}\left(1 + 0.15Re^{0.687}\right) \quad Re \leq 1{,}000 \tag{T.14}$$

$$C_d = 0.44 \quad Re \geq 1{,}000 \tag{T.15}$$

$$Re = \frac{\left|\mathbf{v}_g - \mathbf{v}_s\right| d_p \rho_g \varepsilon_g}{\mu_g} \tag{T.16}$$

Solids Stress Modulus

$$G(\varepsilon_g) = \frac{\partial P_s}{\partial \varepsilon_s} = 10^{-8.76\varepsilon_g + 5.43}\ \mathrm{N/m^2} \tag{T.17}$$

Several parametric studies were conducted in order to understand the formation of clusters in CFB risers (Tsuo and Gidaspow, 1990). Cluster density increased with an increasing solids flux, with a decreasing gas velocity and with a decreasing pipe radius, as expected. The cluster flow appeared to change into a smoother core — annular flow as the particle size was decreased to 100 μm.

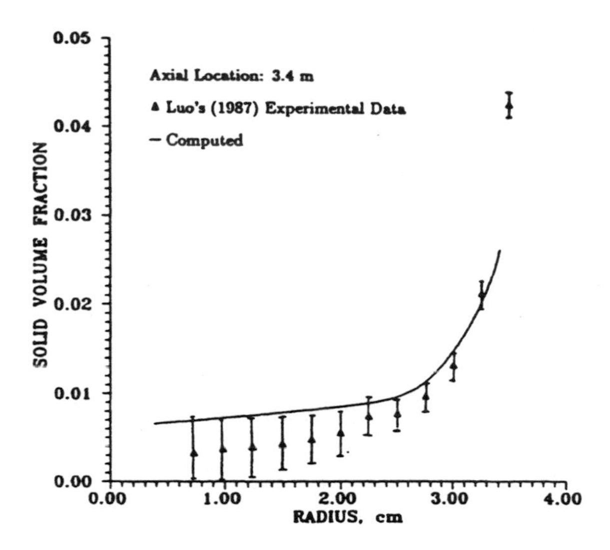

Fig. 8.11 Radial variation of computed and measured solids volume fraction of 520 μm glass beads in the cluster regime. (Tsuo and Gidaspow, 1990. Reproduced by permission of the American Institute of Chemical Engineers. Copyright © 1992 AIChE. All rights reserved.)

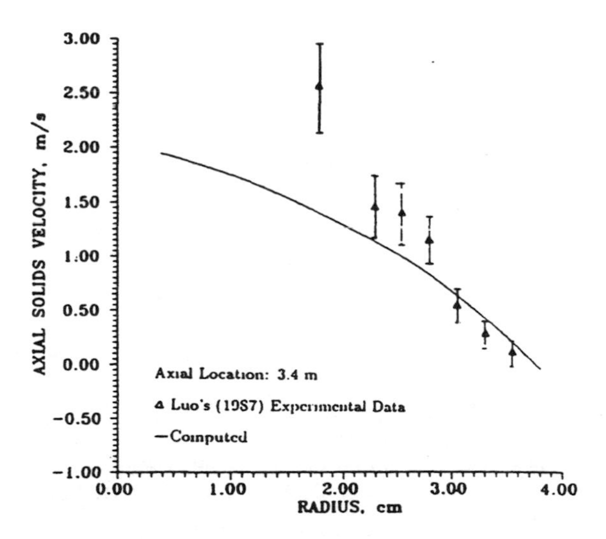

Fig. 8.12 Radial variation of computed and measured solids velocities (averaged from 10 to 15 s for computed results. Experimental velocities obtained from flux measurements using an electrostatic ball probe. (Tsuo and Gidaspow, 1990. Reproduced by permission of the American Institute of Chemical Engineers. Copyright © 1992 AIChE. All rights reserved.)

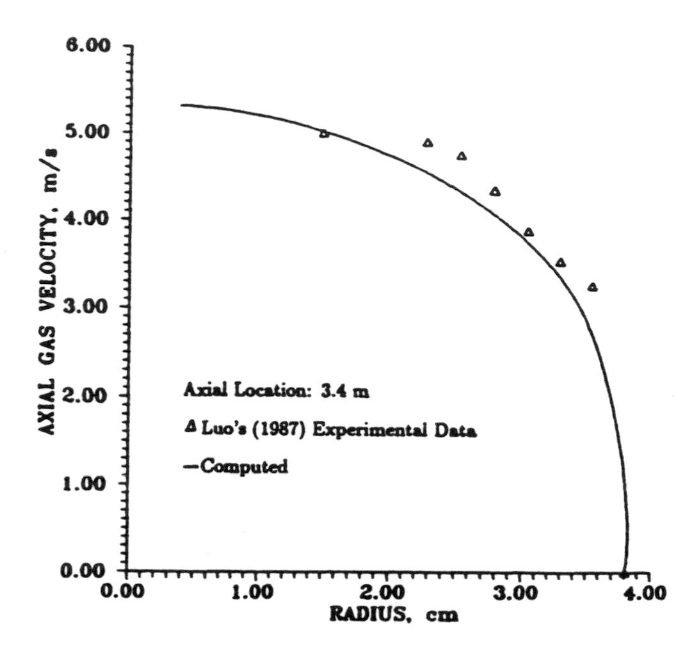

Fig. 8.13 Radial variation of computed and measured gas velocities in the IIT riser (averaged from 10 to 15 s for computed results). (Tsuo and Gidaspow, 1990. Reproduced by permission of the American Institute of Chemical Engineers. Copyright © 1992 AIChE. All rights reserved.)

8.7 Computation of Core–Annular Regime

At the time of computation typical dense CFB data have been obtained by Weinstein *et al.* (1984) at the City College of New York and by Bader *et al.* (1988) at the Institute of Gas Technology (IGT). The experimental conditions of Bader *et al.* were chosen for the computer simulation. The simulations and the experiment were conducted in a 30.5 cm diameter riser with 76 μm FCC catalyst. The superficial gas velocity is 3.7 m/s and the circulating solids flux is 98 kg/m^2-s. Figure 8.14 shows the configuration of the system and the computational conditions. A viscosity coefficient of 0.724 Pa-s was used.

The computer simulation was done for 18 s of the real fluidization time. It took 8 s to fully fill the empty pipe with particles.

The computed volume fraction data were converted into a series of density plots. The number of black dots is proportional to the volume fraction of the solid phase. Figure 8.15 shows the density plots for the solids volume fraction at times of 10 and 16 s after the start-up of particle injection. The empty pipe was fully

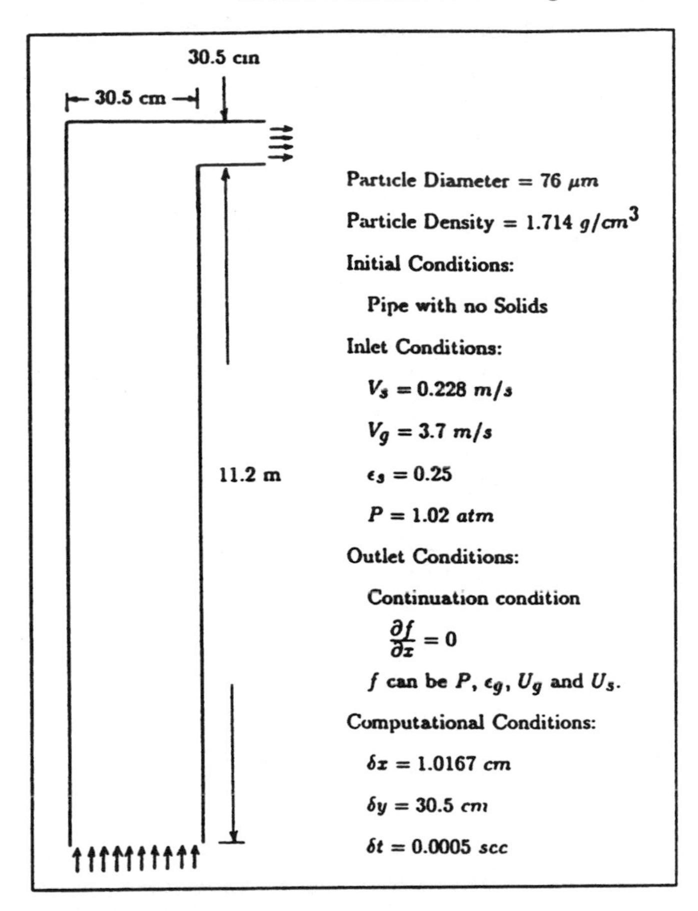

Fig. 8.14 IGT system configuration and computational inlet conditions. (Tsuo and Gidaspow, 1990. Reproduced by permission of the American Institute of Chemical Engineers.)

filled with particles after 8 s of real fluidization time. From the dot density plots, the non-homogeneous distribution of solid particles can be seen in both the axial and the radial directions. There is a dense annular layer at the wall and a dilute core at the center of the riser. Typical computed distributions of porosity in the riser at two axial locations equal to 4.1 and 9.1 m above the solids entry port are shown in Figs. 8.16 and 8.17, respectively. The experimental data of Bader *et al.* (1988) are also included in Figs. 8.16 and 8.17. The match between the computed and the measured porosities is reasonably good. Both the theory and the computation show high solids concentrations near the wall of the riser

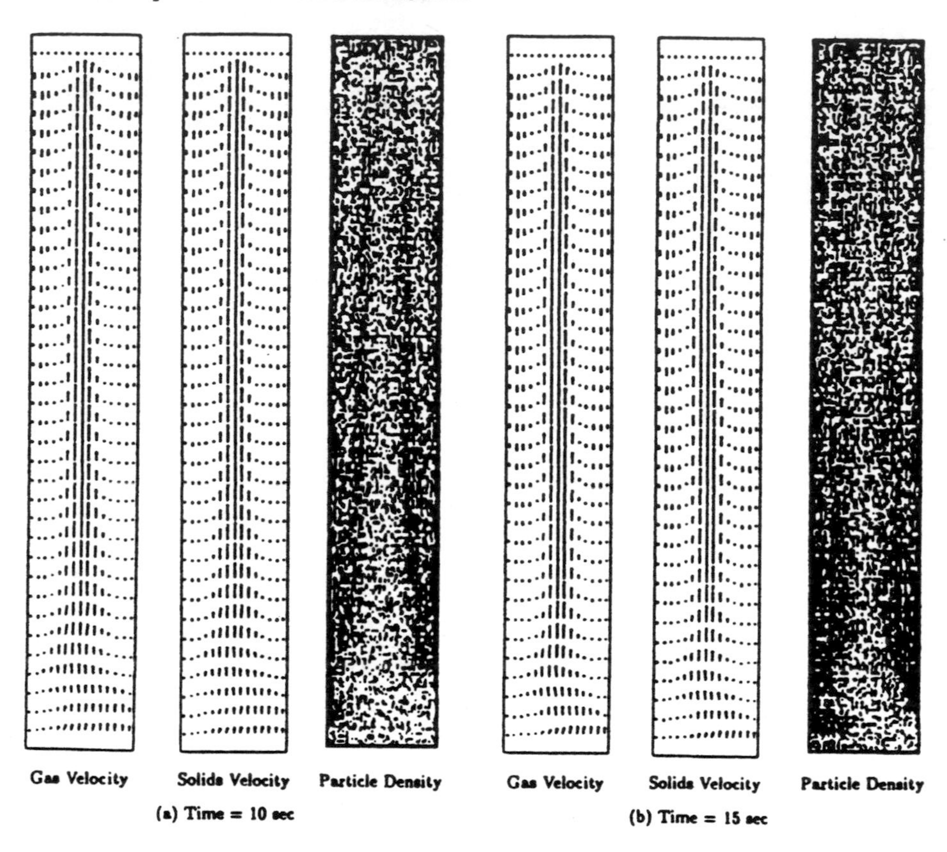

Fig. 8.15 Computed distributions of gas velocities, solids velocities, and particle density of 76 μm FCC particles in the IGT riser. (Tsuo and Gidaspow, 1990. Reproduced by permission of the American Institute of Chemical Engineers. Copyright © 1992 AIChE. All rights reserved.)

corresponding to porosities of 0.85 and 0.75, but much higher porosities, 0.95 to 0.98, at the center of the riser.

Figures 8.18 and 8.19 show typical radial distributions of axial time-averaged solids velocities at axial locations of 4.1 and 9.1 m, respectively. The particle velocity in the center of the riser was computed to be two to three times the superficial gas velocity in the riser.

This sharp rise of the center velocity is caused by the fact that the thick annular region seen in Fig. 8.15 blocks the upward motion of the gas and the particles. By conservation of mass the smaller effective flow area leads to a correspondingly larger gas, and through the drag, a larger central particle velocity. This effect is also seen in all of the measurements obtained by Miller (1991). For example, Fig.

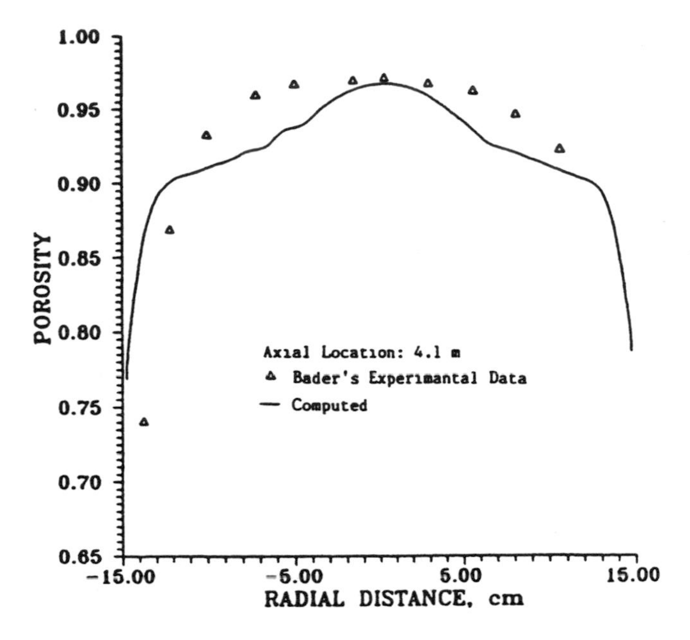

Fig. 8.16 Radial variation of computed and measured void fraction in the IGT riser at 4.1 m. Averaged from 12 to 18 s for computed results. (Tsuo and Gidaspow, 1990. Reproduced by permission of the American Institute of Chemical Engineers. Copyright © 1992 AIChE. All rights reserved.)

8.7 shows the particle velocity near the center for the middle section of the riser to be about 6 m/s for an inlet gas velocity of 2.9 m/s. Miller's upper section shows a lower velocity due to the bend close to the upper section. In his lowest section the particles had not fully accelerated as yet. Miller's calculations showed a significant acceleration effect in this lowest section.

Figures 8.18 and 8.19 also show the experimental data of Bader *et al.* (1988). They determined their particle velocities directly by use of a Pitot tube, averaging the wild fluctuations in pressure. The flux was obtained by the extraction probe already discussed in the previous paragraph. The porosity plotted in Figs. 8.17 and 8.18 was obtained indirectly from the flux and the solids velocity. In Figs. 8.18 and 8.19, the downward and the upward refer to the upward or downward position of the Pitot tube. The difference between the data shows the order of fluctuations of velocity. Such fluctuations were observed in Miller's experiments visually. In view of these experimental difficulties and the use of only a rough value of the solids viscosity in the model, highly idealized inlet conditions and a slice approximation for radial geometry, the comparison between Bader *et al.'s* (1988) experiment and the theory is very reasonable.

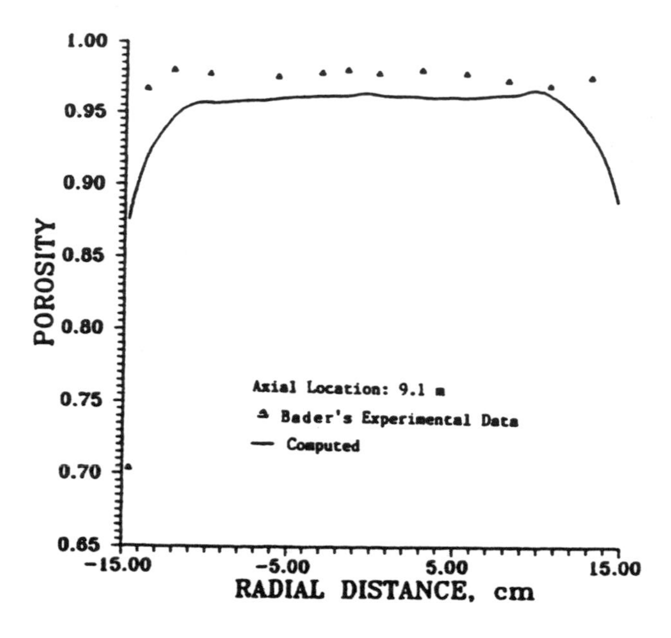

Fig. 8.17 Radial variation of computed and measured void fraction in the IGT riser at 9.1 m (averaged from 12 to 18 s for computed results). (Tsuo and Gidaspow, 1990. Reproduced by permission of the American Institute of Chemical Engineers. Copyright © 1992 AIChE. All rights reserved.)

Figure 8.20 shows typical distributions of axial gas velocity at the two locations. For 76 μm FCC particle flow, the gas flow pattern is similar to the solids flow patterns.

Figure 8.21 shows radial distributions of radial solids velocities at 15 s of real fluidization time at the two locations of 4.1 and 9.1 m. The particles at the 4.1 m level moved from the wall to the center and at the 9.1 m level moved in the reverse direction, from the center to the wall in the left-hand side of the pipe at 15 s after start-up. These velocities oscillate in a turbulent manner. The radial velocities are small compared with the axial velocities.

8.8 Radial Profiles and Turbulence

The objective of comparing the viscous model radial particle concentration profiles to experiments was to determine whether the model could predict the radial inhomogeneities observed in real systems and in experiments. An alternate approach was to use an extension of semi-empirical turbulence models widely

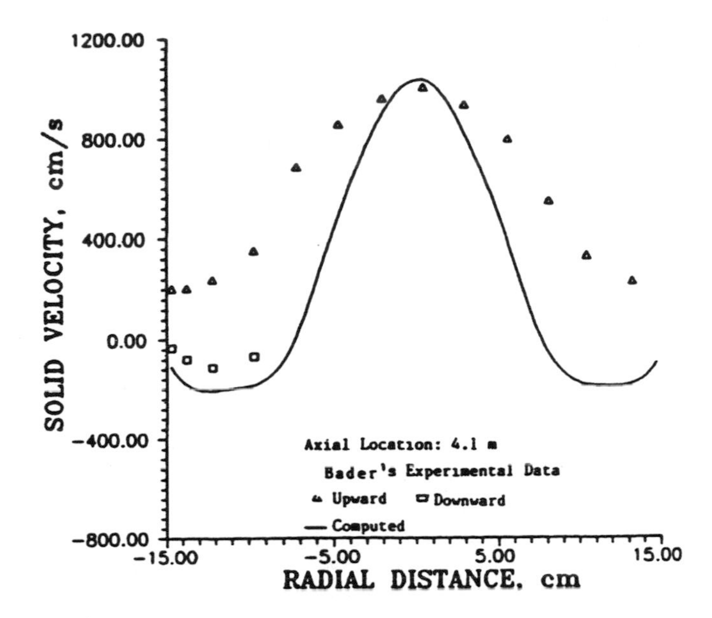

Fig. 8.18 Radial variation of computed and measured axial solids velocity in the IGT riser at 4.1 m. (Tsuo and Gidaspow, 1990. Reproduced by permission of the American Institute of Chemical Engineers. Copyright © 1992 AIChE. All rights reserved.)

used in single-phase flows. Such an approach involves the substitution of average plus fluctuating components of all dependent variables into the conservation laws, performing a time averaging and then using some semi-empirical equations for the unknown Reynolds stresses. Although such an approach had a considerable success in single phase flow, the low-frequency turbulence normally observed in dense two-phase flow makes it possible to directly compute the large-scale turbulence. The small-scale turbulence related to the isotropic part of the Reynolds stresses is computed in the kinetic theory chapter using our equation for an oscillating kinetic energy. Hence, there is no need to introduce two-phase turbulence. As in low Reynolds number single-phase flow, turbulence is computed by direct numerical simulation. However, as in single-phase flow, such direct numerical simulations take longer, since all the oscillations must be resolved.

At the time of the simulation, the only radial concentration data that were available were those of Weinstein *et al.* (1986) from CCNY. Their system was approximated as shown in Fig. 8.22. Two simulations were performed by feeding particles into an initially empty pipe. A viscosity coefficient of 7.24 poises was used. Figure 8.23 shows the particle velocity and concentrations 12 seconds after injection for the high flux simulation. There exists a dilute core of rapidly rising

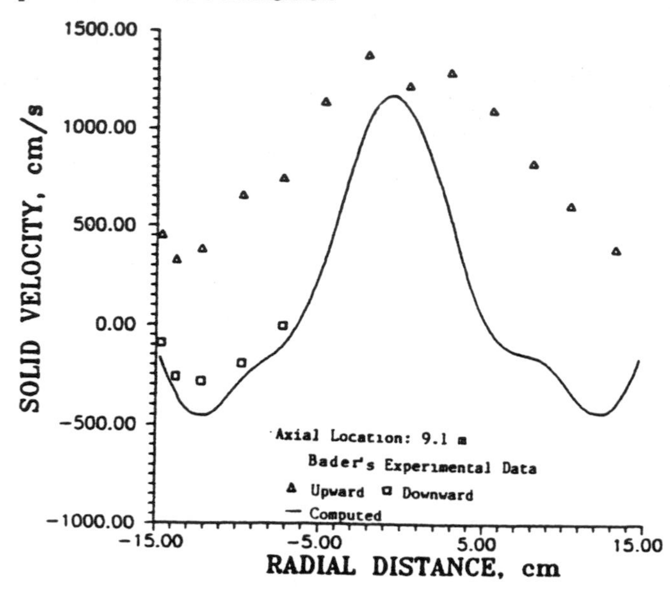

Fig. 8.19 Radial variation of computed and measured axial solids velocity in the IGT riser at 9.1 m. (Tsuo and Gidaspow, 1990. Reproduced by permission of the American Institute of Chemical Engineers. Copyright © 1992 AIChE. All rights reserved.)

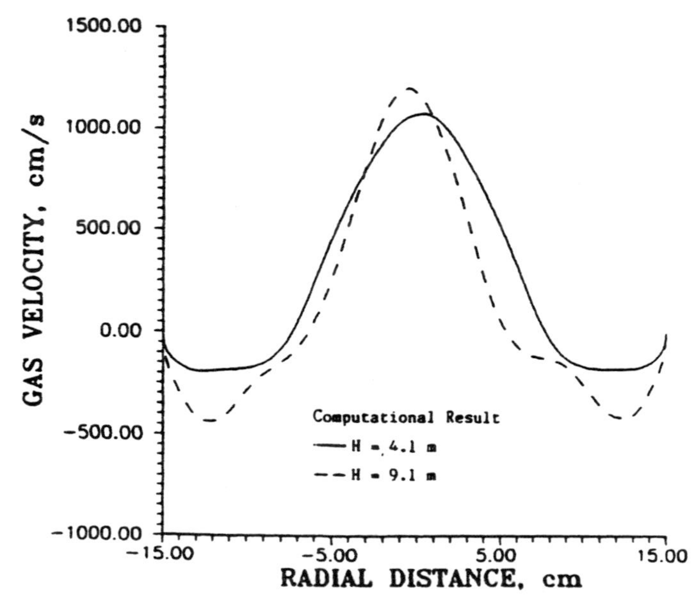

Fig. 8.20 Computed radial variation of axial gas velocity. (Tsuo and Gidaspow, 1990. Reproduced by permission of the American Institute of Chemical Engineers. Copyright © 1992 AIChE. All rights reserved.)

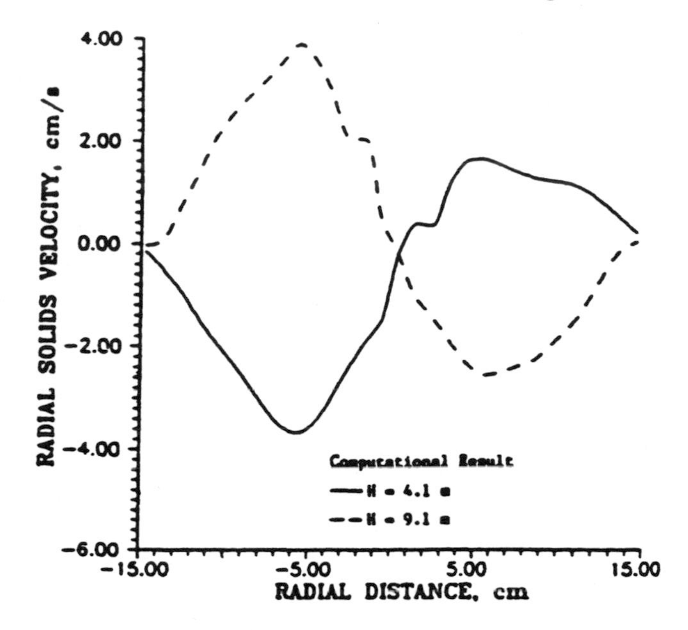

Fig. 8.21 Radial variation of computed radial solids velocity at 15 s real fluidization time in the IGT riser. (Tsuo and Gidaspow, 1990. Reproduced by permission of the American Institute of Chemical Engineers.)

particles with a dense slowly descending annulus. For the conditions of the experiment, flow never develops. The bottom half of the pipe is much denser than the top half. In Fig. 8.23, the first meter of the pipe does not correspond to the experiment because of an unrealistic assumption of constant inlet conditions. Figure 8.24 shows a comparison of time averaged particle concentration to Weinstein *et al.*'s (1986) data. The agreement is very good. The model predicts the observed behavior.

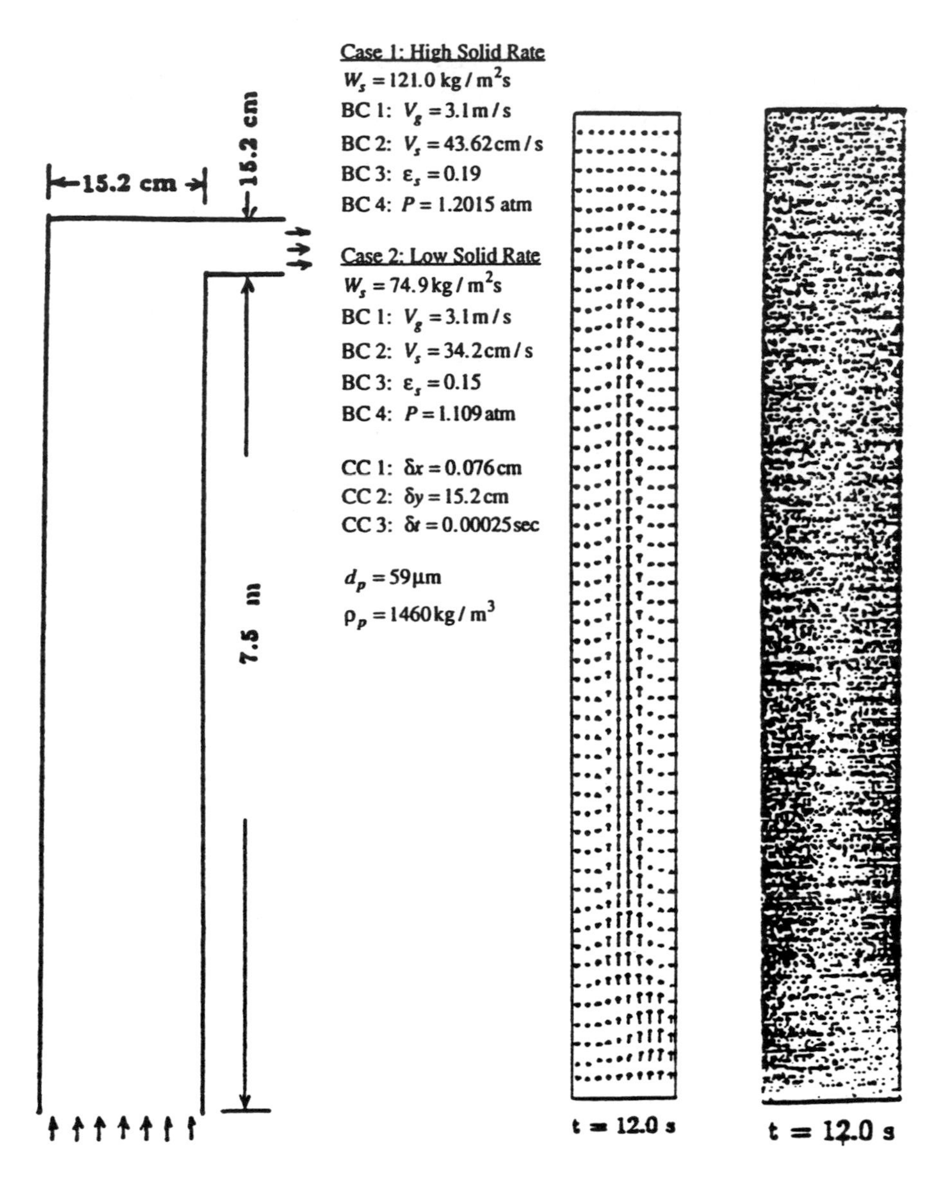

Fig. 8.22 Geometry of the City College of New York fast fluid bed for computer modeling with an idealized inlet condition. (From Gidaspow *et al.*, 1991. Reproduced by permission of Hemisphere Publishing, Washington D.C.)

Fig. 8.23 Computed velocities and concentrations for CCNY riser for $W_s = 121$. (From Gidaspow *et al.* 1991. Reproduced by permission of Hemisphere Publishing, Washington D.C.)

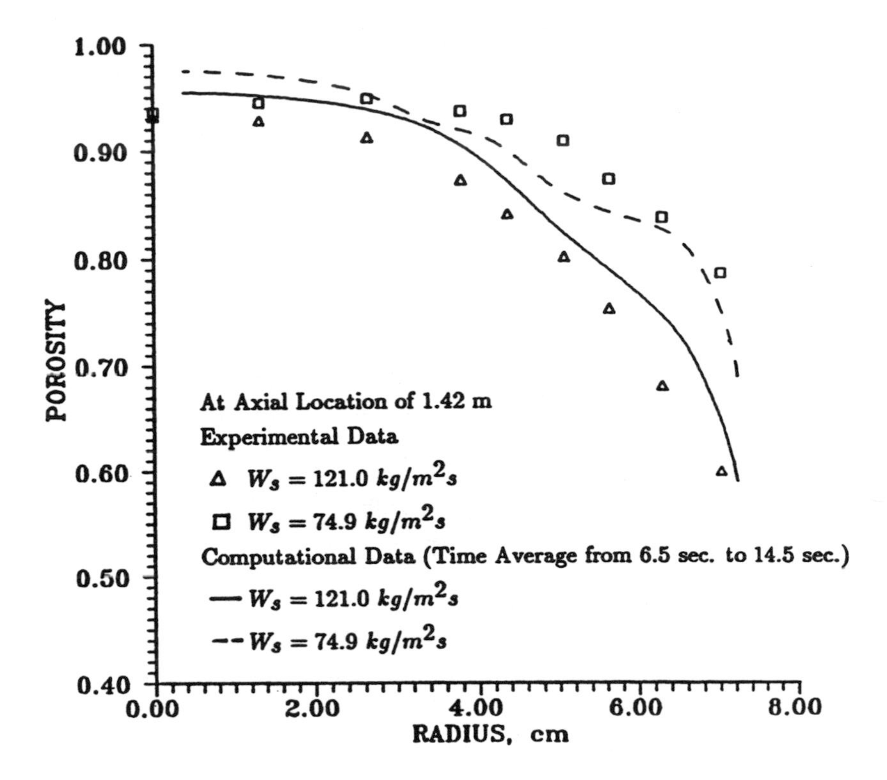

Fig. 8.24 A comparison of computed and measured porosities in the CCNY riser; $d_p = 59\ \mu m$, $\rho_p = 1460\ kg/m^3$, viscosity coefficient = 7.24 poises. (From Gidaspow *et al.*, 1991. Reproduced by permission of Hemisphere Publishing, Washington, D. C.)

Exercises

1. *Scale Factors and Flow Regimes for Homogeneous Two Phase Flow*

Consider homogeneous; that is, equal velocity, two phase particulate flow.

(a) Show that for the incompressible case, analogous to single phase flow, there exists a natural scale factor, called Reynolds number, which divides high-velocity and creeping flow cases.

However, for the two phase homogeneous flow the Reynolds number is

$$Re = \frac{\rho_m U_o d_p}{\mu_s + \mu_f},$$

where the mixture density is

$$\rho_m = \varepsilon_s \rho_s + \varepsilon_f \rho_f$$

and where the other symbols are as in Eq. (8.41).

(b) Show that for dense gas–solid flow,

$$\mathrm{Re} = \frac{\varepsilon_s \rho_s U_o d_p}{\mu_s} \quad \text{for} \quad \varepsilon_s \rho_s >> \varepsilon_f \rho_f .$$

In other words, the gas turbulence has no effect on the particulate flow.

(c) Show that for dilute flow, e.g., $\varepsilon_s < 0.001$, gas-phase turbulence will govern the flow.

2. *Boundary Layer on a Plate in a Two Phase Flow "Wind Tunnel"*

Consider gas–particle flow over a flat plate located in the center of a pneumatic conveyor in the developed flow region, where the pressure drop equals the weight of the bed and buoyancy is balanced by drag (see Fig. 8.25). Since $\rho_s >> \rho_g$, the mixture momentum boundary layer equation for a constant solid kinematic viscosity $\nu_s = \mu_s / \rho_s$ becomes, in terms of $C_s = \rho_s \varepsilon_s$, as follows. The boundary layer equation for C_s in terms of a solids diffusion coefficient D_s also follows:

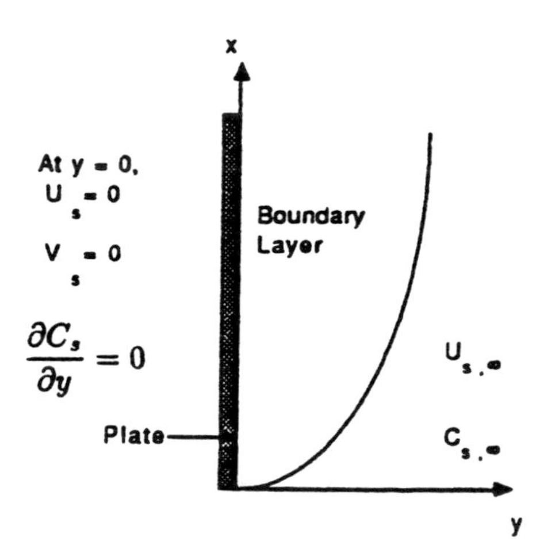

Fig. 8.25 Momentum boundary layer due to particulate viscosity.

Mixture Momentum Boundary Layer

$$C_s U_s \frac{\partial U_s}{\partial x} + C_s V_s \frac{\partial U_s}{\partial y} = \nu_s \frac{\partial^2 U_s}{\partial y^2},$$

Solids Concentration Boundary Layer

$$U_s \frac{\partial C_s}{\partial x} + V_s \frac{\partial C_s}{\partial y} = D_s \frac{\partial^2 C_s}{\partial y^2}.$$

Find:
(a) A solids stream function ψ_s.
(b) A similarity variable η that will reduce the preceding partial differential equations into ordinary differential equations.
(c) The ordinary differential equations for the boundary layer.
(d) The drag on the plate.
(e) The thickness of the concentration boundary layer in terms of the thickness of the momentum boundary layer.
(f) Note that the time-average drag on this plate suspended in the two phase flow will be zero for a zero solids viscosity. Hence, suggest a new method of measuring viscosity in particulate two-phase flow. Discuss potential problems and the approximation made in deriving the solids concentration boundary layer equations (see Schlichting, 1960, for very similar mathematics).

3. *Developed Time-Averaged Two Phase Flow*

In developed two phase flow the variations of velocity in the basic direction of flow are set to zero. The equations reduce themselves to ordinary differential equations. For single phase incompressible flow in a pipe or between parallel plates such solutions give a parabolic velocity distribution. Reduce the two-phase flow equations for such a case. For a kinetic theory approach, see Sinclair and Jackson (1989).

Literature Cited

Abed, R. (1984). "The Characterizaiton of Turbulent Fluid-Bed Hydrodynamics," in H. Kunii and R. Toei, Eds. *Fluidization*. New York: AIChE.

Arastoopour, H., and D. Gidaspow (1979). "Analysis of IGT Pneumatic Conveying Data and Fast Fluidization Using a Thermodynamic Model," *Powder Technol.* 22, 77–87.

Bader, R., J. Findlay, and T. M. Knowlton (1988). "Gas–Solid Flow Patterns in a 30.5 cm Diameter Circulating Fluidized Bed," pp. 123–127 in P. Basu and J. F. Large, Eds. *Circulating Fluidized Bed Technology II*. New York: Pergamon Press.

Blake, T. R., H. Webb, and P. B. Sunderland (1990). "The Nondimensionalization of Equations Describing Fluidization with Application to the Correlation of Jet Penetration Height," *Chem. Eng. Sci.* **45**, 365–371.

Bird, R. B., W. E. Stewart, and E. N. Lightfoot (1960). *Transport Phenomena*. New York: John Wiley & Sons.

Drew, D. A., and R. T. Lahey, Jr. (1979). "Application of General Constitutive Principles to the Derivation of Multidimensional Two Phase Flow Equations," *Int'l. J. Multiphase Flow* **5**, 243–264.

Fitzgerald, T. J. (1985). "Course Particle Systems," pp. 413–435 (Ch. 12) in J. F. Davidson, R. Clift, and D. Harrison, Eds. *Fluidization*. New York: Academic Press.

Galtier, P. A., R. J. Pointer, and T. E. Patureaux (1989). "Near Full-Scale Cold Flow Model for the R2R Catalytic Cracking Process," pp. 17–24 in J. R. Grace, L. W. Shemilt, and M. A. Bergougnou, Eds. *Fluidization VI*. New York: Engineering Foundation.

Geldart, D., and M. J. Rhodes (1986). "From Minimum Fluidization to Pneumatic Transport: A Critical Review of the Hydrodynamics," pp. 21–32 in P. Basu, ed. *Circulating Fluidized Bed Technology*.

Gidaspow, D. (1986). "Hydrodynamics of Fluidization and Heat Transfer: Supercomputer Modeling," *Appl Mech. Rev.* **39** (1), 1–23.

Gidaspow, D., Y. P. Tsuo, and K. M. Luo (1989a). "Computed and Experimental Cluster Formation and Velocity Profiles in Circulating Fluidized Beds," pp. 81–88 in J. C. Grace, L. W. Schemilt, and M. A. Bergougnou, eds. *Fluidization VI*. New York: Engineering Foundation.

Gidaspow, D., Y.-T. Shi, J. Bouillard, and D.Wasan (1989b). "Hydrodynamics of a Lamella Electrosettler," *AIChE J.* **35**, 714–724.

Gidaspow, D., Y. P. Tsuo and J. Ding (1989c). "Hydrodynamics of Circulating and Bubbling Fluidized Beds," presented at Materials Issues in Circulating Fluidized Bed Combustors, EPRI–Argonne National Lab Symposium, EPRI GS-6747, pp. 4-1 – 4-27, Feb. 1990. Preprinted in Gidaspow *et al.* (1991).

Gidaspow, D., Y. P. Tsuo, and J. Ding (1991). "Hydrodynamics of Circulating and Bubbling Fluidized Beds," pp. 485–512 in J. H. Kim, J. M. Hyun, and C.-O. Lee, Eds. *Fluids Engineering: Korea–U.S. Progress*. Washington, D.C.: Hemisphere Publishing Co.

Glicksman, L. R. (1984). "Scaling Relationships for Fluidized Beds," *Chem. Eng. Sci.* **39**, 1373–1379.

Glicksman, L. R. (1990). "The Use of Scaling Laws to Study Erosion in Circulating Fluidized Beds," presented at the Materials Issues in Circulating Fluidized Bed Combustors, pp. 13-1 to 13-5 in *EPRI–Argonne National Lab Symposium*, EPRI GS-6747.

Jayaswal, U. K., D. Gidaspow, and D. Wasan (1990). "Continuous Separation of Particles in Nonaqueous Media in a Lamella Electrosettler," *Separations Technology* **1**, 3–17.

Jeffrey, D. J. and A. Acrivos (1976). "The Rheological Properties of Suspensions of Rigid Particles," *AIChE J.* **22**, 417–432.

Jeffreys, H. (1961). *Cartesian Tensors*. Cambridge, U.K.: Cambridge University Press.

Luo, K. M. (1987). *Experimental Gas–Solid Vertical Transport*. Ph.D. Thesis. Chicago, Illinois: Illinois Institute of Technology.

Lyzczkowski, R. W., D. Gidaspow, and C. W. Solbrig (1982). "Multiphase Flow-Models for Nuclear, Fossil and Biomass Energy Conversion," pp. 198–351 in A. S. Majumdar and R. A. Mashelkar, Eds. *Advances in Transport Processes*, Vol. II. Wiley-Eastern Publisher.

Miller, A. (1991). *Dense, Vertical Gas–Solid Flow in a Pipe*, Ph.D. Thesis. Chicago, Illinois: Illinois Institute of Technology.

Miller, A., and D. Gidaspow (1992). "Dense, Vertical Gas–Solid Flow in a Pipe," *AIChE J.* **38**, 1801–1815.

Rhodes, M. J., T. Hirama, G. Gerutti, and D. Geldart (1989). "Non-uniformities of Solids Flow in Risers of Circulating Fluidized Beds," pp. 733–780 in J. C. Grace, L. W. Schemilt, and M. A. Bergougnou, Eds. *Fluidization VI*. New York: Engineering Foundation.

Roy, R., J. F. Davidson, and V. G. Tuponogiv (1990). "The Velocity of Sound in Fluidized Beds," *Chem. Eng. Sci.* **45**, 3233–3245.

Schlichting, H. (1960). *Boundary Layer Theory*. New York: McGraw-Hill.

Schügerl, K. (1971). "Rheological Behavior of Fluidized Systems," Ch. 6 in J. F. Davidson, R. Clift, and D. Harrison, Eds. *Fluidization*. New York: Academic Press.

Sinclair, J. L., and R. Jackson (1989). ""Gas–Solid Flows in a Vertical Pipe with Particle–Particle Interactions," *AIChE J.* **35**, 1473–1486.

Squires, A. M., M. Kwauk, and A. A. Avidan (1985). "Fluid Beds: At Last, Challenging Two Entrenched Practices," *Science* **230**, 4732, 1329–1337.

Syamlal, M., and D. Gidaspow (1985). "Hydrodynamics of Fluidization: Prediction of Wall to Bed Heat Transfer Coefficients," *AIChE J.* **31**, 127–135.

Tsuo, Y. P. (1989). *Computation of Flow Regimes in Circulating Fluidized Beds*. Ph.D. Thesis. Chicago, Illinois: Illinois Institute of Technology.

Tsuo, Y. P. and D. Gidaspow (1990). "Computation of Flow Patterns in Circulating Fluidized Beds," *AIChE J.* **36**, 885–896.

Weinstein, H., M. Meller, M. J. Shao, and R. J. Parisi (1984). "The Effect of Particle Density on Holdup in a Fast Fluid Bed," in T. M. Knowlton, Ed. "Fluidization and Fluid Particle Systems: Theories and Applications," *AIChE Symposium Series* **234** (80), 52–59.

Weinstein, H., M. Shao, M. Schnitzlein, and R. A. Graff (1986). "Radial Variation in Void Fraction in a Fast Fluidized Bed," pp. 329–336 in K. Ostergaards and A. Sorensen, Eds. *Fluidization V: Proceedings of the Fifth Engineering Foundation Conference on Fluidization*. New York: AIChE.

Yerushalmi, J. (1986). "High Velocity Fluidized Beds," Ch. 7 in D. Geldart, Ed. *Gas Fluidization Technology*. New York: John Wiley & Sons.

9

KINETIC THEORY APPROACH

9.1 Introduction 239
9.2 Maxwellian Distribution for Particles 240
9.3 Properties of the Maxwellian State 242
9.4 Dynamics of an Encounter between Two Particles 244
9.5 The Frequency of Collisions 247
9.6 Mean Free Path 252
9.7 Elementary Treatment of Transport Coefficients 253
9.8 Boltzmann Integral–Differential Equation 256
9.9 Maxwell's Transport Equation 259
9.10 Conservation Laws with No Collisions 261
9.11 Second Approximation to the Frequency Distribution 263
9.12 Integral Equation Solver Strategy 272
9.13 Viscous Kinetic Stress Tensor 274
9.14 Dense Transport Theorem 276
9.15 Particulate Momentum Equation 279
9.16 Fluctuating Kinetic Energy Equation 281
9.17 Viscosity—Collisional Momentum Transfer 284
9.18 Granular Conductivity 290
Literature Cited 294

9.1 Introduction

Bagnold (1954) is generally credited with starting the kinetic theory approach of granular flow. By using a very primitive expression for collision frequency of particles, he derived an expression for the repulsive pressure of particles for uniform shear flow. His repulsive pressure is proportional to the square of the velocity gradient and particle diameter and directly proportional to the particle density. This same dependence is found in the modern theories discussed in this chapter. In 1980, Ogawa, Umemura and Oshima (1980) suggested that the mechanical energy of granular flow is first transformed into random particle motion and then dissipated into internal energy (see Fig. 9.1). They formulated a conservation of energy equation for this random kinetic energy in which this kinetic energy was

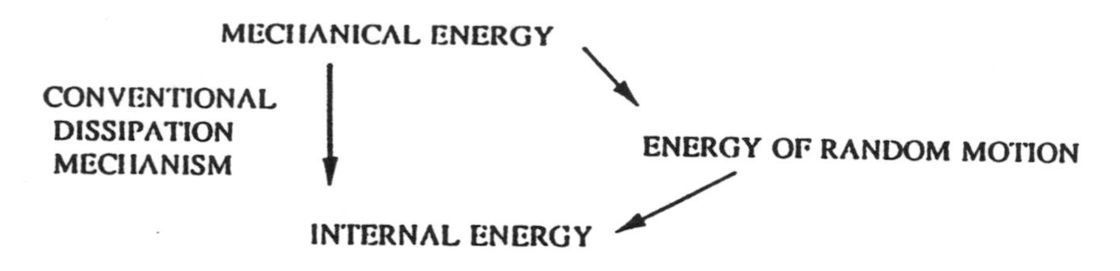

Fig. 9.1 Energy flow in granular and particulate multiphase flows.

produced by shear and dissipated by a heatlike flux vector and a sink. Savage and Jeffrey (1981) related this fluctuating velocity to the absolute value of the shear gradient by means of a dimensionless group and observed that the dense phase kinetic theory, as described in the classical book by Chapman and Cowling (1961), could be put to use. Earlier Soo (1967) had used a similar approach. Jenkins and Savage (1983) continued the development of the kinetic theory approach by modifying the classical treatment of the dynamics of encounter of two particles by introducing the concept of the restitution coefficient and by using the Boltzmann transport equation for dense gases. Lun *et al.* (1984) continued this development. Shahinpour and Ahmadi (1983) and Johnson and Jackson (1987) also contributed to the development of this approach. A number of investigators including the group at the Morgantown Energy Technology Center (Syamlal, 1987), tried to include the drag into these equations to generalize the treatment to multiphase flow. The following description is based on the work of Ding and Gidaspow (1990).

9.2 Maxwellian Distribution for Particles

The frequency distribution of velocities f of particles is a function of time t, position $\mathbf{r}$ and instantaneous velocity $\mathbf{c}$ consistent with the Newton's law of motion:

$$f = f(t,\ \mathbf{r},\ \mathbf{c}). \tag{9.1}$$

The number of particles per unit volume n is

$$n = \int f\ d\mathbf{c}. \tag{9.2}$$

The mean value of a quantity ϕ is defined to be

$$n < \phi > = \int \phi f\ d\mathbf{c}. \tag{9.3}$$

Hence, the hydrodynamic velocity **v** becomes

$$\mathbf{v} = \frac{1}{n} \int \mathbf{c} f \, d\mathbf{c}. \tag{9.4}$$

Based on limited experimental measurements of instantaneous velocities of particles in a liquid fluidized by Carlos and Richardson (1968), it is reasonable to assume that the particles oscillate about mean values in a chaotic manner. Hence, the velocity distribution follows a normal distribution about its mean. In the theory of probability and statistics the normal or Gaussian probability function f_N is expressed as follows about its mean μ with a standard deviation σ:

$$f_N(x) = \frac{1}{\sqrt{2\pi\sigma^2}} \exp\left[-\frac{(x-\mu)^2}{2\sigma^2} \right]. \tag{9.5}$$

Its properties can be better understood when one observes that it is also the unit source solution of the Green's function in an infinite medium for the diffusion equation (Carlslaw and Jaeger, 1959):

$$\frac{\partial f_N}{\partial \sigma^2} = \frac{\partial^2 f_N}{\partial x^2}, \tag{9.6}$$

where σ^2 plays the role of time in diffusion. Thus, the unit source condition is

$$\int_{-\infty}^{\infty} f_N \, dx = 1. \tag{9.7}$$

The variance σ^2 is given in statistics by a formula similar to Eq. (9.3):

$$\sigma^2 = < (x-\mu)^2 > = \int_{-\infty}^{\infty} (x-\mu)^2 f_N \, dx. \tag{9.8}$$

For three-dimensional space equipartition of energy is assumed. This means the probability distributions are multiplicative. Then the variance of the square of the velocity about zero mean is

$$\sigma_{\mathbf{c}}^2 \equiv n < \mathbf{c}^2 > = \iiint \mathbf{c}^2 f_x f_y f_z \, d\mathbf{c}, \tag{9.9}$$

where

$$\mathbf{c}^2 = c_x^2 + c_y^2 + c_z^2. \tag{9.10}$$

When

$$c_x^2 = c_y^2 = c_z^2, \tag{9.11}$$

$$< \mathbf{c}^2 > = 3 < c_x^2 >. \tag{9.12}$$

Hence, the normal distribution for velocities, called the Maxwellian distribution, can be written as

$$f_x = n(3/(2\pi < \mathbf{c}^2 >))^{1/2} \exp[(-3c_x^2)/(2< \mathbf{c}^2 >)]. \tag{9.13}$$

In granular flow, Jenkins and Savage (1983) and those who continued their work called one-third the mean square velocity the granular temperature, called Θ here to distinguish it from T, the thermal temperature:

$$\Theta = \tfrac{1}{3} < \mathbf{c}^2 >. \tag{9.14}$$

In terms of this temperature, the Maxwellian distribution for particles in three dimensions is

$$f(\mathbf{r},\mathbf{c}) = \frac{n}{(2\pi\Theta)^{3/2}} \exp\left[-\frac{(\mathbf{c}-\mathbf{v})^2}{2\Theta}\right]. \tag{9.15}$$

9.3 Properties of the Maxwellian State

To illustrate the integrations involved in the kinetic theory, the mean speed is computed. The mean value of any function $< \phi >$ is given by Eq. (9.3). Hence, the mean speed $< c >$ about zero hydrodynamic velocity using (9.15) is

$$< c > = \frac{1}{(2\pi\Theta)^{3/2}} \int c \exp\left[-\frac{\mathbf{c}^2}{2\Theta}\right] d\mathbf{c}. \tag{9.16}$$

With complete spherical symmetry,

$$d\mathbf{c} = dc_x\, dc_y\, dc_z = 4\pi c^2\, dc. \tag{9.16a}$$

Hence, (9.16) becomes

$$<c> = \frac{4\pi}{(2\pi\Theta)^{3/2}} \int_0^\infty c^3 \exp\left[-\frac{c^2}{2\Theta}\right] dc. \tag{9.17}$$

Infinite integrals of the type given by Eq. (9.17) frequently occur in kinetic theory.

The simplest of these integrals is the error function integral, *erf*:

$$erf\, y = \frac{2}{\sqrt{\pi}} \int_0^y e^{-x^2}\, dx, \tag{9.18}$$

where

$$erf\, \infty = 1. \tag{9.19}$$

Chapman and Cowling (1961) using gamma functions show that

$$\int_0^\infty e^{-\alpha c^2} c^r\, dc = \tfrac{\sqrt{\pi}}{2} \cdot \tfrac{1}{2} \cdot \tfrac{3}{2} \cdot \tfrac{5}{2} \cdots \tfrac{r-1}{2} \alpha^{-(r+1)/2}, \tag{9.20}$$

where r is an even integer, and

$$\int_0^\infty e^{-\alpha c^2} c^r\, dc = \tfrac{1}{2} \alpha^{-(r+1)/2} \left(\frac{r-1}{2}\right)!, \tag{9.21}$$

where r is an odd integer greater than minus one. Using (9.21),

$$\int_0^\infty c^3 e^{-c^2/(2\Theta)}\, dc = \tfrac{1}{2} \left(\frac{1}{2\Theta}\right)^{-2}. \tag{9.22}$$

Hence, Eq. (9.17) gives

$$<c> = (8\Theta/\pi)^{1/2}. \tag{9.23}$$

This is the same expression as that for the mean value of the molecular speed given by Chapman and Cowling's Eq. (4.11-2) when one lets the ratio of the Boltzmann constant to the molecular mass equal one.

The mean value of c^2 can also be obtained using the same formulas:

$$<c^2> = \frac{4\pi}{(2\pi\Theta)^{3/2}} \int_0^\infty c^4 \exp\left[-\frac{c^2}{2\Theta}\right] dc. \tag{9.24}$$

Integration using Eq. (9.20) gives

$$< c^2 > = 3\Theta. \tag{9.25}$$

Hence, as in kinetic theory of gases,

$$< c^2 >^{1/2} = < c > \left(\frac{3\pi}{8}\right)^{1/2} = 1.086 < c >. \tag{9.26}$$

The two average velocities differ by less than 10 percent.

9.4 Dynamics of an Encounter between Two Particles

The following analysis differs from the classical kinetic theory by the introduction of a restitution coefficient first introduced by Jenkins and Savage (1983) into the kinetic theory of granular flow. As in classical theory, consider the dynamics of an encounter of two particles of mass m_1 and m_2. Let $\mathbf{r}_1$ and $\mathbf{r}_2$ be vectors in the rest frame from a fixed origin to the two particles. Introduce the vector $\mathbf{r}_c$ from the origin to the center of mass of particles:

$$\mathbf{r}_c = \frac{m_1\mathbf{r}_1 + m_2\mathbf{r}_2}{m_1 + m_2}. \tag{9.27}$$

The equations of motion for the particles are

$$m_1 \frac{d^2\mathbf{r}_1}{dt^2} = \mathbf{F}_1(r), \tag{9.28}$$

$$m_2 \frac{d^2\mathbf{r}_2}{dt^2} = -\mathbf{F}_1(r), \tag{9.29}$$

where the law of action and reaction has been used. On differentiation of Eq. (9.27) twice with respect to time and using Eqs. (9.28) and (9.29), the result is

$$(m_1 + m_2)^{-1}\left(m_1 \frac{d^2\mathbf{r}_1}{dt^2} + m_2 \frac{d^2\mathbf{r}_2}{dt^2}\right) = 0. \tag{9.30}$$

Since Eq. (9.30) shows that the acceleration of the center of mass vanishes, the center of mass is in uniform rectilinear motion. As in Chapman and Cowling (1961), let $\mathbf{G}$ be this center of mass velocity and m_o be the sum of m_1 and m_2. If the velocities of the particles before collision are $\mathbf{c}_1$ and $\mathbf{c}_2$ and after collision $\mathbf{c}_1'$ and $\mathbf{c}_2'$, then conservation of momentum shows that

$$m_o\mathbf{G} = m_1\mathbf{c}_1 + m_2\mathbf{c}_2 = m_1\mathbf{c}_1' + m_2\mathbf{c}_2'. \tag{9.31}$$

Let $\mathbf{c}_{12}$ be the relative velocity of particles. The relative velocity of the particles after collision is

$$\mathbf{c}_{12}' = \mathbf{c}_1' - \mathbf{c}_2'. \tag{9.32}$$

Let $\mathbf{k}$ be a unit vector directed form the center of the first particle to that of the second upon contact. Consider the particles to be smooth but inelastic with a restitution coefficient e which ranges between zero and one, depending upon the material. When e equals one, the relative velocities of the center of the particles are reversed upon collision, as described in Chapman and Cowling (1961) in paragraph 3.41. For inelastic particles, Jenkins and Savage (1983) let

$$\mathbf{k}\cdot\mathbf{c}_{12}' = -e(\mathbf{k}\cdot\mathbf{c}_{12}). \tag{9.33}$$

The individual particle velocities can be expressed in terms of the relative and the center of mass velocity as in Chapman and Cowling (1961) by solving Eq. (9.31) together with the definition of the relative velocity. For example, to obtain $\mathbf{c}_1$ and $\mathbf{c}_2$, solution of the equations

$$m_1\mathbf{c}_1 + m_2\mathbf{c}_2 = m_o\mathbf{G}, \tag{9.34}$$

$$\mathbf{c}_1 - \mathbf{c}_2 = \mathbf{c}_{12} \tag{9.35}$$

obtained by determinants gives

$$\mathbf{c}_1 = G + \left(m_2/m_o\right)\mathbf{c}_{12}, \tag{9.36}$$

$$\mathbf{c}_2 = G - \left(m_1/m_o\right)\mathbf{c}_{12}. \tag{9.37}$$

Similarly, using Eqs. (9.31) and (9.32),

$$\mathbf{c}_1' = G + \left(m_2/m_o\right)\mathbf{c}_{12}', \tag{9.38}$$

$$\mathbf{c}_2' = G + (m_1/m_o)\mathbf{c}_{21}'. \tag{9.39}$$

Using Eqs. (9.36) and (9.37) it can be shown that the sum of the kinetic energies of the particles is as follows:

$$\tfrac{1}{2}\left(m_1 c_1^{\,2} + m_2 c_2^{\,2}\right) = \tfrac{1}{2} m_o \left(G^2 + m_1 m_2 / m_o^{\,2} c_{12}^{\,2}\right). \tag{9.40}$$

Similarly, Eqs. (9.38) and (9.39) give

$$\tfrac{1}{2}\left(m_1 c_1'^2 + m_2 c_2'^2\right) = \tfrac{1}{2} m_o \left(G^2 + m_1 m_2 / m_o^{\,2} c_{12}'^{\,2}\right). \tag{9.41}$$

Then the translational kinetic energy change during collision ΔE is

$$2\Delta E = m_1 c_1'^2 + m_2 c_2'^2 - m_1 c_1^{\,2} - m_2 c_2^{\,2} = \frac{m_1 m_2}{m_o}\left(c_{12}'^{\,2} - c_{12}^{\,2}\right), \tag{9.42}$$

since the center of mass velocity of the two particles stays the same. Then using Eq. (9.33),

$$\Delta E = \tfrac{1}{2}\frac{m_1 m_2}{m_o}\left(e^2 - 1\right)\left(\mathbf{k}\cdot\mathbf{c}_{12}\right)^2. \tag{9.43}$$

For the case of equal mass particles,

$$m_1 = m_2 = m. \tag{9.44}$$

Equation (9.43) is Jenkins and Savage's (1983) Eq. (8), written as

$$\Delta E = \tfrac{1}{4} m\left(e^2 - 1\right)\left(\mathbf{k}\cdot\mathbf{c}_{12}\right)^2. \tag{9.45}$$

Two other relations will be used in further development of the theory. Using Eqs. (9.36), (9.38) and then (9.33), it follows that

$$\mathbf{c}_1' - \mathbf{c}_1 = m_2/m_o\left(\mathbf{c}_{12}' - \mathbf{c}_{12}\right) = -m_2/m_o\,(1+e)\left(\mathbf{k}\cdot\mathbf{c}_{12}\right)\mathbf{k}, \tag{9.46}$$

which for equal masses becomes

$$\mathbf{c}_1' - \mathbf{c}_1 = -\tfrac{1}{2}(1+e)(\mathbf{k}\cdot\mathbf{c}_{12})\mathbf{k}. \tag{9.47}$$

Similarly, using Eqs. (9.37) and (9.39) and then (9.33) for equal masses, it follows that

$$\mathbf{c}_2' - \mathbf{c}_2 = \tfrac{1}{2}(1+e)(\mathbf{k}\cdot\mathbf{c}_{12})\mathbf{k}. \tag{9.48}$$

9.5 The Frequency of Collisions

The objective of this section is to derive the classical binary frequency of collisions corrected for the dense packing effect, as done with the factor χ in Chapter 16 in Chapman and Cowling (1961). Analogously to the single frequency distribution function given by Eq. (9.1), a collisional pair distribution function, $f^{(2)}$ is introduced:

$$f^{(2)} = f^{(2)}(\mathbf{c}_1, \mathbf{r}_1, \mathbf{c}_2, \mathbf{r}_2). \tag{9.49}$$

It is defined such that

$$f^{(2)} d\mathbf{c}_1 d\mathbf{c}_2 d\mathbf{r}_1 d\mathbf{r}_2 \tag{9.50}$$

is the probability of finding a pair of particles in the volume $d\mathbf{r}_1 d\mathbf{r}_2$ centered on points $\mathbf{r}_1, \mathbf{r}_2$ and having velocities within the ranges $\mathbf{c}_1$ and $\mathbf{c}_1 + d\mathbf{c}_1$ and $\mathbf{c}_2$ and $\mathbf{c}_2 + d\mathbf{c}_2$. Figure 9.2 shows the geometry of collisions. It is that given by Savage and Jeffrey (1981) with a generalization to two rigid spheres of unequal diameters. Figure 9.3 shows the spherical geometry and the solid angle dk.

The number of binary collisions per unit time per unit volume N_{12} can be expressed as follows:

$$N_{12} = \iiint f^{(2)}(\mathbf{c}_1, \mathbf{r}, \mathbf{r}+d_{12}\mathbf{k}, \mathbf{c}_2)\cdot \mathbf{c}_{21}\cdot \mathbf{k} d_{12}{}^2 d\mathbf{k}\, d\mathbf{c}_1\, d\mathbf{c}_2, \tag{9.51}$$

where

$$\mathbf{c}_{21}\cdot \mathbf{k}\, d\mathbf{k} = c_{21}\cos\theta \sin\theta\, d\theta\, d\phi, \tag{9.52}$$

since the volume of the collision cylinder is

$$d\mathbf{r} = d_{12}{}^2 \sin\theta\, d\theta\, d\phi \cdot c_{21}\cos\theta \cdot dt$$

as seen in Fig. 9.4.

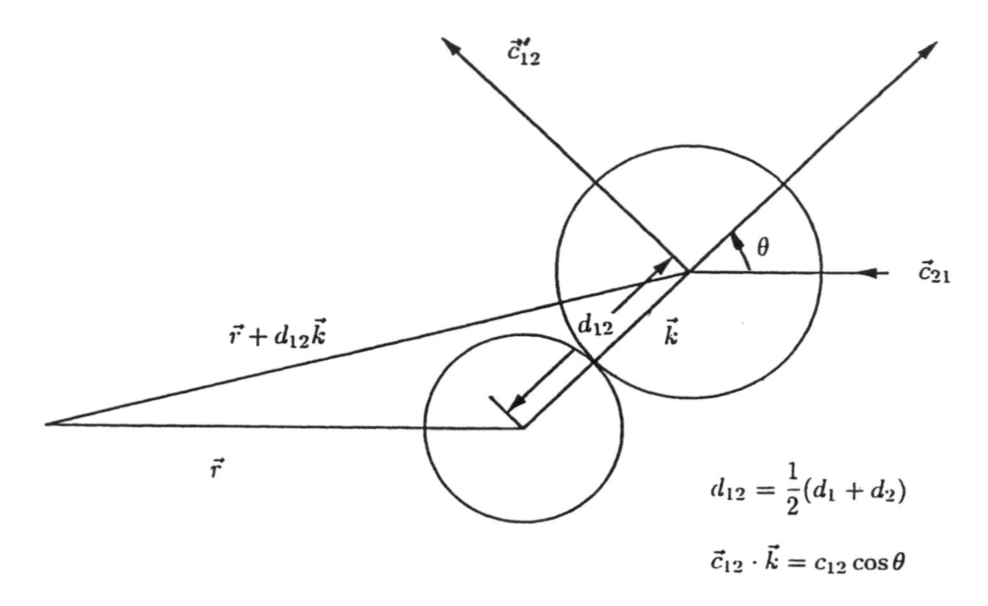

Fig. 9.2 Geometry of a collision of two spheres of diameters, d_1 and d_2.

Chapman and Cowling (1961) treat dense gases by assuming that

$$f^{(2)}(\mathbf{c}_1, \mathbf{r}, \mathbf{r}+d_{12}\mathbf{k}, \mathbf{c}_2) = \chi\left(\mathbf{r}+\tfrac{1}{2}d_{12}\mathbf{k}\right) \cdot f(\mathbf{r}, \mathbf{c}_1) \cdot f(\mathbf{r}+d_{12}\mathbf{k}, \mathbf{c}_2), \quad (9.53)$$

where χ is a quantity that is unity for ordinary gases, and increases with increa-sing density, becoming infinite as the gas approaches the state in which the molecules are packed so closely that motion is impossible. Savage and Jeffrey (1981) have associated χ with the radial distribution function in statistical mechanics of liquids.

The dependence on position indicated in Eq. (9.53) can be related to the volume fraction of the solid. A form successfully used by Ding and Gidaspow (1990) is

$$\chi\left(\mathbf{r}+\tfrac{1}{2}d_{12}\mathbf{k}\right) = g_o(\varepsilon_s) = \left[1 - \left(\frac{\varepsilon_s}{\varepsilon_{s\max}}\right)^{1/3}\right]^{-1}. \quad (9.53a)$$

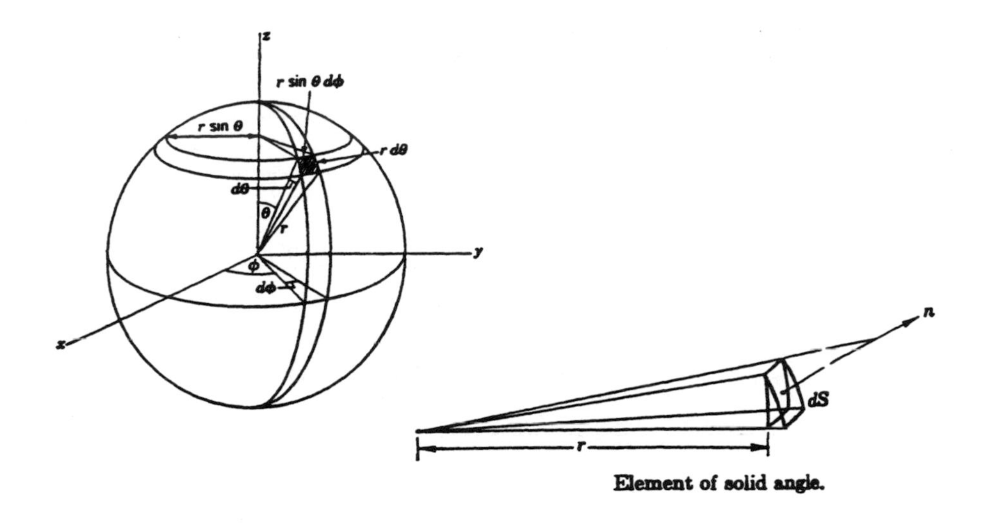

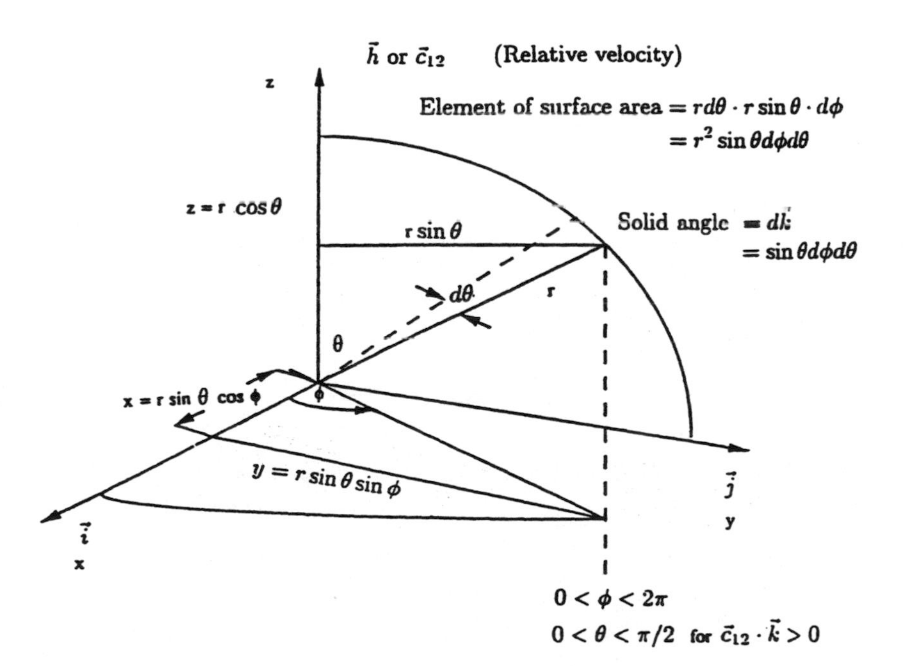

$0 < \phi < 2\pi$

$0 < \theta < \pi/2$ for $\vec{c}_{12} \cdot \vec{k} > 0$

Unit Vector $\vec{k} = \vec{h}\cos\theta + \vec{i}\sin\theta\cos\phi + \vec{j}\sin\theta\sin\phi$

Fig. 9.3 Spherical coordinates and solid angle, dk.

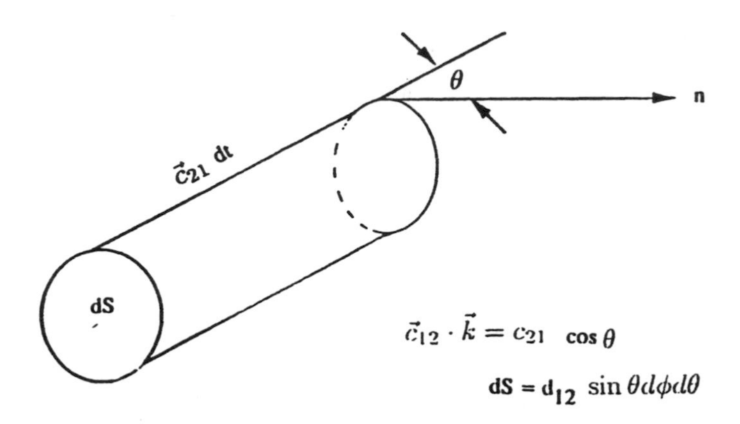

Fig. 9.4 Geometry of collisions.

This form requires, as an empirical input, the maximum packed solids fraction, $\varepsilon_{x\max}$. A better form might be that obtained from a consolidometer directly. Then $f^{(2)}$ becomes

$$f^{(2)} = g_o \frac{n_1 n_2}{(2\pi\Theta)^3} \exp\left[-\frac{(\mathbf{c}_1 - \mathbf{v})^2 + (\mathbf{c}_2 - \mathbf{v})^2}{2\Theta}\right]. \tag{9.53b}$$

For a uniform, steady state, **v** is zero. The integrals over **k** as seen from Eq. (9.52) give

$$\int_o^\pi \cos\theta \sin\theta \, d\theta = \frac{1}{2} \tag{9.54a}$$

and

$$\int_o^{2\pi} d\phi = 2\pi. \tag{9.54b}$$

Thus, the number of collisions per unit time per unit volume using (9.51) to (9.54) in Eq. (9.51) is as follows:

$$N_{12} = \frac{\pi d_{12}{}^2 g_o n_1 n_2}{(2\pi\Theta)^3} \iint c_{21} \exp\left[-\frac{c_1{}^2 + c_2{}^2}{2\Theta}\right] d\mathbf{c}_1 \, d\mathbf{c}_2. \tag{9.55}$$

In view of the definition of granular temperature given by Eq. (9.14) or equivalently with the use of the Maxwellian distribution without the mass of the particles, the present theory is restricted to collision of equal mass particles. Hence, it does not give an expression for N_{12} as general as that found in classical kinetic theory of gases. Hence, for $m_1 = m_2$, Eq. (9.40) becomes

$$c_1^2 + c_2^2 = 2G^2 + \tfrac{1}{2}c_{12}^2. \tag{9.56}$$

In Eq. (9.55) the coordinates are transformed from $\mathbf{c}_1, \mathbf{c}_2$ to $\mathbf{G}, \mathbf{c}_{12}$. As in Chapman and Cowling (1961), the Jacobian of transformation using Eq. (9.36) is

$$\frac{\partial(\mathbf{G}, \mathbf{c}_{12})}{\partial(\mathbf{c}_1, \mathbf{c}_2)} = \frac{\partial(\mathbf{c}_1 + m_2/m_o \mathbf{c}_{21}, \mathbf{c}_2 - \mathbf{c}_1)}{\partial(\mathbf{c}_1, \mathbf{c}_2)} = \frac{\partial(\mathbf{c}_1, \mathbf{c}_2)}{\partial(\mathbf{c}_1, \mathbf{c}_2)} = 1. \tag{9.57}$$

From spherical symmetry,

$$d\mathbf{G} = 4\pi G^2\, dG \tag{9.58}$$

and

$$d\mathbf{c}_{21} = 4\pi c_{21}^2\, dc_{21}. \tag{9.59}$$

Hence, using (9.56) through (9.59), Eq. (9.55) becomes

$$N_{12} = \frac{2n_1 n_2\, d_{12}^2 g_o}{\Theta^3} \int_0^\infty \int_0^\infty \exp\left[-\frac{2G^2 + \frac{1}{2}c_{21}^2}{2\Theta}\right] G^2\, dG c_{21}^3\, dc_{21}. \tag{9.60}$$

Using Eq. (9.21), the integral over c_{21} becomes

$$\int_0^\infty c_{21}^3 \exp\left[-\frac{c_{21}^2}{4\Theta}\right] dc_{21} = \frac{1}{2}\left(\frac{1}{4\Theta}\right)^{-2} = 8\Theta^2, \tag{9.61}$$

and using (9.20), the integral over G yields

$$\int_0^\infty G^2 \exp\left[-\frac{G^2}{\Theta}\right] = \frac{\sqrt{\pi}}{4}\Theta^{3/2}. \tag{9.62}$$

Collecting all terms, the collision frequency becomes

$$N_{12} = 4n_1 n_2\, d_{12}^2 g_o \sqrt{\pi\Theta}. \tag{9.63}$$

This checks with the classical kinetic theory of gases result, Chapman and Cowling's (1961) Eq. (5.21-4), given below:

$$N_{12} = 2n_1 n_2\, d_{12}{}^2 \left(\frac{2\pi k_B T m_o}{m_1 m_2} \right)^{1/2} \tag{9.64}$$

for dilute gases, since

$$m_o / m_1 = 2 \tag{9.65}$$

and

$$k_B / m = 1 \tag{9.66}$$

in the definition of the granular temperature Θ.

Equation (9.66) shows the difference between the granular and the thermal temperature. In the latter, the Boltzmann constant k_B serves to assign the scale of the temperature. In Eq. (9.63) as g_o becomes infinite at maximum packing, the number of collisions becomes infinite, but at this point the granular temperature is expected to approach zero, leaving N_{12} undefined.

9.6 Mean Free Path

In dilute kinetic theory of gases, the intuitive concept of the mean free path plays a very important role. Through the use of this concept, transport coefficients such as viscosity are obtained that are surprisingly close to those obtained from the exact theory.

The mean time between successive collisions, called the collision time τ, is obtained as shown below for dilute flow where g_o is one:

$$\tau = n_1 / N_{11} . \tag{9.67}$$

Hence,

$$\tau = 1 \Big/ \left(4 n_1\, d_{12}{}^2 \sqrt{\pi \Theta} \right). \tag{9.68}$$

The mean free path ℓ is the product of the average velocity and the collision time:

$$\ell = < c > \tau. \tag{9.69}$$

Using Eqs. (9.23) and (9.68), the mean free path becomes independent of Θ:

$$\ell = \frac{1}{\pi\sqrt{2}n_1 d_p^2} = \frac{0.707}{\pi n\, d_p^2 g_o}. \qquad (9.70)$$

Using the relation

$$\varepsilon_s = \tfrac{1}{6}\pi\, d_p^3 n, \qquad (9.71)$$

one obtains a formula for the mean free path that gives the intuitively correct dependence of the mean free path on the particle diameter. Equation (9.70) becomes

$$\ell = \frac{1}{6\sqrt{2}}\frac{d_p}{\varepsilon_s}. \qquad (9.72)$$

Table 9.1 shows some typical values of collision frequencies and mean free paths for particles and molecules.

9.7 Elementary Treatment of Transport Coefficients

It is well known in kinetic theory of gases that a simple, non-rigorous treatment of transport phenomena produced surprisingly accurate values of transport coefficients. Such a treatment is presented below.

Consider some flowing quantity Q, such as mass flux, momentum flux, or current, at two points x and $x + \ell$, where ℓ is the mean free path. By Taylor series,

$$Q(x+\ell) = Q(x) + \ell \cdot \frac{dQ}{dx}, \qquad (9.73)$$

where the derivative of Q is evaluated at some mean value. Then the change in this quantity Q is

$$\Delta Q = \ell\frac{dQ}{dx}. \qquad (9.74)$$

Let Q be the mass flux of species A:

$$Q = \rho_A \langle v_A \rangle \equiv Q_A, \qquad (9.75)$$

Table 9.1 Typical Velocities, Collision Frequencies and Mean Free Paths for Molecules and Particles

Molecule: CO_2, $T = 273$ K	Particle: 500 μm Glass Beads
$<C^2>^{1/2} = \left(\frac{3k_BT}{m}\right)^{1/2} = \left(\frac{3RT}{M}\right)^{1/2} = 158\left(\frac{T}{M}\right)^{1/2}$, m/s For $M_{CO_2} = 44$, $<C^2>^{1/2} = 394$ m/s. Note: Sonic velocity $\equiv \left(\frac{\gamma RT}{M}\right)^{1/2} = 258$ m/s	$<C^2>^{1/2} = \sqrt{3\Theta}$ Exp. discharge velocity ≈ 1 m/s. $<C^2>^{1/2} \le 1$ m/s
$N_{11} = 4n^2 d_{12}^2 \left(\frac{\pi k_B T}{m}\right)^{1/2}$ Diameter of molecule $= d_{12} \cong 3\times10^{-10}$ m $n = \frac{6.023\times10^{23}\text{ molecules/mole}}{22.41\text{ l/mole}} = 2.68\times10^{25}$ molecules/m³ Hence, $N_{11} = 1.06\times10^{35}$ molecules/m³·s Collision frequency $= \frac{1}{2}N_{11}/n \cong 2\times10^9\,\text{s}^{-1}$	$N_{12} = 4n_1n_2\,d_p^2\sqrt{\pi\Theta}$ $n = (6\varepsilon_s)/(\pi d_p^3)$ $N_{11} = \frac{144\sqrt{\pi}}{\pi^2}\frac{\varepsilon_s^2}{d_p^4}\sqrt{\Theta}$ $N_{11} \cong 10^{14}$ particles/m³·s Collision frequency $= \frac{1}{2}\frac{N_{11}}{n_1} =$ $6.77\left(\frac{\varepsilon_s}{d_p}\right)\sqrt{\Theta} \cong 7\times10^3\,\text{s}^{-1}$ down to 1000 s⁻¹
Mean free path $= l$ $= <C^2>^{1/2}/(\text{Collision frequency}) \cong 2\times10^{-7}$ m	$l = \frac{1}{6\sqrt{2}}\left(\frac{d_p}{\varepsilon_s}\right)$ At minimum fluidization, $l = 118$ μm For pneumatic transport, $\varepsilon_s = 0.001$, $l \cong 6$ cm
Conclusion: Since $l >> d_{12}$, dilute kinetic theory is valid.	**Conclusions:** For fluidization, we need dense theory; for transport, wall collisions are essential.

where in Eq. (9.75) a collisional interpretation was used. Hence,

$$\Delta Q_A = l <v_A> \frac{d\rho_A}{dx} \tag{9.76}$$

The Fick's law of diffusion in mass units is

$$\rho_A(v_A - v) = -D\frac{d\rho_A}{dx}, \tag{9.77}$$

where D is the diffusion coefficient and v, the mass average velocity, is zero in this case. Comparing Eqs. (9.76) and (9.77) it is clear that

$$D = l < v_A >. \tag{9.78}$$

The diffusivity is simply the product of an average velocity and the distance between collisions.

Using Eqs. (9.72) and (9.23),

$$D = \frac{1}{3\sqrt{\pi}} \frac{\sqrt{\Theta} d_p}{\varepsilon_s}. \tag{9.79}$$

To estimate a value for this diffusion coefficient one must first know the granular temperature Θ. This "temperature" or kinetic oscillation energy can be found from an oscillation kinetic energy to be derived in later sections.

To obtain the viscosity of the particulate phase, let the momentum flux be

$$Q = \rho v < v >. \tag{9.80}$$

Then for a constant density ρ, the change in momentum flux is

$$\Delta Q = l\rho < v > \frac{dv}{dx}. \tag{9.81}$$

The viscosity μ for fully developed incompressible flow is defined by

$$\text{Shear} = \mu \frac{dv}{dx}. \tag{9.82}$$

The momentum transport ΔQ equals the force per unit area:

$$\Delta Q = \text{Shear}. \tag{9.83}$$

Therefore the viscosity assumes the form

$$\mu = l\rho < v >. \tag{9.84}$$

From Eq. (9.78) it can be seen that

$$\mu = \rho D, \tag{9.85}$$

as in the kinetic theory of gases. Therefore, using Eq. (9.79), a simple formula for the collisional viscosity is

$$\mu = \left(\frac{1}{3\sqrt{\pi}}\right)\rho_p d_p \sqrt{\Theta}. \tag{9.86}$$

When the granular temperature, Θ, is obtained from a kinetic oscillation energy balance, Eq. (9.86) can then be used to obtain a rough value of the collisional viscosity of the particulate phase. Table 9.2 shows an estimate of the particulate viscosity using a more exact formula from Chapman and Cowling (1961). The difference between the more exact formula and that given by Eq. (9.86) is a factor of two. The particulate viscosity is of the order of poise, 100 times that of water.

9.8 Boltzmann Integral–Differential Equation

The Boltzmann equation for the frequency distribution *f*, as given by Eq. (9.1), can be derived, as any conservation equation, using the Reynolds transport theorem. The conservation principle is as follows:

$$\frac{d}{dt}\int_{\mathbf{r}}\int_{\mathbf{c}} f(t,\mathbf{r},\mathbf{c})\, d\mathbf{r}\, d\mathbf{c} = \int_{\mathbf{r}}\int_{\mathbf{c}}\left(\frac{\partial f}{\partial t}\right)_{\text{coll}} d\mathbf{r}\, d\mathbf{c}. \tag{9.87}$$

Rate of change of *f* for a system of a number of particles moving with a mean generalized "velocity" $\left(\frac{d\mathbf{r}}{dt},\frac{d\mathbf{c}}{dt}\right)$ = Rate at which *f* is altered by encounters

Then an application of the usual Reynolds transport theorem, as presented in Chapter 1 and illustrated in population balance problems of that chapter, gives the well-known Boltzmann equation

$$\frac{\partial f}{\partial t} + \mathbf{c}\cdot\frac{\partial f}{\partial \mathbf{r}} + \mathbf{F}\frac{\partial f}{\partial \mathbf{c}} = \left(\frac{\partial f}{\partial t}\right)_{\text{coll}}, \tag{9.88}$$

where **c** and **r** were regarded to be as independent coordinates and where Newton's law of motion

Table 9.2 SOLIDS VISCOSITY FOR DILUTE FLOW

For smooth rigid spherical molecules of diameter d_p (Chapman and Cowling, p. 218),

$$[\mu_1] = \frac{5}{16d_p^2}\left(\frac{k_B mT}{\pi}\right)^{1/2}, \text{ where } m = \rho_p \frac{\pi}{6} d_p^3.$$

To convert T to granular temperature Θ, let $\frac{k_B}{m} = 1$. Thus,

$$\text{DILUTE } \mu_s = \frac{5\sqrt{\pi}}{96}\rho_p d_p \Theta^{1/2}. \tag{T9.1}$$

Interpretation

$$\text{DILUTE } \mu_s = \frac{5\sqrt{\pi}}{96}\cdot(\rho_p \varepsilon_s)\cdot\left(\frac{d_p}{\varepsilon_s}\right)\cdot\Theta^{1/2}$$

VISCOSITY = (Constant) · (Bulk density) · (Mean free path) · (Oscillation velocity)

Numerical Values

$\Theta \approx 1$ m/s from critical flow experiments

Kinematic viscosity= $\nu_s = \mu_s/\rho_p$ = (Constant) · d_p m/s

$\text{Constant} = \frac{5\sqrt{\pi}}{96} = 0.0923$ (Eq. (T9.1)); Constant = 0.186 (approximate value), (Eq. (9.86))

For 500 μm particles, $\nu_s \cong 0.1 \times 5 \times 10^{-4} = 5 \times 10^{-5}\,\text{m}^2/\text{s}$ (close to gas diffusivity)

For $\rho_p = 2000\,\text{kg/m}^3$, $\mu_s \cong 5 \times 10^{-5}\,\text{m}^2/\text{s} \cdot 2000\,\text{kg/m}^3 = 0.1\,\text{kg/m-s}$

1 poise = 1 g/cm-s = 0.1 kg/m-s

Particulate viscosity $\mu_s \cong 1$ poise (recall $\mu_{\text{water}} = 1$ cp, μ_{CO_2} (Table 9.1) $\cong 10^{-2}$ cp)

$$\frac{\text{Force}}{\text{Unit mass}} \equiv \mathbf{F} = \frac{d\mathbf{c}}{dt} \tag{9.89}$$

with the force **F** independent of **c** was substituted to replace the instantaneous acceleration. This derivation is equivalent to using the Eulerian type balance presented in Chapman and Cowling (1961), Chapter 3, and in most other kinetic theory texts. In such a derivation one considers velocities in the range **c** + **F** *dt* equivalent to Eq. (9.89). Chapman and Cowling (1961) extend the Boltzmann equation to a mixture by merely assigning a new label to f. For type S particles, it becomes f_s and the force becomes $\mathbf{F}_s$.

For binary collisions of rigid particles, the right-hand side of the Boltzmann equation (9.88) assumes the form

$$\left(\frac{\partial f}{\partial t}\right)_{\text{collbinary}} = \iint \left(f^{(2)\prime}\mathbf{c}_{12}' \cdot \mathbf{k} - f^{(2)}\mathbf{c}_{12} \cdot \mathbf{k}\right) d_{12}^2 \, d\mathbf{k} \, d\mathbf{c}_1, \tag{9.90}$$

where the primes again indicate the quantities after collision and $f^{(2)}$ is the product of the respective single particle distributions as given by Eq. 9.53). Hence, the Boltzmann equation is an integral–differential equation. Because of its non-linearity it must be solved by iteration. For the first approximation, one takes the Maxwellian distribution. The second approximation, as shown in detail in Chapter 7 in Chapman and Cowling (1961), will give rise to a Navier–Stokes type equation. This is done efficiently using an altered form of the Boltzmann operator, as presented below.

Chapman and Cowling (1961) introduce a relative velocity that they and others call the peculiar velocity, which is the difference between the velocity $\mathbf{c}$ and some reference velocity $\mathbf{c}_o$. For convenience of presentation, equate this velocity $\mathbf{c}$ to the hydrodynamic velocity $\mathbf{v}$. Let

$$\mathbf{C} = \mathbf{c} - \mathbf{v}(t, \mathbf{r}). \tag{9.91}$$

Change the coordinates from $\mathbf{c}$ to $\mathbf{C}$. Then

$$f(t, \mathbf{r}, \mathbf{c}) = f_c(t, \mathbf{r}, \mathbf{C}). \tag{9.92}$$

By chain rules of calculus,

$$\frac{\partial f}{\partial \mathbf{c}} = \frac{\partial f_c}{\partial \mathbf{C}}, \tag{9.93}$$

$$\frac{\partial f}{\partial t} = \frac{\partial f_c}{\partial t} - \frac{\partial f_c}{\partial \mathbf{C}} \frac{\partial \mathbf{v}}{\partial t}, \tag{9.94}$$

$$\frac{\partial f}{\partial x} = \frac{\partial f_c}{\partial x} - \frac{\partial f_c}{\partial \mathbf{C}} \cdot \frac{\partial \mathbf{v}}{\partial x}. \tag{9.95}$$

$1 \times 3 \cdot 3 \times 1$ SCALAR

In view of (9.95), differentiation gives

$$\frac{\partial f}{\partial \mathbf{r}} = \frac{\partial f_c}{\partial \mathbf{r}} - \frac{\partial f_c}{\partial \mathbf{C}} \frac{\partial \mathbf{v}}{\partial \mathbf{r}}. \tag{9.96}$$

$$1\times 3 = 1\times 3 - 1\times 3 \cdot 3\times 3 \text{ Vector}$$

Then, multiplication of Eq. (9.96) by **C** produces the double dot sum useful in the derivation of the Navier–Stokes equation:

$$\mathbf{C}\cdot\frac{\partial f}{\partial \mathbf{r}} = \mathbf{C}\cdot\frac{\partial f_c}{\partial \mathbf{r}} - \frac{\partial f_c}{\partial \mathbf{C}}\left(\mathbf{C}\frac{\partial}{\partial \mathbf{r}}\right)\mathbf{v}. \tag{9.97}$$

Using the notation of Chapman and Cowling (1961) for the convective derivative following the hydrodynamic velocity,

$$\frac{D}{Dt} = \frac{\partial}{\partial t} + \mathbf{v}\frac{\partial}{\partial \mathbf{r}}, \tag{9.98}$$

the Boltzmann equation takes the modified form

$$\frac{Df}{Dt} + \mathbf{C}\frac{\partial f}{\partial \mathbf{r}} + \left(\mathbf{F} - \frac{D\mathbf{v}}{Dt}\right)\frac{\partial f}{\partial \mathbf{C}} - \frac{\partial f}{\partial \mathbf{C}}\mathbf{C} : \frac{\partial \mathbf{v}}{\partial \mathbf{r}} = \left(\frac{\partial f}{\partial t}\right)_{\text{coll}}, \tag{9.99}$$

where the subscript on the function was dropped following the usual convention in physics where two functions are given the same symbol when their meaning stays the same.

9.9 Maxwell's Transport Equation

A transport equation for a quantity ψ can be obtained starting with the Boltzman equation (9.88) by multiplying it by ψ and integrating over **c** as shown below:

$$\int \psi\left(\frac{\partial f}{\partial t} + \mathbf{c}\cdot\frac{\partial f}{\partial \mathbf{r}} + \mathbf{F}\cdot\frac{\partial f}{\partial \mathbf{c}}\right)d\mathbf{c} = \int \psi\left(\frac{\partial f}{\partial t}\right)_{\text{coll}} d\mathbf{c}. \tag{9.100}$$

As in Chapman and Cowling (1961), Chapter 3, these integrals can be transformed by means of equations such as

$$\int \psi \frac{\partial f}{\partial t}\, d\mathbf{c} = \frac{\partial}{\partial t}\int \psi f\, d\mathbf{c} - \int f \frac{\partial \psi}{\partial t} d\mathbf{c} = \frac{\partial n < \psi >}{\partial t} - n < \frac{\partial \psi}{\partial t} >, \tag{9.101a}$$

$$\int \psi c_x \frac{\partial f}{\partial x}\, d\mathbf{c} = \frac{\partial}{\partial x}\int \psi c_x f\, d\mathbf{c} - \int f c_x \frac{\partial \psi}{\partial x} d\mathbf{c} = \frac{\partial n < \psi c_x >}{\partial x} - n < c_x \frac{\partial \psi}{\partial x} >, \tag{9.101b}$$

$$\int \psi \frac{\partial f}{\partial c_x}\, d\mathbf{c} = \int\int [\psi f]_{c_x = -\infty}^{c_x = \infty}\, dc_y\, dc_z - \int f \frac{\partial \psi}{\partial c_x} d\mathbf{c} = -n < \frac{\partial \psi}{\partial c_x} >. \tag{9.101c}$$

In (9.101b) as $\mathbf{c}$ is independent of $\mathbf{r}$, the variable c_x is not included in the differentiation with respect to x; in (9.101c), an integration by parts is performed, and the integrated part vanishes because, by hypothesis, ψf tends to zero as c_x tends to infinity in either direction.

In this way it is found that

$$\frac{\partial n < \psi >}{\partial t} + \frac{\partial}{\partial \mathbf{r}} \cdot n < \psi \mathbf{c} > - n\left[< \frac{\partial \psi}{\partial t} > + < \mathbf{c} \cdot \frac{\partial \psi}{\partial \mathbf{r}} > + \mathbf{F} \cdot < \frac{\partial \psi}{\partial \mathbf{c}} > \right]$$
$$= \int \psi \left(\frac{\partial f}{\partial t} \right)_{\text{coll}} d\mathbf{c} \tag{9.102}$$

Maxwell's equation refers to a function $\psi(\mathbf{c})$ only, so that the terms $(\partial\psi)/(\partial t)$ and $(\partial\psi)/(\partial \mathbf{r})$ do not occur in it. The left-hand side of the balance is then simply an accumulation plus a modified divergence or net rate of outflow of ψ with respect to $\mathbf{r}$ and $\mathbf{c}$.

This equation can be again written in a form more convenient for some applications by changing ψ into the peculiar velocity $\mathbf{C}$ dependence given by Eq. (9.91). Using the chain rules for ψ as given by Eqs. (9.93) to (9.97), Chapman and Cowling's (1961) Eq. (2), paragraph 3.13, becomes as follows for $\psi = \psi < \mathbf{C} >$ only:

$$\frac{Dn < \psi >}{Dt} + n < \psi > \frac{\partial \mathbf{v}}{\partial \mathbf{r}} + \frac{\partial}{\partial \mathbf{r}} n < \psi \mathbf{C} > - n\left(\mathbf{F} - \frac{D\mathbf{v}}{Dt} \right) < \frac{\partial \psi}{\partial \mathbf{C}} >$$

$$- n < \frac{\partial \psi}{\partial \mathbf{C}} \mathbf{C} > : \frac{\partial \mathbf{v}}{\partial \mathbf{r}} = n\psi_c. \tag{9.103}$$

9.10 Conservation Laws with No Collisions

By substituting for ψ mass, momentum and energy, the corresponding conservation laws are easily obtained from the Maxwell's Eq. (9.103).

Case 1. Conservation of Mass

Let $\psi = m$. Then (9.103) becomes the conservation of particles equation shown below:

$$\frac{D(nm)}{Dt} + nm\frac{\partial \mathbf{v}}{\partial \mathbf{r}} = 0. \tag{9.104}$$

Since $\rho_s \varepsilon_s = nm = \rho$, Eq. (9.104) is the standard continuity equation with no phase change, which in this case follows more simply from Eq. (9.102):

$$\frac{\partial(\varepsilon_s \rho_s)}{\partial t} + \frac{\partial(\varepsilon_s \rho_s \mathbf{v}_s)}{\partial \mathbf{r}} = 0. \tag{9.105}$$

The subscript S was added to emphasize that the velocity in Eq. (9.104) is the solid particle velocity.

Case 2. Conservation of Momentum

Let $\psi = m\mathbf{C}$, the relative velocity of the particles relative to its hydrodynamic velocity. It is clear that

$$< \mathbf{C} > = < \mathbf{c} > - \mathbf{v} = 0 \tag{9.106}$$

by definition of **v**, Eq. (9.4).

Hence, the first two terms in Eq. (9.103) vanish. The tensor $\rho < \mathbf{CC} >$ is defined to be the kinetic part of the stress (see Table 9.3):

$$\mathbf{P}_k = \rho < \mathbf{CC} >. \tag{9.107}$$

The last term in Eq. (9.103) gives

$$< \frac{\partial m\mathbf{C}}{\partial \mathbf{C}} \mathbf{C} > = m < \mathbf{C} > = 0. \tag{9.108}$$

Therefore, in this case we recover the inviscid momentum balances of the particles shown below. Table 9.4 shows an alternate derivation:

$$\frac{\partial \mathbf{P}_k}{\partial \mathbf{r}} - \rho\left(\mathbf{F} - \frac{D\mathbf{v}}{Dt}\right) = 0. \tag{9.109}$$

Case 3. Conservation of Energy

Let $\psi = \frac{1}{2}m\mathbf{C}^2$, the relative kinetic energy of the particles, where $\frac{1}{3}\mathbf{C}^2$ is the granular temperature Θ. The heat flux $\mathbf{q}_K$ is defined as

$$\mathbf{q}_K = \frac{1}{2}nm < \mathbf{C}^2\mathbf{C} > = \frac{3}{2}n < \Theta\mathbf{C} >. \tag{9.110}$$

The fourth term in Eq. (9.103) vanishes because

$$< \left(\partial \tfrac{1}{2}m\mathbf{C}^2\right) / (\partial \mathbf{C}) > = m < \mathbf{C} > = 0. \tag{9.111}$$

Table 9.3 KINETIC STRESS TENSOR

$$\mathbf{P}_k \equiv \rho < \mathbf{CC} > = \begin{pmatrix} \rho < C_x^2 > & \rho < C_x C_y > & \rho < C_x C_z > \\ \rho < C_y C_x > & \rho < C_y^2 > & \rho < C_y C_z > \\ \rho < C_z C_x > & \rho < C_z C_y > & \rho < C_z^2 > \end{pmatrix}$$

Since $n < \mathbf{CC} > = \int \mathbf{CC}f\, d\mathbf{C}$,

$$\mathbf{P}_k = \begin{pmatrix} \rho\int C_x^2 f\, d\mathbf{C} & \rho\int C_x C_y f\, d\mathbf{C} & \rho\int C_x C_z f\, d\mathbf{C} \\ \rho\int C_y C_x f\, d\mathbf{C} & \rho\int C_y^2 f\, d\mathbf{C} & \rho\int C_y C_z f\, d\mathbf{C} \\ \rho\int C_z C_x f\, d\mathbf{C} & \rho\int C_z C_y f\, d\mathbf{C} & \rho\int C_z^2 f\, d\mathbf{C} \end{pmatrix},$$

where ρ = BULK DENSITY = $\varepsilon_s \rho_s$ and where $f = f(\mathbf{C})$.

NOTE that if $f(\mathbf{C}) = f_x(C_x)\cdot f_y(C_y)\cdot f_z(C_z)$ as in Maxwellian distribution, then

$$P_{xy} = \rho\int C_x C_y f(\mathbf{C})\, d\mathbf{C} = \rho\int_{-\infty}^{\infty} f_z\, dC_z \int_{-\infty}^{\infty} C_y f_y\, dC_y \int_{-\infty}^{\infty} C_x f_x\, dC_x.$$

But $\int_{-\infty}^{\infty} C_x f_x\, dC_x = < C_x > = < c_x - v_x > = v_x - v_x = 0$.

Hence, for a Maxwellian distribution, $P_{xy} = P_{zx} = P_{ij} = 0 \ (i \neq j)$.

Hence, Eq. (9.103) becomes

$$\frac{3}{2}\left[\frac{D\rho\Theta}{Dt}+\rho\Theta\frac{\partial \mathbf{v}}{\partial \mathbf{r}}\right]+\frac{\partial \mathbf{q}_K}{\partial \mathbf{r}}+\rho<\mathbf{CC}>:\frac{\partial v}{\partial \mathbf{r}}=0. \tag{9.112}$$

Using the continuity Eq. (9.104) and the definition of the kinetic pressure, Eq. (9.107), the equation for the granular temperature becomes

$$\frac{3}{2}\rho\frac{D\Theta}{Dt}=-\frac{\partial \mathbf{q}_K}{\partial \mathbf{r}}-\mathbf{P}_K:\frac{\partial \mathbf{v}}{\partial \mathbf{r}}. \tag{9.113}$$

This is the granular or oscillating energy equation for granular flow without energy dissipation. It shows that the source of production of oscillation is the shear double dotted with the gradient of velocity. This equation can be compared to the corresponding thermal energy balance with a constant specific heat, C_v, written as

$$C_v\rho\frac{DT}{Dt}=-\frac{\partial \mathbf{q}}{\partial \mathbf{r}}-\mathbf{P}:\frac{\partial \mathbf{v}}{\partial \mathbf{r}}. \tag{9.114}$$

Thus, the apparent constant volume heat capacity for granular flow is $\frac{3}{2}$ rather than $\frac{3}{2}k_B/\mathrm{m}$, where k_B is the Boltzmann constant.

9.11 Second Approximation to the Frequency Distribution

The Boltzmann equation, following Chapman and Cowling (1961), for binary collision of rigid particles can be written as

$$\varsigma(f)=J(f\,f_1)+Df=0, \tag{9.115}$$

where the differential operator Df is defined to be

$$Df\equiv\frac{\partial f}{\partial t}+\mathbf{c}\frac{\partial f}{\partial \mathbf{r}}+\mathbf{F}\frac{\partial f}{\partial \mathbf{c}} \tag{9.116}$$

and the integral operator for dilute particles is

Table 9.4 CONSERVATION EQUATIONS USING THE TRANSPORT EQUATION

$$\frac{\partial n<\psi>}{\partial t}+\nabla n<\psi \mathbf{c}>-n\mathbf{F}<\frac{\partial \psi}{\partial \mathbf{c}}>=0, \text{ with no collisional contribution}$$

Conservation of Mass

Let $\psi = m$, since $nm = \varepsilon_s \rho_s$:

$$\frac{\partial(\varepsilon_s \rho_s)}{\partial t}+\nabla \varepsilon_s \rho_s \mathbf{v}_s = 0.$$

Conservation of Momentum

Let $\psi = m\mathbf{c}$:

$$n<\psi \mathbf{c}>= nm<\mathbf{cc}>=\rho<(\mathbf{C}+\mathbf{v})(\mathbf{C}+\mathbf{v})>=\rho\int\left[\mathbf{CC}+2\mathbf{Cv}+\mathbf{v}^2\right]f\, d\mathbf{c}=\mathbf{P}_k+\rho\mathbf{vv},$$

since $\mathbf{P}_k = \rho<\mathbf{CC}>$ and $<\mathbf{C}>=0$,

$$\frac{\partial(\varepsilon_s \rho_s v_s)}{\partial t}+\nabla\left(\mathbf{P}_k+\varepsilon_s\rho_s\mathbf{v}_s\mathbf{v}_s\right)=\varepsilon_s\rho_s\mathbf{F}.$$

Conservation of Energy

Let $\psi = \frac{1}{2}mc^2$:

$$n<\psi \mathbf{c}>=\tfrac{nm}{2}<\mathbf{c}^2\mathbf{c}>=\tfrac{1}{2}\rho<\left(\mathbf{CC}+2\mathbf{Cv}+\mathbf{v}^2\right)(\mathbf{C}+\mathbf{v})>=\tfrac{1}{2}\rho\int\left[\mathbf{C}^2\mathbf{C}+\left(\mathbf{v}^2+\mathbf{C}^2+2\mathbf{CC}\right)\mathbf{v}\right]f\, d\mathbf{c}\,.$$

Note $\mathbf{q}_k = \frac{1}{2}\rho\int \mathbf{C}^2\mathbf{C}f\, d\mathbf{c} = \frac{1}{2}nm<\mathbf{C}^2\mathbf{C}>$ and since $\Theta = \frac{1}{3}<\mathbf{C}^2>$,

$$n<\psi\mathbf{c}>=\mathbf{q}_k+\mathbf{v}\left(\tfrac{1}{2}\rho\mathbf{v}^2+\tfrac{3\Theta\rho}{2}\right)+\mathbf{P}_k\mathbf{v}, \text{ hence}$$

$$\frac{\partial\left(\frac{3}{2}\Theta\rho+\frac{1}{2}\rho\mathbf{v}^2\right)}{\partial t}+\nabla\cdot\left[\mathbf{q}_k+\mathbf{v}\left(\tfrac{3}{2}\Theta\rho+\tfrac{1}{2}\rho\mathbf{v}^2\right)+\mathbf{P}_k\mathbf{v}\right]=\rho\mathbf{F}\cdot\mathbf{v}$$

Subtract $\mathbf{v}\cdot$ momentum to obtain Eq. (9.113).

$$J(FG)\equiv\iint\left(FG-F'G'\right)\mathbf{c}_{12}\cdot\mathbf{k}\,d_{12}^2 d\mathbf{k}\, d\mathbf{c}_1, \tag{9.117}$$

as in Chapman and Cowling Chapter 7 or as given by the more general equation

(9.90) for dense mixtures of particles. The Boltzmann equation is solved by iteration as follows. Express f as an infinite series:

$$f = f^{(0)} + f^{(1)} + f^{(2)} + \ldots . \tag{9.118}$$

Then

$$\varsigma(f) = \varsigma^{(0)}\left(f^{0}\right) + \varsigma^{(1)}\left(f^{(0)}, f^{(1)}\right) + \varsigma^{(2)}\left(f^{(0)}, f^{(1)}, f^{(2)}\right) + \ldots . \tag{9.119}$$

By (9.115),

$$\varsigma^{(0)}\left(f^{0}\right) = 0, \tag{9.120}$$

$$\varsigma^{(1)}\left(f^{(0)}, f^{(1)}\right) = 0, \tag{9.121}$$

$$\varsigma^{(2)}\left(f^{(0)}, f^{(1)}, f^{(2)}\right) = 0. \tag{9.122}$$

The function $f^{(0)}$ was the Maxwellian distribution. Equation (9.121) determines the second approximation to the Boltzmann equation. The third approximation can then be obtained from (9.122).

A convenient form for Df is given by Eq. (9.99). By (9.121) $D^{(1)}$ is then

$$D^{(1)} = \frac{D_o f^{(0)}}{Dt} + \mathbf{C} \cdot \frac{\partial f^{(0)}}{\partial \mathbf{r}} + \left(\mathbf{F} - \frac{D\mathbf{v}_o}{Dt}\right) \frac{\partial f^{(0)}}{\partial \mathbf{C}} - \frac{\partial f^{(0)}}{\partial \mathbf{C}} \mathbf{C} : \frac{\partial \mathbf{v}_o}{\partial \mathbf{r}}. \tag{9.123}$$

Dividing both sides of Eq. (9.123) by $f^{(0)}$, expressing the derivative in terms of a logarithm and using the inviscid momentum balance, Eq. (9.109), the preceding expression can be written as follows:

$$D^{(1)} / f^{(0)} = \frac{D_o \ln f^{(0)}}{Dt} + \mathbf{C} \cdot \frac{\partial \ln f^{(0)}}{\partial \mathbf{r}} + \frac{1}{\rho} \frac{\partial \mathbf{P}_k}{\partial \mathbf{r}} \frac{\partial \ln f^{(0)}}{\partial \mathbf{C}} - \frac{\partial \ln f^{(0)}}{\partial \mathbf{C}} \mathbf{C} : \frac{\partial \mathbf{v}_o}{\partial \mathbf{r}}. \tag{9.124}$$

Using the Maxwell's distribution, Eq. (9.15),

$$\ln f^{(0)} = \text{const} + \ln \frac{n}{\Theta^{3/2}} - \frac{\mathbf{C}^2}{2\Theta}, \tag{9.125}$$

and its derivative becomes

$$\frac{\partial \ln f^{(0)}}{\partial \mathbf{C}} = -\frac{\mathbf{C}}{\Theta}. \tag{9.126}$$

To determine the convective derivative of $f^{(0)}$, one must first obtain the first approximation to the conservation equation. The first approximation to the convective operation D/Dt is

$$\frac{D_o}{Dt} = \frac{\partial_o}{\partial t} + \mathbf{C}_o \frac{\partial}{\partial \mathbf{r}}. \tag{9.127}$$

Then the continuity equation (9.104) becomes

$$\frac{D_o n}{Dt} = -n \frac{\partial \mathbf{v}_o}{\partial \mathbf{r}}, \tag{9.128}$$

and the momentum equation (9.109) is

$$\frac{D_o \mathbf{v}}{Dt} = \mathbf{F} - \frac{1}{\rho} \frac{\partial \mathbf{P}_k^{(0)}}{\partial \mathbf{r}}, \tag{9.129}$$

where the kinetic pressure is

$$\mathbf{P}_k^{(0)} = \int \rho \mathbf{C}\mathbf{C} f^{(0)}\, d\mathbf{c}. \tag{9.130}$$

From the properties of the Maxwellian distribution it can be shown that the non-diagonal elements of the stress $\mathbf{P}_k^{(0)}$ vanish (see Table 9.3). Consider the mean,

$$< C_x > = \frac{1}{n} \int_{-\infty}^{\infty} f(C_x) C_x \, dc_x. \tag{9.131}$$

When $f = f^{(0)}$,

$$< C_x > = 0, \tag{9.132}$$

since the contribution due to the positive and the negative parts of the integral cancel. In general, for any odd integer K,

$$< C^K > = \frac{1}{n}\int_{-\infty}^{\infty} f^{(0)}(C)C^K \, dc = 0. \tag{9.133}$$

Consider a component of stress in (9.130), $P_{xy}^{(0)}$:

$$P_{xy}^{(0)} = \rho\int_{-\infty}^{\infty} f_z^{(0)} \, dC_z \int_{-\infty}^{\infty} f_y^{(0)} C_y \, dC_y \int_{-\infty}^{\infty} f_x^{(0)} C_x \, dC_x . \tag{9.134}$$

The two integrals over x and y are both zero by (9.131). Hence, for a Maxwellian distribution the viscosity μ is also zero, where for developed flow,

$$P_{xy}^{(0)} = -\mu \frac{dv_x^{(0)}}{dy}. \tag{9.135}$$

The diagonal elements of (9.130) give the kinetic pressure P. We have

$$P = P_{xx} = P_{yy} = P_{zz} \tag{9.136}$$

and

$$P_{xx} + P_{yy} + P_{zz} = \rho\left(\left\langle C_x^2 \right\rangle + \left\langle C_y^2 \right\rangle + \left\langle C_z^2 \right\rangle\right). \tag{9.137}$$

Since

$$I:I = 3, \tag{9.138}$$

let

$$P = \tfrac{1}{3}\mathbf{P}_k^{(0)} : I. \tag{9.139}$$

Furthermore, since $\Theta = \frac{1}{3}\left\langle \mathbf{C}^2 \right\rangle$, Eqs. (9.139) and (9.137) show that

$$P = \rho\Theta . \tag{9.140}$$

Equation (9.140) is the equation of state for granular flow. It differs from the ideal gas equation,

$$P = k_B nT , \tag{9.141}$$

where k_B is the Boltzmann constant and T the thermal temperature by the introduction of the absolute temperature scale in defining T to be 273 K at the freezing point of water and 373 K at its boiling point (see Table 9.5). The heat flux q was defined by Eq. (9.110). Its rth iterate can be written as

$$q^{(r)} = \int E\mathbf{C}f^{(r)}(\mathbf{c})d\mathbf{c}, \tag{9.142}$$

where $E = \frac{1}{2}m\mathbf{C}^2$. But if $f^{(r)} = f^{(0)}$, that is, when f is Maxwellian, formula (9.133) shows that

$$q^{(0)} = 0. \tag{9.143}$$

With these preliminaries, the first approximation to the convective derivative can then be evaluated. The first approximations to the mass, momentum, and energy equations from Eqs. (9.104), (9.109), and (9.113) are as follows:

Mass Conservation

$$\frac{D_o n}{Dt} = -n\frac{\partial \mathbf{v}_o}{\partial \mathbf{r}}, \tag{9.144}$$

Momentum Conservation

$$\frac{D_o \mathbf{v}}{Dt} = \mathbf{F} - \frac{1}{\rho}\frac{\partial P}{\partial \mathbf{r}}, \tag{9.145}$$

since

$$\mathbf{P}_k^{(0)} = \mathbf{I}P; \tag{9.145a}$$

Energy Conservation

$$\frac{3}{2}\rho\frac{D_o\Theta}{Dt} = -\rho\Theta\frac{\partial \mathbf{v}_o}{\partial \mathbf{r}} \tag{9.146}$$

using Eq. (9.140).

Combining the energy conservation equation (9.146) with the conservation of mass equation (9.144), one obtains

$$\frac{3}{2\Theta}\frac{D_o\Theta}{Dt} = \frac{1}{n}\frac{D_o n}{Dt} \tag{9.147a}$$

Table 9.5 IDEAL GAS LAW FOR GASES, POWDERS, AND CONVERSIONS

Molecular	Granular Powder
Definition of thermal temperature T: $k_B T = \frac{1}{3} m < C^2 >$, (1A) where the Boltzmann constant k_B converts kinetic energy into temperature: $k_B = 1.3805 \times 10^{-23}\ J/K$.	Definition of granular temperature Θ: $\Theta = \frac{1}{3} < C^2 >$, (1B) where the 3 is due to motion in three directions. To convert from T to Θ, set $k_B/m = 1$ in standard kinetic theory formulas.
Definition of hydrostatic pressure P: $P = \frac{1}{3} \rho < C^2 >$, (2A) where $\rho = nm$. Hence, eliminating $< C^2 >$ in Eqs. (1) and (2) gives the *Ideal Gas Law* $p = n k_B T$. (3A) With Avogadro's number N and $n = N/V$, where V is the molar volume and $R = N k_B$, Eq. (3A) gives $PV = RT$. (4A) The internal energy per unit molecule of the gas, $\langle U \rangle$, $n \langle U \rangle = \int \frac{1}{2} m C^2 f(t, \mathbf{r}, \mathbf{c}) d\mathbf{c}$. (5A) Hence, using Eq. (1A), $n \langle U \rangle = \frac{3}{2} n k_B T$. (6A) Then Molar $C_v = \frac{3}{2} R$. (7A)	Definition of particulate pressure P: $P = \frac{1}{3} \rho < C^2 >$, (2B) where ρ, the bulk density, is $\rho = \varepsilon_s \rho_s$ and the $\frac{1}{3}$ is due to isotropy and $P_{xx} + P_{yy} + P_{zz} = \rho < C^2 >$. Hence we have the *Particulate Ideal State Equation* $p = \rho \Theta$ (3B) or $p = \varepsilon_s \rho_s \Theta$. (4B) Using Eqs. (5A) and (1B), we obtain the *Powder Internal Energy* $\langle U \rangle$ $\langle U \rangle = \frac{3}{2} m \Theta$ (5B) or $C_{vp} = \left(\frac{\partial \langle U \rangle}{\partial \Theta} \right) = \frac{3}{2} m$. (6B) Since temperature exists, there is an entropy S $dS = \frac{1}{\Theta} dU + \frac{p}{\Theta} dV = C_{vp} \frac{d\Theta}{\Theta} - \frac{d\varepsilon_s}{\varepsilon_s}$ (7B) where $V = (\varepsilon_s \rho_s)^{-1}$. Hence, $S = C_{vp} \ln \Theta - \ln \varepsilon_s + \text{constant}$. (8B)

or

$$\frac{D_o \ln n/\Theta^{3/2}}{Dt} = 0. \qquad (9.147b)$$

Equation (9.147b) shows that the particle density ρ is proportional to the three- half power of the granular temperature, as shown below:

$$\rho \propto \Theta^{3/2}. \qquad (9.148)$$

Using Eq. (9.147b), convective differentiation of (9.125) gives

$$\frac{D_o \ln f^{(0)}}{Dt} = \frac{\mathbf{C}^2}{2\Theta^2}\frac{D_o\Theta}{Dt}, \qquad (9.149)$$

and using (9.146) and (9.144), this equation becomes

$$\frac{D_o \ln f^{(0)}}{Dt} = -\frac{1}{3}\frac{\mathbf{C}^2}{\Theta}\frac{\partial \mathbf{v}_o}{\partial \mathbf{r}}. \qquad (9.150)$$

Then the sum of the first and the last terms in the equation for $D^{(1)}$ in equation (9.124) becomes

$$-\frac{1}{3}\frac{\mathbf{C}^2}{\Theta}\frac{\partial \mathbf{v}_o}{\partial \mathbf{r}} + \frac{\mathbf{CC}}{\Theta}:\frac{\partial \mathbf{v}_o}{\partial \mathbf{r}} = \frac{1}{\Theta}\overset{o}{\mathbf{C}\mathbf{C}}:\frac{\partial \mathbf{v}_o}{\partial \mathbf{r}}. \qquad (9.151)$$

Using (9.126) and (9.150) and the definition of the non-divergent tensor employed by Chapman and Cowling (1961)

$$\overset{o}{\mathbf{W}} = \mathbf{W} - \frac{1}{3}\left(W_{xx} + W_{yy} + W_{zz}\right)\mathbf{I}, \qquad (9.152)$$

the third term in Eq. (9.124) can be expressed as follows:

$$\frac{1}{\rho}\frac{\partial \mathbf{P}_k}{\partial \mathbf{r}}\cdot\frac{\partial \ln f^{(0)}}{\partial \mathbf{c}} = \frac{1}{n}\frac{\partial (n\Theta)}{\partial \mathbf{r}}\cdot\left(-\frac{\mathbf{C}}{\Theta}\right) = -\frac{\mathbf{C}}{n\Theta}\frac{\partial (n\Theta)}{\partial \mathbf{r}} \qquad (9.153)$$

using Eq. (9.140). Using Eq. (9.153), the sum of the two middle terms in Eq. (9.124) can be written as

$$\mathbf{C}\frac{\partial \ln f^{(0)}}{\partial \mathbf{r}} + \frac{1}{\rho}\frac{\partial \mathbf{P}_k}{\partial \mathbf{r}}\frac{\partial \ln f^{(0)}}{\partial \mathbf{C}} = \mathbf{C}\cdot\frac{\partial \ln f^{(0)}/n\Theta}{\partial \mathbf{r}}. \tag{9.154}$$

But the Maxwellian distribution, Eq. (9.15), can be written as

$$\ln\frac{f^{(0)}}{n\Theta} = \ln(2\pi)^{-3/2} + \ln\Theta^{-5/2} - \frac{\mathbf{C}^2}{2\Theta}. \tag{9.155}$$

Differentiation of (9.155) with respect to $\mathbf{r}$ shows that Eq. (9.154) can be expressed as

$$\mathbf{C}\frac{(\partial \ln f^{(0)}/n\Theta)}{\partial \mathbf{r}} = \mathbf{C}\left(\frac{\partial \ln \Theta^{-5/2}}{\partial \mathbf{r}} + \frac{\mathbf{C}^2}{2\Theta^2}\frac{\partial \Theta}{\partial \mathbf{r}}\right) = \mathbf{C}\frac{\partial \ln \Theta}{\partial \mathbf{r}}\left(-\tfrac{5}{2} + \frac{\mathbf{C}^2}{2\Theta}\right). \tag{9.156}$$

Therefore, using (9.151) and (9.156), the final expression for $D^{(1)}$ becomes as follows:

$$\frac{D^{(1)}}{f^{(0)}} = \left(\frac{\mathbf{C}^2}{2\Theta} - \tfrac{5}{2}\right)\mathbf{C}\frac{\partial \ln \Theta}{\partial \mathbf{r}} + \frac{1}{\Theta}\overset{\circ}{\mathbf{C}\mathbf{C}}:\frac{\partial \mathbf{v}_o}{\partial \mathbf{r}}. \tag{9.157}$$

If one lets

$$f^{(1)} = f^{(0)}\phi^{(1)}, \tag{9.158}$$

one obtains a linear nonhomogeneous Fredholm integral equation for $\phi^{(1)}$ as follows. Boltzmann Eq. (9.115) for the second order approximation for f is

$$D^{(1)} + J^{(1)} = 0, \tag{9.159}$$

where $D^{(1)}$ is given by Eq. (9.157) and

$$J^{(1)} = J\left(f^{(0)}f_1^{(1)}\right) + J\left(f^{(1)}f_1^{(0)}\right). \tag{9.160}$$

Chapman and Cowling (1961), p. 85, find it convenient to define an integral I by means of the expression

$$n_1^2 I_1(F) \equiv \iint f_1^{(0)} f^{(0)}\left(F_1 + F - F_1' - F'\right)k_1\, d_p^2\, d\mathbf{k}\, d\mathbf{c}. \tag{9.161}$$

Using (9.161), $J^{(1)}$ becomes

$$J\left(f^{(0)}f_1^{(0)}\phi^{(1)}\right)+J\left(f^{(0)}\phi^{(1)}f_1^{(0)}\right)=$$
$$\iint f^{(0)}f_1^{(0)}\left(\phi^{(1)}+\phi_1^{(1)}-\phi'^{(1)}-\phi_1'^{(1)}\right)k_1\,d_p^2\,d\mathbf{k}\,d\mathbf{c}\equiv n^2 I\left(\phi^{(1)}\right). \tag{9.162}$$

Equation (9.159) shows that this expression is the negative of $D^{(1)}$ given by (9.157). Hence, the linear nonhomogeneous integral equation for $\phi^{(1)}$ is as follows:

$$n^2 I\left(\phi^{(1)}\right)=-f^{(0)}\left[\left(\frac{\mathbf{C}^2}{2\Theta}-\tfrac{5}{2}\right)\mathbf{C}\frac{\partial\ln\Theta}{\partial\mathbf{r}}+\frac{1}{\Theta}\overset{\circ}{\mathbf{C}}\mathbf{C}:\frac{\partial\mathbf{v}_o}{\partial\mathbf{r}}\right]. \tag{9.163}$$

9.12 Integral Equation Solver Strategy

The second approximation to the Boltzmann equation $f^{(1)}$ is given by the product of the Maxwell–Boltzmann distribution and $\phi^{(1)}$. In the preceding paragraph it was shown that $\phi^{(1)}$ is determined by the nonhomogeneous linear integral equation (9.163). As shown in detail in Chapman and Cowling, this equation can be solved as follows.

First, note that $\phi^{(1)}$ is a scalar. Hence only scalar solutions of (9.163) are considered. Second, note that Eq. (9.162) shows that $I(\phi^{(1)})$ is linear in $\phi^{(1)}$. Hence, superposition applies. The solution to the integral equation is the sum of the homogeneous equation $I(\phi^{(1)})=0$ plus a particular solution. The latter consists of (1) a linear combination of the components of $\partial\Theta/\partial\mathbf{r}$: for this to be a scalar, it must be given by a scalar product of $\partial\Theta/\partial\mathbf{r}$ and another vector, (2) a linear combination of the components of $\partial\mathbf{v}/\partial\mathbf{r}$: this must be the double dot product of $\partial\mathbf{v}/\partial\mathbf{r}$ and another second order tensor. Thus, one can write

$$\phi^{(1)}=-\frac{1}{n}\mathbf{A}\cdot\nabla\ln\Theta=\frac{1}{n}\mathbf{B}:\nabla\mathbf{v}+\alpha\cdot\psi, \tag{9.164}$$

where **A** and **B** are the vector and tensor functions of **C**, respectively, while α is independent of the velocity. The components of $\nabla\ln\Theta$ and $\nabla\mathbf{v}$ are equated. Thus, the integral equations for A and B are obtained:

$$nI(\mathbf{A}) = f^{(0)}\left(\frac{\mathbf{C}^2}{2\Theta} - \tfrac{5}{2}\right)\mathbf{C}, \tag{9.165}$$

$$nI(\mathbf{B}) = \left(f^{(0)}/\Theta\right)\overset{o}{\mathbf{C}}\mathbf{C}. \tag{9.166}$$

The conditions of solubility of these integral equations are

$$\int \psi^{(i)} f^{(0)}\left(\frac{\mathbf{C}^2}{2\Theta} - \tfrac{5}{2}\right)\mathbf{C}\,d\mathbf{c} = 0, \tag{9.167}$$

$$\int \psi^{(i)} f^{(0)}\,\overset{o}{\mathbf{C}}\,d\mathbf{c} = 0, \tag{9.168}$$

where $i = 1, 2, 3$.

Now let us briefly consider how to determine A and B. They are functions of n, Θ, and $\mathbf{C}$. Since I is a linear, rotationally invariant operator and the right member of (9.165) is a vector in the direction of $\mathbf{C}$, the vector $\mathbf{A}$ must also be in the direction of $\mathbf{C}$. Thus,

$$\mathbf{A} = A(\mathbf{C})\mathbf{C}, \tag{9.169}$$

where A is a scalar function of n, Θ, and $\mathbf{C}$.

Since the right member of (9.166) is a symmetric traceless tensor, $\mathbf{B}$ must be of the form

$$\mathbf{B} = B(\mathbf{C})\overset{o}{\mathbf{C}}\mathbf{C}, \tag{9.170}$$

where B is some function of n, Θ, and $\mathbf{C}$. The quantities A and B are determined by expressing them in terms of polynomials, as shown in Chapman and Cowling (1961). In Eq. (9.164) α can be set to zero. Here the algebraic details are of little interest. In view of (9.169) and (9.170) $\phi^{(1)}$ becomes

$$\phi^{(1)} = -\frac{1}{n}A(\mathbf{C})\mathbf{C}\cdot\nabla\ln\Theta - \frac{1}{n}B(\mathbf{C})\overset{o}{\mathbf{C}}\mathbf{C}:\nabla\mathbf{v}. \tag{9.171}$$

9.13 Viscous Kinetic Stress Tensor

The equation for the second approximation to the kinetic stress tensor $\mathbf{P}_k^{(1)}$ is as follows (see Table 9.3):

$$\mathbf{P}_k^{(1)} = m\int f^{(1)}\mathbf{CC}\,d\mathbf{c} = m\int f^{(0)}\phi^{(1)}\mathbf{CC}\,d\mathbf{c}, \tag{9.172}$$

in view of the definition of $\phi^{(1)}$ given by Eq. (9.158). Into this, substitute Eq. (9.171). Note that the integrals of odd powers of **C** vanish, since the Maxwellian distribution function $f^{(0)}$ is even. Hence,

$$\mathbf{P}_k^{(1)} = -\frac{m}{n}\int f^{(0)}B(\mathbf{C})\left(\overset{o}{\mathbf{C}}\mathbf{C}:\nabla\mathbf{v}\right)\mathbf{CC}\,d\mathbf{c}. \tag{9.173}$$

The double dot product in parentheses is Eq. (9.173), which can be transformed to give a traceless symmetric gradient of the hydrodynamic velocity, using the identity for two tensors **X** and **Y**:

$$\overset{o}{\mathbf{X}}:\mathbf{Y} = \mathbf{X}:\overset{o}{\mathbf{Y}}{}^{s}, \tag{9.174}$$

where

$$\overset{o}{\mathbf{X}} = \mathbf{X} - \tfrac{1}{3}tr\mathbf{X}\cdot\mathbf{I}, \tag{9.174a}$$

and the superscript s indicates that the tensor **Y** is symmetric. Identity (9.174) can be derived from the basic definition of the double dot product as the double sum of the components of the two tensors. Identity (9.174) applied to the expression in parenthesis of (9.173) gives

$$\overset{o}{\mathbf{C}}\mathbf{C}:\nabla\mathbf{v} = \mathbf{CC}:\overset{o}{\nabla}{}^{s}\,\mathbf{v}, \tag{9.175}$$

where in various notations the *rate of shear tensor* **S** is

$$\overset{o}{\nabla}{}^{s}\,\mathbf{v} \equiv \mathbf{S} = \frac{\partial\overset{o}{\mathbf{C}}}{\partial\mathbf{r}} = \tfrac{1}{2}\left(\nabla\mathbf{v} + \nabla^{s}\mathbf{v}\right) - \tfrac{1}{3}(\nabla\mathbf{v}:\mathbf{I})\mathbf{I} = \tfrac{1}{2}\left(\frac{\partial v_j}{\partial x_i} + \frac{\partial v_i}{\partial x_j}\right) - \tfrac{1}{3}\frac{\partial v_i}{\partial x_i}\delta_{ij}. \tag{9.176}$$

Using (9.173), the viscous kinetic stress tensor is rewritten as

$$\mathbf{P}_k^{(1)} = -\frac{m}{n}\int f^{(0)} B(\mathbf{C})\left(\mathbf{CC}:\overset{o}{\nabla}{}^s\,\mathbf{v}\right)\mathbf{CC}\,dc. \tag{9.177}$$

Thus, an algebraic rearrangement produced the rate of shear tensor that was expected from continuum mechanics. Furthermore, since $\overset{o}{\nabla}{}^s\,\mathbf{v}$ is not a function of **C**, integration and differentiation with respect to the spatial coordinates can be rearranged in Eq. (9.177). This can be accomplished using the following identity, given in Chapman and Cowling (1961), which the interested reader should take the time to derive to understand the theory.

For any tensor **W** independent of **C**,

$$\int F(C)\mathbf{CC}\left(\overset{o}{\mathbf{C}}\mathbf{C}:\mathbf{W}\right)d\mathbf{C} = \tfrac{1}{5}\overset{o}{\mathbf{W}}{}^s\int F(C)\left(\overset{o}{\mathbf{C}}\mathbf{C}:\overset{o}{\mathbf{C}}\mathbf{C}\right)d\mathbf{C} = \int F(C)\left(\mathbf{CC}:\overset{o}{\mathbf{W}}{}^s\right)\mathbf{CC}\,d\mathbf{C}. \tag{9.178}$$

With $\mathbf{W} = \nabla\mathbf{v}$, (9.177) using identity (9.178) becomes as follows:

$$\mathbf{P}_k^{(1)} = -\overset{o}{\nabla}{}^s\mathbf{v}\left[\frac{m}{5n}\int f^{(0)} B(C)\left(\overset{o}{\mathbf{C}}\mathbf{C}:\overset{o}{\mathbf{C}}\mathbf{C}\right)d\mathbf{C}\right], \tag{9.179}$$

in which $B(C)\overset{o}{\mathbf{C}}\mathbf{C}$ can be replaced with the tensor **B** using expression (9.170), while $f^{(0)}\mathbf{CC}$ is replaced with $n\mathbf{I}(\mathbf{B})\Theta$ using Eq. (9.166). Then the viscous kinetic stress tensor becomes the double integral given by

$$\mathbf{P}_k^{(1)} = -\left[\frac{m\Theta}{5}\int \mathbf{B}:\mathbf{I}(\mathbf{B})\,dc\right]\overset{o}{\nabla}{}^s\,\mathbf{v}. \tag{9.180}$$

The viscosity μ is defined by means of

$$\mathbf{P}^{(1)} \equiv -2\mu\,\overset{o}{\nabla}{}^s\mathbf{v}. \tag{9.181}$$

Hence, μ is the scalar

$$\mu = \frac{m\Theta}{10}\int \mathbf{B}:\mathbf{I}(\mathbf{B})\,dc. \tag{9.182}$$

Except for a conversion to granular temperature, this is the expression for viscosity given by the bracket integral (7.41-1) in Chapman and Cowling (1961). Its computation is lengthy and is shown in the reference cited. The first approximation gives

$$\mu_{\text{dilute}} = \frac{5\sqrt{\pi}}{96}\rho_p d_p \Theta^{1/2}. \tag{9.183}$$

Table 9.2 shows that such particular viscosities are of the order of one poise for 500 μm glass beads oscillating at velocities of about one meter per second. Such reasonable values of particulate viscosities provide a motivation for applying the kinetic theory of gases to dispersed multiphase flow. Equation (9.183) is Lun *et al.*'s (1984) Eq. (4.9) for perfectly elastic particles. For non-elastic particles there is a correction involving the restitution coefficient. This correction is small. For example, for a restitution coefficient of one-half, the viscosity is $16/15$ that given by Eq. (9.183). For a restitution coefficient of 0.9 the correction is less than 0.3 percent. In view of the approximations made such as neglect of the charge of particles, rotation, etc., Lun *et al.*'s correction will be neglected.

9.14 Dense Transport Theorem

Jenkins and Savage (1983) and Lun *et al.* (1984) have derived a useful dense fluid transport theorem. It generalizes the transport theorem valid for dilute gases and particles used up to now. Savage (1983) calls this theorem the Maxwell–Chapman transport equation since its essence is found in Chapter 16 in Chapman and Cowling's book (1961).

The objective is to derive a useful expression for the mean change of the collisional property $<\psi_c>$ given by the Maxwell transport equation (9.102). Equation (9.90) suggests that for dilute particle mixtures,

$$<\psi_c>_{\text{dilute}} = \frac{1}{2}\iiint f^{(2)}\Delta\psi\, d_p^2 \mathbf{c}_{12}\cdot\mathbf{k}\, d\mathbf{k}\, d\mathbf{c}_1\, d\mathbf{c}_2, \tag{9.184}$$

where

$$\Delta\psi = \underbrace{\psi_2' + \psi_1'}_{\text{After collision}} - \underbrace{\psi_2 - \psi_1}_{\text{Before collision}}, \tag{9.185}$$

where the one half in (9.184) is necessary so as not to count the property twice in view of (9.185). The difference between the dilute and the dense cases is that $f^{(2)}$

associated with particles one and two is at different locations, as indicated in Fig. 9.2. Hence, the integrals must be evaluated separately for particles one and two, as described below. The collisional rate of change of ψ_1 is

$$<\psi_{c1}>=\int_{\mathbf{c}_{12}\cdot\mathbf{k}}\left(\psi_1'-\psi_1\right)d_p^2\mathbf{c}_{12}\cdot\mathbf{k}f^{(2)}\left(\mathbf{r},\mathbf{c}_1,\mathbf{r}+d_p\mathbf{k},\mathbf{c}_2\right)d\mathbf{k}\,d\mathbf{c}_1\,d\mathbf{c}_2, \quad (9.186)$$

where $d_p{}^2\mathbf{c}_{12}\cdot\mathbf{k}\,d\mathbf{k}$ is the collision cylinder per unit volume, $f^{(2)}$ is the pair distribution function evaluated at the position $\mathbf{r}+d_p\mathbf{k}$ and the integral is evaluated for particles about to collide, i.e., $\mathbf{c}_{12}\cdot\mathbf{k}>0$. To evaluate the collisional rate of change for particle 2, due to symmetry, this can be done by interchanging the labels 1 and 2 and by replacing $\mathbf{k}$ by $-\mathbf{k}$:

$$<\psi_{c2}>=\int_{\mathbf{c}_{12}\cdot\mathbf{k}}\left(\psi_2'-\psi_2\right)d_p^2\mathbf{c}_{12}\cdot\mathbf{k}f^{(2)}\left(\mathbf{r}-d_p\mathbf{k},\mathbf{c}_1,\mathbf{r},\mathbf{c}_2\right)d\mathbf{k}\,d\mathbf{c}_1\,d\mathbf{c}_2. \quad (9.187)$$

Then, as in Eqs. (9.184) and (9.185),

$$<\psi_c>=\tfrac{1}{2}<\psi_{c1}>+\tfrac{1}{2}<\psi_{c2}>. \quad (9.188)$$

The values of $f^{(2)}$ separated by a distance $d_p\mathbf{k}$ in Eqs. (9.186) and (9.187) are related to each other by means of the Taylor series

$$f^{(2)}\left(\mathbf{r}-d_p\mathbf{k},\mathbf{r}\right)=f^{(2)}\left(\mathbf{r},\mathbf{r}+d_p\mathbf{k}\right)-\left(d_p\mathbf{k}\cdot\nabla\right)\sum_{m=0}^{\infty}\frac{\left(-d_p\mathbf{k}\nabla\right)^m}{(m+1)!}f^{(2)}\left(\mathbf{r},\mathbf{r}+d_p\mathbf{k}\right). \quad (9.189)$$

Substituting Eq. (9.189) into (9.187) and adding the result to (9.186), the mean value of the collisional integral as in (9.188) can be written in the form

$$<\phi_c>=-\nabla\cdot P_c+N_c, \quad (9.190)$$

where the collisional stress contribution is

$$P_c=-\frac{d_p^3}{2}\int_{\mathbf{c}_{12}\cdot\mathbf{k}}\left(\psi_1'-\psi_1\right)\mathbf{c}_{12}\cdot\mathbf{kk}\sum_{m=0}^{\infty}\frac{\left(-d_p\mathbf{k}\nabla\right)^m}{(m+1)!}f^{(2)}\left(\mathbf{r},\mathbf{c}_1,\mathbf{r}+d_p\mathbf{k},\mathbf{c}_2\right)d\mathbf{k}\,d\mathbf{c}_1\,d\mathbf{c}_2 \quad (9.191)$$

and the sourcelike contribution is

$$N_c = \frac{d_p^2}{2} \int_{\mathbf{c}_{12}\cdot\mathbf{k}} (\psi_2' + \psi_1' - \psi_2 - \psi_1)\mathbf{c}_{12}\cdot\mathbf{k} f^{(2)}(\mathbf{r} - d_p\mathbf{k}, \mathbf{c}_1, \mathbf{r}, \mathbf{c}_2)\, d\mathbf{k}\, d\mathbf{c}_1\, d\mathbf{c}_2 . \quad (9.192)$$

To first order, the collisional stress contribution simplifies to

$$P_c = -\tfrac{1}{2} d_p^3 \int_{\mathbf{c}_{12}\cdot\mathbf{k}} (\psi_1' - \psi_1)\mathbf{c}_{12}\cdot\mathbf{kk} f^{(2)}\left(\mathbf{r} - \tfrac{1}{2} d_p\mathbf{k}, \mathbf{c}_1, \mathbf{r} + \tfrac{1}{2} d_p\mathbf{k}, \mathbf{c}_2\right) d\mathbf{k}\, d\mathbf{c}_1\, d\mathbf{c}_2 , \quad (9.193)$$

since the mean positions of the two particles are the points $\mathbf{r} \pm \tfrac{1}{2} d_p\mathbf{k}$, while the mean position of the point of impact is the point $\mathbf{r}$. With the decompositions (9.190) the Maxwell transport equation (9.102) can be written as

$$\frac{\partial n < \psi >}{\partial t} + \nabla \cdot (n < \mathbf{c}\psi > + P_c) = < n\mathbf{F}_s \frac{\partial \psi}{\partial \mathbf{c}} > + N_c , \quad (9.194)$$

where $\mathbf{F}_s$ is the force acting on the system in place of the force per unit mass produced by the system in (9.102). Equation (9.194) will be refered to as the dense or the Jenkins–Savage transport theorem for convenience.

A form more convenient for some applications can be obtained using Eqs. (9.190) and (9.103):

$$\frac{Dn < \psi >}{Dt} + n < \psi > \nabla\mathbf{v} + \nabla(n < \psi\mathbf{C} > + nP_c) - n\left(\mathbf{F} - \frac{D\mathbf{v}}{Dt}\right) < \frac{\partial \psi}{\partial \mathbf{C}} > - n < \frac{\partial \psi}{\partial \mathbf{C}}\mathbf{C} > : \nabla\mathbf{v} = nN_c , \quad (9.195)$$

where $\mathbf{C}$ is the peculiar velocity. As pointed out in Chapman and Cowling (1961), in the remark at the bottom of page 279, care must be taken in using ψ so that it is not position dependent. It will be taken to coincide with the motion with the mass average velocity.

Next, a more useful expression for P_c, the collisional part of the vector of flow of ψ, is derived. This is done by expanding f_1 and f_2 in a Taylor series and by dropping terms of order higher than the first. With the pair-distribution function a product of the distribution functions given by Eq. (9.53), expansion in Taylor series gives the expression

$$f^{(2)}\left(\mathbf{r}-\frac{d_p}{2}\mathbf{k},\mathbf{c}_1;\mathbf{r}+\tfrac{1}{2}d_p\mathbf{k},\mathbf{c}_2\right)=g_o(\varepsilon_s)f_1\left(\mathbf{r}-\frac{d_p}{2}\mathbf{k},\mathbf{c}_1\right)\cdot f_2\left(\mathbf{r}+\tfrac{1}{2}d_p\mathbf{k},\mathbf{c}_2\right)$$

$$=g_o(\varepsilon_s)\left[f_1(\mathbf{r},\mathbf{c}_1)\cdot f_2(\mathbf{r},\mathbf{c}_2)+\tfrac{1}{2}d_p\mathbf{k}f_2(\mathbf{r},\mathbf{c}_2)\nabla f_1-\tfrac{1}{2}d_p\mathbf{k}f_1\nabla f_2\right], \tag{9.196}$$

which can be written in a more convenient form as

$$f^{(2)}=g_o\left[f_1\cdot f_2+\tfrac{1}{2}d_pf_1f_2\nabla\ln\frac{f_2}{f_1}\right]. \tag{9.197}$$

Substitution of (9.197) into (9.193) gives a more useful expression for the collisional part of the vector of flow of ψ shown below:

$$P_c=-\tfrac{1}{2}g_o(\varepsilon_s)d_p^3\iiint_{\mathbf{c}_{12}\cdot\mathbf{k}>0}(\psi_1'-\psi_1)f_1f_2\mathbf{k}(\mathbf{c}_{12}\cdot\mathbf{k})d\mathbf{k}\,d\mathbf{c}_1\,d\mathbf{c}_2$$

$$-\tfrac{1}{4}g_o(\varepsilon_s)d_p^4\iiint_{\mathbf{c}_{12}\cdot\mathbf{k}>0}(\psi_1'-\psi_1)f_1f_2\mathbf{k}\nabla\ln\frac{f_2}{f_1}\mathbf{k}(\mathbf{c}_{12}\cdot\mathbf{k})d\mathbf{k}\,d\mathbf{c}_1\,d\mathbf{c}_2. \tag{9.198}$$

Since the second term of Eq. (9.198) already involves derivatives, f_1 and f_2 at the point r can be replaced by the Maxwellian distribution functions.

9.15 Particulate Momentum Equation

With the decomposition of the collision integral as given by Eq. (9.190) the Jenkins–Savage transport theorem conveniently expressed by Eq. (9.195) can be used to obtain the conservation equations for mass, momentum and energy, as already done in Section 9.10 for the case of no collisions. By taking ψ to be m, since m is the same before and after collision, Eq. (9.195) gives the conservation of mass equation (9.104) or (9.105).

To obtain the momentum equation for the particulate phase, let

$$\psi=m\mathbf{C}. \tag{9.190}$$

Then, as in Section 9.10, the first two terms in Eqs. (9.195) vanish. The gradient of the first part of the third term gives the kinetic pressure, as defined by Eq.

(9.107). The last term as in Eq. (9.108) gives zero. N_c also becomes zero since momentum is conserved in a collision.

Hence, (9.195) reduces itself to

$$\rho\frac{D\mathbf{v}}{Dt}+\nabla\left(\mathbf{P}_k+\mathbf{P}_c\right)=\rho\mathbf{F}. \tag{9.200}$$

The sign convention for the kinetic pressure is that used in Chapman and Cowling. It is the negative of that used in Chapter 1 for the traction and that used by Ding and Gidaspow (1990) for the kinetic pressure.

The force **F** in Eq. (9.200) consists of the externally applied force and the force of interaction between phases as given by Eq. (1.8). The latter force consists of the drag between the phases and other forces such as the added mass force clearly illustrated by Birkhoff (1960) for acceleration of a gas bubble injected into a liquid. In such a situation the gas bubble displaces and accelerates a fluid that is 1000 times denser than itself. Following Lamb (1932), Birkhoff adds one half the displaced mass to the acceleration to account for this effect. The relative velocity equation presented in Chapter 2 and derived by Gidaspow (1977) takes care of the extra acceleration. Others, such as Drew and Lahey (1979) and Pauchon and Banerjee (1988), accounted for this effect by adding a convective form of relative acceleration corrected by an empirical coefficient. Another force that may play a significant role in fluids with a neutrally buoyant particle is the Saffman (1965) lift force. Rotational effects recently measured by Ye *et al.* (1990) are perhaps best introduced into this theory by modifying the Maxwell–Boltzmann distribution by the addition of an $I\omega^2$ term and by constructing an energy equation for a rotational temperature, as done by Condiff *et al.* (1964) and others based on Chapman and Cowling's (1961) analysis for rough spherical molecules described in their Chapter 11. For flow of solid particles in a pipe, Roco and Cader (1990) give their evaluation of the importance of some of the forces they came across in modeling flow of slurries in pipelines. Clearly, Eq. (9.200) does not include the static particle-to-particle stress that was discussed in earlier chapters and in Chapter 12. The gradient of this solids stress must be either added to (9.200) or the kinetic theory approach restricted to flows with no dead regions. The former approach is clearly preferable. For dilute suspensions, fluid turbulence must also be included in **F**.

For now the only effects included are the gravity, the drag and the buoyancy effect which appears through the gradient of fluid pressure in Model A. Then

$$\rho\mathbf{F}=\varepsilon_s\rho_s\mathbf{g}+\beta_A\left(\mathbf{v}_f-\mathbf{v}_s\right)-\varepsilon_s\nabla P_f. \tag{9.201}$$

Hence, using (9.201), the particulate momentum balance $\mathbf{v}_s$, Eq. (9.200), can be written as follows in conservative form:

$$\frac{\partial(\varepsilon_s \rho_s \mathbf{v}_s)}{\partial t} + \nabla \varepsilon_s \rho_s \mathbf{v}_s \mathbf{v}_s + \nabla(\mathbf{P}_k + \mathbf{P}_c) = -\varepsilon_s \nabla P_f + \varepsilon_s \rho_s \mathbf{g} + \beta_A(\mathbf{v}_f - \mathbf{v}_s). \tag{9.202}$$

For Model B the first term on the right-hand side of Eq. (9.202) would not be present.

9.16 Fluctuating Kinetic Energy Equation

To obtain the equation for the translational granular temperature equation, let

$$\psi = \tfrac{1}{2} m c^2 \tag{9.203}$$

in the collisional transport theorem equation (9.194).

As in Table 9.4,

$$n < \psi > = \tfrac{1}{2} \rho < \mathbf{c}^2 > = \tfrac{1}{2} \rho\left(< \mathbf{CC} > + v^2\right) = \rho\left(\tfrac{3}{2}\Theta + \tfrac{1}{2} v^2\right) \tag{9.204}$$

and

$$n < \psi \mathbf{c} > = \mathbf{q}_k + \mathbf{v}\left(\tfrac{1}{2}\rho v^2 + \tfrac{3}{2}\rho\Theta\right) + \mathbf{P}_k \mathbf{v}. \tag{9.205}$$

The collisional part of P_c in Eq. (9.194) using the substitution $\mathbf{c} = \mathbf{C} + \mathbf{v}$ decomposes into a collisional heat flux $\mathbf{q}_c$ and $\mathbf{v} \cdot \mathbf{P}_c$ as follows:

$$P_c\left(\tfrac{1}{2} m \mathbf{c}^2\right) = \mathbf{q}_c + \mathbf{v} \cdot \mathbf{P}_c, \tag{9.206}$$

where

$$\mathbf{q}_c = P_c\left(\tfrac{1}{2} m \mathbf{C}^2\right), \tag{9.207}$$

to give

$$\frac{\partial \rho\left(\tfrac{3}{2}\Theta + \tfrac{1}{2} v^2\right)}{\partial t} + \nabla \cdot \left[\mathbf{q}_k + \mathbf{q}_c + \mathbf{v}\rho\left(\tfrac{3}{2}\Theta + \tfrac{1}{2} v^2\right) + \mathbf{v}(\mathbf{P}_k + \mathbf{P}_c)\right] = \rho < \mathbf{F}_s \mathbf{c} > + N_c\left(\tfrac{m}{2}\mathbf{c}^2\right). \tag{9.208}$$

The mechanical energy balance is obtained by multiplying momentum equation (9.200) by $\mathbf{v}$ and rearranging it, as done in single phase fluid mechanics, to give

$$\frac{\partial\left(\frac{1}{2}\rho v^2\right)}{\partial t}+\nabla\cdot\tfrac{1}{2}\rho v^2\mathbf{v}=\nabla\mathbf{v}\left(\mathbf{P}_k+\mathbf{P}_c\right)+\mathbf{v}\cdot\rho\mathbf{F}-\left(\mathbf{P}_k+\mathbf{P}_c\right):\nabla\mathbf{v}. \qquad (9.209)$$

To obtain the fluctuating kinetic energy equation, Eq. (9.209) is subtracted from Eq. (9.208). This gives rise to the production of the fluctuations due to the shear $(\mathbf{P}_k+\mathbf{P}_c):\nabla\mathbf{v}$, already seen in Eqs. (9.113) and (9.146), and due to fluctuations in drag. The latter may be expressed as

$$\begin{aligned}\rho<\mathbf{Fc}>-\rho\mathbf{F}\cdot\mathbf{v}&=\beta_A\left[<\mathbf{C}_p\cdot\left(\mathbf{C}_g-\mathbf{C}_p\right)>-\mathbf{v}_p\cdot\left(\mathbf{v}_g-\mathbf{v}_p\right)\right]\\&=\beta_A<\mathbf{C}_g\cdot\mathbf{C}_p>-3\beta_A\Theta,\end{aligned} \qquad (9.210)$$

where $<\mathbf{C}_g\mathbf{C}_p>$ is the correlation between the velocity fluctuations of the gas and the particles. Louge *et al.* (1990) proposed how to relate it to the drag and the granular temperature. Ding and Gidaspow (1990) neglected this effect. Through this term, gas oscillations produce oscillations of particles.

The final form of the granular temperature equation, therefore, is

$$\frac{3}{2}\left[\frac{\partial}{\partial t}\left(\varepsilon_s\rho_s\Theta\right)+\nabla\cdot\varepsilon_s\rho_s\mathbf{v}_s\Theta\right]=$$

$$\underbrace{\left(\mathbf{P}_k+\mathbf{P}_c\right):\nabla\mathbf{v}}_{\text{Production of fluctuations by shear}}-\underbrace{\nabla\left(\mathbf{q}_k+\mathbf{q}_c\right)}_{\text{Dissipation by kinetic + Collisional heat flow}}$$

$$+\underbrace{N_c\left(\tfrac{m}{2}\mathbf{c}^2\right)}_{\text{Dissipation due to inelastic collisions}}+\underbrace{\beta_A<\mathbf{C}_g\cdot\mathbf{C}_p>}_{\text{Production due to fluid turbulence or due to collision with molecules}}-\underbrace{3\beta_A\Theta,}_{\text{Dissipation due to interaction with fluid}} \qquad (9.211)$$

where, using (9.192), (9.45), and (9.197),

$$N_c\left(\tfrac{m}{2}\mathbf{c}^2\right)=\frac{d_p^2}{2}\int\frac{1}{4}m\left(e^2-1\right)\left(\mathbf{k}\cdot\mathbf{c}_{12}\right)^2\cdot g_0\left[f_1^{(0)}f_2^{(0)}+\frac{1}{2}d_pf_1^{(0)}f_2^{(0)}\nabla\ln\frac{f_2^{(0)}}{f_1^{(0)}}\right]d\mathbf{k}\,d\mathbf{c}_1\,d\mathbf{c}_2. \qquad (9.212)$$

Jenkins and Savage (1983) were the first to evaluate such an integral using the methods of dense phase kinetic theory illustrated for determination of the solids viscosity in the section. Ding and Gidaspow's (1990) result, equal to that of Jenkins and Savage for $\alpha = 0$, is

$$N_c\left(\tfrac{m}{2}\mathbf{c}^2\right) = 3\left(e^2 - 1\right)\varepsilon_s^2 \rho_p g_0 \Theta\left[\tfrac{4}{d_p}\left(\tfrac{\Theta}{\pi}\right)^{1/2} - \nabla \cdot \mathbf{v}_s\right]. \tag{9.213}$$

For elastic particles, that is, for $e = 1$, N_c is zero.

In the fluctuating energy equation (9.211) $\mathbf{P}_k$ is given by Eq. (9.181) as twice the viscosity expressed by Eq. (9.183) times the traceless symmetrical gradient of the velocity. The collisional part of the stress will be evaluated in the next section. The kinetic part of the heat flux from dilute kinetic theory of gases (Chapman and Cowling, 1961) is expressed as follows for elastic particles:

$$\mathbf{q}_k = -\kappa_k \nabla\Theta, \tag{9.214}$$

where

$$\kappa_k = \tfrac{5}{2}\mu C_v = \tfrac{15}{4}\mu, \tag{9.215}$$

since C_v is $\frac{3}{2}$, as found in Table 9.5.

The viscosity is given by Eq. (9.183). The collisional flux for a Maxwellian distribution is

$$\mathbf{q}_c = -\kappa_c \nabla\Theta, \tag{9.216}$$

where

$$\kappa_c = 2\rho_p \varepsilon_s^2 d_p (1+e) g_o \left(\tfrac{\Theta}{\pi}\right)^{1/2}, \tag{9.217}$$

as evaluated by Jenkins and Savage (1983) and Ding and Gidaspow (1990). In both papers the kinetic conductivity was zero, since only the Maxwellian distribution was considered, rather than its second perturbation (see Section 9.18 for a more complete discussion).

Note that Eq. (9.211) is valid for a non-flow situation. Consider small, elastic particles suspended in a liquid. These particles are known to exhibit random-like motion due to collision with the molecules of the fluid. Such a motion is called Brownian movement. In Eq. (9.211) the molecules produce periodic oscillations through the term $\beta_A < \mathbf{C}_g \cdot \mathbf{C}_p >$, where $\mathbf{C}_g$ is the molecular velocity and $\mathbf{C}_p$ is the particulate velocity. For a stationary fluid, all terms in Eq. (9.211) vanish, except this term and the granular accumulation and dissipation terms. Equation

(9.211) becomes a heat conduction equation with a time dependent source. For small particles, equations such as (9.217) for granular conductivity show that this conductivity is small. Hence the granular diffusivity κ_c / ρ_s is small. Therefore, molecular motion will produce a corresponding motion of particles. Large particles have a large granular conductivity, so molecular motion will not affect them and may be neglected.

9.17 Viscosity–Collisional Momentum Transfer

The collisional part of the flow of a quantity ψ was shown to be given by Eq. (9.198). To obtain the collisional stress tensor $\mathbf{P}_c$, let

$$\psi = m\mathbf{C} = m(\mathbf{c} - \mathbf{v}). \tag{9.218}$$

Then (9.198), using Eq. (9.47), becomes the sum of two integrals denoted by $\mathbf{P}_{c_1} + \mathbf{P}_{c_2}$:

$$\mathbf{P}_c = \tfrac{1}{4} d_p^3 g_0 m(1+e) \iiint (\mathbf{k}\cdot\mathbf{c}_{12})^2 \mathbf{kk} f_1 f_2 \, d\mathbf{k}\, d\mathbf{c}_1\, d\mathbf{c}_2$$

$$+ \tfrac{1}{8} d_p^4 g_0 m(1+e) \iiint (\mathbf{k}\cdot\mathbf{c}_{12})^2 \mathbf{kk} f_1^{(0)} f_2^{(0)} \mathbf{k}\cdot\nabla \ln \frac{f_2^{(0)}}{f_1^{(0)}} d\mathbf{k}\, d\mathbf{c}_1\, d\mathbf{c}_2 = \mathbf{P}_{c_1} + \mathbf{P}_{c_2}, \tag{9.219}$$

where the second integral is denoted by $\mathbf{P}_{c_2}$; Maxwellian distributions are used. In $\mathbf{P}_{c_1}$ the second approximation will be used here. Integrations with respect to $\mathbf{k}$ can be carried out using Chapman and Cowling's (1961) identities 16.6–16.8, which are

$$\int \mathbf{kk}(k\cdot\mathbf{c}_{12})^2 d\mathbf{k} = \tfrac{2\pi}{15}\left(2\mathbf{c}_{12}\mathbf{c}_{12} + c_{12}^2\mathbf{I}\right), \tag{9.220}$$

$$\int \mathbf{kk}(\mathbf{v}\cdot\mathbf{k})(\mathbf{c}_{12}\cdot\mathbf{k})^2 d\mathbf{k} = \tfrac{\pi}{12}\left[\mathbf{v}\cdot\mathbf{c}_{12}\left(\mathbf{c}_{12}\mathbf{c}_{12} + c_{12}^2\mathbf{I}\right)/c_{12} + c_{12}\left(\mathbf{v}\mathbf{c}_{12} + \mathbf{c}_{12}\mathbf{v}\right)\right]. \tag{9.221}$$

Using identity (9.220) the first collisional integral becomes

$$\mathbf{P}_{c_1} = \tfrac{\pi}{30}(1+e) d_p^3 g_0 m \iint f_1 f_2 \left(2\mathbf{c}_{12}\mathbf{c}_{12} + \mathbf{c}_{12}^2\mathbf{I}\right) d\mathbf{c}_1\, d\mathbf{c}_2 . \tag{9.222}$$

From the definition of the mean for any function ϕ,

$$\int f_1 \phi d\mathbf{c}_1 = \int f_2 \phi d\mathbf{c}_2 = n < \phi >. \tag{9.223}$$

Since $< \mathbf{C}_1 >=< \mathbf{C}_2 >= 0$ and in view of the integration over $\mathbf{c}_1$ and $\mathbf{c}_2$, these variables are arbitrary and may be replaced by $\mathbf{c}$. Hence,

$$< \mathbf{C}_{12}^2 >=< C_1^2 + C_2^2 - 2C_1C_2 >= 2 < C^2 >. \tag{9.224}$$

Then the first collision integral becomes

$$\mathbf{P}_{c_1} = \tfrac{\pi}{15}(1+e)d_p^3 g_0 \rho n\left(2 < \mathbf{CC} > + C^2\mathbf{I}\right). \tag{9.225}$$

Using the volume fraction relation

$$\varepsilon_s = \tfrac{1}{6} n\pi d_p^3, \tag{9.226}$$

the collision integral gives

$$\mathbf{P}_{c_1} = \frac{2(1+e)}{5} g_0 \rho_p \varepsilon_s^2 \left(2 < \mathbf{CC} > + C^2\mathbf{I}\right). \tag{9.227}$$

For a Maxwellian distribution,

$$< \mathbf{C}^{(0)}\mathbf{C}^{(0)} >= \Theta \mathbf{I}, \tag{9.228}$$

and $C^2 = 3\Theta$; hence, Eq. (9.227) gives the collisional contribution to the solid pressure:

$$\mathbf{P}_{c_1} = 2(1+e) g_0 \rho_p \varepsilon_s^2 \Theta \mathbf{I}. \tag{9.229}$$

When to Eq. (9.229) the kinetic pressure, Eq. (9.140), is added, the result is the solid phase pressure derived by Ding and Gidaspow (1990):

Solid Phase Pressure

$$p_s = \varepsilon_s \rho_p \left[1 + 2(1+e) g_0 \varepsilon_s\right]\Theta. \tag{9.230}$$

Equation (9.230) is identical with the Van der Waal's equation of state for a gas given by Chapman and Cowling's (1961) Eqs. (16.51–2) when a change is made from the thermal to the granular temperature and Chapman and Cowling's χ is interpreted as the product of g_0 and $\frac{1}{2}(1+e)$. Thus, the solids modulus G, which is very important in modeling fluidized beds, becomes

$$\text{Solids modulus} = G = \left(\frac{\partial p_s}{\partial \varepsilon_s}\right) = \rho_p\left[1+4(1+e)g_0\varepsilon_s\right]\Theta. \tag{9.231}$$

The solids modulus has the correct property of becoming very large as the volume fraction of solids approaches the maximum packing where g_0 becomes infinite. The collisional term becomes small as the volume fraction of solids becomes small, again as required by the physics. Its only problem is that it vanishes when Θ is zero, that is, when there is no motion. This is not surprising, since when there is little or no motion the static stress must be added to the solids pressure.

For a non-Maxwellian distribution $<\mathbf{CC}>$ can be expressed in terms of the symmetric gradient of the velocity using the second approximation to the frequency distribution in the manner illustrated in Sections 9.11–9.13 and presented in detail in Chapter 16 in Chapman and Cowling (1961). The result is

$$\rho<\mathbf{CC}>^{(1)} = -\frac{2}{(1+e)g_0}\left[1+\tfrac{4}{5}\varepsilon_s g_0(1+e)\right]2\mu_{\text{dilute}}\overset{o}{\nabla}{}^s\,\mathbf{v} \tag{9.232}$$

for e near one.

Similarly, the kinetic flow of heat is

$$\tfrac{1}{2}\rho<C^2\mathbf{C}> = -\frac{2}{(1+e)g_0}\left[1+\tfrac{6}{5}\varepsilon_s g_0(1+e)\right]\kappa_{\text{dilute}}\nabla\Theta, \tag{9.233}$$

where κ_{dilute} and μ_{dilute} are the conductivity and viscosity for dilute suspensions, corresponding to those for a normal gas.

Equations (9.232) and (9.233) with e equal to one are Chapman and Cowling's (1961) Eqs. (16.34-2) and (16.34-3), respectively, where g_0 is Chapman and Cowling's χ and

$$b\rho = 4\varepsilon_s. \tag{9.234}$$

Note that Eqs. (9.232) and (9.233) do not reduce themselves to the viscosity and the conductivity for dilute suspensions unless the restitution coefficient is set to one. In deriving the corresponding equations, Lun *et al.* (1984) do not use Chapman and Cowling's (1961) rigorous method. They use the more approximate and simpler moment method. In this method they find it necessary to use a third term involving a gradient of n in a trial function for ϕ which is of the form given by Eq. (9.157) and (9.158). The trial function used by them is

$$\phi = a_1 \overset{o}{\mathbf{C}}\mathbf{C}:\nabla\mathbf{v} + a_2\left(\frac{5}{2} - \frac{C^2}{2\Theta}\right)\mathbf{C}\cdot\nabla\ln\theta + a_3\left(\frac{5}{2} - \frac{C^2}{2\Theta}\right)\mathbf{C}\cdot\nabla\ln n. \tag{9.235}$$

Thus, Lun *et al.*'s expressions are somewhat more complicated than those presented here. For simple shear flow, Lun *et al.* (1984) find that their dimensionless shear and normal stresses increase with a decreasing volume fraction at low values of the volume fraction. Because of a division by an energy dissipation which vanishes as the square of the volume fraction and which also vanishes for a restitution coefficient of unity, these stresses become infinite for perfectly elastic particles and for a pure fluid. Lun *et al.* (1984) describe an instability problem which they ascribe to a primitive treatment of the restitution coefficient. In view of these uncertainties the formulas involving solids viscosities and conductivities should be restricted to values of the restitution coefficient near unity, as indicated below Eq. (9.232). A viable simple *ad hoc* approach to extend the treatment by Ding and Gidaspow (1990) to non-Maxwellian distributions and thus to non-dense conditions is to add to their dense phase viscosities and conductivities the dilute gas viscosity given by Eq. (9.183) and (9.215), respectively. Such a simple treatment used by Louge *et al.* (1990) for pneumatic transport amounts to neglecting the second term in brackets in Eqs. (9.232) and (9.233).

The second collision integral in (9.219) will give rise to the dense phase viscosity derived by Ding and Gidaspow (1990). The theory of evaluating the second integral was already presented. The steps are, however, quite involved algebraically.

Using identity (9.221), the second integral becomes

$$\mathbf{P}_{c_2} = \frac{\pi}{96}(1+e)d_p^4 g_0 m \iint f_1^{(0)} f_2^{(0)}\Bigg[\mathbf{c}_{12}\cdot\nabla\ln\frac{f_2^{(0)}}{f_1^{(0)}}$$
$$\cdot\left(\mathbf{c}_{12}\mathbf{c}_{12} + \mathbf{c}_{12}^2\mathbf{I}\right)\Big/ c_{12} + c_{12}\left(\mathbf{c}_{12}\cdot\nabla\ln\frac{f_2^{(0)}}{f_1^{(0)}} + \nabla\ln\frac{f_2^{(0)}}{f_1^{(0)}}\mathbf{c}_{12}\right)\Bigg]d\mathbf{c}_1\,d\mathbf{c}_2. \tag{9.236}$$

For the Maxwellian distributions,

$$\ln \frac{f_2^{(0)}}{f_1^{(0)}} = -\frac{1}{2\Theta}\left(\mathbf{C}_2^2 - \mathbf{C}_1^2\right). \tag{9.237}$$

Hence,

$$\nabla \ln \frac{f_2^{(0)}}{f_1^{(0)}} = \frac{1}{2\Theta^2}\left(\mathbf{C}_2^2 - \mathbf{C}_1^2\right)\nabla\Theta + \tfrac{1}{\Theta}\nabla\mathbf{v}\cdot\left(\mathbf{C}_2 - \mathbf{C}_1\right). \tag{9.238}$$

In the integral (9.236) the terms involving $\nabla\Theta$ vanish in integration since they are odd function's of **C**. Then the integral becomes

$$\begin{aligned}\mathbf{P}_{c_2} = \frac{-\pi}{96\Theta}(1+e)d_p^4 g_0 m \iint f_1^{(0)} f_2^{(0)} \Big[\nabla\mathbf{v}:\mathbf{c}_{12}\mathbf{c}_{12}\left(\mathbf{c}_{12}\mathbf{c}_{12} + \mathbf{c}_{12}^2\mathbf{I}\right)/c_{12} \\ + c_{12}\left(\nabla\mathbf{v}\mathbf{c}_{12}\right)\mathbf{c}_{12} + c_{12}\mathbf{c}_{12}\left(\nabla\mathbf{v}\mathbf{c}_{12}\right)\Big] d\mathbf{c}_1\, d\mathbf{c}_2.\end{aligned} \tag{9.239}$$

To evaluate (9.239), change the variables from $\mathbf{c}_1, \mathbf{c}_1$ to $\mathbf{c}_{12}$ and **G**. Then as in Section 9.5, $d\mathbf{c}_1, d\mathbf{c}_2$ is replaced by $d\mathbf{G}, d\mathbf{c}_{12}$ given by Eqs. (9.58) and (9.59), and

$$f_1^{(0)} f_2^{(0)} = \frac{n^2}{(2\pi\Theta)^3}\exp\left[-\frac{2G^2 + \frac{1}{2}c_{12}^2}{2\Theta}\right]. \tag{9.240}$$

Formula (9.21) with $r = 5$ and Eq. (9.62) produce a typical term in Eq. (9.239) involving a component of the gradient of **v** equal to

$$-\frac{\sqrt{\pi}(1+e)}{3} g_0 d_p^4 n^2 m \Theta^{1/2}. \tag{9.241}$$

The trial function ϕ given by Eq. (9.235) suggest that the terms in (9.239) involving $\nabla\mathbf{v}$ will involve the traceless symmetric gradient of **v** using identity (9.174). Components of the type given by (9.241) must be grouped using integral theorems of the type derived by Chapman and Cowling (1961), paragraph 1.421. The result is (Chapman and Cowling, 1961; Ding and Gidaspow, 1990)

$$\mathbf{P}_{c_2} = -\frac{\sqrt{\pi}(1+e)}{3} g_0 d_p^4 n^2 m \Theta^{1/2}\left[\tfrac{4}{5}\overset{o}{\nabla}{}^s\,\mathbf{v} + \tfrac{2}{3}\nabla\cdot\mathbf{v}\mathbf{I}\right], \tag{9.242}$$

or substituting for m and $nm = \varepsilon_s \rho_p$ and rearranging Eq. (9.242), the result is Ding and Gidaspow's second half of Eq. (A.19):

$$\mathbf{P}_{c_2} = -\frac{4\varepsilon_s^2 \rho_p d_p g_0 (1+e)}{3\sqrt{\pi}} \Theta^{1/2} \left[\frac{6}{5} \overset{o}{\nabla}^s \mathbf{v} + \nabla \cdot \mathbf{v} \mathbf{I} \right]. \quad (9.243)$$

The coefficient of the brackets is Chapman and Cowling's $\overline{\omega}$. The second part of the collisional integral as given by Eq. (9.243) can be written in the form

$$\mathbf{P}_{c_2} = -2\varepsilon_s \mu_{sc} \overset{o}{\nabla}^s \mathbf{v} - \varepsilon_s \xi_s \nabla \cdot \mathbf{v} \mathbf{I}, \quad (9.244)$$

where μ_{sc} is the coefficient of viscosity simply called solids viscosity by Ding and Gidaspow (1990) and ξ_s is the solids bulk phase viscosity. Both are defined in such a way that as the volume fraction ε_s approaches zero, the momentum equation vanishes. Comparing Eqs. (9.244) and (9.243), the viscosity coefficients are as follows:

$$\mu_{sc} = \tfrac{4}{5} \varepsilon_s \rho_p d_p g_0 (1+e) \left(\tfrac{\Theta}{\pi}\right)^{1/2}, \quad (9.245)$$

$$\xi_s = \tfrac{4}{3} \varepsilon_s \rho_p d_p g_0 (1+e) \left(\tfrac{\Theta}{\pi}\right)^{1/2}. \quad (9.246)$$

The solids phase viscosity $\mu_{sc}\varepsilon_s$ can be expressed in terms of the dilute phase viscosity (T9.1) as follows: dense phase viscosity from second collision integral + dilute phase viscosity = $2\mu_{sc}\varepsilon_s / \mu_s(\text{dilute})$:

$$\tfrac{384}{25\pi} g_0 (1+e) \varepsilon_s^2 = 4.889 g_0 (1+e) \varepsilon_s^2. \quad (9.247)$$

Equation (9.247) shows that for a porosity of 0.99, i.e., for a volume fraction of solids of one percent, the contribution of the dense phase integral is negligible. In such a situation it is appropriate to use the dilute phase viscosity.

The value of the collisional stress tensor $\mathbf{P}_c$ is the sum of that given by Eq. (9.227) and (9.243) for $\mathbf{P}_{c_1}$ and $\mathbf{P}_{c_2}$, respectively. To obtain the total stress tensor the kinetic contribution given by $\rho < \mathbf{CC} >$ must be added:

$$\mathbf{P} = \mathbf{P}_{c_1} + \mathbf{P}_{c_2} + \mathbf{P}_k. \quad (9.248)$$

Then the total stress tensor becomes

$$\mathbf{P} = \rho_p \varepsilon_s \left[1 + \tfrac{4}{5}(1+e) g_0 \varepsilon_s\right] < \mathbf{CC} > + \tfrac{2}{5}(1+e) g_0 \rho_p \varepsilon_s^2 C^2 \mathbf{I}$$

$$- \frac{4\varepsilon_s^2 \rho_p d_p g_0 (1+e)}{3\sqrt{\pi}} \Theta^{1/2} \left[\tfrac{6}{5} \overset{o}{\nabla} \mathbf{v} + \nabla \cdot \mathbf{v} \mathbf{I}\right], \tag{9.249}$$

where $\rho < \mathbf{CC} >$ is given by Eq. (9.232) and $C^2 = 3\Theta$. Making these substitutions the total stress becomes

$$Stress = \mathbf{P} = \rho_p \varepsilon_s \Theta \left[1 + 2(1+e) g_0 \varepsilon_s\right] \mathbf{I} - \frac{4\varepsilon_s^2 \rho_p d_p g_0 (1+e)}{3\sqrt{\pi}} \Theta^{1/2} \nabla \cdot \mathbf{v} \mathbf{I}$$

$$- \left\{ \frac{2\mu_s}{(1+e) g_0} \left[1 + \tfrac{4}{5}(1+e) g_0 \varepsilon_s\right]^2 + \frac{4\varepsilon_s^2 \rho_p d_p g_0 (1+e)}{5\sqrt{\pi}} \Theta^{1/2} \right\} 2 \overset{o}{\nabla}{}^s \mathbf{v}, \tag{9.250}$$

where μ_s is the dilute viscosity given by Eq. (T9.1) in Table 9.2. The viscosity of the dense suspension is the expression in the curly brackets in Eq. (9.250). It is the sum of $\varepsilon_s \mu_{sc}$ given by Eq. (9.245) and the dilute phase viscosity corrected for a large solids volume fraction and a non-unit restitution coefficient. The viscosity is still a product of a mean free path times the oscillation velocity, that is, granular temperature, and times a density corrected for the large volume fraction and the non-unit restitution coefficient.

9.18 Granular Conductivity

The granular heat flow vector $\mathbf{q}$ is computed in much the same way as the pressure tensor. It consists of a kinetic part $\mathbf{q}_k$ and two collisional parts $\mathbf{q}_{c_1}$ and $\mathbf{q}_{c_2}$.

With the quantity ψ given by Eq. (9.203), the kinetic heat flux is given by

$$\mathbf{q}_k = \tfrac{1}{2} m \int f C^2 \mathbf{C} d\mathbf{C} = \rho < \tfrac{1}{2} C^2 \mathbf{C} >. \tag{9.251}$$

When f is the Maxwellian distribution given by Eq. (9.15), the integral in Eq. (9.251) becomes

$$\mathbf{q}_k^{(0)} = \tfrac{1}{2} \rho \int_{-\infty}^{\infty} \frac{C^2 \mathbf{C}}{(2\pi\Theta)^{3/2}} \exp\left[-\frac{C^2}{2\theta}\right] d\mathbf{C}. \tag{9.252}$$

Since the integral in (9.252) is odd, it was already shown in Section 9.11 that

$$\mathbf{q}_k^{(0)} = 0. \tag{9.253}$$

The second iteration is obtained by using $\phi^{(1)}$ given by Eq. (9.171). Since this $\phi^{(1)}$ was derived for dilute conditions only, the resulting $\mathbf{q}_k$ will also be valid for dilute flow. Then (9.251) becomes

$$\begin{aligned}\mathbf{q}_{k\,\text{dilute}}^{(1)} &= \tfrac{1}{2}m\int f^{(1)}C^2\mathbf{C}\,d\mathbf{C} = \tfrac{1}{2}m\int f^{(0)}\phi^{(1)}C^2\mathbf{C}\,d\mathbf{C} \\ &= -\tfrac{m}{n}\int f^{(0)}C^2\mathbf{C}A(\mathbf{C})\mathbf{C}\cdot\nabla\ln\theta\,d\mathbf{C}.\end{aligned} \tag{9.254}$$

In (9.254) when using (9.171), the integral involving odd functions of the components of $\mathbf{C}$ vanished. Using Chapman and Cowling's (1961) formula (1.42-2),

$$\int F(C)\mathbf{CC}\,d\mathbf{C} = \tfrac{1}{3}\mathbf{I}\int F(C)C^2\,d\mathbf{C}, \tag{9.255}$$

or their Eq. (1.42-4), the first iterate for the dilute granular conductivity, as given by (9.254), can be expressed as

$$\mathbf{q}_{k\,\text{dilute}}^{(1)} = -\tfrac{1}{3}\tfrac{m}{n}\nabla\ln\theta\int f^{(0)}C^4A(C)\,d\mathbf{C}. \tag{9.256}$$

Then using Eq. (9.168), followed by (9.165) and (9.169), the conductivity becomes as follows:

$$\mathbf{q}_{k\,\text{dilute}}^{(1)} = -\tfrac{1}{3}\tfrac{m}{n}\nabla\theta\cdot\int \mathbf{C}A(C)f^{(0)}\left(\tfrac{C^2}{\theta}-\tfrac{5}{2}\right)\mathbf{C}\,d\mathbf{C} = \tfrac{1}{3}\tfrac{m}{n}\nabla\theta\int \mathbf{A}I(\mathbf{A})\,d\mathbf{C}. \tag{9.257}$$

Then in the analogue of Fourier's law of conduction,

$$\mathbf{q} = -\kappa\nabla\Theta, \tag{9.258}$$

dilute phase granular conductivity κ is given by what is known as a bracket integral, shown in Eq. (9.259):

$$\kappa_{\text{dilute}} = \tfrac{1}{3}m\int \mathbf{A}I(\mathbf{A})\,d\mathbf{C}. \tag{9.259}$$

The integral in (9.259) is evaluated in terms of Sonine polynomials in C^2 (Chapman and Cowling, 1961). A good approximation of the conductivity for restitution coefficient of one is

$$\kappa_{\text{dilute}} = \tfrac{5}{2}\mu_{\text{dilute}} C_v, \tag{9.260}$$

where μ_{dilute} is given in Eq. (9.183) and

$$C_v = \tfrac{3}{2}. \tag{9.261}$$

The granular conductivity can be expressed as an "eddy type" conductivity expressed in meters squared per second as follows:

$$\frac{\kappa_{\text{dilute}}}{\rho_p} = \frac{75\sqrt{\pi}}{384} d_p \Theta^{1/2}. \tag{9.262}$$

Equation (9.262) is Lun *et al.*'s (1984) Eq. (4.10) for a restitution coefficient of one. For non-electric collisions Lun *et al.* (1984) provide a correction involving the restitution coefficient which they associate with the first coefficient in the Sonine's polynomial solution. In view of the highly approximate nature of Lun *et al.*'s solution and the unavailability of the more exact solution using Chapman and Cowling's (1961) method as discussed in Section 9.17, Lun *et al.*'s solution is not presented here. Lun *et al.*'s conductivity given by their Eq. (4.10) is smaller than the elastic conductivity given by Eq. (9.262). This means that dissipation given by the conductivity times the gradient square is smaller for elastic than non-elastic particles. The results to follow will be restricted to restitution coefficients near unity. With this restriction the kinetic flow of heat for a dense suspension is given by Eq. (9.233).

The collisional granular conductivity is obtained by substituting ψ given by Eq. (9.203) into Eq. (9.198). Then the collisional granular heat flux $\mathbf{q}_c$ is

$$\begin{aligned}\mathbf{q}_c &= \tfrac{1}{4} m g_0 d_p^3 \iiint \left(C_1'^2 - C_1^2\right) f_1 f_2 \mathbf{k}(\mathbf{c}_{12}\cdot\mathbf{k})\, d\mathbf{k}\, d\mathbf{c}_1 d\mathbf{c}_2 \\ &+ \tfrac{1}{8} m g_0 d_p^4 \iiint \left(C_1'^2 - C_1^2\right) f_1 f_2 \mathbf{k}\nabla \ln \frac{f_2^{(0)}}{f_1^{(0)}} \mathbf{k}(\mathbf{c}_{12}\cdot\mathbf{k})\, d\mathbf{k}\, d\mathbf{c}_1 d\mathbf{c}_2 .\end{aligned} \tag{9.263}$$

Using Eq. (9.47), it is easy to show that

$$C_1'^2 = C_1^2 + \mathbf{C}_1 \cdot \mathbf{k}(1+e) + \tfrac{1}{4}(1+e)^2(\mathbf{k}\cdot\mathbf{c}_{12})^2. \tag{9.264}$$

Upon substitution of (9.264) into the integrals in (9.263), integrations with respect to **k** are performed as summarized by Chapman and Cowling (1961) and Ferziger and Kaper (1972). The first integral gives

$$\mathbf{q}_{c_1} = \tfrac{3}{5}\varepsilon_s g_0(1+e)\rho_p\varepsilon_s < C^2\mathbf{C} >. \tag{9.265}$$

For a Maxwellian distribution, $\mathbf{q}_{c_1}$ is zero, as in Eq. (9.252). Using the second approximation to the velocity distribution an extension of the analysis to dense flow presented in Section 9.11 and summarized in Chapman and Cowling (1961) for dense gases gives

$$\tfrac{1}{2}\rho_p\varepsilon_s < C^2\mathbf{C} >^{(1)} = -\frac{2}{g_0(1+e)}\left[1+\tfrac{6}{5}\varepsilon_s g_0(1+e)\right]\kappa_{\text{dilute}}\nabla\Theta. \tag{9.266}$$

As explained, Eqs. (9.262), (9.265) and (9.266) are restricted to restitution coefficients near unity. The second integral in (9.263), as evaluated by Ding and Gidaspow (1990), does not have this restriction. Its value using Eq. (9.238) to introduce the gradient of granular temperature is as follows:

$$\mathbf{q}_{c_2} = -\kappa_{c_2}\nabla\Theta, \tag{9.267}$$

where the granular conductivity from the second collision integral is

$$\kappa_{c_2} = 2\rho_p\varepsilon_s^2 d_p(1+e)g_0\left(\tfrac{\Theta}{\pi}\right)^{1/2}. \tag{9.268}$$

The value of the conductivity given by Eq. (9.268) equals the specific heat, which is two-thirds times the value of the coefficient in brackets of (9.243), referred to as $\overline{\omega}$ in Chapman and Cowling (1961). Surprisingly, their conductivity is again smaller for a smaller restitution coefficient. The extra dissipation due to non-elastic collisions must be compensated by the generation of heat through the source term that has a restitution coefficient to the second power.

The total granular heat flux is the sum of the kinetic flux and the collisional flux given by the sum of $\mathbf{q}_{c1}$ and $\mathbf{q}_{c2}$:

$$\mathbf{q} = \mathbf{q}_k + \mathbf{q}_{c_1} + \mathbf{q}_{c_2}, \tag{9.269}$$

where

$$\mathbf{q}_k = \tfrac{1}{2}\rho_p \varepsilon_s < C^2 \mathbf{C} >^{(1)}. \tag{9.270}$$

Hence, using (9.265) and (9.267), the total flux is

$$\mathbf{q} = \tfrac{1}{2}\rho_p \varepsilon_s < C^2 \mathbf{C} >^{(1)} \left[1 + \tfrac{6}{5}\varepsilon_s g_0 (1+e)\right] - \kappa_{c_2} \nabla\Theta. \tag{9.271}$$

Substituting for the approximate average given by Eq. (9.266) the total flux is then expressed in a Fourier's type law of conduction:

$$\mathbf{q} = -\left\{ \frac{2}{g_0(1+e)} \left[1 + \tfrac{6}{5}\varepsilon_s g_0 (1+e)\right]^2 \kappa_{\text{dilute}} + \kappa_{c_2} \right\} \nabla\Theta, \tag{9.272}$$

where κ_{dilute} is given by Eq. (9.262) and κ_{c_2} is expressed by Eq. (9.268). This expression differs from Lun *et al.* (1984) for values of restitution coefficients significantly different from one. The approach taken here was to follow the well-accepted kinetic theory of dense gases and to apply it to particulate flow.

All the theory holds for perfectly elastic particles. The restitution coefficient served the role of a minor correction.

Literature Cited

Bagnold, R. A. (1954). "Experiments on a Gravity-Free Dispersion of Large Solid Spheres in a Newtonian Fluid under Shear," *Proc. Roy. Soc.* **A225**, 49–63.

Birkhoff, G. (1960). *Hydrodynamics*. Princeton, New Jersey: Princeton University Press.

Carlos, C. R., and J. F. Richardson (1968). "Solids Movement in Liquid Fluidized Beds I: Particle Velocity Distribution," *Chem. Eng. Sci.* **23**, 813–824.

Carlslaw, H. S., and J. C. Jaeger (1959). *Combustion of Heat in Solids*. Oxford, U.K.: Clarendon Press.

Chapman, S., and T. G. Cowling (1961). *The Mathematical Theory of Non-uniform Gases*, 2nd ed. Cambridge, U.K.: Cambridge University Press.

Condiff, D. W., W. K. Lu, and J. S. Dahler (1964). "Transport Properties of Polyatomic Fluids: A Dilute Gas of Perfectly Rough Spheres," *J. Chem. Phys.* **42**, 3445–3475.

Ding, J., and D. Gidaspow (1990). "A Bubbling Fluidization Model Using Kinetic Theory of Granular Flow," *AIChE J.* **36**, 523–538.

Drew, D. A., and R. T. Lahey (1979). "Application of General Constitutive Principles to the Derivation of Multidimensional Two-Phase Flow Equations," *Int'l J. Multiphase Flow* **5**, 243–264.

Drew, D., L. Cheng, and R. T. Lahey (1979). "The Analysis of Virtual Mass Effects in Two-Phase Flow," *Int'l J. Multiphase Flow* **5**, 223–242.

Ferziger, J. H., and H. G. Kaper (1972). *Mathematical Theory of Transport Processes in Gases.* New York: Elsevier.

Gidaspow, D. (1977). "Fluid Particle Systems," pp. 115–128 in S. Kakac and F. Mayinger, eds. *Two Phase Flow and Heat Transfer*, Vol. I. Washington, D.C.: Hemisphere Publishing Co.

Jenkins, J. T., and S. B. Savage (1983). "A Theory for the Rapid Flow of Identical, Smooth, Nearly Elastic Spherical Particles," *J. Fluid Mech.* **130**, 187–202.

Johnson, P. C., and R. Jackson (1987). "Frictional-Collisional Constitutive Relations for Granular Materials with Application to Plane Shearing," *J. Fluid Mech.* **176**, 67–93.

Lamb, H. (1932). *Hydrodynamics*, 6th Ed. Cambridge, U.K.: Cambridge University Press.

Louge, M. Y., E. Mastorakos, and J. T. Jenkins (1990). "The Role of Particle Collisions in Pneumatic Transport," *J. Fluid Mech.* **23**, 345.

Lun, C. K. K., S. B. Savage, D. J. Jeffrey, and N. Chepurniy (1984). "Kinetic Theories for Granular Flow: Inelastic Particles in Couette Flow and Singly Inelastic Particles in a General Flow Field," *J. Fluid Mech.* **140**, 223–256.

Ogawa, S., A. Umemura, and N. Oshima (1980). "On the Equations of Fully Fluidized Granular Materials," *Z. angew. Math. Phys.* **31**, 483–493.

Pauchon, C., and S. Banerjee (1988). "Interphase Momentum Interaction Effects in the Averaged Multifield Model, Part II: Kinematic Waves and Interfacial Drag in Bubbly Flows," *Int'l J. Multiphase Flow* **14**, 253–264.

Roco, M. C., and T. Cader (1990). "Energy Approach for Wear Distribution in Slurry Pipelines," *Jap. J. Multiphase Flow* **4**, 2–20.

Saffman, P. G. (1965). "The Lift on a Small Sphere in a Slow Shear Flow," *J. Fluid Mech.* **22**, 385–400.

Savage, S. B. (1983). "Granular Flows at High Shear Rates," pp. 339–358 in R. E. Meyer, Ed. *Theory of Dispersed Multiphase Flow*. New York: Academic Press.

Savage, S. B. (1988). "Streaming Motions in a Bed of Vibrationally Fluidized Dry Granular Material," *J. Fluid Mech.* **194**, 457–478.

Savage, S. B., and D. J. Jeffrey (1981). "The Stress Tensor in a Granular Flow at High Shear Rates," *J. Fluid Mech.* **110**, 255–272.

Shahinpour, M., and G. Ahmadi (1983). "A Kinetic Theory for the Rapid Flow of Rough Identical Spherical Particles and the Evolution of Fluctuation: Advances in the Mechanics and the Flow of Granular Materials II," pp. 641–667 in M. Shahinpoor, Ed. *Trans. Tech. Pub.* Switzerland: Andermannsdorf.

Soo, S. L. (1967). *Fluid Dynamics of Multiphase System.* Waltham, Massachusetts: Blaisdell.

Syamlal, M. (1987). *A Review of Granular Stress Constitutive Relations Technical Report under Contract No. DE-AC21-85MC21353.* Morgantown, West Virginia: U. S. Department of Energy.

Ye, J., M. C. Roco, and D. Stafford (1990). "Particle Rotation in Couette Slurry Flow," pp. 19–27 in O. Furuya, Ed. *Cavitation and Multiphase Flow Forum, 1990, FED. Vol 98.* New York: ASME.

10

APPLICATIONS OF KINETIC THEORY

10.1 Granular Shear Flow **297**
- 10.1.1 Dissipation–Production Balance **298**
- 10.1.2 Granular Temperature for Dense Flow **300**
- 10.1.3 Granular Temperature for Dilute Flow **302**
- 10.1.4 Transition from Viscous to Grain Inertia Regime **303**
- 10.1.5 Grain Inertia Regime **305**

10.2 Flow Down a Chute **307**
10.3 Bubbling Bed: Flow Patterns **311**
10.4 Liquid–Solid Fluidization **318**
10.5 Circulating Fluidized Bed Loop Simulation **324**
10.6 Maximum Solids Circulation in a CFB **330**
Literature Cited **334**

10.1 Granular Shear Flow

The simplest solution to the Navier–Stokes equation for a fluid is that of shear or Couette flow (Schlichting, 1960). It is assumed that the flow is steady, incompressible and fully developed. The latter condition insures that the velocity is nonzero in one direction, say, in the x direction only. Then the continuity equation with constant porosity shows that the gradient of velocity is zero. Hence, all accelerations vanish and the momentum balance for granular flow is simply

$$\nabla \cdot \mathbf{P} = 0, \tag{10.1}$$

where **P** is given by Eq. (9.250). Furthermore, analogous to the fluid pressure, one assumes the solids pressure in the x direction to be constant. The momentum balance then assumes the simple form

$$\mu_s \frac{du_s}{dy} = \text{constant}. \tag{10.2}$$

In the simple shear flow one assumes a no slip boundary condition at one wall. The second wall is assumed to move with velocity U. With the fluid adhering to the wall, the boundary conditions are

$$\text{at } y = 0, u_s = 0 \text{ and at } y = h, u_s = U. \tag{10.2a}$$

The velocity profile is then

$$u_s = Uy/h. \tag{10.3}$$

Although Eq. (10.2) does not permit a direct determination of viscosity as is done in the Poiseuille viscometer in which the pressure in Eq. (10.1) varies with length of flow, it turns out that the granular temperature balance, Eq. (9.211), permits one to estimate the viscosity, as developed in Sections 10.1.1 to 10.1.3.

10.1.1 Dissipation–Production Balance

With the same assumptions made for the simple shear flow discussed in the previous paragraph, the granular temperature, Eq. (9.211), with no production due to fluid turbulence or molecular fluid collisions, is

$$-\mathbf{P}:\nabla \mathbf{v}_s = \mu_s \left(\frac{\partial u_s}{\partial y} \right)^2 = \gamma - 3\beta_a \Theta + \gamma_w. \tag{10.4}$$

$$\text{Production of oscillations by surface forces} = \text{Dissipation due to inelastic collisions} + \text{Dissipation in the fluid} + \text{Dissipation by the wall} \tag{10.4a}$$

In obtaining the dissipation by the wall in Eq. (10.4), it was assumed that the granular heat flux in Eq. (9.211) is a constant and that the granular heat flow was developed, as is done in the thermal forced convection problems. For an elastic particle collision with a non-vibrating wall, the dissipation by the wall vanishes. In Eq. (10.4) γ is the symbol used for the collisional source contribution, defined as

$$-\gamma = N_c \left(\tfrac{m}{2} c^2 \right), \tag{10.5}$$

where N_c was given by Eq. (9.192). Although the value of this integral was given previously by Eq. (9.213), it is instructive to present an approximate derivation of this integral. Physically, the integral given by Eq. (9.192) can be regarded as the product of the loss of the kinetic energy of the particles per collision

$$\Delta\psi = \Delta\left(\tfrac{1}{2}mC^2\right) \tag{10.6}$$

and a mean binary collision frequency given by Eq. (9.51). As a very rough approximation, we can assume that this collision frequency is independent of C^2 in Eq. (10.6). Then it is given by Eq. (9.63), while the change in kinetic energy is given by Eq. (9.45). In terms of particle diameter d_p, and the granular temperature Θ, the latter becomes

$$\Delta E = \tfrac{\pi}{4}\left(e^2 - 1\right)\rho_p d_p^3 \Theta. \tag{10.7}$$

Multiplication of ΔE by N_{12} from Eq. (9.63) gives an approximate average value of γ that is three times the value given by Eq. (9.213).

The use of the exact value obtained from Eq. (9.213) gives

$$-\gamma = \Delta E_{AV} \cdot N_{12} = 12\varepsilon_s^2 \rho_p g_o \left(e^2 - 1\right)\left(\tfrac{\Theta}{d_p}\right)\left(\tfrac{\Theta}{\pi}\right)^{1/2}. \tag{10.8}$$

Equation (10.8) shows that the dissipation is proportional to the cube of the random velocity and to the square of the particle concentration. The latter dependence arose because of the assumptions that the binary probability of collision is the product of the single frequency distributions, the so-called "chaos" assump-tion. The dissipation also strongly depends upon the empirically based value of g_o, which depends upon the collision velocity and the fluid medium.

Shen and Ackerman (1982), in analyzing the energy dissipation due to the collisions, ΔE given by Eq. (10.7), also include a dissipation due to a change of velocity in a direction perpendicular to the normal direction. This gives rise to a coefficient of kinetic friction. Using such a concept, Eq. (10.8) for the dissipation γ would include the restitution coefficient and the kinetic friction. For rough spherical particles, Shen and Ackerman's (1982) kinetic friction should be related to the tangential restitution coefficient of Araki and Tremaine (1986). The latter authors consider particle translation and rotation. They have extended the analysis presented in Chapman and Cowling's Chapter 11 (1961) for rough spherical molecules to include kinetic energy losses due to normal and tangential restitution coefficients.

The dissipation due to the presence of a fluid is given by the product of the friction coefficient and the random velocity which is expressed in terms of the granular temperature in Eq. (10.4). For low particle Reynolds numbers and dense suspensions, the friction coefficient is obtained from the first term in the Ergun equation. It becomes

$$\beta_A = \frac{18\mu_f \varepsilon_s}{d_p^2}. \tag{10.9}$$

It involves the fluid viscosity μ_f, and a reciprocal of a permeability.

10.1.2 Granular Temperature for Dense Flow

For a dense suspension in a fluid of a small viscosity, the production–dissipation balance, Eq. (10.4), reduces itself to

$$-P_{xy}\left(\frac{\partial u_s}{\partial y}\right) = \gamma. \tag{10.10}$$

The shear P_{xy} is a component of the general stress given by Eq. (9.250). For the dense suspension or for a Maxwellian velocity distribution, it is

$$-P_{xy} = \tfrac{4}{5}\varepsilon_s^2 \rho_p d_p g_o (1+e)\left(\tfrac{\Theta}{\pi}\right)^{1/2}\left(\frac{\partial u_s}{\partial y}\right). \tag{10.11}$$

The use of γ given by Eq. (10.8) gives an equation for the granular temperature Θ,

$$(1-e)\Theta = \tfrac{1}{15}\left(\frac{\partial u_s}{\partial y}\right)^2 d_p^2, \tag{10.12}$$

or for the random velocity $<\mathbf{C}^2>^{1/2}$,

$$\sqrt{1-e} <\mathbf{C}^2>^{1/2} = \tfrac{1}{\sqrt{5}}\left|\frac{du_s}{dy}\right| d_p. \tag{10.13}$$

Equation (10.13) gives rise to the dimensionless group R introduced by Savage (1983):

$$R = \frac{d_p\left|\frac{du_s}{dy}\right|}{<\mathbf{C}^2>} = \frac{\text{particle diameter} \cdot \text{shear rate}}{\text{random velocity}} \tag{10.14}$$

Equation (10.13) shows that the value of R depends only upon the restitution coefficient as shown below:

$$R=\sqrt{5(1-e)}. \tag{10.15}$$

This value of R differs from Savage's (1983) value given by his Eq. (4.4) because he used an earlier version of his kinetic theory.

An equation similar to Eq. (10.13) was obtained by Shen and Ackerman (1982). It involves their kinetic friction coefficient, as well as the drag term from Eq. (10.4).

To obtain kinetic coefficients, such as viscosity and conductivity, the granular temperature must be known. However, Eq. (10.13) makes it clear that the random velocity cannot be estimated without a knowledge of the restitution coefficient. The example below shows the problem.

For millimeter-size particle diameters used by Savage and shear rates as high as $10^3 s^{-1}$ with a restitution coefficient of 0.9 used by Savage, we have, using Eq. (10.13),

$$<C^2>^{1/2}=\frac{1}{\sqrt{0.5}}\cdot 10^3 s^{-1}\times 10^{-3}\,m \approx 1\,m/s. \tag{10.16}$$

Such oscillations are quite reasonable in view of hopper, critical flow rates and Carlos and Richardson's (1968) direct measurements of oscillations in a fluidized bed, as simulated by Gidaspow *et al.* (1991). However, for flow of 100 micron particles in a vertical pipe where shear rates are of the order of $10^2 s^{-1}$, we have

$$\sqrt{1-e}<C^2>^{1/2}=\frac{1}{\sqrt{5}}10^2 s^{-1}\times 10^{-4}\,m. \tag{10.17}$$

For an oscillation velocity of one meter per second, we have roughly

$$1-e\cong 10^{-4}. \tag{10.18}$$

In this theory the restitution coefficient must be much closer to one to obtain reasonable values of viscosities. Computations of a restitution coefficient from an analysis of a collision of two slightly non-elastic spheres (Kuwabura and Kono, 1987) shows that the restitution coefficient approaches unity as the velocity of collision is reduced.

10.1.3 Granular Temperature for Dilute Flow

The use of the dilute limit viscosity given by Eq. (9.183) in the balance given by Eq. (10.4), and neglecting the dissipation due to the walls and the fluid, with γ given by Eq. (10.8), produces an expression for the granular temperature shown in the following equation:

$$12\varepsilon_s^2\left(1-e^2\right)\Theta = \left(\frac{5\pi}{96}\right)\left(\frac{\partial u_s}{\partial y}\right)^2 d_p^2. \qquad (10.19)$$

This expression is similar to Eq. (10.12). The essential difference is the presence of the square of the particle concentration that multiplies the granular temperature. Since the right-hand side of Eq. (10.19) is finite as the particle concentration approaches zero, the granular temperature approaches infinity. This apparently strange behavior is present throughout the theory of Lun *et al.* (1984). It also occurs in the molecular dynamics computer simulations of Campbell and Brennan (1983) and Walton and Braun (1986).

To clarify the behavior exhibited by Eq. (10.19), this equation can be rewritten in terms of the mean free path given by Eq. (9.72). The mean free path, except for the constant in Eq. (9.72), is a ratio of the particle diameter to the volume fraction of the solid. Hence (10.19) becomes

$$\left(1-e^2\right)^{1/2}\Theta^{1/2} = \left(\frac{5\pi}{18}\right)^{1/2} \ell \cdot \left(\frac{\partial u_s}{\partial y}\right). \qquad (10.20)$$

Random velocity = Mean free path · Shear rate

The highest mean free path that a particle can achieve equals the diameter of its containing vessel. This restricts the highest value that can be achieved by the granular temperature. Hence, the mathematical singularity in the dilute limit in Eq. (10.19) does not occur. Equation (10.19) should be restricted by the inequality

$$\text{For } \ell = \frac{1}{6\sqrt{2}}\left(\frac{d_p}{\varepsilon_s}\right) \geq D_{\text{pipe}}, \quad \ell = D_{\text{pipe}}. \qquad (10.21)$$

Sinclair and Jackson (1989) use such a concept for modeling developed riser flow. Equation (10.13) for the random velocity for dense flow is then essentially the same equation as the dilute flow Eq. (10.20). In (10.13), the mean free path is the particle diameter divided by a constant, since the porosity is about one-half.

Then, the constant shear viscosity for this dilute limit, using Eq. (9.183) is

$$\mu_s = \frac{5\pi\sqrt{5}}{288\sqrt{2}}\rho_p d_p l\left(\frac{\partial u_s}{\partial y}\right)\frac{1}{\sqrt{1-e^2}}. \tag{10.22}$$

For a constant shear the viscosity increases as the particle concentration decreases or as the mean free path between the particles increases. The viscosity is also non-Newtonian.

In the presence of a fluid, there is also dissipation of oscillations by the fluid, as given by Eq. (10.4). Consider the limiting case of no dissipation between the particles, that is, restitution coefficient of unity then, using Eq. (10.9), the expression for granular temperature becomes

$$\Theta^{1/2} = \frac{5}{864\sqrt{\pi}}\left(\frac{\rho_p}{\mu_f n}\right)\left(\frac{\partial u_s}{\partial y}\right)^2. \tag{10.23}$$

Again, the granular temperature is high for dilute flow, that is, for small number density of particles n. As expected, the particle oscillations become small in a fluid of a high viscosity μ_f. The unexpected result is that the particle viscosity will vary with the square of the shear gradient. Hence, the shear in such a situation will vary with the cube of the shear rate. Again, in Eq. (10.23), the number density n cannot be too low. Below some low value of n, the dissipation by the wall, as given by Eq. (10.4), becomes dominant. In such a situation the granular temperature is finite for a given shear due to the finite γ_W in (10.4). Hence in a complete analysis of dilute transport, dissipation by the wall must be considered.

10.1.4 Transition from Viscous to Grain Inertia Regime

For dilute viscous flow of elastic particles, Savage's dimensionless R given by Eq. (10.14) can be evaluated by substituting Eq. (10.23) for the oscillation velocity. The result is

$$R = \frac{5184\varepsilon_s}{5\sqrt{3\pi}}\frac{\mu_f}{\rho_p\left|\frac{du_s}{dy}\right|d_p^2}. \tag{10.24}$$

It is a ratio of the fluid viscosity to a particular phase viscosity due to collisions. Except for the concentration dependence its inverse is the dimensionless shear strain group found by Bagnold (1954):

$$N_{\text{Bagnold}} = \frac{\rho_p d_p^2 \left| \frac{du_s}{dy} \right|}{\mu} = \frac{\text{inertial stress}}{\text{viscous stress}}. \tag{10.25}$$

Equation (10.25) suggests that for a sufficiently small N_{Bagnold}, fluid shear will dominate the stresses, while for a large N_{Bagnold}, grain inertia will determine the flow processes. N_{Bagnold} can also be looked at as a type of a Reynolds number. In his review, Savage (1983) has also included a concentration effect into the definition of this dimensionless group.

Bagnold (1954) constructed an annular shear cell in which he determined the shear by measuring the force exerted on a stationary wall and determined the normal force of neutrally buoyant millimeter-size particles suspended in the annular compartment by measuring the fluid pressure. The relations he found were

$$\text{shear} = \text{constant}(\lambda d_p)^2 \left(\frac{du_s}{dy} \right)^2, \quad \text{for large } \frac{du_s}{dy}; \tag{10.26}$$

$$\text{shear} = \text{constant } \lambda^{3/2} \mu_f \frac{du_s}{dy}, \quad \text{for small } \frac{du_s}{dy}. \tag{10.27}$$

The grain concentration λ is defined as the ratio of grain diameter to the distance between the grains s, as shown below:

$$\lambda = d_p / s. \tag{10.28}$$

It is related to the radial distribution function g_o by means of the relation

$$g_o = 1 + \lambda. \tag{10.29}$$

Hence, the function g_o can be interpreted as the dimensionless distance between spheres as follows:

$$g_o = \left[1 - \left(\frac{\varepsilon_s}{\varepsilon_{s,\max}}\right)^{1/3}\right]^{-1} = \frac{s + d_p}{s}. \tag{10.30}$$

For uniform spheres $\varepsilon_{s,\max} = \pi / 3\sqrt{2} = 0.7405$. Bagnold also found that the ratio of the shear to the normal force per unit area, the solids pressure, was constant in the grain inertia regime and equal to about 0.3.

10.1.5 Grain Inertia Regime

The Bagnold number, Eq. (10.25), shows that we will always be in the grain inertia regime for suspensions of particles in air, since the gas viscosity is very low. Hence, Savage (1983) and students had constructed an annular shear cell for studying shear and solids pressure for particles suspended in air. Their cell was similar to that of Bagnold (1954), except that the height of the annular compartment could be varied, while the particles were rotated, by putting various weights on top of the cell. The weights gave them a direct measure of the normal pressure. The volume fraction of particles was determined from the initial weight of the particles and the height of the compartment during rotation. Only the average value of the volume fraction of solids could be determined. The shear was measured by means of a spring balance. The shear rate was determined as in Bagnold's experiment by means of the formula

$$\frac{du_s}{dy} = \frac{R_m w}{\delta R}, \tag{10.31}$$

where R_m is the mean radius of the assembly, w is the rotational speed and δR is the gap width.

Figures 10.1 and 10.2 show the measured dimensionless shear and pressure, respectively, and a comparison to various versions of the kinetic theory. Ding and Gidaspow's (1990) theory only considered the dense regime, similarly to that discussed by Savage (1983). For this situation, the dimensionless shear can be obtained by substituting the granular temperature from Eq. (10.12) into the shear given by Eq. (10.11). The result is

$$\frac{P_{xy}}{\rho_p d_p^2 \left(\frac{du_s}{dy}\right)^2} = \frac{4\sqrt{15}}{75\sqrt{\pi}} \varepsilon_s^2 g_o \frac{1+e}{(1-e)^{1/2}}. \tag{10.32}$$

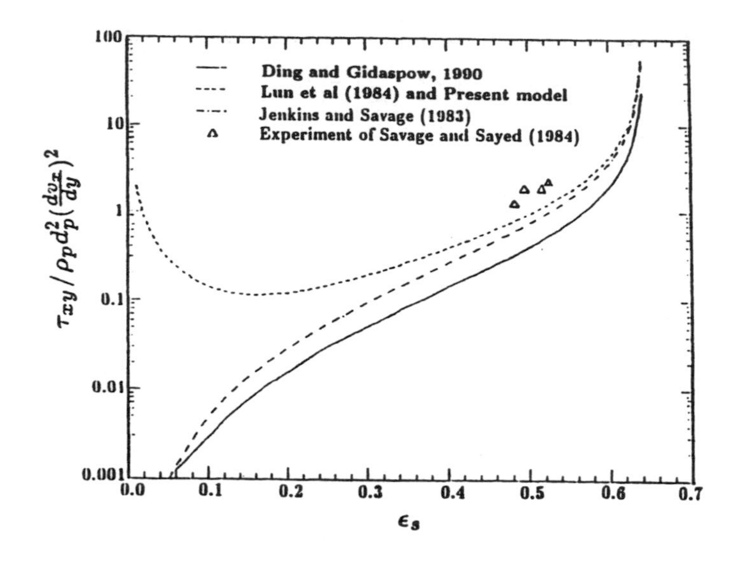

Fig. 10.1 Non-dimensional shear stress versus solid volume fraction for the case of simple shear flow. (From Ding and Gidaspow, 1990. Reproduced by permission of the American Institute of Chemical Engineers. Copyright © 1990 AIChE. All rights reserved.)

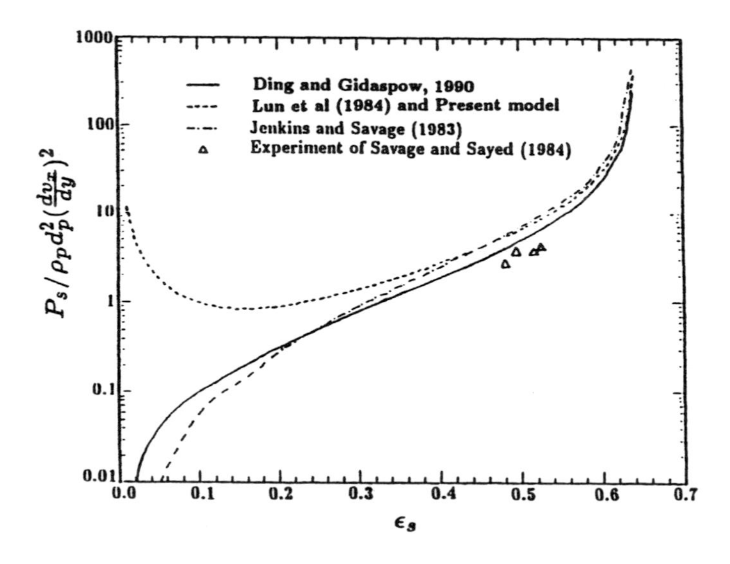

Fig. 10.2 Non-dimensional normal stress versus solid volume fraction for the case of simple shear flow. (From Ding and Gidaspow, 1990. Reproduced by permission of the American Institute of Chemical Engineers. Copyright © 1990 AIChE. All rights reserved.)

Similarly, the pressure is obtained by substituting the granular temperature from Eq. (10.12) into Eq. (9.230):

$$\frac{P_S}{\rho_p d_p^2 \left(\frac{du_s}{dy}\right)^2} = \frac{\varepsilon_s}{15(1-e)}\left[1+2\varepsilon_s g_o (1+e)\right]. \tag{10.33}$$

The data of Savage and Sayed (1984) obtained from the shear cell for millimeter-size particles are close to the values computed from Eqs. (10.32) and (10.33) using restitution coefficients of 0.8 or 0.9. Shen and Ackerman's (1982) expressions give much lower shear and pressures because they did not include g_o into their theory, which is of the order of 10. The shear and the pressures in the more complete theory and that of Lun *et al.* (1984) shown in Figs. 10.1 and 10.2 have a minimum because of the increase of the mean free path and hence granular temp-erature at low volume fraction of solids. For high g_o the ratio of shear to the solids pressure is given by

$$\frac{P_{xy}}{P_s} = \frac{2}{5}\left[15/\pi(1-e)\right]^{1/2} = \frac{2}{5}\left(\frac{3}{\pi}\right)^{1/2} R \quad \text{for} \quad \varepsilon_s g_o >> 1. \tag{10.34}$$

The ratio of the shear to the normal stress can be interpreted to be the tangent of the dynamic angle of repose ϕ_d:

$$P_{xy}/P_s = \tan\phi_d. \tag{10.35}$$

Hence, in this dense regime, this angle depends only upon the restitution coefficient.

10.2 Flow Down a Chute

As a second simple example, consider fully developed two-dimensional, free surface flow of granular material down a rough plane inclined at an angle ξ to the horizontal as shown in Fig. 10.3. The fully developed condition implied that flow properties vary only in the y direction, normal to the surface. The momentum equations, similarly to single-phase flow, are then

Momentum Balance Normal to the Surface

$$\frac{dP_s}{dy} = g\cos\xi\rho_b, \tag{10.36}$$

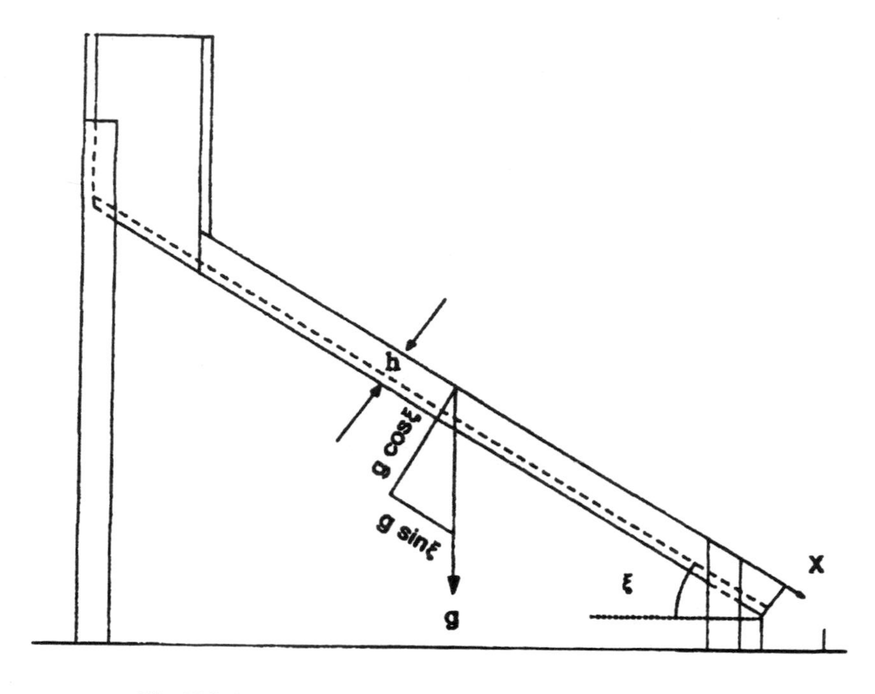

Fig. 10.3 Gravity flow of granular material down an inclined plane.

Momentum Balance Parallel to the Surface

$$\frac{dP_{xy}}{dy} = g \sin \xi \rho_b. \tag{10.37}$$

For an elastic, nonvibrating wall there is no flux of granular temperature into the surface. There is also a zero gradient of granular temperature at the free surface. Hence for granular thermally developed flow the granular flux is zero. For steady developed flow, the convective derivative of the granular temperature also vanishes. Then the granular temperature Eq. (9.211) reduces itself to Eq. (10.10). Therefore the granular temperature Θ is a constant and is given by Eq. (10.12). In view of the constancy of Θ, the shear is also a constant. The velocity gradient is then given by

$$\frac{du_s}{dy} = \frac{\left[15(1-e)\Theta\right]^{\frac{1}{2}}}{d_p}. \tag{10.37}$$

For no slip boundary condition at $y = h$, integration of Eq. (10.37) gives the linear velocity distribution obtained by Savage (1983):

$$u_s = \left[15(1-e)\Theta\right]^{1/2}(h-y)/d_p. \tag{10.38}$$

Similarly, with a zero pressure at the free interphase

$$P_s(0) = 0, \tag{10.39}$$

integration of Eq. (10.36) gives

$$P_s = g\rho_p \cos\xi \cdot \int_o^y \varepsilon_s dy. \tag{10.40}$$

But for constant shear, the solids pressure was found to be given by Eq. (10.33).

Substitution of Eq. (10.38) into (10.33) and use of (10.40) gives the expression for the granular temperature found by Savage (1983):

$$\theta = \frac{g\cos\xi \int_o^y \varepsilon_s dy}{2\varepsilon_s^2 g_o (1+e)} = \text{constant}. \tag{10.41}$$

Then assuming the mean values below are approximately equal

$$\frac{1}{y}\int_o^y \varepsilon_s dy \cong \frac{1}{h}\int_o^h \varepsilon_s dy. \tag{10.42}$$

Savage (1983) finds that the particle concentration distribution across the width of the chute is given by the relation shown below:

$$\frac{\varepsilon_s^2 g_o(\varepsilon_s)}{\varepsilon_{sh}^2 g_o(\varepsilon_{sh})} = \frac{y}{h}. \tag{10.43}$$

To evaluate the concentration profile, a value of the volume fraction at the bottom of the chute, ε_{sh}, must be specified. Figure 10.4 shows the velocity, granular temperature and concentration profiles corresponding to flow down an inclined chute with a 30° inclination angle studied experimentally by Ishida and Shirai (1979). Their particles were glass beads in the diameter range of 350 to 500 μm. Using a fiber optic probe, they obtained velocity distributions for several inclination angles. In constructing the plots in Fig. 10.4, Savage assumed the

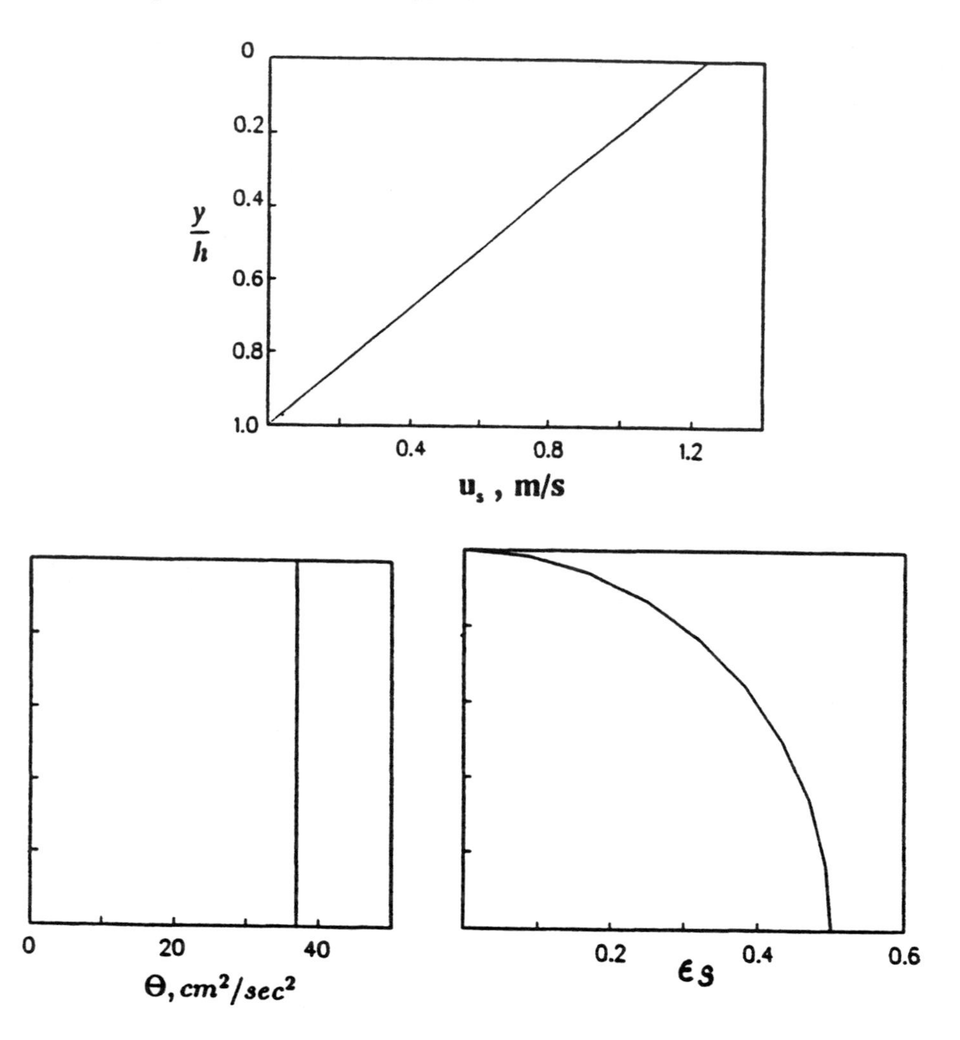

Fig. 10.4 Predicted profiles for velocity, granular temperature and solids fraction during flow down an inclined chute.

restitution coefficient to be 0.8, a steady height of 6 mm and the volume fraction of the solids at the surface to be 0.5. Numerical simulations performed by Ding roughly agree with Savage's analysis. However, flow never develops fully. Hence, the computed behavior is more complex.

A relation between the normal pressure and the shear follows directly from Eqs. (10.36) and (10.37):

$$\frac{dp_s}{dp_{xy}} = \tan\xi. \tag{10.44}$$

Then Eq. (10.34) shows that

$$\tan\xi = \tfrac{2}{5}\left[15/\pi(1-e)\right]^{1/2} = \frac{P_s}{P_{xy}}. \tag{10.45}$$

This implies that for a given value of restitution coefficient there is only one angle of inclination for which there exists steady, fully developed flow. Equation (10.45) also shows that for nearly elastic particles, the solids pressure is small compared with the shear. In view of the incompleteness of the theory, caution must be exercised in interpreting relations such as Eq. (10.45).

10.3 Bubbling Bed: Flow Patterns

Inviscid two-phase flow models were able to predict the formation, the growth and the bursting of bubbles in gas fluidized beds with large jets. In such situations the jet establishes the flow pattern. However, many industrial fluidized beds are built with complex gas distributors that can be approximated by assuming the gas enters the bed uniformly. The gas actually enters the bed through a number of multi-jet tuyers which may have the shape of bubble caps. To prevent these tuyers from being made inoperative by blockage of solids, a distributor having a pressure drop of at least 20% of the bed weight is frequently used (Whitehead, 1971, 1985). Such a two-dimensional bed was constructed at Illinois Institute of Technology to determine the flow patterns as a function of gas velocity and bed height (Gamwo, 1992). The purpose of this study was to determine the predictive capability of the hydrodynamic models. Obstacles were also put into the bed to simulate horizontal heat exchange tubes. The computations were made using the dense phase kinetic theory with the Maxwellian distribution (Ding and Gidaspow, 1990). Table 10.1 summarizes the equations used. For the Maxwellian distribution $\mu_{s,\text{dilute}}$ and $\kappa_{s,\text{dilute}}$ are zero. The restitution coefficient was taken to be 0.8, based on Savage's (1983) annular shear cell studies.

For uniform inlet flow, at low gas velocities or for shallow beds, bubbles form between the center of the bed and the wall. Figure 10.5 shows the observed and the computed bubbles and the flow pattern for a two-dimensional bed with a uniform inlet velocity of air of 100 cm/s. This flow pattern and the bubble forma-

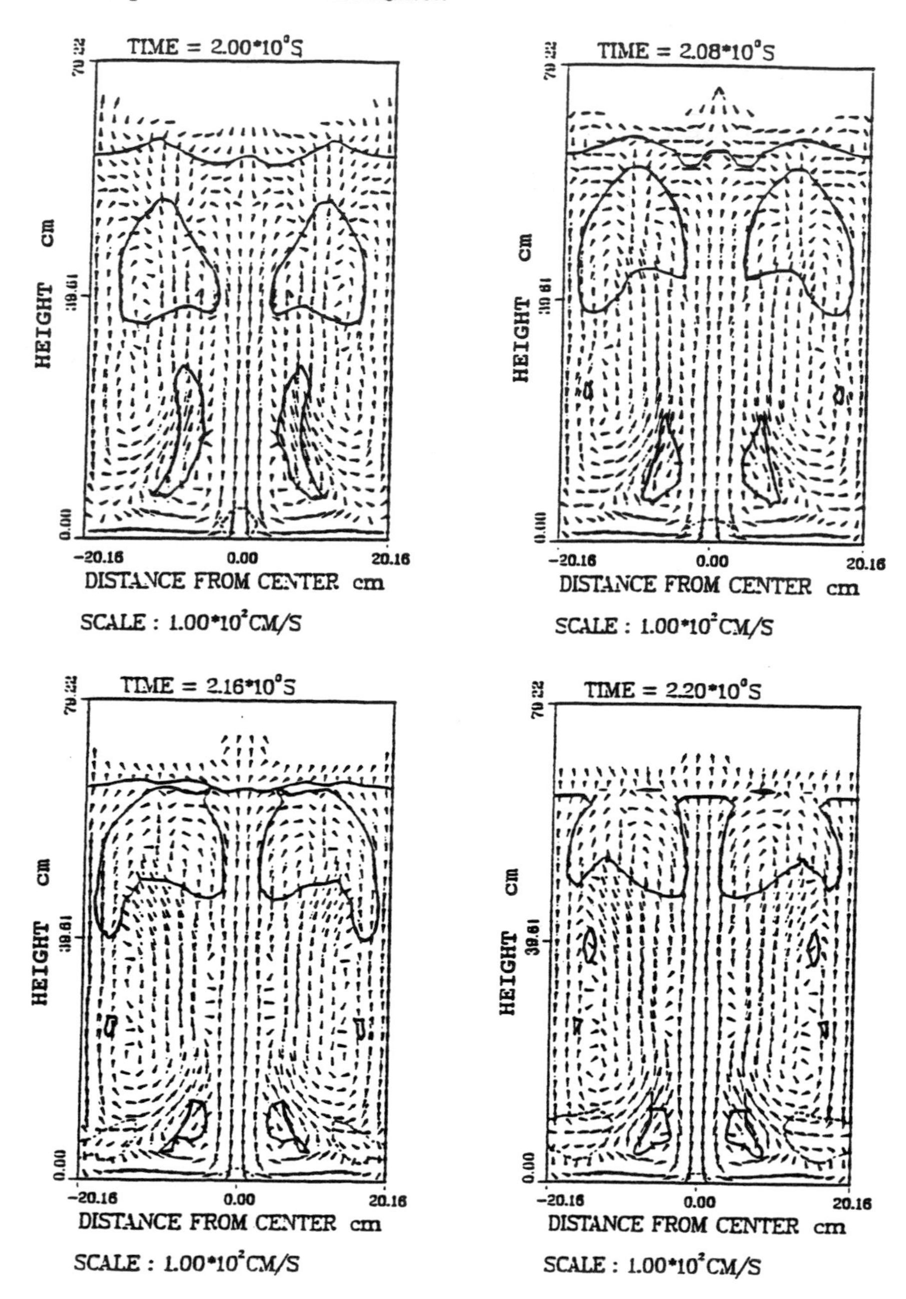

Fig. 10.5 (a) Computed solids velocities and porosity distributions (----, ε = 0.8; - - - -, ε = 0.6) (Ding, Gidaspow, and Gamwo, 1990 Annual AIChE Meeting).

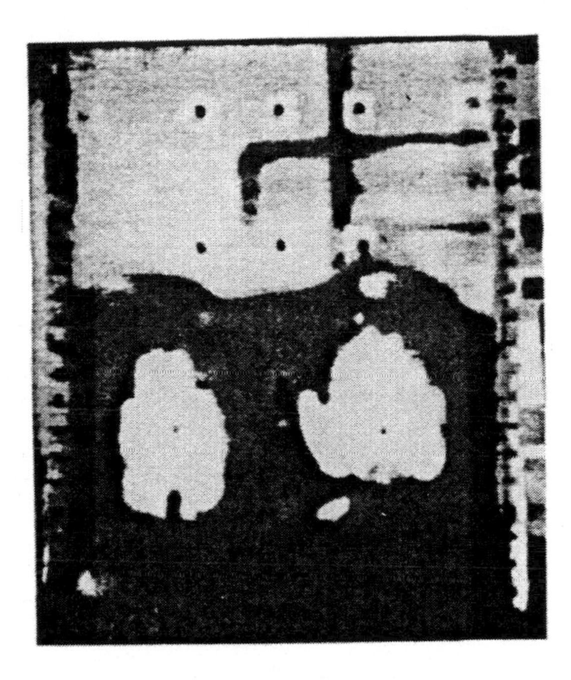

Fig. 10.5 (b) Experimental bubbles at t =2.08 s.

tion can be explained qualitatively based on the inviscid analysis presented in Chapter 7, where an equation for the solids vorticity had been derived. In the absence of viscosity the rate of change of solids vorticity, ζ, can be related to the gradient of porosity as

$$\frac{d\zeta}{dt^s} = -\frac{g}{\left(1-\varepsilon_{mf}\right)}\frac{\partial \varepsilon}{\partial y}, \tag{10.46}$$

where ε_{mf} is the porosity at minimum fluidization. Downward particle motion at the wall sets up the circulation or vorticity. The solids vorticity then gives rise to the vertical porosity gradient, as shown by the above equation. A non-uniform porosity gradient at the bottom of the bed causes non-uniform void propagation. For Geldart type B particles, the void moves faster in the dilute region. There is a catch-up effect, an intersection of void paths. Hence, bubbles form in the region between the walls and the center of the bed, as shown in the flow patterns in Fig. 10.5. The angle that the bubbles make with the horizontal depends on the fluid

Table 10.1 KINETIC THEORY MODEL FOR MULTIPHASE FLOW

Continuity Equation for Phase $k(=g,s)$

$$\frac{\partial}{\partial t}(\varepsilon_k\rho_k)+\nabla\cdot(\varepsilon_k\rho_k\mathbf{v}_k)=\dot{m}_k \qquad \sum_k \varepsilon_k=1 \tag{T10.1}$$

Momentum Equation for Phase $(k=g,s;\ 1=g,s;\ 1\neq k)$

$$\frac{\partial}{\partial t}(\varepsilon_k\rho_k\mathbf{v}_k)+\nabla\cdot(\varepsilon_k\rho_k\mathbf{v}_k\mathbf{v}_k)=\varepsilon_k\rho_k\mathbf{F}_k+\nabla\cdot\tau_k+\beta_B(\mathbf{v}_l-\mathbf{v}_k)+\dot{m}_k\mathbf{v}_k \tag{T10.2}$$

Acceleration of phase k = Gravity + Stress + Drag force + Phase change momentum

Constitutive Equation for Stress and Buoyancy

$$\tau_g=\left[-P_g+\xi_g\nabla\cdot\mathbf{v}_g\right]\mathbf{I}+2\mu_g\mathbf{S}_g \tag{T10.3}$$

where

$$\mathbf{S}_g=\frac{1}{2}\left[\nabla\mathbf{v}_g+(\nabla\mathbf{v}_g)^T\right]-\frac{1}{3}\nabla\cdot\mathbf{v}_g\mathbf{I} \tag{T10.3a}$$

$$\mathbf{F}_g=\mathbf{g}/\varepsilon_g \tag{T10.4}$$

Constitutive Equation for Solid Phase Stress and Buoyancy

$$\tau_s=\left[-P_s+\xi_s\nabla\cdot\mathbf{v}_s\right]\mathbf{I}+2\mu_s\mathbf{S}_s \tag{T10.4a}$$

where

$$\mathbf{S}_s=\frac{1}{2}\left[\nabla\mathbf{v}_s+(\mathbf{v}_s)^T\right]-\frac{1}{3}\nabla\cdot\mathbf{v}_s\mathbf{I} \tag{T10.5}$$

$$\mathbf{F}_s=\mathbf{g}(1-\rho_g/\rho_s) \tag{T10.6}$$

Solids Phase Stress

Kinetic Theory Model (Gidaspow's 1991 extension of Savage (1983) and Ding and Gidaspow's 1990 expressions to dilute and dense flow)

Solids Phase Pressure

$$P_s=\rho_p\varepsilon_s\Theta\left[1+2(1+e)g_o\varepsilon_s\right] \tag{T10.7}$$

Solids Phase Bulk Viscosity

$$\xi_s=\frac{4}{3}\varepsilon_s^2\rho_p d_p g_o(1+e)\left(\frac{\Theta}{\pi}\right)^{1/2} \tag{T10.8}$$

Solids Phase Shear Viscosity

$$\mu_s=\frac{2\mu_{s_{dil}}}{(1+e)g_o}\left[1+\frac{4}{5}(1+e)g_o\varepsilon_s\right]^2+\frac{4}{5}\varepsilon_s^2\rho_p d_p g_o(1+e)\left(\frac{\Theta}{\pi}\right)^{1/2} \tag{T10.9}$$

Table 10.1 (continued)

Solids Phase Stress (cont.)

Solids Phase Dilute Viscosity

$$\mu_{s_{dil}} = \frac{5\sqrt{\pi}}{96}\rho_p d_p \Theta^{1/2} \tag{T10.10}$$

Radial Distribution Function

$$g_o = \frac{3}{5}\left[1 - \left(\frac{\varepsilon_s}{\varepsilon_{s_{max}}}\right)^{1/3}\right]^{-1} \tag{T10.11}$$

Fluctuating Energy $\left(\Theta = \frac{1}{3} < C^2 >\right)$ ***Equation***

$$\frac{3}{2}\left[\frac{\partial}{\partial t}\left(\varepsilon_s\rho_s\Theta\right) + \nabla\cdot\left(\varepsilon_s\rho_s\mathbf{v}_s\Theta\right)\right] = \boldsymbol{\tau}_s : \nabla\mathbf{v}_s - \nabla\cdot\mathbf{q} - \gamma \tag{T10.12}$$

Collisional Energy Dissipation γ

$$\gamma = 3\left(1 - e^2\right)\varepsilon_s^2\rho_s g_o\Theta\left(\frac{4}{d_p}\left(\frac{\Theta}{\pi}\right)^{1/2} - \nabla\cdot\mathbf{v}_s\right) \tag{T10.13}$$

Flux of Fluctuating Energy q

$$\mathbf{q} = -\kappa\nabla\Theta \tag{T10.14}$$

Conductivity of Fluctuating Energy

$$\kappa = \frac{2}{(1+e)g_o}\left[1 + \frac{6}{5}(1+e)g_o\varepsilon_s\right]^2\kappa_{dil} + 2\varepsilon_s^2\rho_p d_p g_o(1+e)\left(\frac{\Theta}{\pi}\right)^{1/2} \tag{T10.15}$$

Dilute Phase (Eddy Type) Granular Conductivity

$$\kappa_{dil} = \frac{75}{384}\sqrt{\pi}\rho_p d_p \Theta^{1/2} \tag{T10.16}$$

Gas–Solid Drag Coefficients

For $\varepsilon_g < 0.8$ *(based on Ergun equation)*

$$\beta_B = 150\frac{\varepsilon_s^2\mu_g}{\varepsilon^2 d_p^2} + 1.75\frac{\rho_g\varepsilon_s\left|\mathbf{v}_g - \mathbf{v}_s\right|}{\varepsilon d_p} \tag{T10.17}$$

Table 10.1 (Continued)

Gas–Solid Drag Coefficients (cont.)

For $\varepsilon_g > 0.8$ *(based on empirical correlation)*

$$\beta_B = \tfrac{3}{4} C_d \frac{\varepsilon_s \rho_g \left| \mathbf{v}_g - \mathbf{v}_s \right|}{d_p} \varepsilon^{-2.65} \qquad \text{(T10.18)}$$

where

$$C_d = \tfrac{24}{Re_p}\left[1 + 0.15 Re_p^{0.687}\right] \text{for } Re_p < 1000 \qquad \text{(T10.19)}$$

$$C_d = 0.44 \text{ for } Re_p > 1000 \qquad \text{(T10.20)}$$

$$Re_p = \left(\varepsilon \rho_g \left| \mathbf{v}_g - \mathbf{v}_s \right| d_p\right) / \mu_g \qquad \text{(T10.21)}$$

velocity, the height of the bed, and the geometry. In the case of fluidization with a jet, the bubble always forms at the jet. A strength of the present model is that it can correctly predict bubbling fluidization with a uniform injection velocity.

The downward flow at the center of the bed is due to conservation of mass of the particles and symmetry. Lun *et al.* (1984) measured such time-averaged flow patterns using radioactive tracer particles. As the bubbles form at the bottom, rise and break on top, there is, of course, some variation in the flow patterns, as shown in Fig. 10.5.

Figure 10.6 shows the computed time-averaged granular temperature. The computed values seem reasonable. The random velocity is higher in the top half of the bed due to the presence of larger bubbles. The peaks are caused by the bubbles breaking at the surface, as seen in Fig. 10.5. The time-averaged solids viscosity shown in Fig. 10.7 is not unreasonable, although it appears to be somewhat low. This may be due to the use of a restitution coefficient that is too low.

The second basic flow pattern for uniform inlet flow occurs for higher flow rates and taller beds. As the bed is filled with particles to a higher and higher level, the bubbles coalesce to a single central bubble. Such a situation is illustrated in Fig. 10.8 for a cylindrical fluidized bed. This simulation corresponds to a radioactive particle tracking experiment of Lin *et al.* (1985). Figure 10.9 is a comparison of the computed and measured time averaged velocities of the particles. Both the simulation and the experiment show the centrally moving upward region caused by the bubble shown in Fig. 10.8. Conservation of mass then requires that there exist a downward flow near the wall. Lin *et al.*'s (1985) experiments show that as the velocity is reduced to 45 cm/s and less, the flow pattern changes to that described for the two-dimensional bed simulation at a velocity of

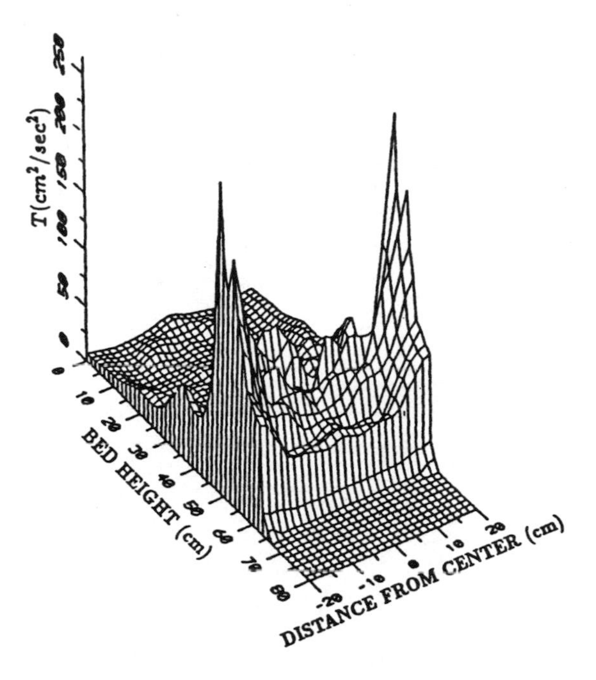

Fig. 10.6 Time-averaged granular temperature cm^2/s^2 in the bed (Ding, Gidaspow and Gamwo, 1990 Annual AIChE Meeting).

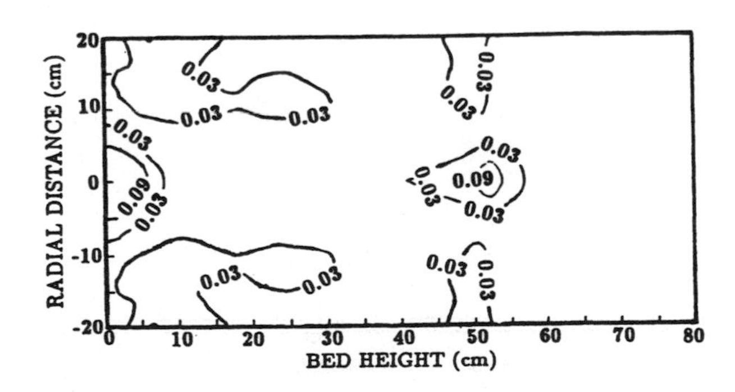

Fig. 10.7 Time-averaged solids viscosity (Pa·s) in the bed.

100 cm/s. The differences in the velocities at which the flow pattern changes for the two-dimensional and the cylindrical simulations are probably due to the wall effect.

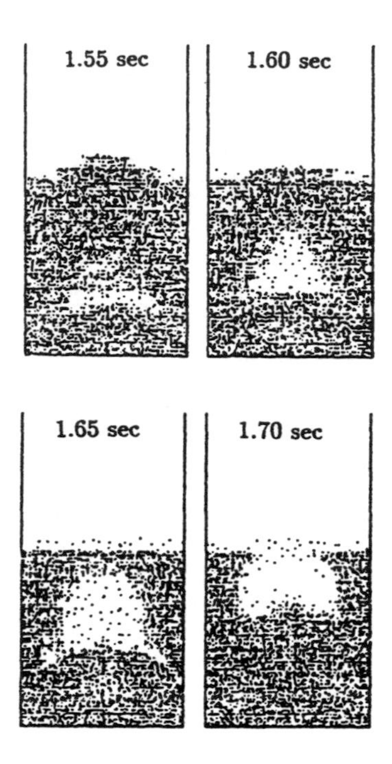

Fig. 10.8 Computed bubbles in a cylindrical bed at several times, inlet air velocity 64.1 cm/s. Bed diameter, 13.8 cm; static bed height, 11.3 cm; particle diamter, 500 μm; particle density, 2.5 g/cm^3. (From Ding and Gidaspow, 1990. Reproduced by permission of the American Institute of Chemical Engineers.

The two basic flow patterns — central downflow at low velocities and low bed levels, and central upflow and bubble coalescence at high velocities and high bed heights — are also observed for bubbling gas–liquid flow. These patterns can be explained by two concepts: (1) vorticity generation due to wall effects and (2) large bubble formation due to shock formation caused by the intersection of characteristic paths.

An effort to determine the circulation patterns in bubble columns using a radioactive particle tracking facility is underway (Moslemian *et al.*, 1990). Their preliminary observations are consistent with the general description given earlier.

10.4 Liquid–Solid Fluidization

Carlos and Richardson (1968) measured particle velocities in a liquid–solid fluidized bed with a uniform distribution by photographing the movement of

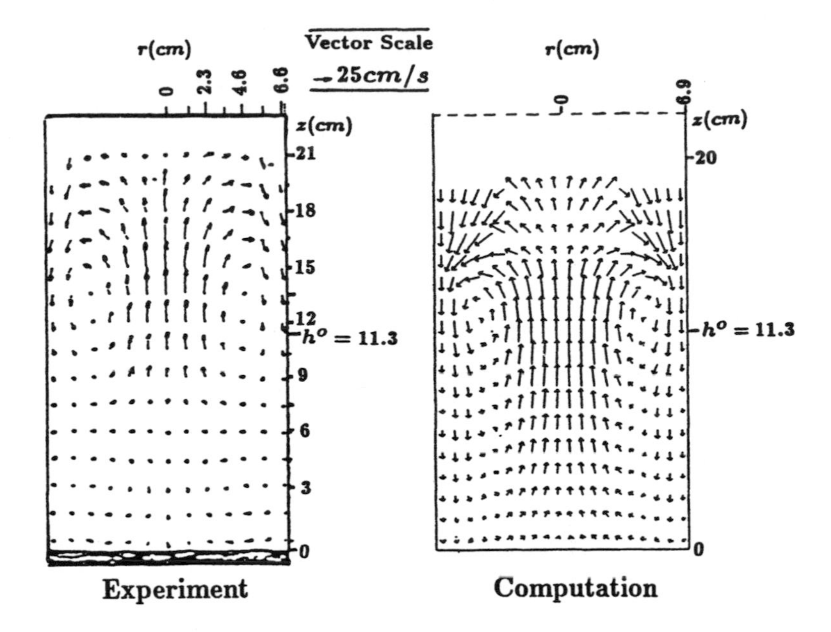

Fig. 10.9 Experimental and computed solids velocities in cylindrical bed, inlet air velocity 64.1 cm/s (Lin *et al.*, 1985).

tracer particles in a completely transparent bed of glass particles fluidized with dimethyl phthalate. The latter liquid has the same refractive index as the particles. Hence, only a blackened tracer particle is visible. Photographs of the tracer particles were taken at equally spaced periods. From the coordinates of the tracer particles, the velocity was determined. In constructing histograms of radial, tangential and axial velocities, Carlos and Richardson neglected the position dependence of velocities. In view of this and to illustrate the technique, an artificial example is shown in Fig. 10.10, which corresponds to the data in Carlos and Richardson's paper (1968) for a ratio of inlet to fluidization velocities of two. At a given position, such data can be obtained by use of a laser-doppler velocity meter. In Fig. 10.10, the hypothetical velocity data has been subdivided into intervals of 0.6 cm/s.

The ordinate is the frequency, f, expressed as the fraction of the total number of velocities falling in the interval, divided by the size of the interval. The area under the histogram is unity. A normal distribution with a standard deviation of unity fits the data. Since the standard deviation is the square root of the granular temperature, these measurements give a granular temperature of 1 cm^2 / s^2 at the location indicated in Fig. 10.10. Unfortunately, Carlos and Richardson lumped

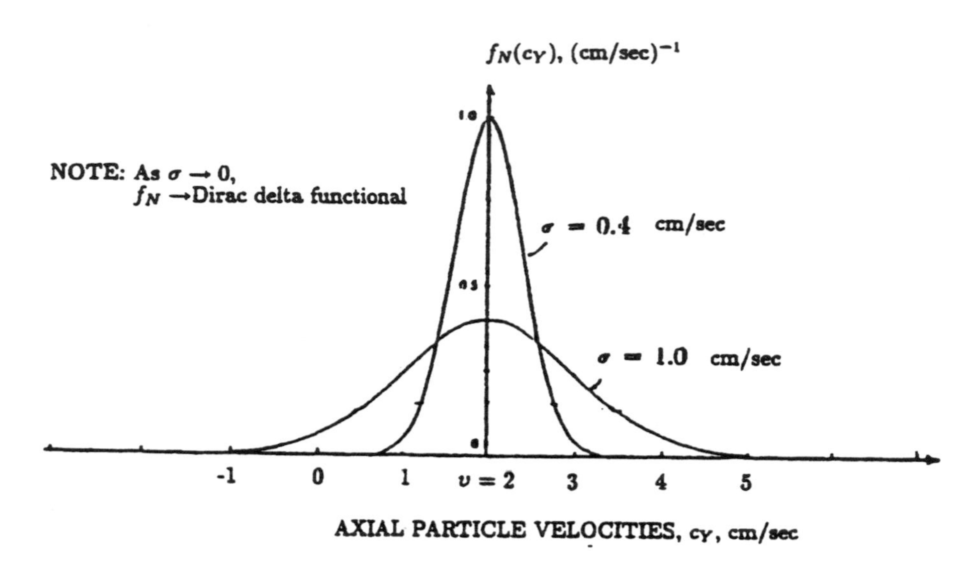

Fig. 10.10 An idealized fit of a histogram of mean axial velocities of particles in a circular liquid fluidized bed at the center of the bed, 8 cm above a uniform distributor to a Maxwellian or normal distribution, f_N:

$$f_N(c_Y) = \left(1/\sqrt{(2\pi)}\sigma\right)\exp\left[-(c_Y - v)^2/2\sigma^2\right],$$

where σ = standard deviation = $\sqrt{(\Theta)}$,

Θ = granular temperature and the hydrodynamic velocity is v,

$$v = \int_{-\infty}^{\infty} c_Y f_N \, dc_Y; \quad \Theta = \int_{-\infty}^{\infty} (c_Y - v)^2 f_N \, dc_Y$$

$$\text{at } c_Y = v; \quad f_N = 0.4 = \frac{1}{\sqrt{(2\pi)}\sigma}$$

all their position-dependent data into overall histograms where mean velocities are zero.

Numerical simulations were performed for a 122 cm long, 10 cm diameter column filled with 0.889 cm diameter glass beads of density of 2.49 g/cm^3. Table 10.2 summarizes the flow conditions used in the simulation. The density of the liquid is 1.19 g/cm^3.

The computations were made using the dense phase kinetic theory with the Maxwellian or the normal velocity distribution (Gidaspow *et al.*, 1991). Table 10.1

Table 10.2 FLOW CONDITIONS:
U_{mf} 4.8 cm/s; Height of fluidized bed = 26 cm

Superficial liquid velocity U (cm/s)	U/U_{mf}	Voidage
7.6	1.6	0.53
9.4	2.0	0.58
11.7	2.4	0.63
15.1	3.1	0.70

10.1 summarizes the equations used, where for the Maxwellian distribution $\mu_{s,\text{dilute}}$ and $\kappa_{s,\text{dilute}}$ are zero, as in the previous gas-fluidized bed example. Angular symmetry was assumed to exist.

Both the experiment and the computations show that the flow pattern consists of an upward movement of the particles in the center and downward movement at the walls. Figure 10.11 shows a comparison of the computed and the measured mean axial velocities. Figure 10.12 shows marked inward radial movement near the bottom of the bed and outward radial movement higher in the bed. Figure 10.13 shows the computed time-averaged granular temperature. There is a great deal of stirring in the lower section of the bed. In view of this, both the computed

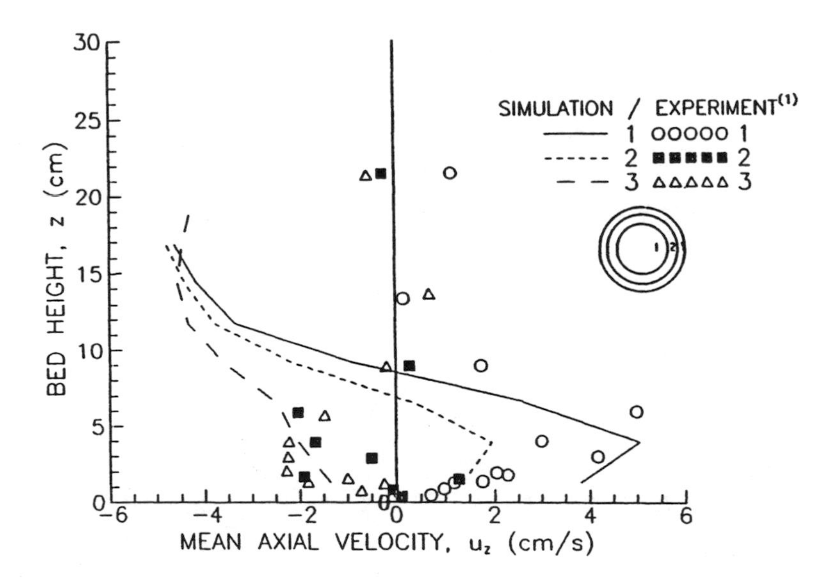

Fig. 10.11 Mean axial velocities (U/U_{mf} = 2.0), #1 – Carlos and Richardson (1968). (Reprinted with permission from Gidaspow *et al.*, 1991.)

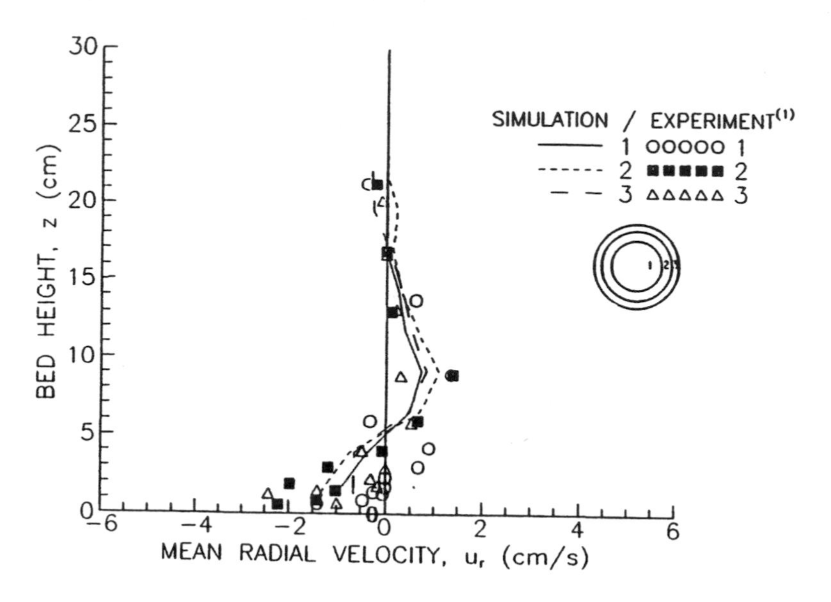

Fig. 10.12 Mean radial velocities ($U/U_{mf} = 2.0$), #1 – Carlos and Richardson (1968). (Reprinted with permission from Gidaspow *et al.*, 1991.)

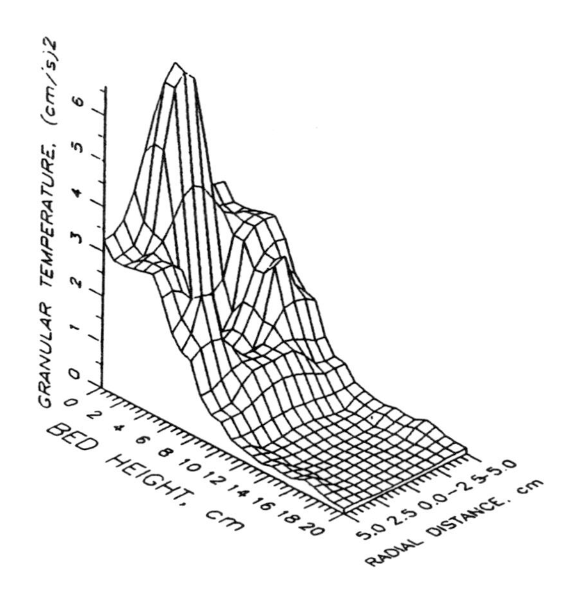

Fig. 10.13 Computed granular temperature in the bed ($U/U_{mf} = 2.0$). (Reprinted with permission from Gidaspow *et al.*, 1991.)

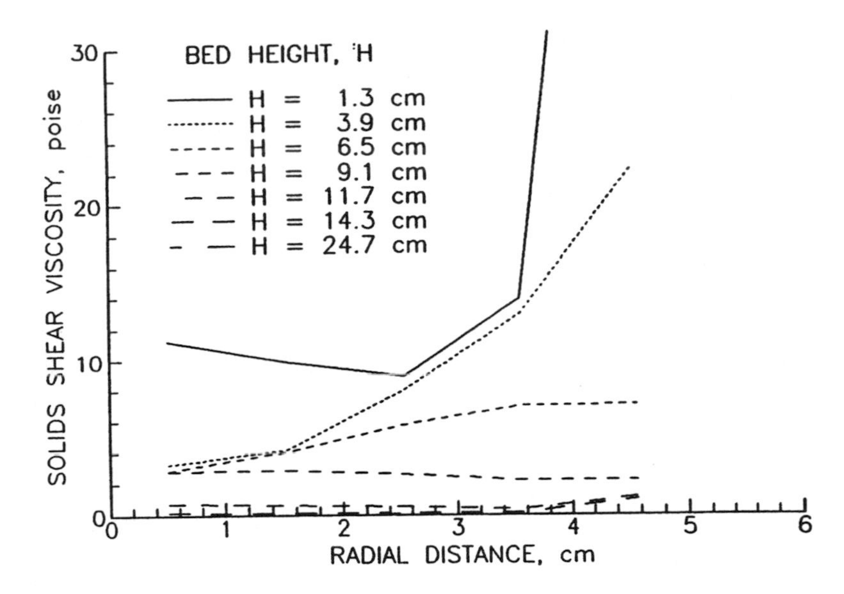

Fig. 10.14 Computed solids shear viscosity as a function of radial distance at various bed heights ($U/U_{mf} = 2.0$). (Reprinted with permission from Gidaspow *et al.*, 1991.)

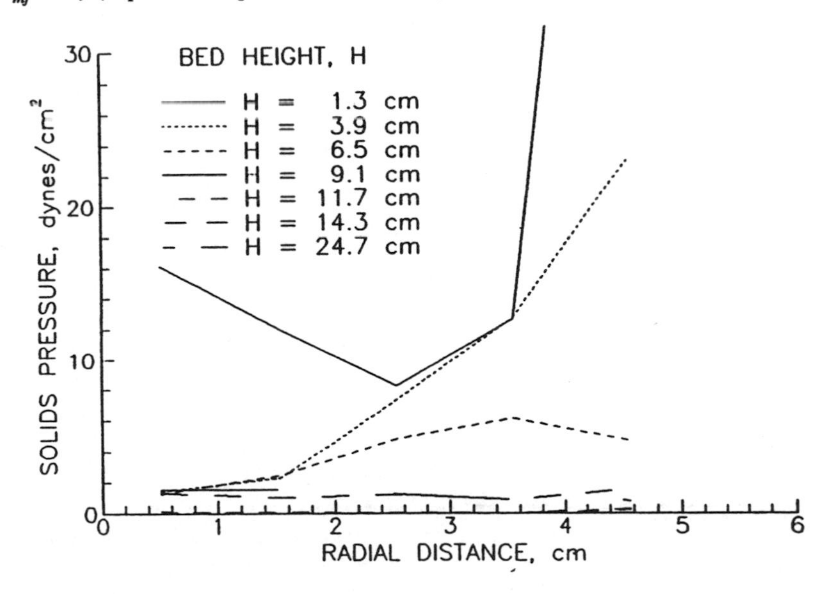

Fig. 10.15 Computed solids pressure as a function of radial distance at various bed heights ($U/U_{mf} = 2.0$). (Reprinted with permission from Gidaspow *et al.*, 1991.)

viscosity shown in Fig. 10.14 and the granular pressure shown in Fig. 10.15 are high in the lower section of the bed. The high value of viscosity, 1 to 20 poises, is expected in such systems because of the large particle diameter.

10.5 Circulating Fluidized Bed Loop Simulation

The complete set of equations shown in Table 10.1 were solved for the circulating fluidized bed loop shown in Fig. 10.16 (Gidaspow *et al.*, 1990; Gidaspow *et al.*, 1992; Bezbaruah, 1991). This loop is a simplified version of an actual PYROFLOW circulating fluidized bed system built for Goodrich Co. In Henry, Illinois (Johnk and Wietske, 1989). The main difference is the replacement of the hot gas cyclone by a screen type gas–solid separator that allows the gas to leave but not the solids. To prevent the separators from plugging up, it was extended beyond the dimensions shown in the PYROFLOW system. An ideal circulating fluidized bed (CFB) normally operates at velocities of 2 to 10 m/s, with loadings of 10 to 15% solids by volume and with a flux of 400 kg/m^2-s or larger. Since the pressure drop in a CFB is approximately equal to the weight of the solids, such operation requires the use of high-pressure blowers. Thus, some CFB boilers using inexpensive blowers operate at loadings of only a few percent solids in the riser. In the simulation we assumed that the CFB operates in an ideal mode. To obtain the desirable high loadings in the riser, we assumed that at time zero, the CFB was filled with 150 μm particle solids to a height of 4.5 m, as shown in Fig. 10.16. The solids were assumed to be at minimum fluidization. This height of solids gives a pressure drop of 0.455 atm. To allow for some losses due to friction, the inlet pressure was increased by another 5%. The exit pressure was assumed to be 1 atm. These requirements give an inlet pressure of 0.153 MPa. Simulations using dense phase kinetic theory and turbulence in the riser were done at three velocities ranging from 2 to 5 m/s (Gidaspow *et al.*, 1990). To permit operation at high fluxes, the solids in the standpipe were fluidized. To obtain dense flow in the standpipe, the fluidization velocity was kept as low as possible. It was at its minimum fluidization. In the simulation presented by Gidaspow *et al.* (1990) using dense phase kinetic theory and turbulence in the dilute phase, the results were not sensitive to the restitution coefficient. Thus, a restitution coefficient of 0.8 gave satisfactory results. However, with no imposed turbulence and the use of the complete kinetic theory model, the results become sensitive to the restitution coefficient. To obtain values of solids viscosities in agreement with Miller's measurements (1991), a restitution coefficient of 0.999 was used.

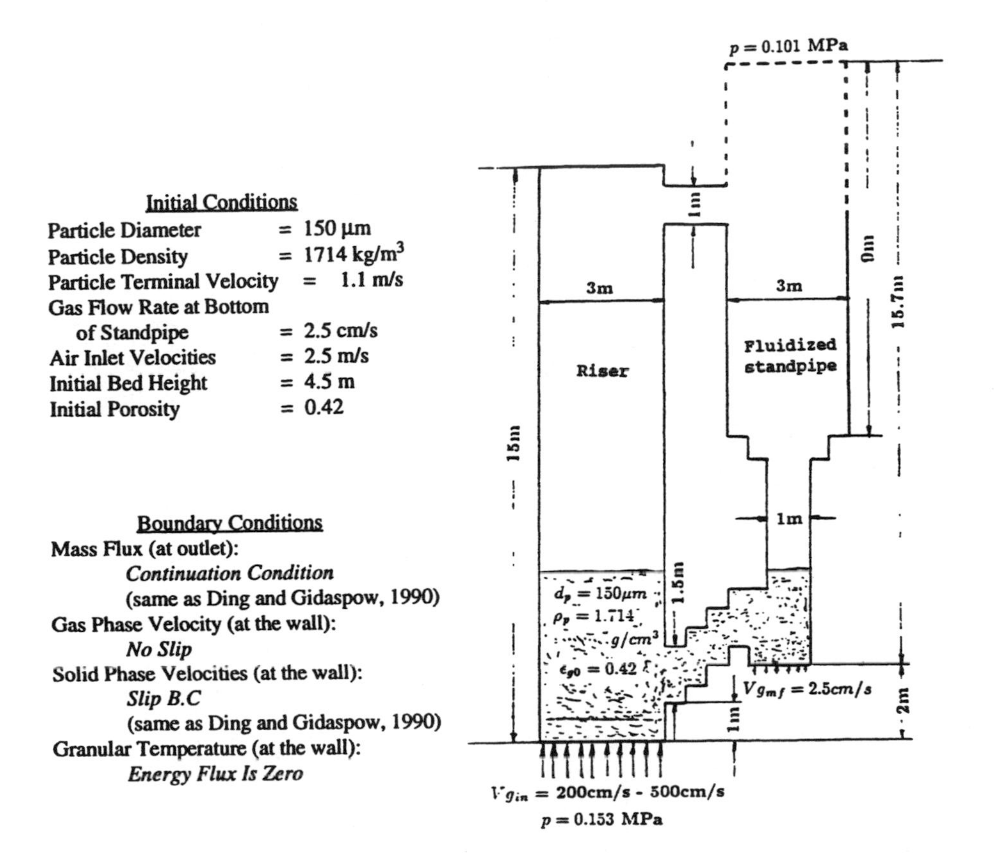

Fig. 10.16 Schematic of CFB loop with initial and boundary conditions.

Figure 10.17 shows the particle concentrations, velocities, the granular temperature and the solids viscosities 5 s after start-up. There is the expected dense flow in the lower section of the downcomer, strong downflow at both walls of the riser, reasonable values of the granular temperature and solids viscosity. A color representation of the particle concentrations shown in Fig. 10.17 clearly shows the film like dense annular region in the riser. It also shows the density distributions in the downcomer which cannot be seen in the dot plot representation in Fig. 10.17. A color video for an inlet velocity of 2 m/s shows the turbulent structure of the particles as a function of time. Cluster like dense regions are discharged from

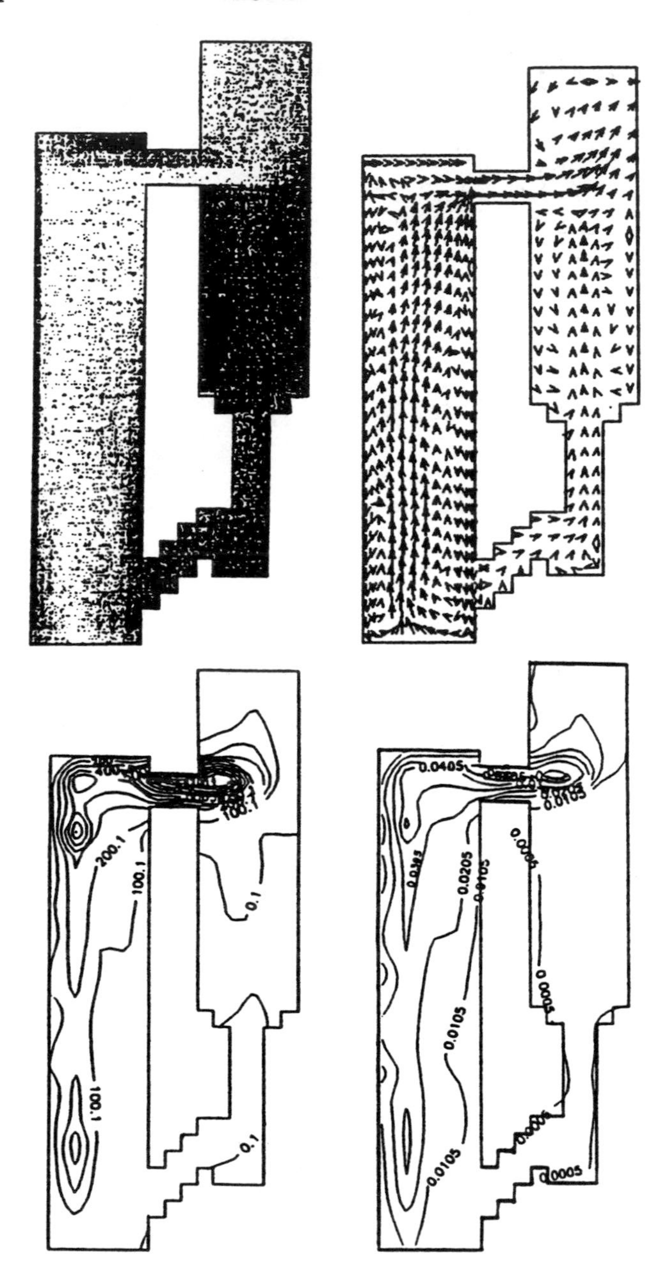

Fig 10.17 Particle concentration, velocities, granular temperatures (cm/s)^2 and solid viscosities (poise) for $V_{gin} = 5$ m / s.

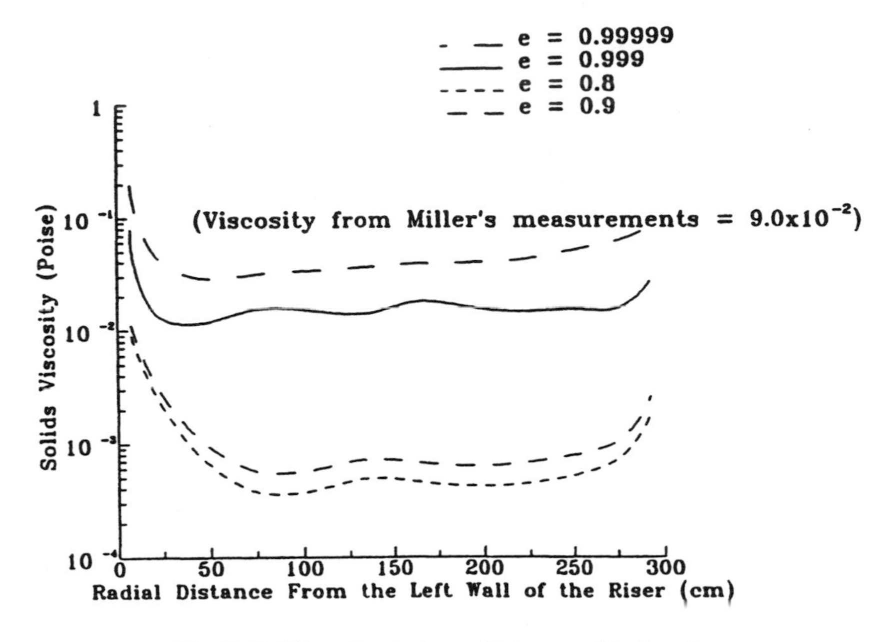

Fig. 10.18 Effect of restitution coefficient on solids viscosity.

the downcomer. They move up in the riser. Velocity plots show a vortex formation at the bottom of the riser. The granular temperature and hence the viscosity was adjusted to have reasonable values by varying the restitution coefficient. Both the granular temperature and the viscosity are very high, however, in the dilute regions. Probably the restitution coefficient must be made velocity-dependent to prevent the occurrence of such high oscillations. Clearly, the restitution coefficient must decrease, as the oscillations rise. Figure 10.18 shows how the solids viscosity varies with the restitution coefficient for an inlet velocity of 2 m/s. It is interesting to observe that the viscosity is constant in the core of the riser. Experimental measurements by Miller support such an observation. The experimental and the computed solids velocities, at the wall undergo cyclic oscillations.

Figure 10.19 shows this behavior at the right wall of the riser. Figure 10.20 displays the discharge from the standpipe to the riser. The discharge velocity oscillates about the 1 m/s velocity typical of discharge velocities from hoppers. This computed velocity is independent of the gas velocity into the riser (Gidaspow *et al.*, 1990). It represents the critical discharge velocity. Thus, the theory

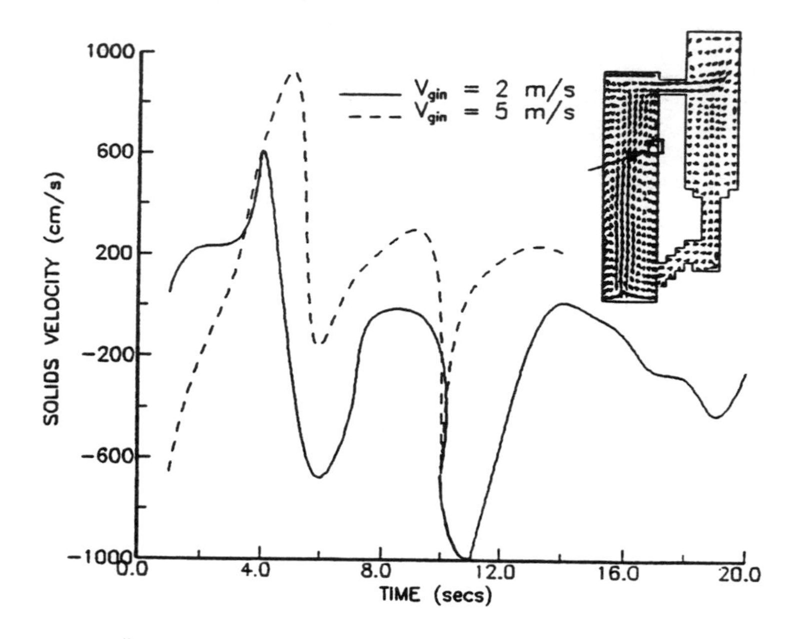

Fig. 10.19 Particle velocity at the right wall of the riser at 10 m.

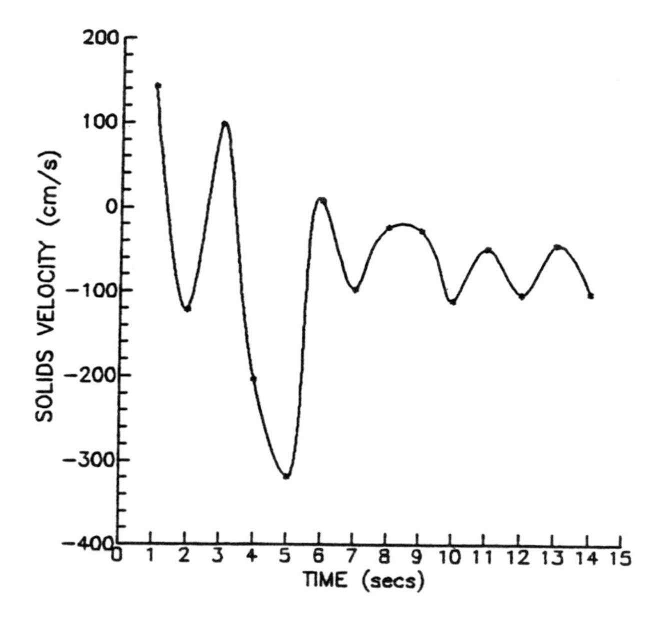

Fig. 10.20 Solids discharge velocity from standpipe to riser for $V_{gin} = 5$ m/s.

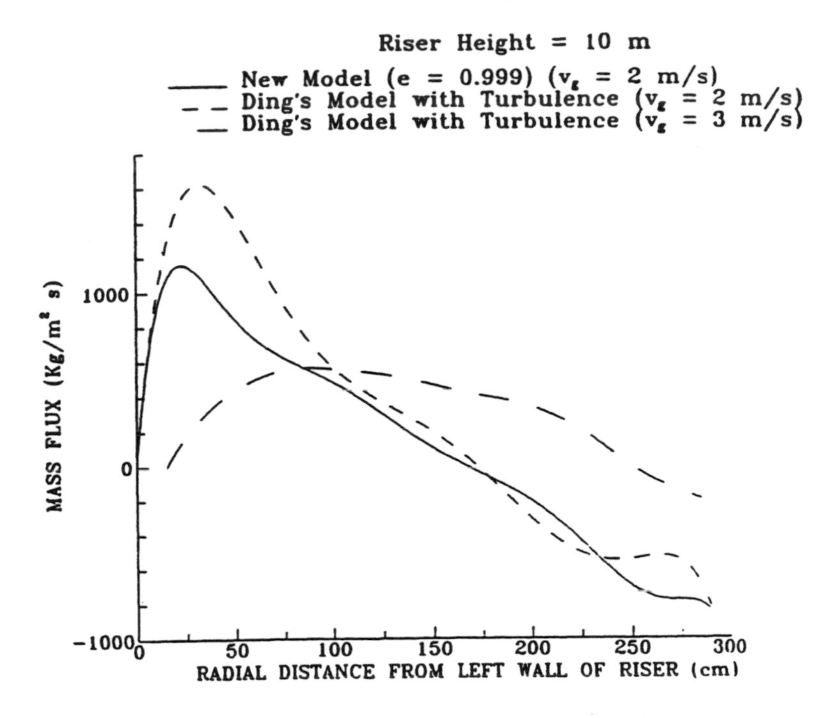

Fig. 10.21 Time averaged radial mass flux profile.

suggests that a restriction in the dense phase of the loop will limit the circulation velocity in the CFB. The next section gives a derivation for the maximum circulation rate which is not limited by this critical velocity.

Figure 10.21 shows a comparison of the kinetic theory model to Ding's model (Gidaspow *et al.*, 1992) with turbulence for the mass flux as a function of radial distance. The viscosity does not affect the system behavior greatly once its value is reasonable. The strong asymmetry disappears at a higher gas velocity. An average value of the solids flux at velocities of 2 to 3 m/s is about 350 kg/m^2-s, a value used in some commercial CFB units. Figure 10.22 displays the solids velocities for the same operating conditions and at the same location as in Fig. 10.21. The particle velocities are much higher than the inlet velocities because of area blockage by solids. Such behavior is found to be true experimentally (Miller, 1991).

Figure 10.23 shows a comparison of the area-averaged solids volume fractions for the two models and for two inlet gas velocities. In each case the average volume fraction is higher near the solids inlet, as expected. At the two meter inlet velocity there is no fully developed region. At a velocity of 5 m/s the

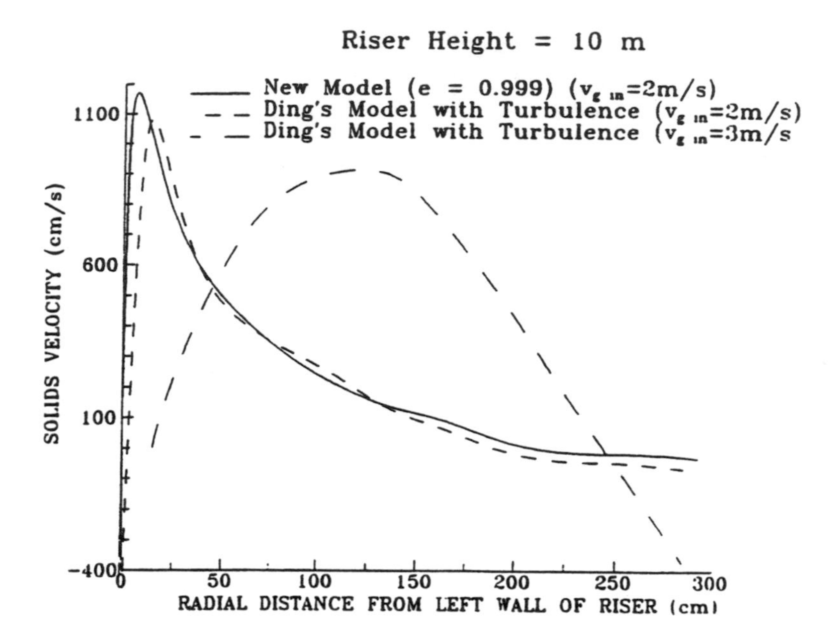

Fig. 10.22 Time-averaged radial solids velocity profile.

riser particle concentration is smaller and there exists an approximate developed region.

The simulation of the CFB has shown the power of the method to make predictions. The method should be useful for making design improvements.

10.6 Maximum Solids Circulation in a CFB

Squires *et al.* (1985) have called a circulating fluidized bed a fast bed and have pointed out the advantages of operating in this flow regime. Hence, it is of great interest to determine the maximum possible circulation rate. It was already pointed out that if there is an area restriction in the flow, the velocity will reach a critical flow condition. Hence, the total rate will be restricted by this maximum possible velocity. It is nevertheless useful to derive another maximum velocity in the absence of a critical flow restriction.

To understand the behavior of solids circulation it is useful to obtain a simplified analytical solution to the problem. To do this a number of drastic assumptions must be made in the multidimensional two phase flow equations. Based on order of magnitude estimates it is possible to:

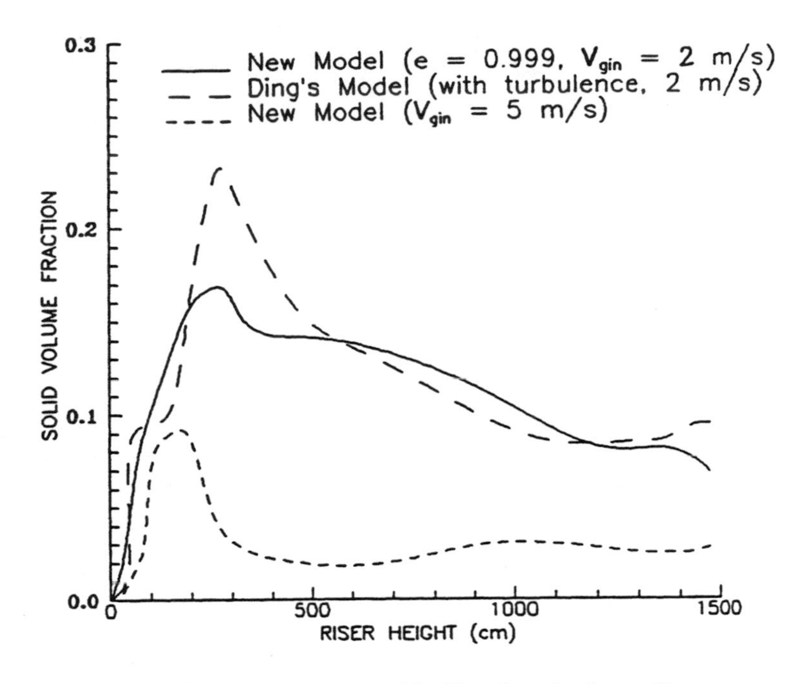

Fig. 10.23 Area-averaged axial solids volume fraction profile.

(1) Neglect gas momentum with respect to solids momentum because of the much larger density of the solid compared with that of the gas.
(2) Neglect the momentum head in the downcomer region because of small solids velocities.
(3) Neglect wall friction and solids pressure.

With these assumptions the mixture momentum equations become

Mixture Momentum in the Riser Region in Conservative Form

$$\frac{d}{dx}\left(\rho_s \varepsilon_{sr} u_{sr}^2\right) = -\frac{dP_r}{dx} - g\rho_s \varepsilon_{sr}, \tag{10.47}$$

Mixture Momentum in the Downcomer

$$\frac{dP_d}{dx} = -g\rho_s \varepsilon_{sd}. \tag{10.48}$$

Away from the entrance of the downcomer and the riser region, where the solids are accelerated to be carried pneumatically in the riser region, the momentum equations can be expressed in integrated form. With constant solids volume fraction in each region, we then have the balances

$$\rho_s \varepsilon_{sr} u_{sr}^2 + \Delta P_r + g \rho_s \varepsilon_{sr} x = 0, \tag{10.49}$$

$$\Delta P_d + g \rho_s \varepsilon_{sd} x = 0. \tag{10.50}$$

Equation (10.49) is somewhat similar to Yang and Keairns' (1974) equation (16) for the draft tube. In Eq. (10.49) the wall friction has been neglected. Also, the kinetic head does not have a factor of one-half in Eq. (10.49). This is because the mixture momentum equation has been integrated in the conservation form given by Eq. (10.47). Integrating it in the form $\rho_s \varepsilon_{sr} U_{sr} (dUsr / dx)$ leads to the one-half in the Yang and Keairn's formula. However, this involves an additional assumption of constancy of ε_{sr} throughout the riser. Integration of the mixture momentum equation in the conservative form only assumes that the solid entered the region with a zero velocity. Thus, in Eq. (10.49) only the value of ε_{sr} in the gravity term has been assumed to be a constant. This constant may further be interpreted as an average value over the distance x.

Equation (10.50) states that the pressure drop in the downcomer region is all due to the weight of the solids, as is well known to be approximately true near minimum fluidization. When the downcomer is not fluidized, part of the weight is supported by the solids pressure. Furthermore, the boundary conditions suggest that

$$\Delta P_r = \Delta P_d . \tag{10.51}$$

Elimination of the pressure drop in Eq. (10.49) gives

$$\rho_s \varepsilon_{sr} u_{sr}^2 = g \rho_s s (\varepsilon_{sd} - \varepsilon_{sr}). \tag{10.52}$$

From continuity equation for the solid we know that

$$\rho_s \varepsilon_s u_s = \text{constant} = W_s, \tag{10.53}$$

where W_s is the recirculation rate of solids per unit area. Then multiplication of Eq. (10.52) by $\rho_s \varepsilon_{sr}$ and solution for W_s gives the desired formula:

$$\text{Solids circulation} = W_s = \rho_s \sqrt{\varepsilon_{sr} g x (\varepsilon_{sd} - \varepsilon_{sr})}. \tag{10.54}$$

This formula for solids circulation flux, W_s, states that circulation is caused by the differences in densities or solids volume fractions, $\varepsilon_{sd} - \varepsilon_{sr}$, between the two regions. Qualitatively, this has been known to be the cause of circulation in fluidized beds for many years. Davidson and Harrison's group has taken advantage of this cause to suggest various designs involving blowing gases at different flow rates at various bed positions. The solids circulation formula (10.54) is also related to the orifice formula, Eq. (39) of Yang and Keairns (1974) and to the empirical correlations, Eqs. (2) and (3) of Yang and Keairns. All the formulae involve a driving head due to a column of a solids mixture.

In the solids circulation formula the void fractions in the two regions can be estimated similarly to that discussed by Yang and Keairns (1974). ε_{sd} can be obtained from the pressure drop, which can be related to the friction given by the Ergun type equations. ε_{sr} can be approximately obtained from a drag or friction between the phases type expression used by Yang and Keairns.

Approximate Solids Circulation Formula

Away from the entrance region, gradients of relative velocity vanish and we can simply balance gravity with drag between the phases. Then for constant slip,

$$\beta(u_g - u_s) = g, \tag{10.55}$$

where the friction coefficient β is related to the standard drag coefficient C_d by the formulas given in Table 10.1.

Then substituting $W_s = \varepsilon_{sr} \rho_s u_{sr}$ into Eq. (10.55), the volume fraction of solids in the riser region is given by

$$\varepsilon_{sr} = \frac{W_s}{\rho_s} \left(\frac{1}{u_{gr} - g/\beta} \right) \tag{10.56}$$

and the solids circulation rate is

$$W_s = \frac{\rho_s \varepsilon_{sd} g x (u_{gr} - g/\beta)}{g x + (u_{gr} - g/\beta)^2}. \tag{10.57}$$

Although in Eq. (10.57) β is implicitly a function of W_s through a dependence of β on C_d, as in Arastoopour and Gidaspow (1979), the formula shows the dependence of circulation on riser velocity u_{gr}, bed height x, and particle diameter.

For small slip velocity or large bed height,

$$W_s = \rho_s \varepsilon_{sd}\left(u_{gr} - g/\beta\right). \tag{10.58}$$

Thus, for constant β, the circulation rate increases linearly with riser velocity under these conditions. Indeed, Gidaspow *et al.* (1990) find such an approximate rise in the CFB loop simulation. For the other extreme there is a decrease of circulation with riser velocity. For constant β, there is a maximum circulation rate at the velocity

$$u_{gr,\text{optimum}} = \frac{g}{\beta} + \sqrt{gx} \tag{10.59}$$

given by

$$W_{s,\max} = \frac{\rho_s \varepsilon_{sd}}{2}\sqrt{gx}. \tag{10.60}$$

For a downcomer height of 4.5 meters with particles at minimum fluidization,

$$\sqrt{gx} = 6.64\,\text{m/s}. \tag{10.61}$$

Since this value is much higher than the expected 1 m/s critical velocity, the huge discharge rate of 3320 kg/m^2-s given by Eq. (10.62) will not materialize. However, with sufficient aeration the critical velocity will jump to the order of magnitude higher homogeneous two-phase flow velocity. Using such a strategy the circulation rates will be very large and limited by the driving head of the solids, as approximately derived here.

Literature Cited

Araki, S., and S. Tremaine (1986). "Dynamics of Dense Particle Discs," *ICARUS* **65**, 83–109.

Arastoopour, H., and D. Gidaspow (1979). "Vertical Pneumatic Conveying Using Four Hydrodynamic Models," *I&E Fundamentals* **18**, 123–130.

Bagnold, R. A. (1954). "Experiments on a Gravity-Free Dispersion of Large Solid Spheres in a Newtonian Fluid under Shear," *Proc. Roy. Soc.* **A255**, 49–63.

Bezbaruah, R. (1991). "Determination of Cohesive and Normal Stresses and Simulation of Fluidization using Kinetic Theory," M.S. Dissertation in Chemical Engineering. Chicago, Illinois: Illinois Institute of Technology.

Campbell, C. S., and C. E. Brennen (1983). "Computer Simulation of Shear Flows of Granular Material," in J. T. Jenkins and M. Satare, eds. *Mechanics of Granular Materials: New Models and Constitutive Relations*. New York: Elsevier Science.

Carlos, C. R., and J. F. Richardson (1968). "Solids Movement in Liquid Fluidized Beds I: Particle Velocity Distribution," *Chem. Eng. Sci.* **23**, 813–824.

Chapman, S., and T. G. Cowling (1961). *The Mathematical Theory of Non-Uniform Gases*, 2nd ed. Cambridge, U.K.: Cambridge University Press.

Ding, J., and D. Gidaspow (1990). "A Bubbling Fluidization Model Using Kinetic Theory of Granular Flow," *AIChE J.* **36**, 523–538.

Gamwo, I. (1992). Ph.D. Thesis. Chicago, Illinois: Illinois Institute of Technology.

Gidaspow, D., J. Ding, and U. K. Jayaswal (1990). "Multiphase Navier-Stokes Equation Solver," pp. 47–56 in *Numerical Methods for Multiphase Flows, FED–Vol. 91*. New York: ASME.

Gidaspow, D., U. K. Jayaswal, and J. Ding (1991). "Navier–Stokes Equation Model for Liquid–Solid Flows using Kinetic Theory" pp. 165–178 in *Liquid Solid Flows, FED-Vol. 118*. New York: ASME.

Gidaspow, D., R. Bezbaruah, and J. Ding, (1992). "Hydrodynamics of Circulating Fluidized Beds: Kinetic Theory Approach," pp. 75–82 in O. E. Potter and D. J. Nicklin, Eds. *Fluidization VII*. Engineering Foundation.

Ishida, M., and T. Shirai (1979). "Velocity Distributions in the Flow of Solid Particles in an Inclined Open Channel," *J. Chem. Eng. of Japan* **12** (1), 46–50.

Johnk, C., and D. Wietzke (1989). "Ahlstrom PYROFLOW Circulating Fluidized Bed Boiler: Tube Erosion Experience," pp. 15-1 in *Material Issues in Circulating Fluidized-Bed Combustors Workshop*. EPRI GS-6747.

Kuwabara, G., and K. Kono (1987). "Restitution Coefficient in a Collision Between Spheres," *Japanese J. Appl. Phys Part I* **26**, 1230–1233.

Lin, J. S., M. M. Chen, and B. T. Chao (1985) "A Novel Radioactive Particle Tracking Facility for Measurement of Solids Motion in Gas Fluidized Beds," *AIChE J.* **31**, 465–473.

Lun, C. K. K., S. B. Savage, D. J. Jeffrey, and N. Chepurniy (1984). "Kinetic Theories for Granular Flow: Inelastic Particles in Couette Flow of Singly Inelastic Particles in a General Flow Field," *J. Fluid Mech.* **140**, 223–256.

Miller, A. (1991). Ph.D. Thesis. Chicago, Illinois: Illinois Institute of Technology.

Moslemian, D., N. Devanathon and M. P. Dudukovic (1990). "Liquid Circulation in Bubble Columns via the CAOPT Facility: Effects of Gas Superficial Velocity, and Column Diameter," *ASME Winter Annual Meeting Preprint*.

Savage, S. B. (1983). "Granular Flows at High Shear Rates," pp. 339–358 in R. E. Meyer, Ed. *Theory of Dispersed Multiphase Flow*. New York: Academic Press.

Savage, S. B., and M. Sayed (1984). "Stresses Developed by Dry Cohesionless Granular Materials Sheared in an Annular Shear Cell," *J. Fluid Mech.* **1142**, 391–430.

Schlichting, H. (1960). *Boundary Layer Theory*. Translated by J. Kestin. New York: McGraw-Hill.

Shen, H., and N. L. Ackermann (1982). "Constitutive Relationships for Fluid–Solid Mixtures," *J. Eng. Mech. Div. Am. Soc. Div. Eng.* **108**, 748–763.

Sinclair, J. L., and R. Jackson (1989). "Gas–Particle Flow in a Vertical Pipe with Particle–Particle Interactions," *AIChE J.* **35**, 1473–1486,

Squires, A. M., M. Kwauk, and A. A. Avidan (1985). "Fluid Beds: At Last, Challenging Two Entrenched Practices," *Science* **230**, 1329–1337.

Walton, O. R., and R. L. Braun (1986). "Stress Calculations for Assemblies of Inelastic Spheres in Uniform Shear," *Acta Mechanics* **63**, 73–86.

Whitehead, A. B. (1971). "Some Problems in Large-Scale Fluidized Beds," pp. 781–814 in J. F. Davidson and D. Harrison, Eds. *Fluidization*. New York, Academic Press.

Whitehead, H. R. (1985). "Distributor Characteristics and Bed Properties," pp. 173–199 in J. J. Davidson, R. Clift, and D. Harrison, eds. *Fluidization*, 2nd Ed. New York: Academic Press.

Yang, W. C., and D. L. Keairns (1974). "Recirculating Fludized Bed Reactor Data Utilizing a Two Dimensional Cold Model," *AIChE Symposium Series* **70** (141), 27–40.

11

KINETIC THEORY OF GRANULAR MIXTURES

11.1 Empirical Input: Restitution Coefficients **337**
11.2 Boltzmann Equations for a Mixture **339**
11.3 Dense Transport Theorem **339**
11.4 Granular Temperatures and Applications **341**
11.5 Particle-to-Particle Drag **343**
11.6 Summary **345**
Exercises for Kinetic Theory (Chapters 9 – 11) **348**
1. Collision Dynamics **348**
2. Mean Values of *i*th Granular Temperature **348**
3. Collision Frequency for a Binary Mixture **349**
4. Mean Free Path: Diffusion Coefficient and Viscosity **349**
5. Multiphase and Multicomponent Collisionless Balances for Mass, Momentum and Energy **349**
6. Homogeneous Multiphase Mixture Balances for Frictionless Flow **349**
7. Momentum Equations for Phase *i* and the Fick's Law for Dense Mixtures **350**
8. Viscosity of Phase *i* **352**
9. Collisional Stress, Momentum Source, and Energy Dissipation **353**
10. Phase *i* Granular Temperature Equation **353**
Literature Cited **353**

11.1 Empirical Input: Restitution Coefficients

The first two kinetic theory chapters have shown how to derive mass, momentum and granular temperature balances and the constitutive relations for flow of a mixture of uniform size and density particles. Real mixtures consist of particles of various sizes and densities. These particles may segregate by size and density. Hence, mixtures must be studied.

The usual input into the hydrodynamic equations, as illustrated in Chapter 8, is the viscosity of the mixture. The kinetic theory of granular flow allows the computation of the rheology of the mixture. However, this theory also requires empirical input. It consists of the normal and the tangential restitution coefficients. The latter arise when one considers rotation of rough particles.

Figure 10.18 shows the dramatic effect of the normal restitution coefficient on the computed viscosity. Hence, a brief discussion of how the restitution coefficient varies with the elastic and dynamic properties is needed. A more complete treatment is found in the book by K. L. Johnson (1985). During impact of particles the work of deformation can be expressed in terms of the elastic pressure P_e and the plastic pressure P_p, and the deformation volume by means of the usual relation, as

$$\text{Work of deformation} = \int_0^V \left(P_e + P_p\right) dV. \tag{11.1}$$

This work equals the relative velocity of impact squared times half the mass. However, the relative kinetic energy equals only the integral of the elastic pressure. Hence, the restitution coefficient e can be expressed as

$$e^2 = \frac{v'^2}{v^2} = \frac{\int_0^V P_e dV}{\int_0^V \left(P_e + P_p\right) dV}, \tag{11.2}$$

where v is the relative velocity before impact and v' is the rebound velocity. Equation (11.2) suggests that for low velocities, where the plastic deformation is small, the restitution coefficient will be nearly one. It also clearly shows that the restitution coefficient is a function of the material properties, as well as the dynamic properties associated with plastic flow. Indeed, data summarized by Johnson (1985) show that for hard materials the restitution coefficient is nearly one for impact velocities of 0.1 m/s and less (see Fig. 11.1). The granular temperatures

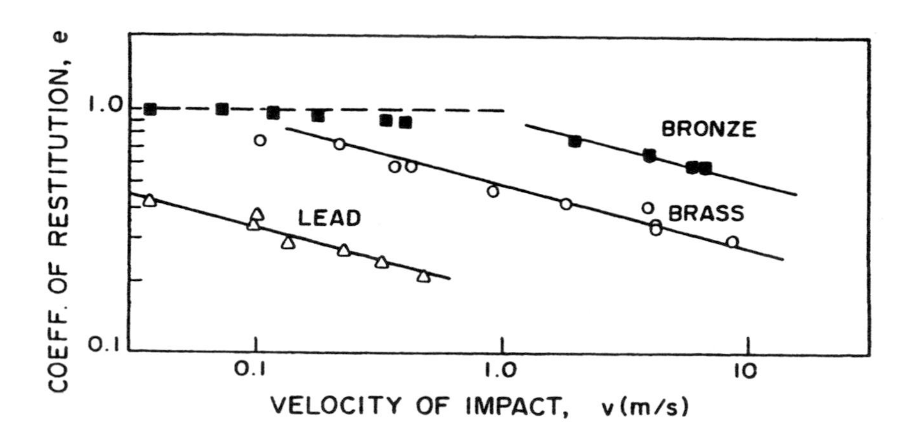

Fig. 11.1 Restitution coefficients for various materials (based on data cited in Johnson, 1985).

in Fig. 10.17 give impact velocities of this order of magnitude. Hence, the choice of $e = 0.999$ made in that example was reasonable, as also demonstrated by a comparison to measured viscosities.

For higher impact velocities Johnson's (1985) theory shows that e is proportional to $v^{-1/4}$. It is supported by some experimental measurements. Hence, in the simulation of the circulating fluidized beds, the restitution coefficient may be made velocity-dependent to eliminate the small regions where the granular temperature may be too high.

11.2 Boltzmann Equations for a Mixture

Boltzmann equations can be derived for each component or each phase, as done in Section 9.8, Eq. (9.87). Considering binary collisions only, the equations for the frequency distributions f_i are

$$\frac{\partial f_i}{\partial t} + \mathbf{c}_i \frac{\partial f_i}{\partial \mathbf{r}} + \mathbf{F}_i \frac{\partial f_i}{\partial \mathbf{c}_i} = \sum_{k=1}^{r} \iint \left(f_i' f_k' - f_i f_k \right) \mathbf{c}_{ik} \cdot \mathbf{k} d_{ik}^2 \, d\mathbf{c}_k \, d\mathbf{k}. \qquad (11.3)$$

$$i = 1, 2, \ldots r$$

Equation (11.3) generalizes the binary frequency distributions for hard sphere models given by Chapman and Cowling's (1961) Eqs. (4.3.1) and (8.1.1), where the prime denotes the functions before collisions, with all symbols maintaining the same meaning. Dahler and Sather (1963) have extended this treatment to include rotation and have provided an alternate explanation for the collision integral in Eq. (11.3). Jenkins and Mancini (1987) have based their derivations of balance laws and constitutive equations for a binary granular mixture on Dahler and Sather's study of kinetic theory of molecules. This paper by Dahler and his later paper (Condiff *et al.*, 1965) are important for future development of granular flow theory. Here only an outline can be given to (one hopes) stimulate future research and to show the relation to the derivations given previously in greater detail.

11.3 Dense Transport Theorem

Dense transport theorems for a binary for a property $\psi_i = \psi_i(\mathbf{c}_i)$ have been given by Farrell *et al.* (1986) and by Jenkins and Mancini (1987) for a two-dimensional geometry. The expressions quoted below are based on their papers.

A generalization of the transport equation (9.194) for component i or phase i is

$$\frac{\partial n_i < \psi_i >}{\partial t} + \nabla \cdot \left(n_i < \mathbf{c}_i \psi_i > + \sum_{k=i,j} P_{c_{ik}} \right) = n_i < \mathbf{F}_{s_i} \frac{\partial \psi_i}{\partial \mathbf{c}_i} > + \sum_{k=i,j} N_{c_{ik}}, \quad (11.4)$$

where $i = i$ or j.

The collisional pressure or energy contribution corresponding to Eq. (9.193) is given by

$$P_{c_{ik}} = -\frac{d_{p_{ik}}^3}{2} \int_{\mathbf{k}\cdot\mathbf{c}_{12}>0} \left(\psi'_{1_i} - \psi_{1_i} \right) (\mathbf{k}\cdot\mathbf{c}_{21}) \mathbf{k} f_{ik}^{(2)} \left(\mathbf{r} - \tfrac{1}{2} d_{p_{ik}} \mathbf{k}, \mathbf{c}_1 ; \mathbf{r} + \tfrac{1}{2} d_{p_{ik}} \mathbf{k}, \mathbf{c}_2 \right) d\mathbf{k}\, d\mathbf{c}_1 d\mathbf{c}_2. \quad (11.5)$$

The sourcelike contribution corresponding to Eq. (9.192) is

$$N_{c_{ii}} = \frac{d_{p_{ik}}^2}{2} \int_{\mathbf{k}\cdot\mathbf{c}_{12}>0} \left(\psi'_{1i} + \psi'_{2i} - \psi_{1i} - \psi_{2i} \right) (\mathbf{k}\cdot\mathbf{c}_{12}) \mathbf{k} f^{(2)} d\mathbf{k}\, d\mathbf{c}_1 d\mathbf{c}_2, \quad (11.6)$$

where the binary collisional frequency distribution $f^{(2)}$ is evaluated as in Eq. (9.192) with the particle diameter replaced by the mean particle diameter $d_{p_{ik}}$. The contribution due to the collisions between unlike particles is

$$N_{c_{ij}} = \frac{d_{p_{ik}}^2}{2} \int_{\mathbf{k}\cdot\mathbf{c}_{12}>0} \left(\psi'_{1i} - \psi_{1i} \right) (\mathbf{k}\cdot\mathbf{c}_{12}) f^{(2)} \left(\mathbf{r}, \mathbf{c}_1 ; \mathbf{r} + d_{p_{ik}} \mathbf{k}, \mathbf{c}_2 \right) d\mathbf{k}\, d\mathbf{c}_1 d\mathbf{c}_2$$

$$+ \frac{d_{p_{ik}}^2}{2} \int_{\mathbf{k}\cdot\mathbf{c}_{12}>0} \left(\psi'_{2i} - \psi_{2i} \right) (\mathbf{k}\cdot\mathbf{c}_{12}) f^{(2)} \left(\mathbf{r} - d_{p_{ik}} \mathbf{k}, \mathbf{c}_1 ; \mathbf{r}, \mathbf{c}_2 \right) d\mathbf{k}\, d\mathbf{c}_1 d\mathbf{c}_2, \quad (11.7)$$

$$i \neq j,$$

where the mean particle diameter was defined by

$$d_{p_{ik}} = \frac{d_{p_i} + d_{p_k}}{2}. \quad (11.8)$$

The reader is asked to derive the conservation of species, momentum and energy equations for components *i*, and phase *i*, as well as the mixture equations, as exercises.

11.4 Granular Temperatures and Applications

In kinetic theory of gases, equipartition of energy (Chapman and Cowling, 1961) permits the use of a single temperature for a mixture. Farrell *et al.* (1986) point out that in granular flow this may not be true because of different restitution coefficients between different particle types. However, a definition of the mixture temperature for a binary, as given below,

$$nT = n_A T_A + n_B T_B, \tag{11.9}$$

where T is the mixture temperature and n_A is the number of A particles per unit volume, shows that one can equate T to T_A when n_B / n_A is small. Thus, as is physically evident, for diffusion of a small quantity of B into A, we can use the temperature of A as the system temperature. Jenkins and Mancini (1987) have apparently used such an approximation in their derivation of a constitutive equation for a granular binary mixture. They let the granular temperature for phase i, Θ_i, be defined by

$$\Theta_i = \tfrac{1}{3} m_i < \mathbf{C}_i^2 >. \tag{11.10}$$

Then the Maxwellian distribution becomes essentially that used in gas theory with the Boltzmann constant equal to unity, since the temperature in Eq. (11.10) has the units of kinetic energy:

$$f_i^{(0)} = n_i \left(\frac{m_i}{2\pi\Theta_i} \right)^{3/2} \exp\left[-\frac{m_i}{2\Theta_i} (\mathbf{c}_i - \mathbf{v}_i)^2 \right]. \tag{11.11}$$

Equation (11.11) is a generalization of Eq. (9.15).

The pair distribution function $f_{ik}^{(2)}$ is again obtained by assuming chaos. It can be given by the product

$$f_{ik}^{(2)} = f_i^{(1)} f_k^{(1)} g_{ik}, \tag{11.12}$$

where the zeroth approximation to $f_i^{(1)}$ is given by the Maxwellian distribution $f_i^{(0)}$ and where the radial distribution function g_{ik} is given by Eq. (10.30) with the particle diameter taken to be the average of the diameter for i and k particles. For a Maxwellian distribution function the integrals for P_{Cik} and N_{Cij} can now be evaluated when we substitute momentum and energy for ψ. To carry out the integrations the individual particle velocities must be related to the relative and to the center of mass velocities, as done in Chapter 9. For different particles considered here this is left as an exercise. Jenkins and Mancini (1987) give the final relation. We expect that the resulting phase viscosities and conductivities will be valid only for dense flow. For the complete flow regime, non-Maxwellian effects must be included. Jenkins and Mancini (1987) generalized the distribution by means of a Taylor series approach, while Chapman and Cowling (1961) present the traditional perturbation approach discussed in Chapter 9. Chapman and Cowling's result applied to particles, with the same approximations made in their Chapter 8, is

$$f_i = f_i^{(0)}\left[1 + a_1\mathbf{C}_i \cdot \frac{\partial \ln \Theta_i}{\partial \mathbf{r}} + a_2\mathbf{C}_i \cdot \mathbf{d}_{ik} + a_3\mathbf{C}_i \cdot \mathbf{C}_i : \nabla \mathbf{v}_i\right], \tag{11.13}$$

where a_1, a_2 and a_3 are constants similar to those in Eq. (9.235). The new group that appears is the diffusion gradient $\mathbf{d}_{ik}$. Its standard form (Hirschfelder *et al.*, 1954) is

$$\mathbf{d}_{ik} = \nabla\left(\frac{n_i}{n}\right) + \left(\frac{n_i}{n} - \frac{n_i m_i}{\rho}\right)\nabla \ln P - \frac{\rho_i \rho_k}{P\rho}\left(\mathbf{F}_i - \mathbf{F}_k\right). \tag{11.14}$$

This new group leads to diffusion of phase i in phase j. Particularly interesting is the analogue of thermal diffusion, that is, the outcome of such an analysis. One obtains an analogue of Fick's law:

$$\mathbf{v}_i - \mathbf{v}_j = -\frac{n^2}{n_i n_j}\left[D_{ij}\mathbf{d}_{ij} + \frac{k}{\Theta_i}\nabla\Theta_i\right], \tag{11.15}$$

where D_{ij} is the binary diffusion coefficient given by

$$D_{ij} = \frac{3}{8nd_{p_{ij}}^2}\left[\frac{\Theta_i\left(m_i + m_j\right)}{2\pi m_i m_j}\right]^{1/2}. \tag{11.16}$$

Equation (11.16) is very close in numerical value to the more approximate expression given by Eq. (9.79) for equal masses. Approximate equations such as Eq. (10.13) suggest that Eq. (11.15) can be interpreted as

$$\text{Flux of phase } i = \text{Diffusivity} \times \text{Concentration gradient} + \text{Constant} \times \text{Gradient of shear}\,. \tag{11.17}$$

Hence this theory suggests that segregation takes place because of the presence of shear. Further quantitative analysis is needed.

11.5 Particle-to-Particle Drag

In pneumatic conveying of a binary mixture, particle-to-particle drag has long been recognized as a momentum source that needs to be included in the separate phase momentum balances to properly describe the observed segregation (Nakomura and Capes, 1976; Arastoopour *et al.*, 1980). This particle drag is due to the head-on collision of particles. Muschelknautz (1959) appears to have been the first to give it a primitive kinetic theory derivation. The more exact treatment based on the work of Jenkins and Mancini (1987) starts with the collision integral given by Eq. (11.7) with ψ_i replaced by $m_i\mathbf{c}_i$. For a uniform granular temperature and particle concentration, this term corresponds to Jenkins and Mancini's ϕ_i. Disregarding the location where $f^{(2)}$ is evaluated, ϕ_i can be written as

$$\phi_i = \sum_{k=1}^{2} \frac{d_{p_{ik}}^2}{2} \int \left(m_i\mathbf{c}_i' - m_i\mathbf{c}_i\right)\left(\mathbf{k}\cdot\mathbf{c}_{12}\right) f^{(2)} d\mathbf{k}\, d\mathbf{c}_1 d\mathbf{c}_2\,. \tag{11.18}$$

The mechanics of binary collisions shows that

$$\mathbf{c}_i' - \mathbf{c}_i = \left(\frac{m_k}{m_i + m_k}\right)\left(1 + e_{ik}\right)\cdot\left(\mathbf{k}\cdot\mathbf{c}_{21}\right)\mathbf{k}. \tag{11.19}$$

Inspection of the integral given by Eq. (11.18) shows that it can be interpreted as

$$\phi_i = \text{Frequency of collision} \times \text{Average change of momentum per collision}\,. \tag{11.20}$$

The frequency of collisions is the integral in (11.18) with momentum equal to unity. The value is given by Eq. (9.64) with the Boltzmann constant equal to one. The average change of momentum per collision is

$$< \mathbf{c}_i' - \mathbf{c}_i > \cong \left(\frac{m_k}{m_i + m_k} \right) (1 + e_{ik})(\mathbf{v}_1 - \mathbf{v}_2). \tag{11.21}$$

The product of (11.21) and N_{12} given by Eq. (9.64) gives the value of ϕ_i obtained by Jenkins and Mancini, except for a constant made in approximating the integral in (11.18). The exact value of ϕ_i is then

$$\phi_i = \tfrac{4}{3} g_{ik} (1 + e_{ik}) d_{p_{ik}}^2 n_i n_k \left(\frac{2\pi m_i m_k \Theta_i}{m_i + m_k} \right)^{1/2} (\mathbf{v}_k - \mathbf{v}_i). \tag{11.22}$$

The particle–particle drag in Eq. (11.22) is of the form of friction coefficient β_{ik} times the relative velocity, as shown below:

$$\phi_i = \beta_{ik} (\mathbf{v}_k - \mathbf{v}_i). \tag{11.23}$$

The friction coefficient in the theoretical expression is in terms of the granular temperature Θ_i. To make it determinate one needs a kinetic energy equation for Θ_i. Before the invention of kinetic theory of granular flow, this friction coefficient had to be evaluated in a more empirical form. The square root of the granular temperature, a velocity had to be evaluated in a different way. This was done by evaluating the frequency of collision by dividing the relative velocity of the particles by the distance between collisions.

Arastoopour (1986) presents correlations based in part on experiments done at the Institute of Gas Technology. For flow of coarse particles and sand in the range of 1% by volume and less, he recommends

$$\beta_{ik} = c_f \frac{\rho_{s_i} \varepsilon_{s_i}}{d_{p_k}} \left| v_{s_k} - v_{s_i} \right|, \tag{11.24}$$

with

$$c_f = 0.7 d_{p_k}^{-0.146}, \tag{11.25}$$

where the coarse particle diameter d_{p_k} is expressed in meters. This correlation is valid in the range of 7 to 16 m/s. For co-current flow of solids in the velocity range of 0.3 to 4.6 m/s, the coefficient c_f in Eq. (11.24) was found to be

$$c_f = d^{34.239} \frac{\rho_{s_k}^{1.34} d_{p_k}^{3.37} \varepsilon_{s_i}^{1.30}}{Re_{ik}}, \tag{11.26}$$

where the Reynolds number was defined as

$$Re_{ik} = \frac{\varepsilon_{s_i} \rho_{s_i} d_{p_k} \left| v_{s_k} - v_{s_i} \right|}{\mu_g}, \tag{11.27}$$

and where ρ_{s_i} is in kilograms per cubic meter and d is in meters.

For fluidization of a binary mixture of solids, Gidaspow *et al.* (1986) used Nakamura and Capes' (1976) semi-theoretical expression of the type of Eq. (11.22), with granular temperature replaced by the relative velocity divided by the mean free path corrected for a packing effect. There is clearly a need to determine better theoretically based correlations, especially for dense flow and fluidization of mixtures.

11.6 Summary

Chapter 9 has shown how the kinetic theory of gases can be modified to describe granular flow of spherical inelastic particles of uniform size and density, and how it can be extended to flow of such particles in a fluid by including drag between the fluid and the particles. The major novelty is the introduction of the concept of granular temperature, which is proportional to the square of a random velocity with which the particles oscillate about a mean hydrodynamic velocity. The result of the analysis is two coupled Navier–Stokes type equations. The kinematic viscosity of the particulate phase can be computed as a product of the mean free path, which is the particle diameter divided by the volume fraction of the particulate phase times the random velocity of the particles. This random velocity is computed by means of a fluctuating kinetic energy equation which in the kinetic theory of gases corresponds to the thermal energy equation. The generation of the random motion is due to shear. The dissipation is due to the inelastic collisions of the particles and due to the drag with the continuous fluid. Molecules always undergo random motion, while particles will cease to oscillate when there is no more shear. The relation between production of random motion and dissipation is explicitly treated in Chapter 10. Then the theory is applied to several technologically important flows.

Table 11.1 GAS MIXTURES: DEFINITIONS AND STATE EQUATIONS

Number density of ith constituent $= n_i$:

$$n_i = \int f_i(\mathbf{r}_i, \mathbf{c}_i, t) d\mathbf{c}_i; \quad n = \sum_i n_i;$$

$$\rho_i = n_i m_i: \qquad \rho = \sum_i \rho_i.$$

For any property ϕ_i,

$$n_i < \phi_i > \equiv \int f_i \phi_i dc_i \quad ; \quad n < \phi > = \sum_i \int f_i \phi_i \, dc_i .$$

Since all components flow at the same average velocity (homogeneous flow: $\mathbf{v}_i = \mathbf{v}_j$), $< \mathbf{c}_i > \equiv \mathbf{v}_i = \mathbf{v}$. Thus, the peculiar velocity $= \mathbf{C}_i = \mathbf{c}_i - \mathbf{v}$:

$$\rho < \mathbf{c}_i > = \int f_i m_i \mathbf{c}_i \, d\mathbf{c}_i \quad ; \quad \rho \mathbf{v} = \sum_i \int f_i m_i \mathbf{c}_i \, d\mathbf{c}_i .$$

The temperature T is defined as $T_i = T_j = T$:

$$kT_i \equiv \tfrac{1}{3} m_i < \mathbf{C}_i^2 > .$$

The kinetic contribution to the ith pressure tensor is

$$\mathbf{P}_i = \rho_i < \mathbf{C}_i \mathbf{C}_i >; \quad \mathbf{P} = \sum_i \rho_i < \mathbf{C}_i \mathbf{C}_i > = \sum_i \mathbf{P}_i .$$

Define the mean partial pressure as $p_i \equiv \frac{1}{3} \mathbf{P}_i : \mathbf{I}$. Then from the definition of temperature and $P_i = \frac{1}{3} n_i m_i < \mathbf{C}_i^2 >$ we obtain the component ideal gas law $P_i = kn_i T_i$. Using the ideal gas law for the mixture $P = nkT$, the partial pressure is related to total pressure by means of

$$P_i = \left(\frac{n_i}{n} \right) P.$$

Chapter 11 is an introduction to the theory of multiphase flow of particles of various sizes and densities. It deals with mixing, and the flow and segregation of such mixtures. Although segregation by shaking has been used since Biblical times and is the basis of many industrial and natural processes, a complete quantitative theory has not yet been developed. The ash agglomerator for removing large beads of ash from the bottom of a coal gasifier is an example of a potential industrial process requiring further understanding. The upward movement of manganese

Table 11.2 BINARY GRANULAR MIXTURES' CONSTITUTIVE RELATIONS[1]

Let collisional pressure for phase $i \equiv P_{c_{ik}}$:

$$P_{cik} = \tfrac{\pi}{3} g_{ik} d_{p_{ik}}^3 n_i n_k (1 + e_{ik}) \Theta_i,$$

$$\text{Hydrostatic pressure (mixture pressure)} = \sum_{i=1}^{2} \sum_{k=1}^{2} P_{cik} + \sum_{i=1}^{2} n_i \Theta_i,$$

where temperature $= \Theta_i = \Theta_k = \tfrac{1}{3} m_i < C_i^2 >$.

Collisional Stress for Phase i

$$\mathbf{P}_{c_{ik}} = P_{cik}\left[\mathbf{I} - \tfrac{4}{5} d_{p_{ik}} \left(\frac{2 m_i m_k}{\pi m_{ik} \Theta_i}\right)^{1/2} \left(\nabla^s \mathbf{v}_i + \tfrac{1}{2} \nabla \cdot \mathbf{v}_i \mathbf{I}\right)\right]$$

Phase Momentum Source (Drag)

$$\phi_i = P_{c_{ik}}\left[\frac{m_k - m_i}{m_{ik}} \nabla \ln \Theta_i + \nabla \ln \frac{n_i}{n_k} + \frac{4}{d_{p_{ik}}}\left(\frac{2 m_i m_k}{\pi m_{ik} \Theta_i}\right)^{1/2} (\mathbf{v}_k - \mathbf{v}_i)\right] \quad ; \quad \text{for } i \neq k$$

Phase i Heat Flux

$$q_{ik} = \frac{3 P_{c_{ik}}}{\pi m_{ik}} \left\{ -\tfrac{2}{3} d_{p_{ik}} \left(\frac{2\pi m_i m_k \Theta_i}{m_{ik}}\right)^{1/2} \nabla \ln \Theta_i + \tfrac{\pi}{3}(m_i \mathbf{v}_i + m_k \mathbf{v}_k) - \frac{m_k}{m_{ik}}(1 - e_{ik}) \cdot \left[\tfrac{1}{4} d_{p_{ik}} (m_k - m_i) \left(\frac{2\pi m_{ik} \Theta_i}{m_i m_k}\right)^{1/2} \nabla \ln \Theta_i + \tfrac{1}{6} d_{p_{ik}} \left(\frac{2\pi m_{ik}^3 \Theta_i}{m_i m_k}\right)^{1/2} \nabla \ln \frac{n_i}{n_k} + \frac{\pi}{2} m_{ik} (\mathbf{v}_k - \mathbf{v}_i)\right]\right\}$$

Energy Dissipation

$$\gamma = \sum_{k=1,2} \tfrac{6}{\pi} P_{c_{ik}} \left\{ \tfrac{\pi}{6}\left(\frac{m_i - m_k}{m_{ik}}\right) \nabla \cdot \mathbf{v}_i + \frac{m_k (1 - e_{ik})}{m_{ik}} \left[\frac{1}{d_{p_{ik}}} \left(\frac{2\pi m_{ik} \Theta_i}{m_i m_k}\right)^{1/2} - \tfrac{\pi}{2} \nabla \cdot \mathbf{v}_i\right]\right\}$$

Notation

m_{ik} is defined as $m_{ik} = (m_i + m_k)$.

e_{ik} is the restitution coefficient for collision between particle i and particle k.

g_{ik} is the contact radial distribution function for a mixture of spheres equal to χ_{ij} in Chapman and Cowling (1961), page 292.

Θ_i is defined by Eq. (11.10) with other symbols the same as in Chapter 9.

[1]Based on Jenkins and Mancini, 1987, for equal phase temperature, and Chapman and Cowling, 1961.

nodules at the bottom of the sea is an example of a naturally occurring process in need of more quantitative understanding. Separation of minerals, such as pyrites from finely ground coal, toxic heavy metal oxides from dusts that cannot be buried safely due to water pollution, or someday, enrichment of ores on the moon or Mars, can be accomplished by means of an application of a direct electric field in view of the differences of work functions of the constituent minerals. Invention and design of efficient processes for such separations will be helped by a better understanding of the granular motion of such mixtures. Chapman and Cowling (1961) have a chapter on the kinetic theory in ionized gases. However, there exists no such analysis for fine particles which naturally carry a surface charge allowing them to be separated. For example, pyrites are negatively charged, while coal and other carbon bearing materials in shale are positively charged relative to copper (Gupta *et al.*, 1993).

As an aid to further development of the kinetic theory of mixtures of particles, Tables 11.1 and 11.2 are presented. The first table summarizes the elementary theory of gas mixtures. In terms of multiphase flow, the major assumptions are that of homogeneous flow and equal component temperatures. Table 11.2 gives some of the constitutive relations for binary granular mixtures, based primarily on those given by Jenkins and Mancini (1987), with their effects of unequal temperature deleted. Except for the energy dissipation that is new, their expressions are essentially those found in Chapman and Cowling (1961) when one corrects for the presence of the restitution coefficient. In many ways Chapman and Cowling's (1961) constitutive relations are more complete.

Exercises for Kinetic Theory (Chapters 9–11)

1. *Collision Dynamics*

(a) Derive a generalization of Eq. (9.48) for collision of particles of masses m_1 and m_2 (see Eq. (11.19)).

(b) Show that

$$m_1(\mathbf{c}_1 - \mathbf{c}_1') = -m_2(\mathbf{c}_2 - \mathbf{c}_2').$$

Explain the meaning of this equation.

2. *Mean Values of ith Granular Temperature*

Use the definition of the granular temperature of phase i, given by Eq. (11.10) and the Maxwellian distribution given by Eq. (11.11) to find

(a) $< c_i >$ in terms of Θ_i; and

(b) $< c_i^2 >^{1/2}$ in terms of Θ_i.

3. *Collision Frequency for a Binary Mixture*
Use the Maxwellian distribution given by Eq. (11.11) to find the collision frequency for a binary granular mixture. Compare your result to Eq. (9.64).

4. *Mean Free Path: Diffusion Coefficient and Viscosity*
(a) Generalize Eq. (9.70) to the collision of particles of masses m_1 and m_2 using the collision frequency found in Exercise 3.
(b) Generalize Eq. (9.79) to the case of diffusion of particles of masses m_1 and m_2.
(c) Generalize Eq. (9.86).

5. *Multiphase and Multicomponent Collisionless Balances for Mass, Momentum and Energy*
Use the generalization of phase i or component i transport Eq. (9.103) and the generalization of the collisionless transport equation in Table 9.4 to derive the following balances.

(a) Conservation of phase i equation. Compare your result to that derived in Chapter 1.
(b) Conservation of momentum of phase i. See Table 11.1 for an analogous definition of pressure of phase i.
(c) Energy equation for phase i.
(d) Develop an equation of state for phase i, that is, relate P_i and Θ_i.
(e) Repeat (a) through (c) for a multicomponent single-phase mixture. The temperature T_i is defined by

$$k_B T_i = \frac{1}{2} m_i < \mathbf{C}_i^2 >,$$

where k_B is the Boltzmann constant and the heat flux q_i is defined by

$$q_i = \frac{1}{2} \rho_i \int \mathbf{C}_i^2 \mathbf{C}_i f_i \, d\mathbf{c}_i .$$

6. *Homogeneous Multiphase Mixture Balances for Frictionless Flow*
Use the summation of Eq. (11.4) to derive the following equations for equal velocity, equal granular temperature flow of a mixture.

(a) For $\psi_i = 1$, obtain the homogeneous flow mixture equation

$$\frac{dn}{dt}+n\nabla\cdot\mathbf{v}=0$$

with

$$n=\sum_i n_i .$$

(b) For $\psi_i=m_i\mathbf{c}_i$, obtain the homogeneous momentum equation

$$\rho\frac{d\mathbf{v}}{dt}+\nabla P_m=0,$$

where

$$P_m=\sum_{i=1}^{r}\sum_{j=1}^{r}n_i\Theta_i\left(1+b_{ij}g_{ij}\right)$$

with

$$\rho=\sum_{i=1}^{r}\rho_i \quad ; \quad b_{ij}=\tfrac{\pi}{3}\left(1+e_{ij}\right)d_{p_{ij}}^3 n_j .$$

See Eq. (9.109) and Table 11.2.

(c) For $\psi_i=\frac{1}{2}m_i c_i^2$, obtain the energy equation with no conduction

$$\frac{1}{\Theta_i}\frac{d\Theta_i}{dt}+\frac{2P_m}{3n\Theta_i}\nabla\cdot\mathbf{v}=0,$$

where

$$\frac{d}{dt}=\frac{\partial}{\partial t}+\mathbf{v}\cdot\nabla .$$

(d) Show that these equations are valid for the case of a Maxwellian distribution.

7. ***Momentum Equations for Phase i and the Fick's Law for Dense Mixtures***

(a) Let $\psi_i=m_i\mathbf{c}_i$ in Eq. (11.4) and show that the phase i momentum balance becomes

$$\frac{\partial}{\partial t}\left(\rho_i\varepsilon_i\mathbf{v}_i\right)+\nabla\cdot\left(\rho_i\varepsilon_i\mathbf{v}_i\mathbf{v}_i+\Pi_i\right)=\varepsilon_i\rho_i\mathbf{F}_i-\phi_i,$$

where

$$\Pi_i=-\varepsilon_i\rho_i\left[\left(\mathbf{v}_i-\mathbf{v}\right)\left(\mathbf{v}_i-\mathbf{v}\right)+<\varepsilon_i\rho_i\mathbf{C}_i\mathbf{C}_i>+\sum_k P_{c_{ik}}\right.$$

with $\mathbf{C}_i$ defined as

$$\mathbf{C}_i = \mathbf{c}_i - \mathbf{v}.$$

P_{Cik} is given by Eq. (11.5) and has been evaluated for a binary in Table 11.2. $\mathbf{F}_i$ is the external force per unit mass of phase i:

$$\phi_i \equiv N_{c_{ij}}\left(m_i \mathbf{C}_i\right); \quad i \neq j.$$

ϕ_i has been evaluated in Table 11.2 for a binary mixture of granular solids. The density ρ_i is the density of the particle i. The average velocity $\mathbf{v}$ is defined by

$$\rho \mathbf{v} = \sum_i \varepsilon_i \rho_i \mathbf{v}_i .$$

(b) Generalize the momentum balance to include a continuous fluid between the mixture of particles.
(c) Show that the mixture momentum balance becomes

$$\rho \frac{d\mathbf{v}}{dt} = -\nabla \mathbf{P} + \rho \mathbf{F},$$

where

$$\mathbf{P} = \sum_i \left[\Pi_i + (\mathbf{v}_i - \mathbf{v})(\mathbf{v}_i - \mathbf{v})\right] = \sum_i \left(\underbrace{< \varepsilon_i \rho_i \mathbf{C}_i \mathbf{C}_i >}_{\text{Kinetic pressure}} + \underbrace{\sum_k P_{C_{ik}}}_{\text{Collisional pressure}} \right)$$

and

$$\rho \mathbf{F} = \sum_i \varepsilon_i \rho_i F_i .$$

The sum of all the momentum sources is zero.
(d) Show that for no flow $\mathbf{v}_i = 0$, Π_i is like a partial pressure and is given by

$$\Pi_i = \left[n_i + \sum_k \tfrac{\pi}{3} g_{ik} d_{p_{ik}}^3 n_i n_k \left(1 + e_{ik}\right) \right] \Theta_i .$$

(e) What assumptions in the kinetic theory of granular flow make it impossible to treat molecules and granular particles in an identical fashion?

(f) Show that the momentum balance for phase i can be rearranged in terms of a diffusion velocity, $\mathbf{v}_i - \mathbf{v}$ and expressed as follows:

$$\frac{\partial}{\partial t}(\mathbf{v}_i - \mathbf{v}) + (\mathbf{v}_i \cdot \nabla)(\mathbf{v}_i - \mathbf{v}) + (\mathbf{v}_i - \mathbf{v})\nabla \mathbf{v} = \tfrac{1}{\rho}(\nabla P - \rho \mathbf{F}) - \frac{1}{\rho_i \varepsilon_i}\nabla \cdot \Pi_i + \mathbf{F}_i + \frac{1}{\rho_i \varepsilon_i}\phi_i .$$

(g) Ignore the inertia terms, that is all the accelerations in part f and subtract the resulting equation for phase i from that for phase j to obtain

$$0 = -\frac{1}{\varepsilon_i \rho_i}\nabla \cdot \Pi_i + \frac{1}{\varepsilon_i \rho_i}\nabla \Pi_j + \mathbf{F}_i - \mathbf{F}_j + \frac{\phi_i}{\varepsilon_i \rho_i} - \frac{\phi_j}{\varepsilon_j \rho_j}.$$

(h) Use the equation in part (g) and the phase momentum source term in Table 11.2, ϕ_i, to obtain the Fick's law of diffusion for a dense binary mixture. This is essentially the equation developed by Thorne and given in Chapman and Cowling (1961), page 293. The result should be

$$\mathbf{v}_i - \mathbf{v}_j = -\frac{n^2}{n_i n_j} D_{ij}\left\{\frac{\varepsilon_i \rho_i \varepsilon_j \rho_j}{\rho n \Theta_i}(\mathbf{F}_j - \mathbf{F}_i) + \frac{\rho_j}{\rho n \Theta_i}\nabla\left[n_A \Theta_i\left(1 + b_i \varepsilon_i \rho_i g_{ii} + b_j' \varepsilon_j \rho_j g_{ij}\right)\right]\right.$$

$$\left. - \frac{\rho_i}{\rho n \Theta_i}\nabla\left[n_j \Theta_j\left(1 + b_j \varepsilon_j \rho_j g_{jj} + b_i' \varepsilon_i \rho_i g_{ij}\right)\right] + \tfrac{2}{3}\pi \frac{n_i n_j}{n} d_{p_{ij}}^3 g_{ij}\left[\left(\frac{m_i - m_j}{m_i + m_j}\right)\nabla \ln \Theta_i + \nabla \ln \frac{n_j}{n_i}\right]\right\},$$

where the diffusion coefficient D_{ij} is given by the expression

$$D_{ij} = \frac{3}{8 n d_{p_{ij}}^2 g_{ij}}\left[\frac{\Theta_i (m_i + m_j)}{2\pi m_i m_j}\right]^{1/2}$$

and

$$b_i \varepsilon_i \rho_i = \tfrac{2}{3}\pi n_i d_{p_{ii}}^3, \quad i = i, j,$$

$$b_i' \varepsilon_i \rho_i = \tfrac{2}{3}\pi n_i d_{p_{ij}}^3, \quad i = i, j.$$

(i) Interpret the Fick's law expression in terms of driving forces due to particle concentration, shear and effects of external forces, such as electric fields on particles carrying a different surface charge. When does the shear have the largest effect on particle segregation?

8. *Viscosity of Phase i*
(a) Identify the viscosity of phase *i* using Table 11.2.
(b) Compare this expression to the more general viscosity for a uniform particle mixture from Eq. (9.250) and explain the differences.
(c) Find the viscosity of a binary granular mixture using the expression from Table 11.2.

9. *Collisional Stress, Momentum Source, and Energy Dissipation*
Show how the collisional stress, the momentum source, and the energy dissipation in Table 11.2 were obtained.

10. *Phase i Granular Temperature Equation*
In Equation (11.4), let $\psi_i = m_i c_i^2$ and derive an energy equation in terms of the granular temperature Θ_i defined by Eq. (11.10), (from Jenkins and Mancini, 1987).

Literature Cited

Arastoopour, H. (1986). "Pneumatic Conveying of Solids," pp. 349–382 in N. P Cheremisinoff, Ed. *Encyclopedia of Fluid Mechanics, Volume 4: Solids and Gas–Solids Flows*. Houston, Texas: Gulf Publishing.

Arastoopour, H., D. Lin, and D. Gidaspow (1980). "Hydrodynamic Analysis of Pneumatic Transport of a Mixture of Two Particle Sizes," pp. 1853–1971 in T. N. Veziroglu, Ed. *Multiphase Transport Vol. 4.* New York: Hemisphere Publishing Corp.

Chapman, S., and T. G. Cowling (1961). *The Mathematical Theory of Non-uniform Gases,* 2nd Ed. Cambridge, U.K.: Cambridge University Press.

Condiff, D. W., W. K. Lu, and J. S. Dahler (1965). "Transport Properties of Polyatomic Fluids, a Dilute Gas of Perfectly Rough Spheres," *J. Chem. Phys.* **42**, 3445–3475.

Dahler, J. S., and N. F. Sather (1963). "Kinetic Theory of Loaded Spheres I," *J. Chem. Phys.* **38**, 2363–2382.

Farrell, M., C. K. K. Lun, and S. B. Savage (1986). "A Simple Kinetic Theory for Granular Flow of Binary Mixtures of Smooth, Inelastic Spherical Particles," *Acta Mechanica* **63**, 45–60.

Gidaspow, D., M. Symlal, and Y. Seo (1986). "Hydrodynamics of Fluidization of Single and Binary Size Particles: Supercomputer Modeling," in K. Ostergaard and A. Sorensen, Eds. *Fluidization V.* New York: AIChE.

Gupta, R., D. Gidaspow, and D. T. Wasan (1993). "Electrostatic Separation of Powder Mixtures Based on the Work Functions of its Constituents," *Powder Technol.* **75**, 79–87.

Hirschfelder, J. O., C. F. Curtis, and R. B. Bird (1954). *Molecular Theory of Gases and Liquids.* New York: John Wiley & Sons.

Jenkins, J. T., and F. Mancini (1987). "Balances Laws and Constitutive Relations for Plane Flows of a Dense, Binary Mixture of Smooth, Nearly Elastic, Circular Disks," *J. Appl. Mech.* **54**, 27–34.

Johnson, K. L. (1985). *Contact Mechanics.* Cambridge, U.K.: Cambridge University Press.

Muschelknautz, E. (1959). "Theoretical and Experimental Investigations on the Pressure Loss in Pneumatic Conveyers with Special Regard to the Influence of Friction and Weight of Materials to be Conveyed," *VDI-Forschungsheft 476.* Dusseldorf: VDI-Verlag GMBH.

Nakomura, K. and C. E. Capes. (1976). "Vertical Pneumatic Conveying of Binary Particle Mixtures," pp. 159–184 in D. L. Keairns, Ed. *Fluidization Technology, Vol. II.* New York: Hemisphere Publishing Company.

12

SEDIMENTATION AND CONSOLIDATION

12.1 Conservation of Particles **355**
12.2 Settling in a Sedimentation Column: Introduction **357**
12.3 Free Settling **360**
12.4 Compression Settling **366**
12.5 Consolidation: Relation to Osmotic Pressure **370**
12.6 Electrokinetic Phenomenon: Zeta Potential **376**
12.7 Effect of Zeta Potential on Sedimentation **379**
Exercises **382**
1. Hydrodynamics of a Thickener **382**
2. Electrokinetic Phenomena in Porous Media or in Membranes **383**
3. Lamella Electrosettler **383**
Literature Cited **388**

12.1 Conservation of Particles

Sedimentation of particles is a physically simple, common example of multiphase flow. Consider a tall settling column initially containing a uniform dispersed mixture of particles. For settling of non-colloidal particles of a uniform size after some time t, a clear interface will appear at the top of the column. For sedimentation of a dilute suspension of particles there will exist a constant settling zone in the center. At the bottom of the column the particles will slow down and form a sludge layer. Figure 12.1 shows the three regions and the system for which a particle balance is made.

Let n be the number of particles per unit volume. Then a particle balance for an arbitrary element of a volume fixed in space, as shown in Fig. 12.1, referred to in the literature as a "shell" or an "Eulerian" balance, is

$$\frac{d}{dt}\int_x^{x+\Delta x} na\,dx \quad + \quad nva\Big]_x^{x+\Delta x} \quad = \quad 0. \qquad (12.1)$$

Rate of accumulation of particles in the volume element $a\Delta x$ + Net rate of particles Outflow = Zero rate of particle production

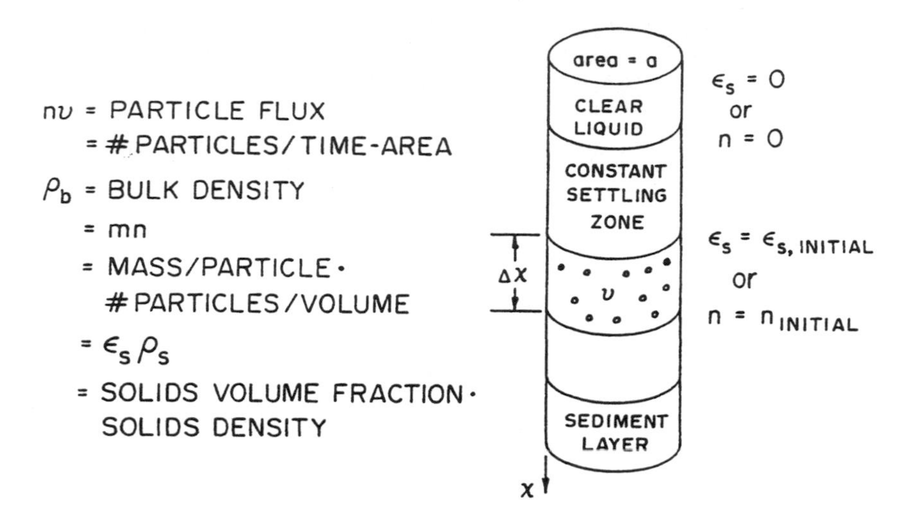

Fig. 12.1 Settling of a suspension of particles of a uniform size.

An application of the mean value theorem of integral calculus to the integral in Eq. (12.1), of the mean value theorem of differential calculus to the difference in fluxes written in terms of Leibnitz notation used in the fundamental theorem of calculus and an application of the limiting process yields the particle balance

$$\frac{\partial(na)}{\partial t}+\frac{\partial(nva)}{\partial x}=0. \tag{12.2}$$

A particle balance made on a system of particle mass moving with a velocity v with a time-dependent volume $ax(t)$, the so-called "Lagrangian" balance, gives

$$\frac{d}{dt}\int_{x(t)}^{x(t)+\Delta x(t)} na\,dx=0. \tag{12.3}$$

Then an application of the Reynolds transport theorem widely used in Chapter 1 produces the same equation (12.2). For a constant area a the particle balance is simply

$$\frac{\partial n}{\partial t}+\frac{\partial(nv)}{\partial x}=0. \tag{12.4}$$

A similar balance can be obtained for bulk density ρ_b where

$$\rho_b = m \cdot n, \tag{12.5}$$

with m equal to the mass of the particle.

The Lagrangian balance, stating the principle of conservation of mass, is

$$\frac{d}{dt}\int_{x(t)}^{x(t)+\Delta x(t)} \rho_b a\, dx = 0. \tag{12.6}$$

Hence,

$$\frac{\partial \rho_b}{\partial t} + \frac{\partial(\rho_b v)}{\partial x} = 0. \tag{12.7}$$

With a constant m, Eq. (12.7) is the same as Eq. (12.4), the conservation of particles. Similarly, the particle conservation equation can be expressed in terms of the particle volume fraction of solids ε_s using the relation

$$\rho_b = \varepsilon_s \rho_s, \tag{12.8}$$

where ρ_s is the density of the solid particles. When the density of the particles remains constant, the equivalent of the conservation of particle density is the volume of particles conservation equation

$$\frac{\partial \varepsilon_s}{\partial t} + \frac{\partial(\varepsilon_s v)}{\partial x} = 0, \tag{12.9}$$

where $v = v_s$, the only velocity recognized so far. The porosity or voidage, that is, the volume fraction between the particles ε, is

$$\varepsilon = 1 - \varepsilon_s. \tag{12.10}$$

This is the common notation for porosity used in chemical engineering literature. Although the derivation was with reference to settling of particles in a fluid, the preceding balances and relations are valid in a vacuum, too.

12.2 Settling in a Sedimentation Column: Introduction

To obtain the concentration of particles in the sedimentation column shown in Fig. 12.1, the particle balance given by Eq. (12.4) must be supplemented by a momentum balance for the velocity v. The objective of this book was to show how to solve such problems. However, many limiting cases can be solved by use of the mass balance alone. In the case of slow, dilute sedimentation of non-colloidal particles, the particle velocity very quickly becomes the velocity of fall of a single particle in an infinite Newtonian fluid, the Stokes velocity, given by

$$v = v_{\text{Stokes'}} = \frac{d_p^2\left(\rho_s - \rho_f\right)g}{18\mu}. \tag{12.11}$$

In such a case, the particle balance is simply the first-order partial differential equation:

$$\frac{\partial n}{\partial t} + v\frac{\partial n}{\partial x} = 0, \tag{12.12}$$

which is equivalent to the ordinary differential equation

$$\frac{dn}{dt} = 0 \tag{12.13}$$

along the path

$$\frac{dx}{dt} = v. \tag{12.14}$$

Equation (12.13) shows that the particle concentration stays constant along the characteristic direction or path given by Eq. (12.14),

$$n = \text{constant}. \tag{12.15}$$

Hence, there exists a constant settling zone having a constant particle concentration, as sketched in Fig. 12.1. In the top zone in Fig. 12.1, this concentration is zero, since it has been implicitly assumed that zero time in this problem is the time when the particles acquired the terminal velocity given by Eq. (12.11) and just began to move away from the top boundary. In the free settling zone, the particle concentration stays at its initial value.

This elementary analysis can be applied to the settling of particles having a wide particle-size distribution. Let n_i be the number of particles per unit volume of size i, with diameter d_i, and density ρ_i. If there is neither formation of size i particles nor destruction, then a balance on size i moving with velocity v_i is

$$\frac{d}{dt}\int_{x(t)}^{x(t)+\Delta x(t)} a n_i \, dx = 0. \tag{12.16}$$

An application of the Reynolds transport theorem in one dimension for a constant cross-sectional area a gives a generalization of Eq. (12.4):

$$\frac{\partial n_i}{\partial t}+\frac{\partial(n_i v_i)}{\partial x}=0, \qquad i=1,2,3,\ldots k \text{ sizes.} \tag{12.17}$$

If each particle-size travels with its own terminal velocity and if particle–particle size interactions due to particle collisions are neglected, then in the Stokes regime each size falls with velocity given by Eq. (12.11) with d_p replaced by d_i and ρ_s replaced by ρ_i. Then the k particle differential equations are

$$\frac{\partial n_i}{\partial t}+v_i\frac{\partial n_i}{\partial x}=0. \tag{12.18}$$

These can be written as k ordinary differential equations as

$$\frac{dn_i}{dt}=0 \tag{12.19}$$

along the respective paths

$$\frac{dx}{dt}=v_i. \tag{12.20}$$

Integration of Eq. (12.19) shows that

$$n_i = \text{constant}, \qquad \text{i}=1,2,3,\ldots k, \tag{12.21}$$

along the paths

$$x-x_o=v_i(t-t_o), \qquad i=1,2,3,\ldots k. \tag{12.22}$$

Hence, the particles separate according to size and density into constant concentration regions, as sketched in Fig. 12.2 for case $\rho_i > \rho_f$. Sedimentation of multisize particles with a more general drag law, with particle–particle interactions and with retarding electrical forces, as well as with the momentum effects to be discussed in this chapter, have been illustrated by Shih *et al.* (1987). The principal effect observed is, however, the segregation discussed earlier. If some particles are lighter than the fluid, they will move to the top, while the denser particles descend. Separation of minerals by froth flotation works on this principle, where air bubbles, acting as *particles*, carry fine tailings to the top of a flotation unit.

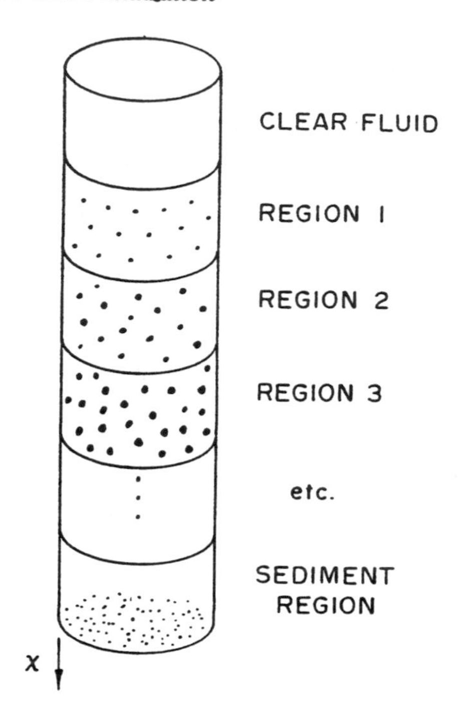

Fig. 12.2 Settling of a mixture of particles of various sizes.

12.3 Free Settling

The introductory treatment of sedimentation neglected fluid motion and the concentration effects caused by the fact that in settling, the particles are not suspended in an infinite fluid, as is assumed in Stokes' law. In a settling column, the fluid velocity can be related to the particle velocity from simple mass balances already used. The particle balance, Eq. (12.4), when expressed in terms of the volume fraction of the solid, ε_s, is given by Eq. (12.9) for an incompressible solid. A similar balance can be written for an incompressible fluid. Since the volume fractions add up to one, the equation for the mixture is simply the sum

$$\frac{\partial\left(\varepsilon_s v_s + \varepsilon v_f\right)}{\partial x} = 0 . \tag{12.23}$$

Integration of Eq. (12.23) shows that

$$\varepsilon_s v_s + \varepsilon v_f = \text{Constant}. \tag{12.24}$$

The integration constant is zero, since at the top of the column both v_f and v_s are zero because of the incompressibility of both the solid and the fluid. Hence, Eq. (12.24) states that the solids flux settling down equals the flux of the liquid rising:

$$\varepsilon_s v_s = -\varepsilon v_f. \tag{12.25}$$

The velocity that is needed in the Stokes' law or in any other constitutive relation for the drag is the relative velocity, since constitutive equations must be independent of the frame of reference. Let the solids velocity relative to the fluid be v_r:

$$v_r = v_s - v_f. \tag{12.26}$$

Then using Eq. (12.25), the relative velocity becomes

$$v_r = v_s / \varepsilon. \tag{12.27}$$

Equation (12.27) already indicates that a packing effect will enter the sedimentation problem. Historically, further progress was made by assuming that the velocity of fall of a particle depends only upon its local concentration (Kynch, 1952). Before discussing Kynch's conclusions based on this assumption, consistent with the aim of this book, a momentum balance is introduced.

For zero particle acceleration, negligible wall effect and zero particle normal stress, the force balance on the volume element $a\Delta x$ in Fig. 12.1 is simply a balance between buoyancy and drag. In terms of the number of particles per unit volume n, a fluid–particle friction coefficient β, and an effective gravitational acceleration $\tilde{g}$ introduced by Batchelor (1988) in a similar derivation, the balance is

$$\underset{\text{Buoyant force}}{am\tilde{g}\int_x^{x+\Delta x} n\,dx} + \underset{\text{Drag force}}{a\int_x^{x+\Delta x} \beta v_r\,dx} = 0, \tag{12.28}$$

Buoyant force + Drag force = 0

where

$$\tilde{g} = \frac{g(\rho_s - \rho_f)}{\rho_s}. \tag{12.29}$$

Simplification of the force balance yields the relation

$$\varepsilon_s g\left(\rho_s - \rho_f\right) = -\beta v_r. \tag{12.30}$$

A more complete momentum balance is derived in Chapter 2. It is given by Eq. (2.17). Although in the balance given by Eqs. (12.28) and (12.30), the drag on the particles has been assumed to be equal to the coefficient β times the relative velocity, a momentum balance on the fluid for developed flow is

$$\varepsilon \frac{dP}{dx} = \beta_A v_r. \tag{12.31}$$

The multiplication by porosity in Eq. (12.31) is due to the assumption that the pressure acts on the fluid portion of the mixture only. Since in Eq. (12.30) βv_r is the dissipation equal to the pressure drop,

$$\beta = \beta_A / \varepsilon. \tag{12.32}$$

The pressure drop can be correlated by the Ergun equation (Bird *et al.*, 1960). This equation is given by Eq. (2.10) in Chapter 2.

For low Reynolds numbers the first term of this equation can be used. It is

$$\beta_A = \frac{150\varepsilon_s^2 \mu}{\varepsilon\left(d_p \phi_s\right)^2}. \tag{12.33}$$

Using Eqs. (12.27), (12.30), (12.32) and (12.33), the solids velocity becomes

$$v_s = -\frac{\varepsilon^3 g \Delta\rho\left(d_p \phi_s\right)^2}{150\varepsilon_s \mu}. \tag{12.34}$$

Although in the form given by Eq. (12.34) a singularity appears in the solids velocity at the dilute limit, this apparent problem vanishes when Eq. (12.34) is substituted into the full conservation of solids volume Eq. (12.9) as follows:

$$\frac{\partial \varepsilon}{\partial t} + \frac{\varepsilon^2 g \Delta\rho\left(d_p \phi_s\right)^2}{50\mu} \frac{\partial \varepsilon}{\partial x} = 0. \tag{12.35}$$

This is the same result as that obtained using Stokes' law for velocity described in the previous section, except that the characteristic direction is concentration dependent, as can be seen by integrating Eq. (12.35):

$$\frac{d\varepsilon}{dt} = 0 \tag{12.36}$$

along the path

$$\frac{dx}{dt} = v_o \varepsilon^2, \tag{12.37}$$

where for convenience we let

$$v_o = \frac{g\Delta\rho\left(d_p\phi_s\right)^2}{50\mu}. \tag{12.38}$$

This form of the drag law can now be used to illustrate the dispersion region in sedimentation first proposed by Kynch (1952) and extended to include the effect of the sediment layer by Tiller (1981) and Font (1990). Kynch considered the motion of the clear interface which can be considered to be a discontinuity. Let H be the height of this interface from the bottom of the container. Then a mass balance across this shock can be written mathematically as

$$\frac{d}{dt}\lim_{\xi\to 0}\int_{H-\xi}^{H+\xi} a\varepsilon_s\rho_s\,dx + \lim_{\xi\to 0} a\varepsilon_s\rho_s v_s\Big]_{H-\xi}^{H+\xi} = 0. \tag{12.39}$$

Rate of accumulation of mass + Net rate of mass outflow = 0

Differentiation of the integral gives

$$\lim_{\xi\to 0}\int_{H-\xi}^{H+\xi}\frac{\partial\left(\varepsilon_s\rho_s\right)}{\partial t}a\,dx + a\varepsilon_s\rho_s\Big]_-^+\frac{dH}{dt} = -a\varepsilon_s\rho_s v_s\Big]_-^+, \tag{12.40}$$

where the notation $]_-^+$ means evaluated *right above H* minus *right below H*, similar to the notation used in the Fundamental Theorem of Integral Calculus. Since the integral in (10.40) vanishes for a constant area, the rate of change of the height is then given by

$$\frac{dH}{dt} = -\frac{\varepsilon_s\rho_s v_s\Big]_-^+}{\varepsilon_s\rho_s\Big]_-^+}. \tag{12.41}$$

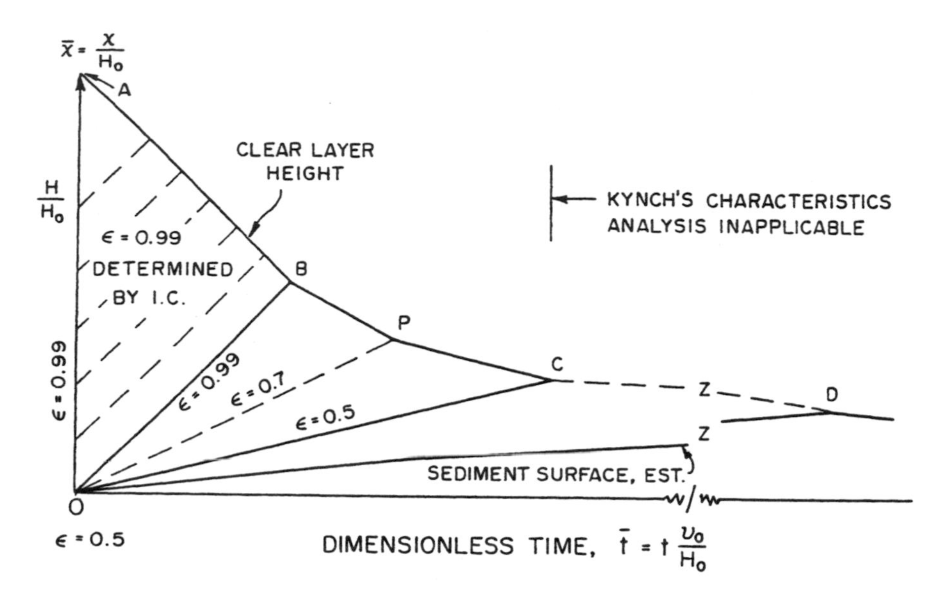

Fig. 12.3 Computation of interface with a concentration-dependent settling rate.

For a constant density and zero volume fraction of solids above the interface, Eq. (12.41) gives the clear layer interface height as

$$\frac{dH}{dt} = -v_s. \tag{12.42}$$

In the example considered,

$$v_s = v_o \varepsilon^2. \tag{12.43}$$

As an example, consider settling of a mixture initially stirred to a uniform volume fraction of one percent, so that at time zero, $\varepsilon = 0.99$. The settling process described by the differential equations (12.36), (12.37), (12.42), and (12.43) is illustrated graphically in Fig. 12.3. Equation (12.36) shows that the porosity is constant along the paths given by Eq. (12.37). To represent the solution of the differential equation, it is best to use dimensionless variables in the differential equations. Let H_o be the initial height of the settling column filled with the constant concentration of the particles. Let

$$\bar{x} = \frac{x}{H_o}, \quad \bar{H} = \frac{H}{H_o}, \tag{12.44}$$

$$\bar{t} = t\left(\frac{v_o}{H_o}\right). \tag{12.45}$$

The equation (12.37) becomes

$$\frac{d\bar{x}}{d\bar{t}} = \varepsilon^2. \tag{12.46}$$

In view of Eq. (12.36), all the dashed lines in Fig. 12.3 in the region ABO are straight and have a slope of approximately unity, since

$$\frac{d\bar{x}}{d\bar{t}} = (0.99)^2. \tag{12.47}$$

The initial constant concentration or void fraction of 0.99 propagates along these lines. The whole region ABO has the constant porosity of 0.99. To determine the solution in the other regions, boundary conditions must be prescribed at the bottom of the container. Suppose at the bottom the porosity is 0.5 at zero time. Then the slope of the line OC is

$$\frac{d\bar{x}}{d\bar{t}} = (0.5)^2 = \frac{1}{4}. \tag{12.48}$$

The porosity is 0.5 only along this line. But between B and C we can draw many straight lines where slopes will vary from $(0.99)^2$ to $(0.5)^2$. The corresponding porosities will be constant along these lines and vary between 0.99 and 0.5. A typical line of slope 0.49 is given by line OP. Let us now consider the motion of the clear layer interface $\bar{H}$:

$$\frac{d\bar{H}}{d\bar{t}} = -\varepsilon^2. \tag{12.49}$$

Since $\varepsilon = 0.99$ up to point B,

$$\bar{H} = 1 - (0.99)^2\bar{t}, \tag{12.50}$$

as shown in Fig. 12.3. But from B to C, the concentration changes. Hence, Eq. (12.49) must be integrated, step by step, to obtain the curve

$$\overline{H} = \overline{H}(\bar{t}). \tag{12.51}$$

Figure 12.3 illustrates a very crude two-increment integration. For the line passing through P of slope 0.49,

$$\frac{d\overline{H}}{d\bar{t}} = -0.49. \tag{12.52}$$

A slope of minus 0.49 has been drawn through the point B to P. Then at P, a slope of 0.25 has been constructed. Clearly between the points B and C, there exists a curved region. The height of the interface is dropping at a decreasing rate. After a time, corresponding to point B, the column will not have a constant concentration region any longer. It consists of a clear layer, a layer of decreasing concentration and a sediment layer. The region to the right and below the line OC will be influenced by the sediment layer. A shock balance given by Eq. (12.41) can also be written for the size of the sediment layer. However, since the physical assumptions made so far are not valid for the use of the sediment, its surface in Fig. 12.3 is merely sketched in to indicate a possible situation for this problem. The drop of the clear layer beyond point C has also not been determined, since its analysis depends upon the rise of the sediment layer, as first pointed out by Tiller (1981). Eventually the sediment will meet the clear layer at point D. The mathematics for this problem will be presented in the next section. Kynch's (1952) analysis of the descent of the interface and dispersion is valid until point C in Fig. 12.3.

12.4 Compression Settling

Kynch's theory was successful in explaining free settling without any effect of the sediment layer. It cannot be used to analyze the sediment layer, since flux is no longer a function of particle concentration only, as given by the force balance Eq. (12.30). An interparticle stress or particle pressure must be added to the buoyancy equals drag balance. Let σ be this particle pressure. This particle pressure is a result either of collision of particles with each other, or of elastic or inelastic contact between particles, causing a repulsive force. A compressive force must be applied to bring the particles closer together. With this additional force, the momentum balance with negligible inertia made on the volume element shown in Fig. 12.1 is

$$am\tilde{g}\int_x^{x+\Delta x} n\,dx + \int_x^{x+\Delta x} a\beta v_r\,dx + a\sigma\Big]_x^{x+\Delta x} = 0. \tag{12.53}$$

Buoyant force + Drag force + Solids pressure force = 0

An application of the mean value theorems of calculus to the expressions in Eq. (12.53) and the limiting process, together with the relations given by Eqs. (12.5), (12.8) and (12.32), gives the force balance below:

$$-g\Delta\rho\varepsilon\varepsilon_s = \beta_A v_r + \varepsilon\frac{d\sigma}{dx}. \tag{12.54}$$

This is the same as Eq. (2.17) in Chapter 2 when wall friction is neglected. Equation (12.54) is also identical with Eq. (10) of Tiller (1981). In place of the friction coefficient Tiller uses the permeability k. The relation between the two is

$$\text{Permeability} = k = \frac{\mu\varepsilon^2}{\beta_A}. \tag{12.55}$$

Use of Stokes' law, Eq. (12.11), for the friction coefficient shows that for this limiting case the permeability becomes

$$k_{\text{Stokes'}} = \frac{d_p^2}{18}\frac{\varepsilon}{\varepsilon_s}. \tag{12.56}$$

The permeability has the units of (length)^2. The length scale is the particle diameter for a medium made up of spheres. The usual capillary flow model for a porous medium (Scheidegger, 1960) gives the slightly modified value shown below:

$$k_{\text{capillary}} = \frac{d_p^2\varepsilon}{32}, \tag{12.57}$$

where d_p is now an average pore diameter. Precise values of permeabilities must be measured, although the Ergun equation discussed in Chapter 2 gives more reliable values of friction coefficients, and thus permeabilities, than either of the two rough expressions (12.56) and (12.57).

Equation (12.54) can be rearranged into the usual form used in Fick's law of diffusion. It reads

$$\text{Relative solids flux} = \varepsilon_s v_r = \frac{-g\Delta\rho\varepsilon\varepsilon_s^2}{\beta_A} - \frac{\varepsilon\varepsilon_s}{\beta_A}\frac{d\sigma}{dx}. \tag{12.58}$$

To make further progress one must assume a constitutive relation for the stress σ. In the absence of the usually small effect due to particle collisions, the stress is a function of the displacement, or

$$\sigma = \sigma(\varepsilon_s). \tag{12.59}$$

Substitution into Eq. (12.58) requires the knowledge of the derivative of this stress. Then we let the modulus of "elasticity," G, be

$$G = \frac{d\sigma}{d\varepsilon_s}. \tag{12.60}$$

Then using the relation between relative and solids velocity, Eq. (12.27), the solids flux relative to the stationary wall $\varepsilon_s v_s$ becomes

$$\varepsilon_s v_s = \frac{-g\Delta\rho\varepsilon^2\varepsilon_s^2}{\beta_A} - \frac{\varepsilon^2\varepsilon_s G}{\beta_A}\frac{\partial\varepsilon_s}{\partial x}. \tag{12.61}$$

Substitution of this flux into the conservation of solids equation (12.9) gives the convective diffusion equation below:

$$\frac{\partial\varepsilon_s}{\partial t} + \frac{\partial(v_{\text{sett}}\varepsilon_s)}{\partial x} = \frac{\partial}{\partial x}\left(\mathcal{D}\frac{\partial\varepsilon_s}{\partial x}\right), \tag{12.62}$$

where v_{sett} is a convective settling velocity

$$v_{\text{sett}} = \frac{-g\Delta\rho\varepsilon^2\varepsilon_s}{\beta_A} \tag{12.63}$$

and $\mathcal{D}$ is a diffusion or consolidation coefficient used in solid mechanics:

$$\mathcal{D} = \frac{\varepsilon^2\varepsilon_s G}{\beta_A}. \tag{12.64}$$

The consolidation coefficient is a ratio of the solids modulus and the friction coefficient; just as in the usual Fick's law, the diffusion coefficient is a ratio of thermal energy to dissipation due to molecular drag given by Stokes' law in terms of the radius of the molecule. Hence the process of sediment rise can be described by a diffusion equation. The solution of this convective diffusion equation requires the knowledge of the particle modulus G. The present state of knowledge of settling is such that this modulus must be determined from the settling experiment itself. After settling is essentially complete, when there is no motion, Eq. (12.58) gives

$$\sigma = g\Delta\rho \int_0^x \varepsilon_s \, dx \,. \tag{12.65}$$

A measurement of the solids distribution, ε_s, as a function of height by an x-ray densitometer or by other means gives this solids pressure as a function of the solids concentration. Attempts to obtain this modulus from osmotic pressure will be discussed in the following section on consolidation.

A typical example of settling in a sedimentation column from the study of Shih *et al.* (1986) is summarized below. Settling of fine and coarse particles in non-aqueous media, which is of interest to the petroleum industry, was carried out in the sedimentation column equipped with a moveable x-ray densitometer for measuring particle concentration and with electrodes for applying an electric field for enhancing settling of colloidal particles as shown in Fig. 12.4.

Figure 12.5 shows typical data obtained for coarse particles with no applied electric field. Initially, the particles were dispersed in the column by an ultrasonic wave mixer. The concentrations were measured with the movable x-ray densitometer at the times shown in Fig. 12.5. For the 63 μm illite particles dispersed in toluene we see the constant settling regions and the build-up of the sediment. The sediment concentration varies from a maximum at the bottom of the column to that present initially in the free settling zone. In this case, the free settling follows the simple theory discussed earlier, except that the settling velocity is not the Stokes velocity but had to be corrected for a packing effect. The solids stress modulus G was obtained as a function of the volume fraction of solids from the data shown in Fig. 12.5 and from other similar data. The convective diffusion Eq. (12.62) was then solved by standard finite difference techniques. The comparison to experiment shows that the model captures the basic features of the settling process. It is not predictive, since neither the drag nor the stress modulus were known before the experiment was conducted. The hope is that the understanding achieved since Kynch's theory will lead to predictive models, at least for some model systems.

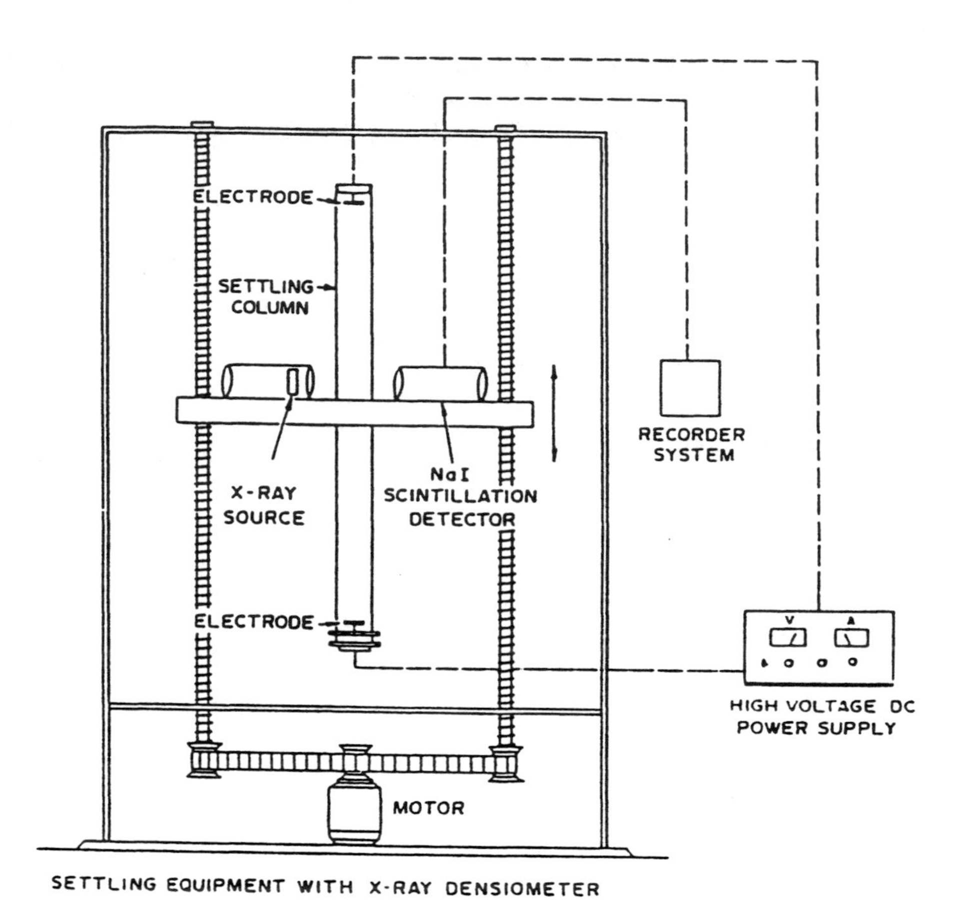

Fig. 12.4 Settling equipment with x-ray densitometer.

12.5 Consolidation: Relation to Osmotic Pressure

onsolidation is an important civil engineering problem, since buildings, bridges, and dams are frequently founded on soils. Subsidence can lead to a complete collapse of such structures. Earthquakes can cause liquifaction and catastrophic collapse of buildings, as has recently occurred to buildings built in the Marina district of San Francisco. Consolidation theory can also be applied to

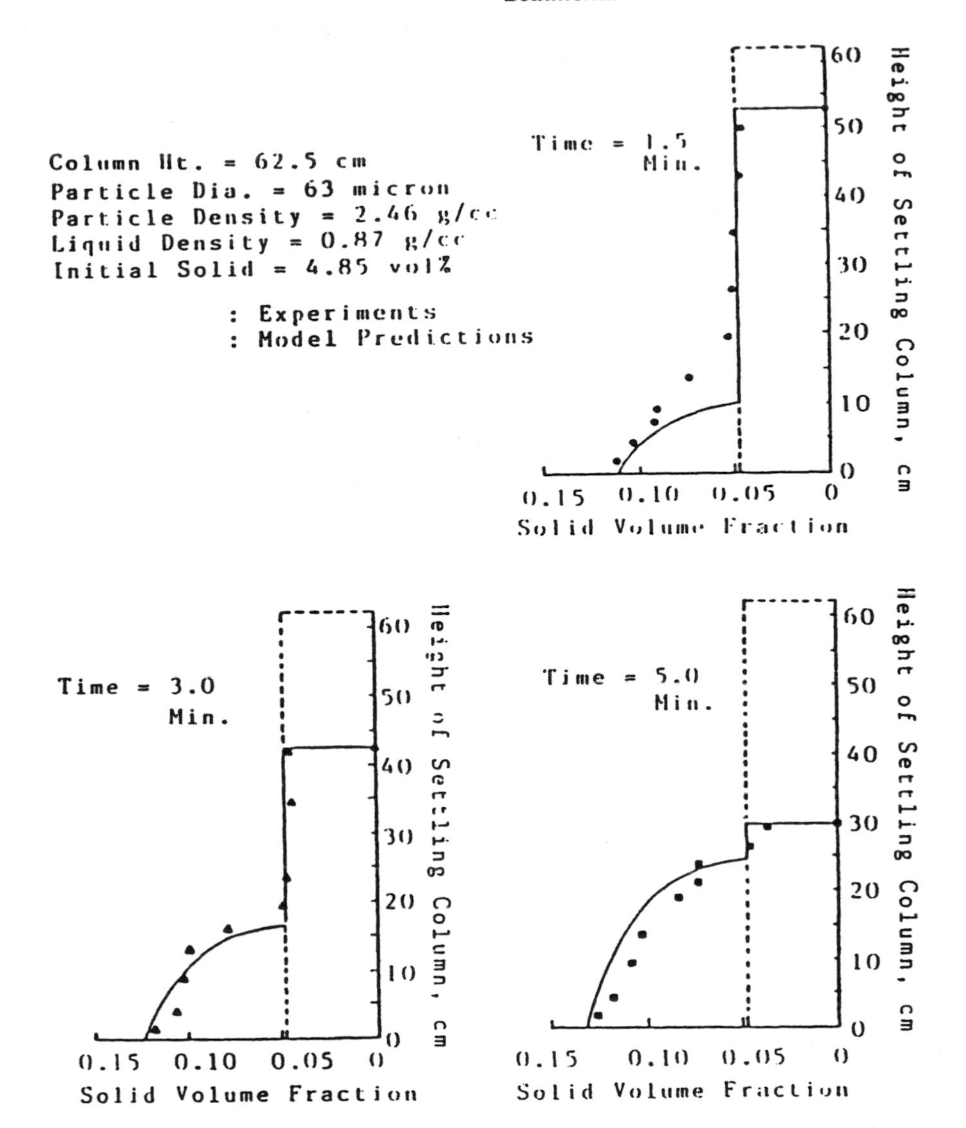

Fig. 12.5 Settling of illite in toluene — comparison of theory to experiments in settling column, Fig. 12.4. (from Shih *et al.*, 1986).

filtration and manufacture of materials by hot pressing of powders. In civil engineering, the basic tool for obtaining consolidation coefficients for various soils or powders is a consolidometer (Wissa *et al.*, 1971; Taylor, 1948; Schiffman *et al.*,

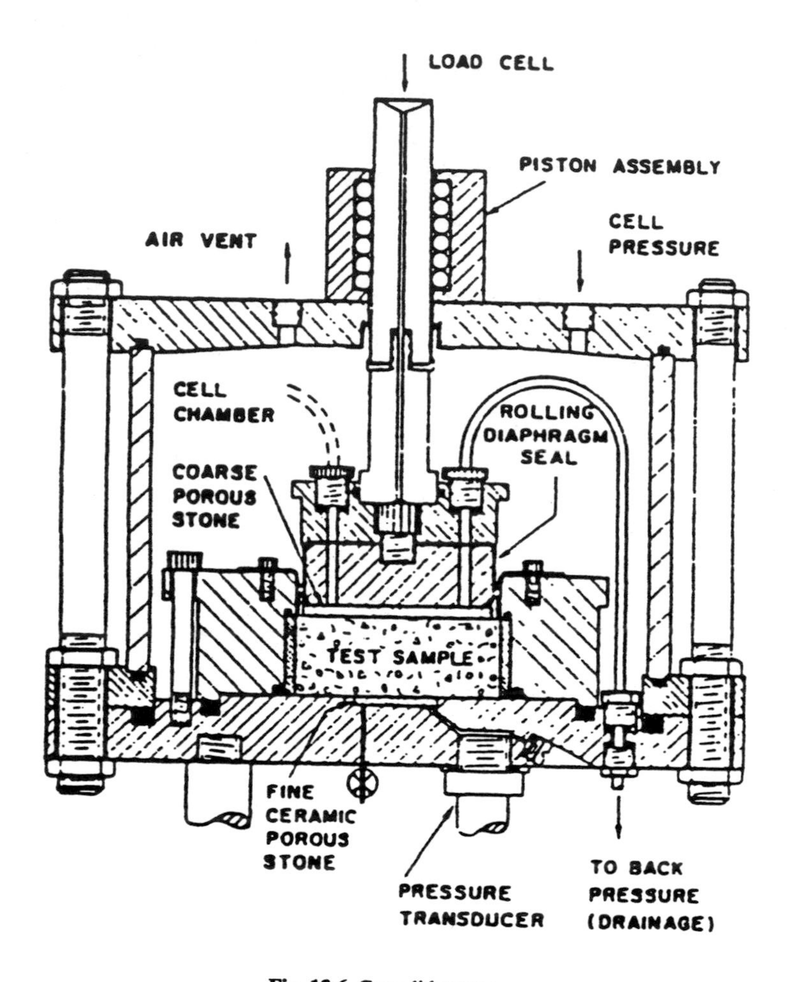

Fig. 12.6 Consolidometer.

1985). It is based on Terzaghi's (1943) consolidation theory that will be reviewed here.

A typical consolidometer is shown in Fig. 12.6. A test sample is put into the machine and a given load is applied. The fluid is forced out of the sample through porous walls. The displacement of the level is measured as a function of time accurately. This gives the porosity as a function of time, as required for the solution of Eq. (12.62). When v_{settle} is zero, Eq. (12.62) is a diffusion equation for the volume fraction of solid or equivalently for porosity. Taylor (1948) shows how a solution to such a diffusion equation for a constant consolidation coefficient and experimental displacement data for a step input in load gives a value of the consolidation coefficient. Although this brief description is the essence of the consolidation theory, a more complete understanding of Terzaghi's theory and its extensions requires the consideration of the momentum balance for the mixture. The origin of this approach dates back to Terzaghi, since the consolidometer, as shown in Fig. 12.6,measures the fluid pressure, as well as the total applied load.

The mixture momentum equation had been derived in Chapters 1 and 2. Table 2.1 shows that in one dimension with zero acceleration and no wall friction, the mixture momentum balance shows that the sum of the gradient of the fluid pressure P and the intergranular or solids stress σ is balanced by the weight of the mixture. Then the mixture momentum balance for the test sample in Fig. 12.6 is

$$\frac{dP}{dx}+\frac{d\sigma}{dx}=-g\left(\varepsilon_s\rho_s+\varepsilon\rho_f\right). \tag{12.66}$$

When the weight of the sample can be neglected compared with the large applied load, the mixture momentum balance shows that

$$\frac{d}{dx}(P+\sigma)=0. \tag{12.67}$$

Integration of Eq. (12.67) shows that

$$P+\sigma=\text{Applied load}=\sigma_t. \tag{12.68}$$

In soil mechanics, σ is called the effective stress. For a constant applied load,

$$dP=-d\sigma. \tag{12.69}$$

Furthermore, for zero applied load, the intergranular stress is simply the negative of the fluid pressure in the pores of the soil which is the osmotic pressure. Hence, under the restrictive conditions,

$$\underset{\text{Intergranular stress}=-\text{Osmotic pressure}}{\sigma=-P.} \tag{12.70}$$

The utility of this relation is that there exist many theoretical and empirical equations for osmotic pressure. Therefore, the consolidation coefficient which involves the derivative of the intergranular stress can be estimated without a measurement, as done in some recent papers (Auzerais *et al.*, 1988; Buscall, 1990).

Furthermore, the relation between the intergranular stress and the osmotic pressure, as given by Eq. (12.70) can be used to explain the strength of materials, as discussed by Taylor (1948). A stiff clay has a cohesive strength σ because of past compression in the ground, where the previous applied load squeezed out the water of osmotic pressure P. Similarly, Taylor explains the strength of steel which he considers to be a supercooled liquid. During cooling of the steel from a molten state, shrinkage occurred, more and more crystals were formed, and the osmotic pressure between crystals increased until when the steel solidified, large internal pressures were captured, giving the steel its great strength.

Large values of osmotic pressures can be obtained from equations, such as used by Auzerais *et al.* (1988) in their sedimentation studies. They cite the relation

$$P = \text{Osmotic pressure} = \frac{k_B T}{\frac{4}{3}\pi r^3}\varepsilon_s Z(\varepsilon_s), \tag{12.71}$$

where

$$Z(\varepsilon_s) = \frac{1.85}{0.64 - \varepsilon_s}, \tag{12.72}$$

and where k_B is the Boltzmann constant, T is the thermal temperature and r is the radius of a colloidal particle in the mixture.

With this perspective, it is now useful to rederive Terzaghi's linear consolidation theory in a few steps. This is normally expressed in terms of a void ratio e, which is the ratio of a volume of void to the volume of the solid phase. Porosity is thus expressed as

$$\varepsilon = \frac{e}{1+e}. \tag{12.73}$$

The derivation will be given in terms of the more convenient volume fraction of solids ε_s. The flux of solids can be obtained from the momentum balance given by Eq. (12.53). Neglecting the gravitational force and using the relation between the relative velocity and the solids velocity, Eq. (12.27), the balance is

$$\text{Flux} = \varepsilon_s v_s = -\frac{\varepsilon^2 \varepsilon_s}{\beta_A}\frac{d\sigma}{dx}, \tag{12.74}$$

where the friction coefficient β_A may be expressed in terms of the more commonly used permeability given by Eq. (12.55). The permeability used in soil mechanics contains the fluid viscosity, so that the friction coefficient is essentially the inverse of the effective permeability used in soil mechanics. Substitution of Eq. (12.74) into the conservation of solids volume, Eq. (12.9), yields

$$\frac{\partial \varepsilon_s}{\partial t} = \frac{\partial}{\partial x}\left(\frac{\varepsilon^2 \varepsilon_s}{\beta_A}\frac{\partial \sigma}{\partial x}\right), \tag{12.75}$$

since $\sigma = \sigma(\varepsilon_s)$:

$$\frac{\partial \sigma}{\partial t} = \frac{\partial \sigma}{\partial \varepsilon_s}\frac{\partial \varepsilon_s}{\partial t} = G\frac{\partial \varepsilon_s}{\partial t}. \tag{12.76}$$

Then for a constant friction coefficient relation in Eq. (12.75), a combination of (12.75) and (12.76) gives the differential equation for the intergranular stress,

$$\frac{\partial \sigma}{\partial t} = \mathcal{D}\frac{\partial^2 \sigma}{\partial x^2}, \tag{12.77}$$

where

$$\text{Consolidation coefficient} = \mathcal{D} = \frac{\varepsilon^2 \varepsilon_s G}{\beta_A} = \frac{k'}{a_v}, \tag{12.78}$$

with coefficient of volume compressibility a_v being

$$a_v = \frac{1}{G} = \frac{d\varepsilon_s}{d\sigma}, \tag{12.79}$$

and with k' an effective soil permeability. For an application of a load μ_i in the consolidometer shown in Fig. 12.6, the initial and boundary conditions are

$$\text{at } t = 0, \quad \sigma = \mu_i; \tag{12.80}$$

$$\text{at } x = 0, \quad \sigma = 0; \tag{12.81}$$

$$\text{at } x = L, \quad \sigma = 0. \tag{12.82}$$

The boundary conditions given by Eqs. (12.81) and (10.82) hold because it was assumed that the sample is drained at the top and at the bottom, resulting in zero fluid pressure and hence no intergranular stress. Standard solutions to the diffusion equation are available for both short and long times (Carslaw and Jaeger, 1959). As discussed in this chapter, the consolidation coefficient may be estimated from the knowledge of the friction coefficient or permeability for the medium and

$$G = \frac{dP}{d\varepsilon_s}, \tag{12.83}$$

where P is the osmotic pressure. Such an approach may aid in the interpretation and better understanding of consolidation.

12.6 Electrokinetic Phenomena: Zeta Potential

Particles suspended in a fluid normally carry a positive or more frequently a negative surface charge. In gases, particles frequently acquire a surface charge by triboelectrification, that is, frictional contact with non-conductive surfaces.

Elements in the periodic table can be grouped according to their electronegativities, which are related to their work functions. Hence, contact of two different chemical substances results in a contact-potential difference. Corona discharge, as in electrostatic precipitation, produces charged particles because of adsorption of electrons. In air, charged particles exert an electric field over distances much larger than their size because of the low electrical conductivity and dielectric constant of air. In fluids, such as water, having a much larger conductivity and dielectric constant, the electric field produced by the particle acts over distances much smaller than the size of the particle. In such situations, the concept of the zeta potential and double-layer thickness pioneered by Debye and Huckel (1923) is very useful.

In liquids, a surface charge on a particle is frequently produced by adsorption and partial ionic dissociation of a chemical substance on the surface of the particle, as shown in Fig. 12.7. In oils processed in the petroleum industry, asphal-tenes disassociate, adsorb on fine particles and produce colloidal particles that are difficult to separate. In water, the dissociation is often simpler, involving the equilibrium

$$H_2O \leftrightarrow OH^- + H^+, \tag{12.84}$$

with the hydrogen ion adsorbed on the surface as shown in Fig. 12.7. Hence, in aqueous solutions, the hydrogen ion concentration, as measured by the pH, often controls the charge of the suspended colloidal particles. The charge can be controlled by simply controlling the pH of the suspension.

In Fig. 12.7, although there is overall electroneutrality, the absorbed charged ions produce an electric field near the surface. The electric field is described by one of the Maxwell's equations of electromagnetism. With a zero magnetic field,

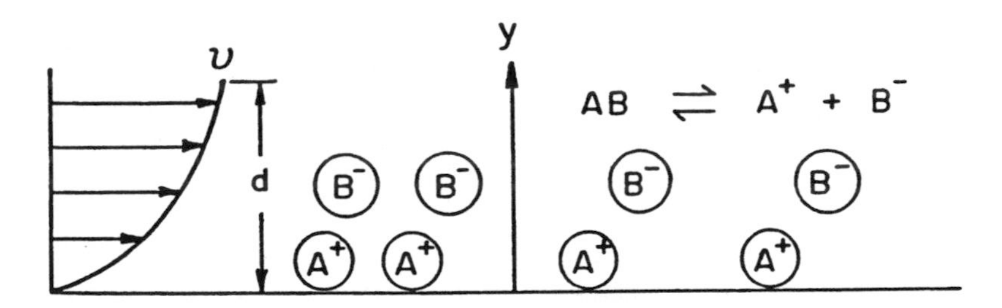

Fig. 12.7 Concept of zeta potential.

the relevant equation is that for the conservation of charge, q. In terms of the electric potential, and the electrical units used in Levich (1962) and in the Zeta-meter manuals (1968), the Poisson equation in one dimension is

$$\frac{d^2\phi}{dy^2} = -\frac{4\pi}{D}q, \tag{12.85}$$

where D is the dielectric constant.

With the fluid flowing past the surface in laminar flow, the steady state fully developed momentum balance is

$$\mu\frac{d^2v}{dy^2} = -qE, \tag{12.86}$$

where μ is the viscosity and E is the electric field strength. The electric force in Eq. (12.86) is clearly the product of the charge and the electric field strength as defined in electrostatics. Elimination of charge between Eqs. (12.85) and (12.86) gives the equation

$$\frac{d^2v}{dy^2} = \left(\frac{DE}{4\pi\mu}\right)\frac{d^2\phi_E}{dy^2}. \tag{12.87}$$

Integration of Eq. (12.87) gives the linear function

$$v = \left(\frac{DE}{4\pi\mu}\right)\phi_E + C_1y + C_2, \tag{12.88}$$

where C_1 and C_2 are the integration constants.

Since far from the wall the velocity is some constant, C_1 is set to zero. If ϕ_s is the potential when v becomes zero, at or near the wall, then

$$C_2 = -\left(\frac{DE}{4\pi\mu}\right)\phi_s. \quad (12.89)$$

Far from the wall, the potential ϕ_E has the constant value ϕ_o equal to the potential of the bulk of the solution and the wall. Here the velocity is taken to be v_o. Therefore,

$$v_o = \left(\frac{DE}{4\pi\mu}\right)(\phi_o - \phi_s). \quad (12.90)$$

The difference in the potential across the mobile portion of the double layer is called ζ-potential:

$$\zeta = \phi_o - \phi_s. \quad (12.91)$$

In colloidal chemistry, the zeta-potential is usually determined by measuring electrophoretic mobility, *EM* (*Zetameter Manual*, 1968):

$$EM = \frac{v_o}{E}. \quad (12.92)$$

This measurement is made by determining, under a microscope, the time it takes a colloidal particle to travel a fixed distance under the application of a constant electric field of strength E, where E is simply the voltage divided by the distance between two electrodes.

Sophisticated instruments make such a measurement automatically. Hence the relation between the electrophoretic mobility and the zeta potential is

$$EM = \frac{D\zeta}{4\pi\mu}. \quad (12.93)$$

In the literature Eq. (12.93) is known as the Helmholtz–Smoluchowski formula.

Colloidal stability (Davies and Riedeal, 1961) is produced when small suspended particles have a sufficiently high zeta potential such that they repel each other during inevitable collisions. The collisions are produced by Brownian motion in stationary fluids caused by molecular non-uniform bombardment of the suspended particles. Shear caused by bulk motion also produces similar random fluctuations and collisions of particles. In stationary fluids colloidal stability is achieved when (Davies and Riedeal, 1961)

$$\begin{gathered}\text{Work of repulsions} >> \text{Thermal energy} \\ d_p D \zeta^2 >> k_B T.\end{gathered} \tag{12.94}$$

In Eq. (12.94) the work of repulsion was roughly estimated by approximately integrating the charge of the particle, as obtained from Eq. (12.85), times the product of the electric field over the particle diameter. The thermal energy is the product of the Boltzmann constant, k_B, times the temperature.

Equation (12.94) shows that for a sufficiently low zeta potential or a low dielectric constant, a colloidal suspension will lose its stability. This means that the particles will not repel each other upon collision. Instead, the attractive van der Waals forces will cause them to stay together. They may form flocs and settle as particle agglomerates rather than as individual particles.

Since the effective size of such a floc is many times larger than the individual particle, settling will be quite rapid. Figure 12.8 shows the settling rates in the constant settling regime of 5 μm illite particles suspended in toluene with various concentration of Athabosca asphaltene obtained in the sedimentation column depicted in Fig. 12.4. The zeta potential of the particles was varied by using various concentrations of asphaltene. At a 10% asphaltene concentration, the zeta potential was −22.2 mV. The same effect can be produced by using parts per million concentration of commercial surfactants. Figure 12.8 shows that below a zeta potential of about 8 mV, the settling rate increased drastically. Microscopic examination of the sediment withdrawn from the settling column showed that the particles had settled as large flocs. Dilution of the suspension to achieve such a low zeta potential of particles is not economically feasible, nor is charge neutralization because of the large quantities of the positively charged asphaltenes, hence electrophoresis was tried, as depicted in Fig. 12.8 (Shih *et al.*, 1986).

12.7 Effect of Zeta Potential on Sedimentation

Non-equilibrium thermodynamics (de Groot *et al.*, 1952) can be used to obtain qualitative as well as quantitative effects of the zeta potential on sedimentation. In non-equilibrium thermodynamics, fluxes are expressed as linear functions of forces. Charged particles are subjected to two forces: gravity and electric field. Flow of charged particles produces a current. In terms of mobilities b_{ij}, which can be related to the Onsager friction coefficients, we can write

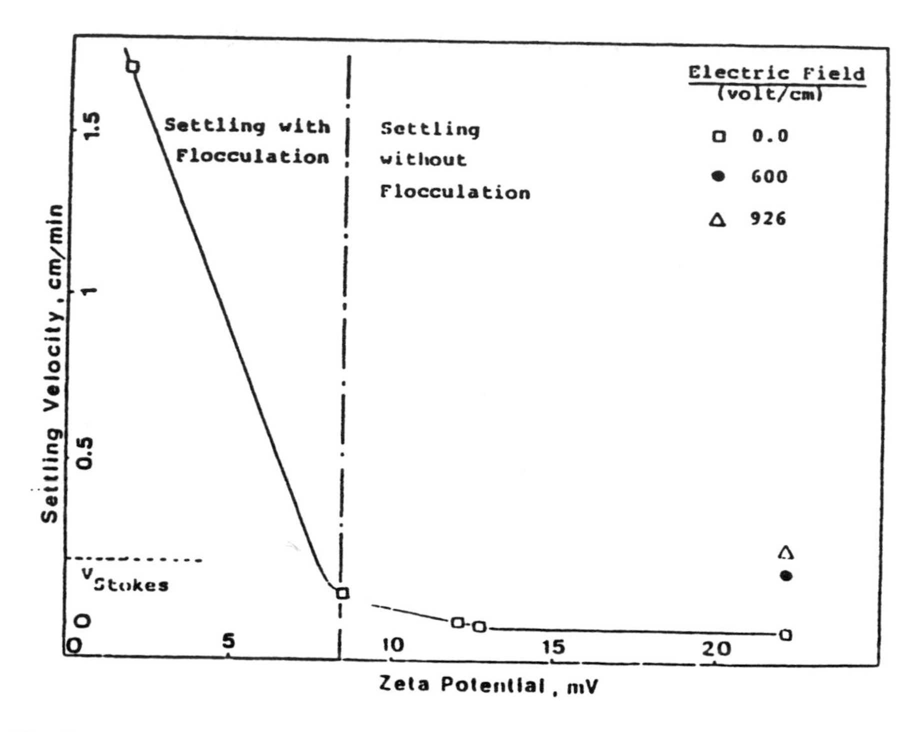

Fig. 12.8 Variation of settling velocity with zeta potential and enhancement of settling velocity by external electric field (from Shih *et al.* 1986).

$$\text{Current} = I = b_{11}E + b_{12}g, \tag{12.95}$$

$$\text{Mass flux} = J_m = b_{21}E + b_{22}g. \tag{12.96}$$

For a dilute suspension, b_{22} can be obtained from Stokes' law, Eq. (12.11) as

$$b_{22} = \frac{\rho d_p^2 \left(\rho_s - \rho_f\right)}{18\mu}. \tag{12.97}$$

The effect of concentration of particles can then be included as in Eq. (12.33) since b_{22} is essentially the permeability. Clearly these equations apply for flow through porous media where Δp is replaced by the pressure drop using the simplified momentum balance. The coefficient b_{11} is the electrical conductivity. We can relate b_{12} or b_{21} to the zeta-potential, ζ using the well-known relation between the electrophoretic mobility and the zeta potential Eq. (12.93). This gives, using Onsager's relation,

$$b_{12} = b_{21} = \frac{\rho D\zeta}{4\pi\mu}. \tag{12.98}$$

For zero current flow, Eq. (12.95) shows that we can write the ratio of settling velocity with electric field, v to the settling velocity with zero zeta potential, v_o, as

$$\frac{v}{v_o} = 1 - \frac{b_{12}^2}{b_{12}b_{22}}, \tag{12.99}$$

where

$$b_{12} \le \sqrt{(b_{11}b_{22})}, \tag{12.100}$$

according to the Clausius–Duhem inequality. The use of Stokes' law for b_{22} and Eq. (12.98) gives

$$\frac{v}{v_o} - 1 - \frac{18\rho}{k\mu(\rho_s - \rho_f)}\left(\frac{D\zeta}{\pi d_p}\right)^2. \tag{12.101}$$

This equation differs from that given by Davies and Riedeal (1961) by the presence of the numerical value of 18 and the ratio of densities. Its derivation from thermodynamics also permits a ready extension to include concentration effects. Eq. (12.101) shows that to speed up settling we can

(a) decrease the dielectric constant, as is done in adding a promoter liquid of a high characterization factor (Gorin *et al.*, 1977);
(b) decrease the zeta potential by use of some surfactant;
(c) increase the electrical conductivity. It was shown that the retardation of sedimentation by surface charges is small for sufficiently large particle sizes. The dimensionless group in Eq. (12.101) says what is a large particle size.

Equation (12.96) also shows that the mass flux or the settling rate can be substantially increased by applying an external field E (see Fig. 12.8). This technique will be effective whenever the electrical conductivity, b_{11}, is small.

Exercises

1. *Hydrodynamics of a Thickener*

Consider steady state operation of a thickener shown in Fig. 12.9. The volumetric feed rate is F m^2 / hr. The inlet volume fraction of the feed is ε_{SF}. The underflow sludge is withdrawn at a volumetric rate S m^3 / hr with a volume fraction of solids of ε_{SU}. As a first approximation it can be assumed that in the critical zone, the particles do not contact each other, while in the compression zone the relative velocity between the liquid and the solid is zero. Using such an assumption, estimate the concentrations in the two zones. The step-by-step procedure is

(a) Find the relation between the volume fractions and $\dot{F}$ and $\dot{S}$.

(b) Apply the buoyancy equals drag balance discussed in the drift flux chapter. This analysis assumes the settling velocity as a function of concentration is known for a given sludge. Explain why in practice it is necessary to obtain such a relation experimentally.

(c) Sketch a drift flux versus volume fraction of fluid curve and the operating line involving $\dot{F}$ and $\dot{S}$. Show the location of the critical zone concentration.

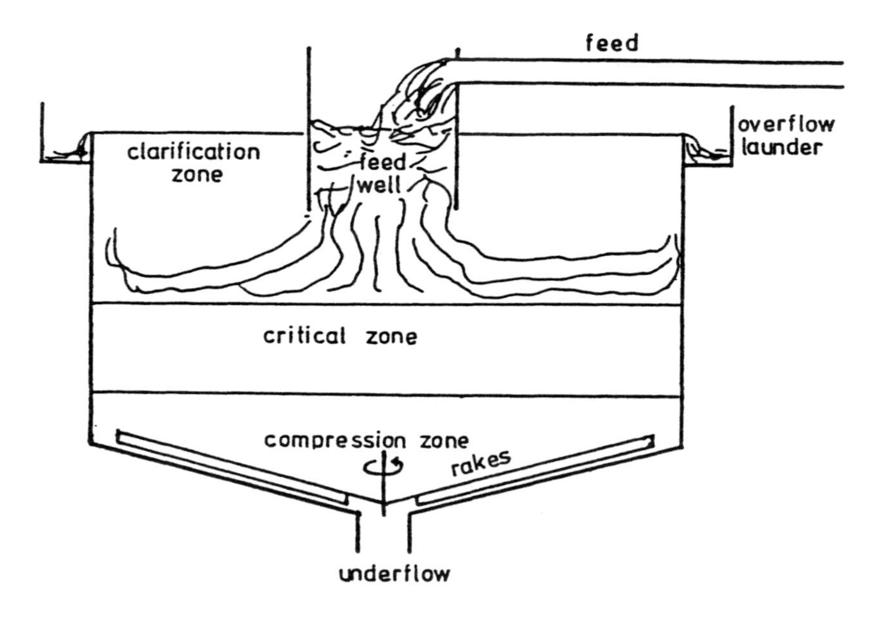

Fig. 12.9 The zones in a thickener.

(d) Note that in the compression zone Eq. (12.62) becomes

$$\mathcal{B}\frac{d\varepsilon_s}{dx} = \text{Constant}. \qquad (12.102)$$

For a constant consolidation coefficient, find an approximate relation between the sludge concentration and the thickness of the sludge zone.

2. *Electrokinetic Phenomena in Porous Media or in Membranes*

Salt water flows through a section of a porous medium or through a membrane. A manometer records a pressure drop ΔP and an electrometer shows a voltage difference $\Delta\psi$. The equations for flow of mass and current, analogous to Eqs. (12.95) and (12.96), are

$$I = b_{11}\Delta\psi + b_{12}\Delta P, \qquad (12.103)$$

$$J = b_{21}\Delta\psi + b_{22}\Delta P. \qquad (12.104)$$

Darcy's law is obtained when b_{21} is zero in the second equation. Ohm's law follows, for b_{12} is zero in Eq. (12.103) and I becomes the streaming current when a short-circuit is created by setting $\Delta\psi$ to zero. Electro-osmosis is obtained in Eq. (12.104) for zero pressure drop.

(a) Find an expression for b_{22}.
(b) Show that for zero current

$$J = b_{22}\left(1 - \frac{b_{12}^2}{b_{11}b_{22}}\right)\Delta P \qquad (12.105)$$

when $b_{12} = b_{21}$.
(c) What is the effect of electrokinetics on the permeability of the medium (McGregor, 1974)?

3. *Lamella Electrosettler*

Lamella gravity settlers are widely used to separate solid particles from slurries as described by Mace and Laks (1978). Their kinematics was analyzed by Acrivos and Herbholzheimer (1979). Their practical operation was given by Probstein *et al.* (1981). They are, however, not useful for separating colloidal particles, since the inherent settling velocity of such particles is essentially zero. In non-aqueous liquid suspensions, this settling velocity can be made large by imposing a large electrostatic field across the suspension, in effect, substituting electrophoresis for gravity settling. A lamella settler based on this principle was invented and built at Illinois Institute of Technology. Jayaswal *et al.* (1990) demonstrated that it could continuously separate a slurry of colloidal particles into a clear liquid and into a thick suspension.

Figures 12.10 to 12.12 show the apparatus, the model equations and a comparison of particle concentrations measured with an x-ray densitometer to the computed steady values obtained from the model.

(a) Explain the equations in Figure 12.11. Why is the particle viscosity much higher than that of the liquid? Suggest a better model.
(b) Show that during transient operation the dimensionless height of the interface can be given by

$$\overline{H}(t) = -\frac{b}{H_o}\sin(\alpha) + \left(1 + \frac{b}{H_o}\sin(\alpha)\right)\exp\left(-\frac{E_m E t}{b}\right), \tag{12.106}$$

Fig. 12.10 Lamella electrosettler at IIT (Jayaswal *et al.*, 1990). Overflow exit, top right; inlet feed, top left; thickened slurry underflow, bottom left; electrodes, top and bottom plates connected to a high voltage supply.

1. Continuity equation for phase k (= f, s)

$$\frac{\partial}{\partial t}(\varepsilon_k \rho_k) + \nabla \cdot (\varepsilon_k \rho_k \mathbf{v}_k) = 0 \tag{T.1}$$

2. Momentum equation for phase k (= f, s)

$$\frac{\partial}{\partial t}(\varepsilon_k \rho_k \mathbf{v}_k) + \nabla \cdot (\varepsilon_k \rho_k \mathbf{v}_k \mathbf{v}_k) = \varepsilon_k \rho_k \mathbf{g} + q_k \mathbf{E} + \nabla \cdot [\tau_k] + \beta_{kf}(\mathbf{v}_f - \mathbf{v}_k) \tag{T.2}$$

3. Constitutive equation for stress

$$[\tau_k] = [-P_k][\mathbf{I}] + 2\varepsilon_k \mu_k [\mathbf{S}_k] \tag{T.3a}$$

$$[\mathbf{S}_k] = \frac{1}{2}\left[\nabla \mathbf{v}_k + (\nabla \mathbf{v}_k)^T\right] - \frac{1}{3}\nabla \cdot \mathbf{v}_k [\mathbf{I}] \tag{T.3b}$$

3A. Empirical solids viscosity and stress model

$$\nabla p_s = G(\varepsilon_s)\nabla \varepsilon_s \tag{T.3c}$$

$$G(\varepsilon_s) = 10^{8.76\varepsilon_s^{-0.27}} \text{ dynes/cm}^2 \tag{T.3d}$$

$$\mu_s = 69.3 \text{ centipoise (example)} \tag{T3.e}$$

4. Fluid–solid drag coefficients for $\varepsilon_f \leq 0.8$ (based on empirical correlation)

$$\beta_{sf} = \left(150\frac{(1-\varepsilon_f)\varepsilon_s \mu_f}{(\varepsilon_f d_p \phi_s)^2} + 1.75\frac{\rho_f \varepsilon_s |\mathbf{v}_f - \mathbf{v}_s|}{\varepsilon_f d_p \phi_s}\right)\frac{\rho_s}{(\rho_s - \rho_f)} \tag{T.4a}$$

for $\varepsilon_f > 0.8$ (based on empirical correlation)

$$\beta_{sf} = \frac{3}{4} C_d \frac{\varepsilon_s \rho_f \rho_s |\mathbf{v}_f - \mathbf{v}_s|}{d_p \phi_s (\rho_s - \rho_f)} \varepsilon_f^{-2.65} \tag{T.4b}$$

where

$$C_d = \frac{24}{Re_s}\left[1 + 0.15(Re_s)^{0.687}\right], \text{ for } Re_s < 1{,}000, \tag{T.4c}$$

$$C_d = 0.44, \text{ for } Re_s \geq 1{,}000, \tag{T.4d}$$

$$Re_s = \frac{\varepsilon_f \rho_f |\mathbf{v}_f - \mathbf{v}_s| d_p \phi_s}{\mu_f} \tag{T.4e}$$

5. Surface charge of particles

$$q_v = \frac{9}{2} \cdot \frac{\mu_f \cdot E_M}{(d_p/2)^2} \varepsilon_s \tag{T.5a}$$

Electrophoretic mobility

$$E_M = 1.02 \times 10^{-4} \cdot E + 6.04 \times 10^{-4} \, (\mu\text{m/s})/(\text{V/cm}) \tag{T.5b}$$

Fig. 12.11 Equations used in modeling electrosettling of colloidal suspensions of alumina particles in kerosine stabilized with a surfactant.

I.C.1	at $t=0$;	$u_s = u_f = 0,\quad v_s = v_f - 0$	
		$\varepsilon_s = \varepsilon_{s,o}$	
B.C.1	at $y=0$;	$v_s = v_f = V_{\text{feed}}$	
		$\varepsilon_s = \varepsilon_{\text{feed}}$	$0 < x < x_1$
		$v_s = v_f = 0$	$x_1 < x < x_2$
		$v_s = v_f = -V_{uf}$	
		$\varepsilon_s = \varepsilon_{uf}$	$x_2 < x < b$
B.C.2	at $y=L$;	$v_s = v_f = 0$	for all x
B.C.3	at $x=0$;	$u_s = u_f = 0$	$0 < y < y_1$
		$u_s = u_f = -U_{of}$	$y_1 < y < L$
B.C.4	at $x=b$;	$u_s = u_f = 0$	for all y

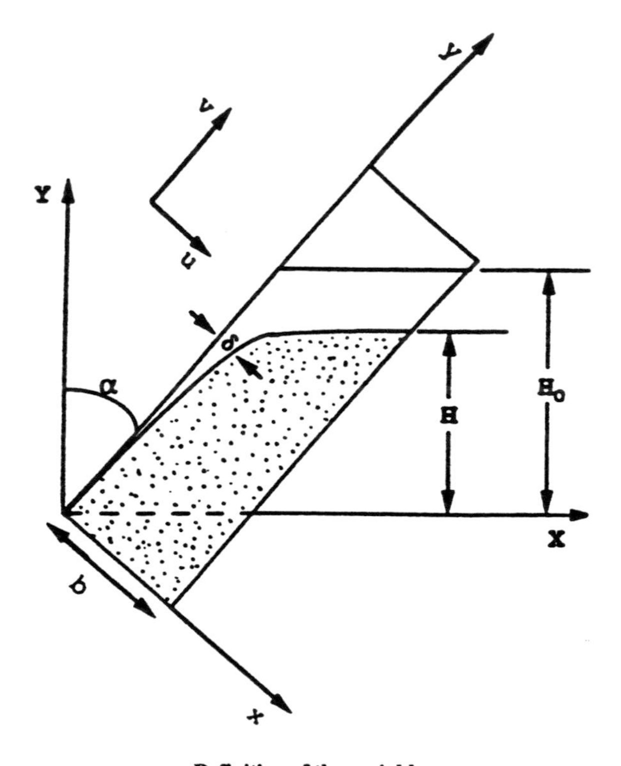

Definition of the variables

Fig. 12.11 (continued)

Where H_o is the initial height, E_m is the electrophoretic mobility (particle velocity divided by the electric field E), and other symbols are defined in Fig. 12.11, and where

$$\overline{H} = \frac{H}{H_o}, \quad H = H_o \text{ at } t = 0. \tag{12.107}$$

Hint: If we assume the motion of the interface is governed by the motion of the particles at the interface, the volume of clarified liquid produced per unit of time is

$$\frac{d}{dt} \int_{V(t)} dV = \int_{S(t)} \mathbf{u}_s \cdot \mathbf{n} \, dS, \tag{12.108}$$

where $\mathbf{u}_s$ is the slip velocity and is given by either U_o or $E_m \cdot E$. In terms of the variables defined in Fig. 12.11, the volume balance given by Eq. (12.108) can be written as

$$-\frac{b}{\cos(\alpha)} \frac{dH(t)}{dt} + \int_{S(t)} \mathbf{u}_s \cdot \mathbf{n} \, dS = 0. \tag{12.109}$$

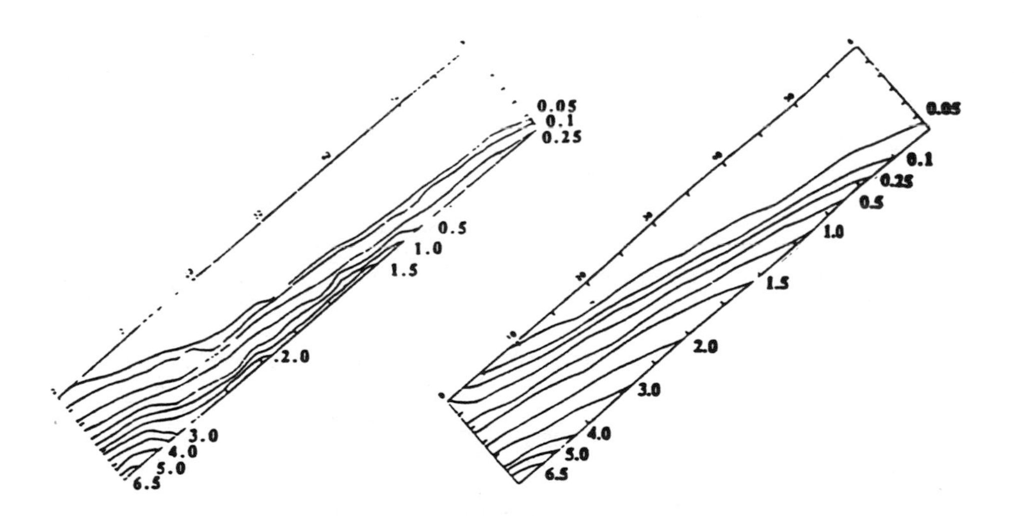

Fig. 12.12 Experimental (left) and computed (right) alumina particle concentrations expressed in volume percent in the lamella electrosettler at a steady state with a feed rate of 112 cm^3/min, at an electric field strength of 1000 volts/cm (3 μm diameter alumina particles, with a zeta potential of 50 mV, a liquid viscosity of kerosene of 1.49 centipoises and a solids viscosity of 69 centipoises. $\alpha = 50°$. (Jayaswal *et al.*, 1990.)

Noting that $\mathbf{u}_s$ is identically zero at the walls, Eq. (12.109) integrated over the top flat and thin electroboundary-layer interfaces gives Eq. (12.106) when gravitational settling is neglected.
(c) In the lamella electrosettler the particles, as shown in Fig. 12.12, continuously move down to the right electrode, forming the sludge at the bottom. What is the cause of this motion? Suggest a more precise way to model this sliding motion assuming that the particles are spheres that lost their charge upon contact with the electrode.

Literature Cited

Acrivos, A., and E. Herbholzheimer (1979). "Enhanced Sedimentation in Settling Tanks with Inclined Walls," *J. Fluid Mech.* **92**, Part 3, 435–437.

Auzerais, F. M., R. Jackson, and W. B. Russel (1988). "The Resolution of Shocks and the Effect of Compressible Sediments in Transient Settling," *J. Fluid Mech.* **195**, 437–462.

Batchelor, G. K. (1988). "A New Theory of the Instability of a Uniform Fluidized Bed," *J. Fluid Mech.* **193**, 75–100.

Bird, R. B., W. E. Stewart, and E. N. Lightfoot (1960). *Transport Phenomena.* New York: John Wiley & Sons.

Buscall, R. (1990). "The Sedimentation of Concentrated Colloidal Suspensions," *Colloids and Surfaces* **43**, 33–53.

Carslaw, H. S., and J. C. Jaeger (1959). *Conduction of Heat in Solids*, 2nd ed. London, U.K.: Oxford University Press.

Davies, J. T., and E. K. Riedeal (1961). *Interfacial Phenomena.* New York: Academic Press.

Debye, P., and E. Hueckel (1923). "On the Theory of Electrolytes," *Physikalische Zeitschrift* **24** (9), 185–206.

de Groot, S. R., P. Mazur, and J. Th. G. Overbeek (1952). "Non-equilibrium Thermodynamics of the Sedimentation Potential and Electrophoresis," *J. Chem. Physics* **20**, 1825–1829.

Font, R. (1990). "Calculation of the Compression Zone Height in Continuous Thickeners," *AIChE J.* **36**, 3–12.

Gorin, E., C. J. Kulik, and H. E. Lebowitz (1977). "Deashing of Coal Liquifaction Products via Partial De-asphalting," *Ind. Eng. Chem.: Process Des. Dev.* **16**, 95–105.

Jayaswal, U. K., D. Gidaspow, and D. T. Wasan (1990). "Continuous Separation of Fine Particles in Nonaqueous Media in a Lamella Electrosettler," *Separations Technology* **1**, 3–17.

Kynch, G. J. (1952). "Theory of Sedimentation," *Trans. Faraday Soc.* **48**, 166–181.

Levich, V. G. (1962). *Physiochemical Hydrodynamics*. Englewood Cliffs, N.J.: Prentice Hall, Inc.

Mace, G. R., and R. Laks (1978). "Developments in Gravity Sedimentation," *CEP* **74** (7), 77–83.

McGregor, R. (1974). *Diffusion and Sorption in Fibers and Films*. New York: Academic Press.

Probstein, R. F., D. Yung, and E. D. Hicks (1981). "A Model for Lamella Settlers," pp. 53–92 in M. P. Freeman and J. A. Fitzpatrick, Eds. *Physical Separations*. New York: Engineering Foundation.

Scheidegger, A. E. (1960). *The Physics of Flow through Porous Media*. Toronto, Canada: University of Toronto Press.

Schiffman, R. L., V. Pane, and V. Sunara (1985). "Sedimentation and Consolidation," pp. 57–121 in B. M. Moudgil and P. Somasundaran, eds. *Flocculation, Sedimentation and Consolidation*, Proceedings of the Engineering Foundation Conference at Sea Island, Georgia. New York: Engineering Foundation.

Shih, Y. T., D. Gidaspow, and D. T. Wasan (1986). "Sedimentation of Fine Particles in Nonaqueous Media, Part I–Experimental; Part II–Modeling," *Colloids and Surfaces* **21**, 393–429.

Shih, Y. T., D. Gidaspow, and D. T. Wasan (1987). "Hydrodynamics of Sedimentation of Multisized Particles," *Powder Technology* **50**, 201–215.

Taylor, D. W. (1948). *Fundamentals of Soil Mechanics*. New York: John Wiley & Sons.

Terzaghi, K. (1943). *Theoretical Soil Mechanics*. New York: John Wiley & Sons.

Tiller, F. M. (1981). "Revision of Kynch Theory," *AIChE J.* **27**, 823–829.

Wissa, E. Z., J. T. Christian, E. H. David, and S. Heiberg (1971). "Consolidation at Constant Rate of Strain," *J. Soil Mechanics and Foundation Div. ASCE* **SM10**, 1393–1411.

Zeta-Meter Manual (1968). New York: Zeta-Meter Inc.

APPENDICES

FORMULATION OF CONTINUUM PROBLEMS: INTRODUCTION

The objective of the following presentation is to illustrate the basic mathematical concepts of formulation of continuum problems for a reader whose knowledge of mathematics does not extend beyond the normal one year of calculus and an introduction to differential equations course normally required for a bachelor's degree in engineering. Such a person is a typical first-year graduate student in chemical engineering encountered by the author in the last 30 years of teaching.

Only one key idea is used in the formulation of continuum problems, that of conservation of some countable quantity, such as mass, chemical component, momentum, energy or entropy. For a system of constant mass or constant species, the rate of increase of such a countable quantity equals its rate of production. Such a logical construction is applied to continuous physical problems or to discrete socio-economic problems, as in the Club of Rome study of population pollution or in growth of money in a bank, illustrated by the compound interest formula or by the oscillation of national income in the classical Samuelson equation (Goldberg, 1958). Some physicists (Spielberg and Anderson, 1987) consider the conservation principles as one of the seven ideas along with relativity, quantum mechanics, and the second law of thermodynamics, that shook the universe of physics. While physicists are primarily concerned with the basic building blocks of matter, it is the engineers who have perfected the skills of making proper balances.

There are two equivalent ways to derive conservation equations in space and time. We can choose our system to be a constant volume into which some quantity, such as mass, momentum or energy, is flowing in and out. This approach is called the Eulerian view or the shell balance approach in the chemical engineering literature. In the second approach we consider the system to be a constant mass or a constant mass of species i moving with its own velocity. Since the mass or species is conserved, there is no net outflow across the boundary. However, there may exist internal production, such as formation of species i from species j. This approach is called the Lagrangian view. Both points of view produce identical equations. The Lagrangian approach is more useful for making momentum balances, since Newton's second law of motion, the rate of change of momentum equals the forces acting on the system, applies to a constant mass system. The Eulerian approach is often simpler to apply in non-rectangular geometries for energy and species balances. Hence, the two view points must be understood.

In analyzing complex particle motion problems, physicists often use Hamiltonian mechanics in the place of the Newton's second law. The equivalent formulation, which involves minimizing an integral using the methods of calculus of variation, may one day find an application in fluidization and multiphase flow, but has not been developed to a sufficient extent to merit presentation in the text. Gidaspow (1978) has derived a relative velocity equation by minimizing the rate of entropy production. Li *et al.* (1988, 1992) have used an energy minimization principle to explain some fluidization characteristics, such as cluster formation.

APPENDIX A

OVERALL (MACROSCOPIC) BALANCES

In engineering thermodynamics we choose as our system a unit—for example, a fuel cell or a compressor. We make balances on this unit. Everything outside the unit is called the surroundings. Since detailed processes are not considered, we call such balances overall or macroscopic balances. In such balances the only independent variable is time, t.

A general conservation scheme for balances of mass, species, momentum, energy, entropy, money, etc., is

A.1 Conservation Scheme

$$\text{Accumulation} = \text{Inflow} - \text{Outflow} + \text{Creation} \qquad \text{(A.1)}$$

$$\text{(Crossing the boundary of the system)} + \text{(Inside the system).}$$

Thus, accumulation of some quantity such as mass is said to occur because of either excess or net rate of inflow and because of possible creation of that quantity inside the system (see Fig. A.1).

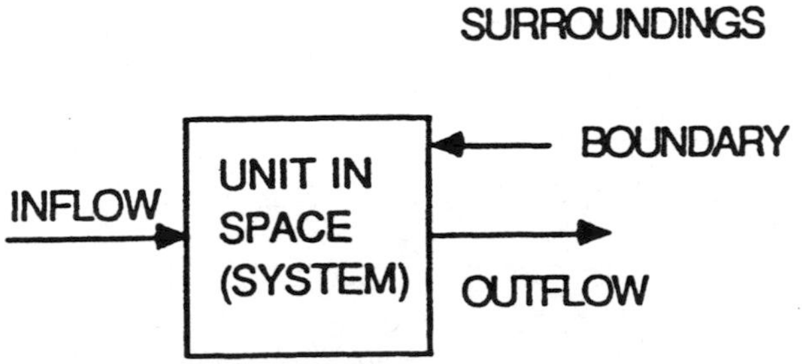

Fig. A.1 System for overall balances.

A.2 Conservation of Mass

Let m = mass, kg;

W_{in} = mass flow rate into the unit above, kg / s;

W_{out} = mass flow rate out of the unit above, kg / s;

R = rate of creation of mass in the unit, kg / s.

Then in time increment $t_f - t$, the conservation scheme gives the equation

$$m(t_f) - m(t) = W_{\text{in}}(t')\cdot(t_f - t) - W_{\text{out}}(t'')\cdot(t_f - t) + R(t''')\cdot(t_f - t), \qquad \text{(A.2)}$$
$$t < t' < t_f \qquad t < t'' < t_f \qquad t < t''' < t_f$$

where () following m, W_{in}, W_{out} means substitution.

We assume the functions W_{in}, W_{out} and R are continuous and differentiable. We divide both side of Eq. (A.2) by $t_f - t$ and take the limit as $t_f \to t$:

$$\lim_{t_f \to t} \frac{m(t_f) - m(t)}{t_f - t} = \lim_{t_f \to t}\left[W_{\text{in}}(t') - W_{\text{out}}(t'') + R(t''')\right]. \qquad \text{(A.3)}$$

As $t_f \to t, t', t''$ and $t''' \to t$. The left side of Eq. (A.3) is the derivative of m with respect to time, by definition:

$$\frac{dm}{dt} = W_{\text{in}} - W_{\text{out}} + R. \qquad \text{(A.4)}$$

Mass is *conserved*, that is, for mass, $R = 0$. Thus,

$$\frac{dm}{dt} = W_{\text{in}} - W_{\text{out}}. \qquad \text{(A.5)}$$

We have a *steady state* when $(dm)/(dt) = 0$. Then $W_{\text{in}} = W_{\text{out}}$.

A.3 Conservation of Species

For each chemical species we obtain a conservation of species equation in the same way as that for mass. For species i,

$$\frac{dm_i}{dt} = W_{\text{in},i} - W_{\text{out},i} + R_i. \tag{A.6}$$

If i were to refer to an element rather than to species such as H_2O, a conservation of element principle would apply to a chemical reaction since elements are not created or destroyed in a chemical reaction. Otherwise, R_i is not zero for a species.

A.3.1 Fuel Cell: Example of a Species Balance

The reactions taking place in a fuel cell with an acid electrolyte are indicated in Fig. A.2.

The rate of creation of H_2O or the rate of depletion (reaction) of H_2 in Eq. (A.6) can be expressed in terms of current I using Faraday's law. Faraday's law states that in the electrochemical reaction, F coulombs of charge are produced per an equivalent of species. F is approximately 96,500. The current I is the rate of change of charge per unit time. By definition, 1 ampere = 1 coulomb/s.

Suppose m_A moles of gas A (in this case A is H_2) react in the fuel cell per unit time. Then Faraday's law allows one to equate

$$m_A nF = I, \tag{A.7}$$

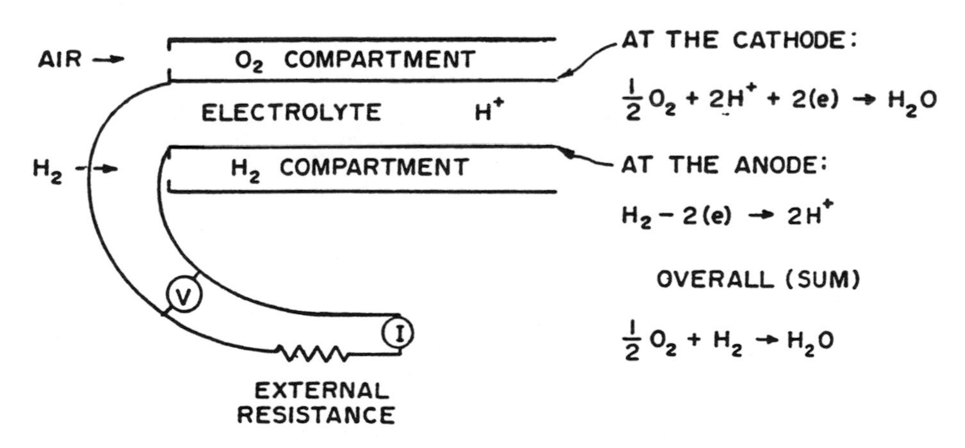

I = current, measured with an ammeter, as indicated.
V = voltage, measured with a voltmeter, as indicated.

Fig. A.2 Hydrogen-air fuel cell.

where n is the number of equivalents per mole. To check Eq. (A.7) look at the units:

$$m_A \quad \cdot \quad n \quad \cdot \quad F \quad = \quad I$$

$$\frac{\text{moles } H_2}{\text{s}} \quad \frac{\text{equiv } H_2}{\text{moles } H_2} \quad \frac{\text{coulombs}}{\text{equiv}} = \frac{\text{coulombs}}{\text{s}}$$

In Eq. (A.6) the rate of creation is expressed in mass units. To convert to molar units in Eq. (A.7) we use the molecular weight of the reacting species, H_2 in this case. Let M be the molecular weight of H_2. Then, since hydrogen is being reacted (depleted) rather than being produced, we introduce a negative sign and obtain

$$R_i = -m_A \cdot M$$

$$\frac{g}{\text{s}} = \frac{\text{moles}}{\text{s}} \cdot \frac{\text{g}}{\text{mole}}. \tag{A.8}$$

With this introduction, let us make a *hydrogen balance* on the hydrogen compartment of a fuel cell shown in Fig. A.3.

The inlet gas consists of a mixture of hydrogen and carbon dioxide. Such gases can be produced by reaction of natural gas with steam. The source of hydrogen may also be coal or biomass, say, garbage or peanut shells. The hydrogen balance becomes

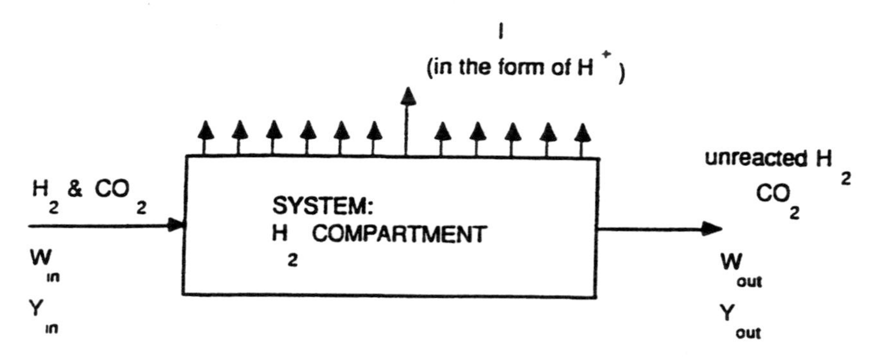

Y = weight fraction of H_2 and the subscripts in and out denote inlet and outlet conditions, respectively.

Fig. A.3 System for an overall hydrogen balance.

$$W_{\text{in}}Y_{\text{in}} - W_{\text{out}}Y_{\text{out}} = \frac{IM}{nF},$$
$$\frac{\text{g}}{\text{sec}} = \frac{\text{coulombs/s} \times \frac{\text{g}}{\text{mole}}}{\frac{\text{equiv}}{\text{mole}} \times \frac{\text{coulombs}}{\text{equiv}}} \tag{A.9}$$

Note that Eq. (A.9) is a special case of the more general balance Eq. (A.6). As long as all hydrogen is consumed by the electrochemical reactions shown in Fig. A.2, it is exact. Any imbalance would indicate either leaks, hydrogen burning or production by some other means, accumulation or instrument errors. Hence, the importance of the balance made.

A.4 Conservation of Energy

Let us choose a *constant mass* system as shown in Fig. A.4. Its total energy is E. The energy of the surroundings is E'.

Since energy is *additive*, we have

$$\begin{aligned} &\text{at time } t: \quad E_{\text{total}}(t) = E(t) + E'(t), \\ &\text{at time } t_o: \quad E_{\text{total}}(t_o) = E(t_o) + E'(t_o), \\ &\text{subtracting: } \Delta E_{\text{total}} = \Delta E + \Delta E', \end{aligned} \tag{A.10}$$

where Δ means difference, as shown earlier.

Since energy is defined to be a function that is always conserved,

$$\Delta E_{\text{total}} = 0.$$

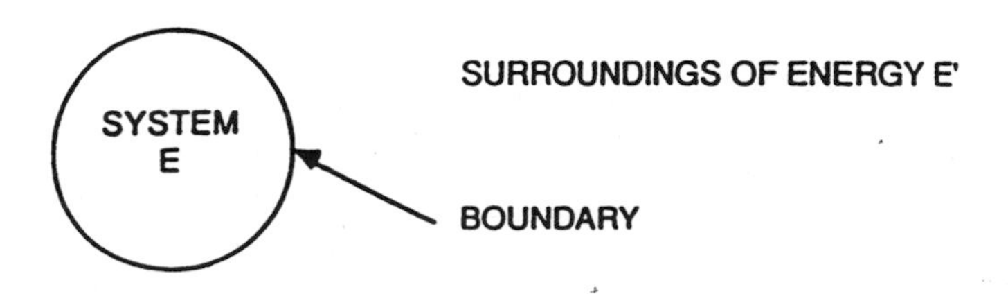

Fig. A.4 System for an overall energy balance.

Divide both sides of Eq. (A.10) by Δt and take the limit as $\Delta t \to 0$. Using the definition of a derivative we obtain

$$\frac{dE}{dt} = -\frac{dE'}{dt}. \tag{A.11}$$

Equation (A.11) says that the energy of our constant mass system, called a "closed" system, changes because of a corresponding change in the energy of the surroundings. But surroundings can affect the system only through flow of energy through its boundary. We say that energy can flow across the boundary in two forms, as work, W, or as heat, Q. Therefore, we let

$$-\frac{dE'}{dt} = \frac{dQ}{dt} - \frac{dW}{dt}, \tag{A.12}$$

or using the conservation of energy principle, Eq. (A.11),

$$\frac{dE}{dt} = \frac{dQ}{dt} - \frac{dW}{dt}. \tag{A.13}$$

The convention often used for heat and work is

	Heat	Work
into the system:	+	−
out of the system:	−	+

Thus, the net rate of heat *outflow* is $-\frac{dQ}{dt}$. Work done by the system results in a decrease of energy of the system for zero net rate of heat outflow.

Work is defined in mechanics. Equation (A.13) says that whatever is not work will be heat.

A.4.1 Constant Pressure Closed System

Make the following simplifications. Let

(1) $E = U =$ internal energy,
(2) $\frac{dW}{dt} = P\frac{dV}{dt}$, where $P =$ pressure $= \frac{\text{force}}{\text{area}}$,

where V = volume of the system. This formula is obtained from the formula in mechanics:

$$W = \text{work} = \int_{x_0}^{x_1} (\text{force})\, dx, \quad \text{where } x \text{ is distance.} \tag{A.14}$$

$$\text{Let area} = A, \text{ then } W = \int_{x_0}^{x_1} PA\, dx = \int_{V_0}^{V_1} P\, dV. \tag{A.15}$$

In terms of the rate of change of volume with respect to time,

$$dW = \int_{t_0}^{t_1} P \frac{dV}{dt} dt = \int_{V_0}^{V_1} P\, dV,$$

using the formula from calculus for changing the limits of integration. With these simplifications Eq. (A.13) gives

$$\frac{dU}{dt} + P\frac{dV}{dt} = \frac{dQ}{dt}.$$

Let the enthalpy = $H = U + PV$. In thermodynamics (Callen, 1960) it is shown that the enthalpy is the Legendre transform of energy with respect to volume and is the natural energy to use for constant pressure processes. Thus, for constant P we have the equation

$$\frac{dH}{dt} = \frac{dQ}{dt}. \tag{A.16}$$

Let m = mass of the closed system, and h = specific enthalpy, J/kg = (H/m). For a single component substance we know that $h = h(P,T)$, where T is absolute temperature. For a constant pressure process, $h = h(T)$ only. Therefore,

$$\frac{dH}{dt} = m\frac{dh}{dt} = m\left(\frac{\partial h}{\partial T}\right)_P \frac{dT}{dt}, \tag{A.17}$$

by chain rule for composite function.

We define the specific heat at constant pressure as

$$C_P = \left(\frac{\partial h}{\partial T}\right)_P. \tag{A.18}$$

Therefore, at a constant pressure the energy balance, Eq. (A.16), becomes

$$mC_P \frac{dT}{dt} \quad - \quad \frac{dQ}{dt} \quad = 0. \tag{A.19}$$

Accumulation of energy (enthalpy) + Net rate of outflow of heat = 0

A.5 Heat and Work Produced by a Fuel Cell

A simple, nontrivial example of an energy balance is that of a fuel cell, which is becoming a practical device for producing electricity. See Fig. A.5.

In addition to mechanical work needed to push the gases through the fuel cell, a fuel cell also produces electrical work. The power (rate of work) produced by the fuel cell is the voltage *V* time the current *I*. Thus, Eq. (A.16) must include an extra work term. Rather than deriving a general formula, consider a steady state fuel cell operation. The temperature of the fuel cell is constant. Reactants enter the fuel cell at this constant temperature. The pressure drop is assumed to be negligible. Thus, the pressure is also constant. In words, the *energy balance* is

Rate of enthalpy (outflow − inflow) = Rate of heat flow into the fuel cell

−Power produced by the fuel cell (A.21)

At a constant pressure and temperature, the reaction

$$H_2 + \tfrac{1}{2}O_2 \rightarrow H_2O \tag{A.22}$$

yields a change of enthalpy, $-\Delta H$ per mole of hydrogen reacted. At one atmosphere at 25° C this is the tabulated heat of reaction of hydrogen. In terms of

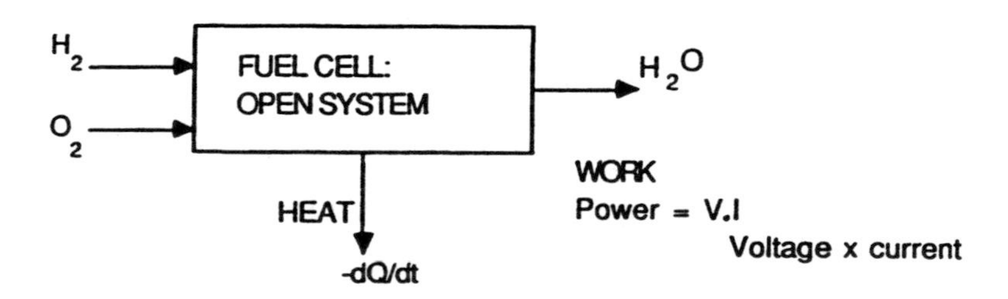

Fig. A.5 Open system example.

moles of hydrogen reacted per unit time given by Eq. (A.7), the energy balance is

$$+\Delta H \cdot m_A = \frac{dQ}{dt} - V \cdot I. \tag{A.23}$$

Substituting the species balance, Eq. (A.7) gives the rate of heat produced by the fuel cell:

$$-\frac{dQ}{dt} = \left(\frac{-\Delta H}{nF} - V \right) \cdot I. \tag{A.24}$$

There are two useful limits that may be imposed upon Eq. (A.24).

Limit 1: No power production, V = 0 (short-circuited)

$$-\frac{dQ}{dt} = -\Delta H(I/nF) \tag{A.25}$$

All of the heat of reaction results in heat.

Limit 2: Reversible Operation
Practically this can be achieved only for very small I. Thus, in this mode of operation, power = $V \cdot I$ is again very small. The reversible voltage of an electrochemical cell can be calculated from the free energy of the reaction from the formula

$$\Delta G = -nFE, \tag{A.26}$$

where

E = reversible potential, volts,
ΔG = free energy change,
F = Faraday's constant,
n = number of equivalents per mole.

From the definition of G as $G = H - TS$, at a constant temperature we have $\Delta G = \Delta H - T\Delta S$. Substituting this relation and Eq. (A.26) into Eq. (A.24) we find that

$$-\frac{dQ}{dt} = -\frac{T\Delta S}{nF} \cdot I. \tag{A.27}$$

For a reaction such as $C + O_2 \rightarrow CO_2$, ΔS is nearly zero. Therefore, the heat produced by a hypothetical fuel cell burning carbon electrochemically is zero.

Since $\frac{dQ}{dt}$ is zero, Eq. (A.23) shows that it is theoretically possible to convert all of the heat of reaction of carbon into electrical work.

A.5.1 Alternate Form of Fuel Cell Energy Balance

Using the definition of the reversible potential *E*, Eq. (A.26), we can rearrange the fuel cell energy balance as

$$-\frac{dQ}{dt} = (E - V) \cdot I - \frac{T\Delta S}{nF} \cdot I. \tag{A.28}$$

Rate of heat production by the fuel = cell (given off to the surroundings) | Production due to + irreversible operation | Heat changes due to reversible reaction

Equation (A.28) shows that a fuel cell produces heat because of irreversible operation. Particularly for molten carbonate fuel cells or for fuel cells with a solid electrolyte, we can express the difference between reversible and actual operating potentials in terms of an electrical resistance, *R*, times the current:

$$E - V = I \cdot R. \tag{A.30}$$

Thus, we find

$$-\frac{dQ}{dt}_{\text{(irreversible only)}} = I^2 \cdot R. \tag{A.31}$$

Thus, a fuel cell operating with a nonzero current will convert some of the stored enthalpy into heat. One of the objects of ongoing fuel cell development is to minimize *R*.

A.6 Open System Energy Balance

Open system energy balances are useful in single phase and in homogeneous multiphase flow. In the latter the energies and the enthalpies are simply the mixture quantities. The derivation of the open system energy balance is made on the closed system shown in Fig. A.6.

Consider closed systems at times t_o and t. At time t_o, boundaries are a–a; at time *t*, boundaries are b–b.

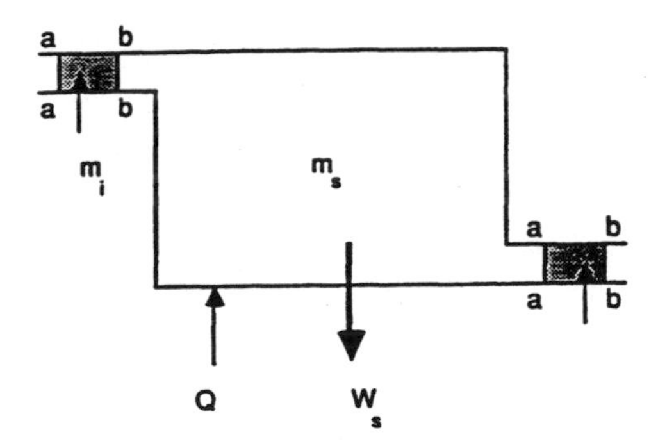

Fig. A.6 Open system using a closed system approach.

	Mass of System	Energy of System
at time t:	$m_s + m_o$	$e_s m_s + e_o m_o$
at time t_o:	$m_s + m_i$	$e_s m_s + e_i m_i$

The change in energy of the system between t and t_o is

$$\Delta E = e_s m_s(t) - e_s m_s(t_o) + e_o m_o - e_i m_i. \tag{A.32}$$

The work done by the system at this time interval consists of the work done by the shaft, the work of expansion at the inlet and the work at the outlet:

$$W = W_s - P_i \rho_i^{-1} m_i + P_o \rho_o^{-1} m_o. \tag{A.33}$$

The energy balance is

$$\Delta E = Q - W_s + P_i \rho_i^{-1} m_i - P_o \rho_o^{-1} m_o. \tag{A.34}$$

By the definition of a derivative in the limit as t approaches t_o, Eq. (A.34) becomes

$$\frac{dE}{dt} = \frac{dQ}{dt} - \frac{dW_s}{dt} + P_i \rho_i^{-1} \frac{dm_i}{dt} - P_o \rho_o^{-1} \frac{dm_o}{dt}. \tag{A.35}$$

When the total energy E consists of the kinetic, potential and internal energy, Eq. (A.35) becomes the conventional macroscopic balance

$$\frac{d(e_s m_s)}{dt} + \left(\tfrac{1}{2}v_o^2 + gz_o + U_o + P_o\rho_o^{-1}\right)\frac{dm_o}{dt} - \left(\tfrac{1}{2}v_i^2 + gz_i + U_i + P_i\rho_i^{-1}\right)\frac{dm_i}{dt} = \frac{dQ}{dt} - \frac{dW_s}{dt} \tag{A.36}$$

In terms of enthalpy, at steady state using the dot notation to represent the time derivative, this balance reduces itself to the most useful expression,

$$\Delta h + g\Delta z + \tfrac{1}{2}\Delta v^2 = \frac{\dot{Q}}{\dot{m}} - \frac{\dot{W}_s}{\dot{m}}. \tag{A.37}$$

Hence, for steady state single or multiphase homogeneous flow in the absence of gravitational or kinetic energy changes, we have the following useful relations.

For heat exchanger:

$$\dot{Q} = \dot{m}\Delta h. \tag{A.38}$$

For an adiabatic compressor:

$$\dot{W}_s = -\dot{m}\Delta h. \tag{A.39}$$

For a throttling valve:

$$h_i = h_o. \tag{A.40}$$

APPENDIX B

EULERIAN APPROACH: ONE-D ENERGY BALANCE

The shell balance approach of formulating continuum problems can be best illustrated for an energy balance discussed in Appendix A. In this method, called the Eulerian approach in the mechanics literature, the system is a fixed small arbitrary volume element contained in the macroscopic system discussed in Appendix A.

To begin with, consider the example of heat generation in the rod depicted in Fig. B.1. The system is shown in Fig. B.2.

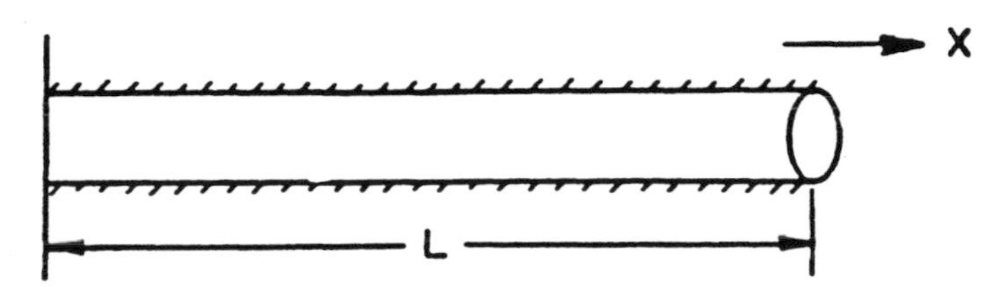

Physical Situation: A rod of cross-sectional area a starts to generate heat at $t = 0$ at a rate of Ψ Btu/hr-ft^3. What is its temperature distribution at any time t, if it is insulated except for its two ends?

System for derivation of the differential equation:
Element: $a \cdot \Delta x$ fixed in space; x, arbitrary.

Fig. B.1 Insulated rod.

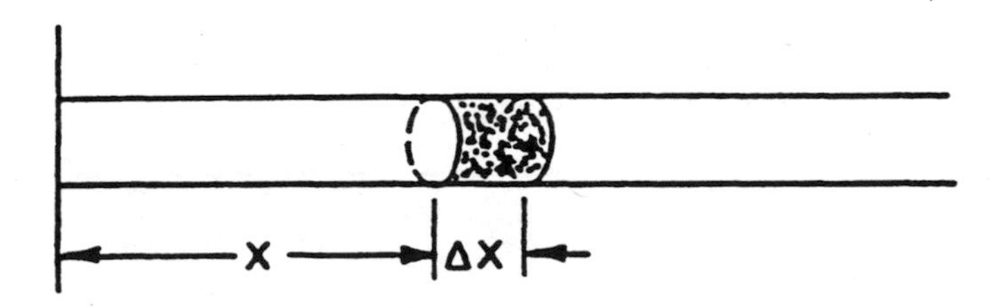

Fig. B.2 Eulerian system for the rod.

The energy balance at a constant pressure involves the enthalpy and is

$$\begin{aligned}&\textit{Scheme:}\ \text{Rate of energy (enthalpy) increase} + \text{Rate of energy outflow}\\&-\text{Rate of energy inflow} = \text{Rate of energy generation}\end{aligned} \tag{B.1}$$

The individual quantities in the balance are

$$\text{Total enthalpy in element } a\cdot\Delta x\text{:}\, H = \int_x^{x+\Delta x} h\rho\, dV, \tag{B.2}$$

where h is the specific enthalpy and ρ the density.

$$\text{Net rate of energy outflow} = q(t, x+\Delta x)\cdot a - q(t,x)\cdot a \equiv q\cdot a\big]_x^{x+\Delta x}, \tag{B.3}$$

where q is the heat flux, defined + for outflow.

Outflow minus inflow is represented by the Fundamental Theorem of Integral Calculus notation on the right side of Eq. (B.3).

Energy generation in volume element $a\cdot\Delta x$

$$= \int_x^{x+\Delta x} \psi a\, dx. \tag{B.4}$$

The balance becomes

$$\frac{d}{dt}\int_x^{x+\Delta x} h\rho\, dx + q\big]_x^{x+\Delta x} = \int_x^{x+\Delta x} \psi\, dx. \tag{B.5}$$

Since the limits of integration are independent of time, the Leibnitz rule for differentiating an integral gives simply

$$\frac{d}{dt}\int_x^{x+\Delta x} h\rho\, dx = \int_x^{x+\Delta x} \frac{\partial (h\rho)}{\partial t}\, dx. \tag{B.6}$$

Conservation of mass principle for this system is

$$\frac{d}{dt}\int_x^{x+\Delta x} \rho a\, dx = 0 \quad \text{or} \quad \int_x^{x+\Delta x} \frac{\partial \rho}{\partial t}\, dx = 0, \tag{B.7}$$

since there is no mass flow. Therefore, product differentiating yields

$$\int_x^{x+\Delta x} \frac{\partial(h\rho)}{\partial t}dx = \int_x^{x+\Delta x} \rho\frac{\partial h}{\partial t}dx. \tag{B.8}$$

Since $h = h(T)$ only and $C_p \equiv (\frac{\partial h}{\partial T})_p, h(t,x) = h(T(t,x))$, by chain rule of differentiation:

$$\frac{\partial h}{\partial t} = \left(\frac{\partial h}{\partial T}\right)_p \cdot \frac{\partial T}{\partial t} = C_p\frac{\partial T}{\partial t}. \tag{B.9}$$

Therefore,

$$\frac{d}{dt}\int_x^{x+\Delta x} h\rho\, dx = \int_x^{x+\Delta x} \rho C_p\frac{\partial T}{\partial t}dx. \tag{B.10}$$

Extensions of the mean value theorems shown in Figs. B.3 and B.4 to multiplace functions are needed to complete the derivation.

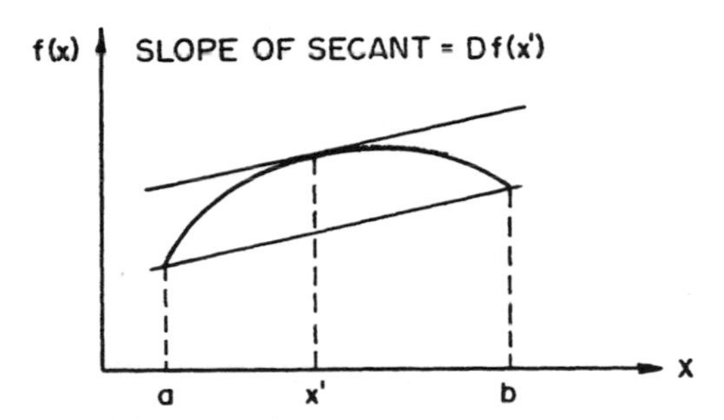

Fig. B.3 Differential calculus mean value theorem.

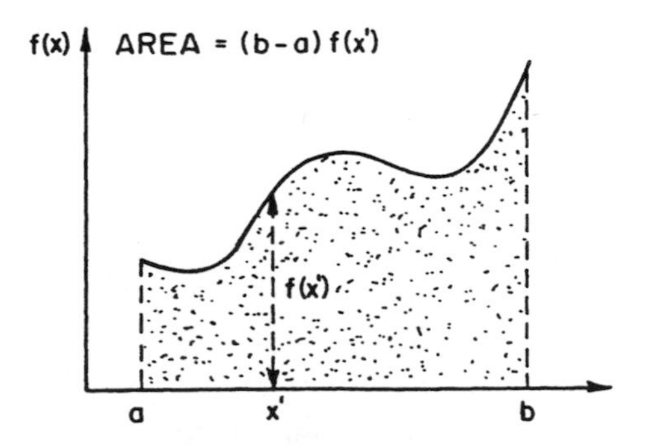

Fig. B.4 Integral calculus mean value theorem.

The Mean Value Theorem of Differential Calculus
If the function f is (a) continuous in the closed interval and (b) has a derivative at every point, then at some point x' of the interval,

$$\frac{f(b)-f(a)}{b-a}=Df(x'), \qquad a<x'<b.$$

The Mean Value Theorem of Integral Calculus
If f is integrable in (a, b) and has m, M as closest bounds,

$$\int_a^b f(x)dx=\mu(b-a), \quad m\leq\mu\leq M.$$

Further, if $f(x)$ is continuous, $\mu=f(x'), a<x'<b$, then

$$\int_a^b f(x)dx=(b-a)f(x').$$

Applying the mean value theorem of I.C. to the integral, and the mean value theorem of D. C. to the difference, we obtain

$$\rho C_p\frac{\partial T}{\partial t}(t,x')\cdot\Delta x+\frac{\partial q(t,x'')}{\partial x}\cdot\Delta x=\psi(t,x''')\cdot\Delta x. \tag{B.11}$$

$$x<x'<x+\Delta x \qquad x<x''<x+\Delta x \qquad x<x'''<x+\Delta x$$

Divide by Δx and let $\Delta x\to 0$. Then $x', x'', x'''\to x$. Therefore, for any x in the open interval (note: boundary not included),

$$\rho C_p\frac{\partial T}{\partial t}+\frac{\partial q}{\partial x}=\psi.$$

By Fourier's law of heat conduction (Carslaw and Jaeger, 1959),

$$q=-k\frac{\partial T}{\partial x}. \tag{B.12}$$

Substituting, we obtain the diffusion equation with heat generation:

$$\rho C_p \frac{\partial T}{\partial t} = \frac{\partial}{\partial x}\left(k \frac{\partial T}{\partial x}\right) + \psi . \tag{B.13}$$

To complete the problem we also need two boundary conditions, and an initial condition.

APPENDIX C

LEIBNIZ FORMULA AND RELATION TO TRANSPORT

Formulas for differentiation integrals are so useful in transport that a brief review of their calculus is necessary. The formula for differentiating an integral with variable limits given below is usually called the Leibniz formula:

$$\frac{d}{dt}\int_{a(t)}^{b(t)} G(t,x)\,dx = \int_{a(t)}^{b(t)} \frac{\partial G}{\partial t}\,dx + G(t,b)\frac{db}{dt} - G(t,a)\frac{da}{dt}. \tag{C.1}$$

To comprehend it fully, it is useful to present its derivation. Since in the integration t acts as a parameter, we can let G be a partial derivative with respect to the second place denoted by D_2. Hence, let

$$G(t,x) = D_2 F(t,x). \tag{C.2}$$

Using the convenient D notation to represent one place differentiation the integral on the left side of (C.1) by the Fundamental Theorem of Integral Calculus becomes

$$D\int_a^b D_2 F(t,x) = D\left[F(t,b) - F(t,a)\right]. \tag{C.3}$$

Differentiation of the right side of (C.3) and the use of the fundamental theorem of Integral Calculus in reverse gives the three parts of the Leibniz formula in (C.1):

$$\begin{aligned} &D_1 F(t,b) + D_2 F(t,b)\cdot Db - D_1 F(t,a) - D_2 F(t,a)\cdot Da \\ &= \int_a^b D_1\left[D_2 F(t,x)\right] + D_2 F(t,b)\cdot Db - D_2 F(t,a)\cdot Da. \end{aligned} \tag{C.4}$$

In the Lagrangian approach to be discussed in greater detail in Appendix D, the system chosen is one of constant mass that moves through space. Hence, the limits of integration on the volume are a function of time, unlike in the Eulerian approach discussed in Appendix B. In one dimension conservation of mass for a unit area states that for an arbitrary volume element $1 \cdot \Delta x$, we have

$$\frac{d}{dt}\int_{x(t)}^{x(t)+\Delta x(t)} \rho(t,x)\,dx = 0. \tag{C.5}$$

Application of the Leibniz formula (C.1) to (C.5) gives the equation

$$\int_{x(t)}^{x(t)+\Delta x(t)} \frac{\partial \rho}{\partial t}dx + \rho(t, x+\Delta x)\frac{d}{dt}(x+\Delta x) - \rho(t,x)\frac{dx}{dt} = 0. \tag{C.6}$$

Since the velocity is defined to be

$$v = \frac{dx}{dt}, \tag{C.7}$$

the balance is

$$\int_{x(t)}^{x(t)+\Delta x(t)} \frac{\partial \rho}{\partial t}dx + \rho v(t, x+\Delta x) - \rho v(t,x) = 0. \tag{C.8}$$

An application of the mean value theorems of calculus to (C.8) gives

$$\frac{\partial \rho(t,x')}{\partial t}\cdot \Delta x + \frac{\partial(\rho v)(t,x'')}{\partial x}\cdot \Delta x = 0. \tag{C.9}$$

$$x < x' < x + \Delta x \qquad x < x'' < x + \Delta x$$

The shrinking of the arbitrary element to zero gives the one-dimensional single phase continuity equation

$$\frac{\partial \rho}{\partial t} + \frac{\partial(\rho v)}{\partial x} = 0 \tag{C.10}$$

This technique can be clearly extended to multiple integrals, providing an alternate derivation to the Reynolds transport theorem illustrated in subsequent appendices.

APPENDIX D

LAGRANGIAN APPROACH: ONE-D CONSERVATION OF SPECIES AND POPULATION BALANCE

The purpose of this section is to illustrate the derivation of conservation laws using the Lagrangian approach. In this method the system chosen is a constant mass of some species which moves through space. The space may be real or abstract, as in the population balance.

D.1 Conservation of Species

As an illustration of one-dimensional balances, consider conservation of species.

Model: For i components we assume the existence of i distinct continua within any arbitrary volume. Each space point in the multi-component continuum has n velocities, v_i, $i = 1, 2, 3,...n$, one velocity for each coexistent continuum. All functions are assumed to be differentiable.

In contrast to the multiphase balances, the molecules are assumed not to occupy any space. At the same position, at the same time, all i components are present. Such an assumption is reasonable in view of the small size of the molecules compared with a sample volume.

Approach: Take a constant mass of species i and move with the element of mass i at a velocity v_i (see Fig. D.1).

At time t_o, the "particle" is at x_o. This identifies the particle. At time t, the particle is at x and $x = x(t, x_o)$ as we follow the particle. Since the function x is single-valued and differentiable, its inverse exists and is expressed as

$$x_o = x_o(t, x). \tag{D.1}$$

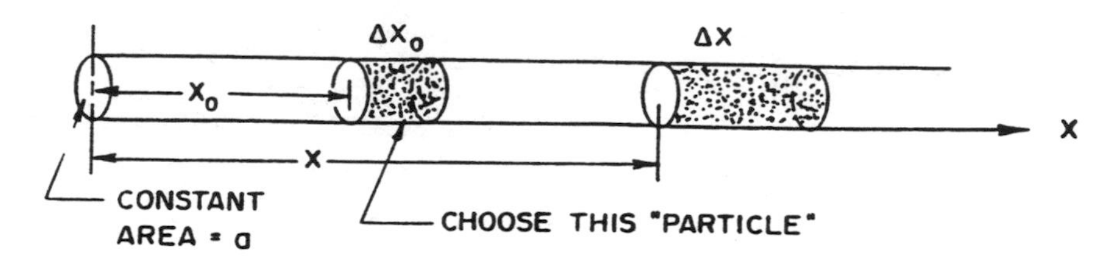

Fig. D.1 Element for Lagrangian balance.

The species balance states

$$\frac{d}{dt}\int_x^{x+\Delta x} a\rho_i(t,x)\,dx \qquad = \qquad -\int_x^{x+\Delta x} a r_i M_i\, dx. \tag{D.2}$$

Rate of change of mass of species i where ρ_i = Rate of consumption of species i by molar rate of

is the density of species i, kg/m^3 reaction r_i, where M_i is the molecular weight of i

The flow velocity or "particle" velocity for the continuum is defined as

$$v(t,x) = \frac{dx}{dt} = \frac{\partial x(t,x_o)}{\partial t}. \tag{D.3}$$

The mathematical problem now is to differentiate the integral with variable limits of integration. The technique is to transform the integral to one with constant limits of integration. Transform

$$(x,t) \rightarrow (x_o,t) \text{ coordinates}, \tag{D.4}$$

where x_o is fixed for the particular particle. We have

$$\frac{d}{dt}\int_{x(t,x_o)}^{x(t,x_o)+\Delta x(t,x_o)} \rho_i(t,x)\,dx. \tag{D.5}$$

Since we have only partial integration and the limits are functions of x_o, we recall that the formula for changing the limits is

$$\int_{g(a)}^{g(b)} h(x)\,dx = \int_a^b h(g)\cdot\frac{dg}{dx}\cdot dx. \tag{D.6}$$

Proof: Let $h = Df$, since such an f must exist to integrate.

$$\int_{g(a)}^{g(b)} Df = f(g)(b) - f(g)(a) \tag{D.7}$$

by the Fundamental Theorem of Integral Calculus, but

$$\int_a^b D[f(g)] = f(g)(b) - f(g)(a). \tag{D.8}$$

Also

$$\int_a^b D[f(g)] = \int_a^b Df(g) \cdot Dg \quad \text{by chain rule.} \tag{D.9}$$

This completes the proof.

Therefore, since in our integral t is suppressed,

$$I = \frac{d}{dt}\int_{x(t,x_o)}^{x(t,x_o)+\Delta x(t,x_o)} \rho_i(t,x)\,dx = \frac{d}{dt}\int_{x_o}^{x_o+\Delta x_o} \rho_i[x(x_o,t),t]\frac{\partial x}{\partial x_o}dx_o. \tag{D.10}$$

Since now the integral is over x_o only, which is not a function of time, only the integrand is differentiated:

$$I = \int_{x_o}^{x_o+\Delta x_o}\left\{\frac{\partial \rho_i[x(x_o,t),t]}{\partial t}\cdot\frac{\partial x}{\partial x_o} + \rho_i \cdot \frac{\partial}{\partial t}\left[\frac{\partial x}{\partial x_o}\right]\right\}dx_o. \tag{D.11}$$

Note that

$$\frac{\partial}{\partial t}\frac{\partial x(t,x_o)}{\partial x_o} = \frac{\partial}{\partial x_o}\frac{\partial x(t,x_o)}{\partial t} = \frac{\partial v_i(t,x_o)}{\partial x_o}. \tag{D.12}$$

Interchanging order of differentiation   By definition of velocity for i-th particle

Strictly speaking, every x should carry the subscript i for ith particle.

But v_i itself changes as the particle moves:

$$v_i = v_i(t,x), \tag{D.13}$$

$$v_i(t,x_o) = v_i(t,x(x_o,t)), \tag{D.14}$$

$$\frac{\partial v_i}{\partial x_o} = \frac{\partial v_i}{\partial x} \cdot \frac{\partial x}{\partial x_o}, \tag{D.15}$$

$$I = \int_{x_o}^{x_o+\Delta x_o} \left[\frac{d\rho_i[t, x(x_o, t)]}{dt} + \rho_i \frac{\partial v_i}{\partial x} \right] \frac{\partial x}{\partial x_o} dx_o \tag{D.16}$$

Write d/dt to emphasize that x_o is held constant and that differentiation has not yet been performed. Next differentiate using the chain rule and change the limits back to x:

$$\int_x^{x+\Delta x} \left(\frac{\partial \rho_i}{\partial t} + \frac{\partial \rho_i}{\partial x} \cdot v_i + \rho_i \frac{\partial v_i}{\partial x} \right) dx. \tag{D.17}$$

Combine the expression in (D.17) using the product rule of differentiation:

$$I = \int_x^{x+\Delta x} \left(\frac{\partial \rho_i}{\partial t} + \frac{\partial (\rho_i v_i)}{\partial x} \right) dx. \tag{D.18}$$

Then the balance becomes

$$\int_x^{x+\Delta x} \left[\frac{\partial \rho_i}{\partial t} + \frac{\partial (\rho_i v_i)}{\partial x} + r_i M_i \right] dx = 0. \tag{D.19}$$

Since the integral above vanishes for an arbitrary volume and the integrand is continuous, the integrand itself must vanish. This follows from a proof by contradiction. Suppose the integrand did not vanish for some point P but were positive. Then since the integrand is continuous it would be positive for some neighborhood of P. We may take Δx within this neighborhood entirely, and for this the integral would not vanish. There is a contradiction. Therefore,

$$\frac{\partial \rho_i}{\partial t} + \frac{\partial (\rho_i v_i)}{\partial x} = -r_i M_i. \tag{D.20}$$

Example: *Combustion in a Channel*

A dilute mixture of hydrogen in air is burning in a long tube. The reaction is first order. The flow is highly turbulent. Find the steady state concentration profile.

In this system, mass is conserved and therefore mass units should be used. Also, since the flow is very rapid we may assume that the hydrogen moves with the total stream velocity, that is, $v = v_i$.

At steady state, $\frac{\partial \rho}{\partial t} = \frac{\partial \rho_i}{\partial t} = 0$ and the conservation of mass and species equations become

$$\frac{\partial(\rho v)}{\partial x} = 0, \tag{D.21}$$

$$\rho v \frac{\partial Y_i}{\partial x} + Y_i \frac{\partial(\rho v)}{\partial x} = -r_i M_i, \tag{D.22}$$

where Y_i is the weight fraction of species so that $\rho_i = Y_i \rho$. Using the conservation of mass equation we see that

$$\rho v \frac{dY_i}{dx} = -r_i M_i \tag{D.23}$$

for each reacting species.

For the first order reaction we may express the rate of combustion of hydrogen as (Gidaspow and Ellington, 1964)

$$r_{H_2} = k_p P_t y, \tag{D.24}$$

where k_p is the first order rate constant in partial pressure units (for example, moles H_2 / hr − m^3 − atm), P_t is the total pressure and y the mole fraction of hydrogen. The stoichiometry gives the following relation between conversion, mole and weight fractions for the reaction below:

$$H_2 + \tfrac{1}{2}O_2 \rightarrow H_2O. \tag{D.25}$$

Table D.1*

	Feed	At Any Position
H_2	n_o	$n_o - n$
O_2	$1 - n_o$	$1 - n_o - 0.5n$
H_2O	0	n
Total	1.0	$1 - 0.5n$

*Basis: 1 mole of feed

In Table D.1, n is the number of moles of H_2 reacted per mole of feed, and n_o is the number of moles of H_2 in feed per mole of feed. From Table D.1 it is clear that

$$y = \frac{n_o - n}{1 - 0.5n} \tag{D.26}$$

and

$$Y = \frac{(n_o - n) M_{H_2}}{M_{mo}}, \tag{D.27}$$

where M_{mo} is the average molecular weight of feed.

Substitution of expressions for mole and weight fractions into the simplified rate expression yields

$$\left(\frac{\rho v}{M_{mo}} \right) \frac{dn}{dx} = k_p P_t \frac{n_o - n}{1 - 0.5n}. \tag{D.28}$$

Integration yields

$$\frac{k_p P_t M_{mo} \Delta x}{(\rho v)} = \ln \frac{n_o}{n_o - n} - 0.5 \left(n_o \ln \frac{n_o}{n_o - n} - n \right). \tag{D.29}$$

For systems with low hydrogen conversion or dilute feeds, $0.5n << 1$. Then the preceding expression simplifies to

$$-\frac{k_p P_t M_{mo} \Delta x}{(\rho v)} = \ln \frac{y}{y_o}, \tag{D.30}$$

where y_o is the inlet hydrogen mole fraction. Then

$$y = y_o \exp \left[-\frac{k_p P_t M_{mo} \Delta x}{(\rho v)} \right]. \tag{D.31}$$

The conservation of species equation was derived in terms of weight units. Weight or mass units are most useful when mass is conserved. When the total number of moles are conserved or often when chemical reactions occur, molar quantities are more appropriate. The molar concentration of species i, C_i, is related to the partial density, ρ_i, by

$$C_i = \rho_i / M_i \,. \tag{D.32}$$

The conservation of species in terms of C_i becomes

$$\frac{\partial C_i}{\partial t} + \frac{\partial (C_i v_i)}{\partial x} = -r_i . \tag{D.33}$$

The total molar density C is

$$C = \sum_{n=1}^{n} C_i \,, \tag{D.34}$$

and the average molar velocity becomes

$$v^* = \sum_{i=1}^{n} C_i v_i \Big/ \sum_{i=1}^{n} C_i \,. \tag{D.35}$$

When moles are conserved, Eq. (D.31) can be derived more quickly starting with the molar balance, Eq. (D.33).

D.2 Reynolds Transport Theorem

Starting with the balance expressed by Eq. (D.2) the integral could also have been differentiated using the Leibniz rule, as illustrated in Appendix C. Both results can be summarized by stating the one-dimensional version of the so-called Reynolds Transport Theorem:

$$\frac{d}{dt^i} \int_{x(t)}^{x(t)+\Delta x(t)} \Im(t,x)\, dx = \int_{x(t)}^{x(t)+\Delta x(t)} \left(\frac{\partial \Im}{\partial t} + \frac{(\Im v_i)}{\partial x} \right) dx \,, \tag{D.36}$$

where

$$v_i = \frac{dx}{dt^i} . \tag{D.37}$$

Its extension to more than one dimension can be done using either of the techniques of Appendices C or D.

D.3 Population Balance

As an application of the Reynolds transport theorem, Eq. (D.36), consider a population balance made on a well-mixed crystallizer, depicted by Fig. D.2. The number of crystals in the size range r to $r+\Delta r$, N, is expressed as follows in terms of the frequency distribution $f(t,r)$:

$$N = \int_o^V \int_r^{r+\Delta R} f\, dr\, dV. \tag{D.38}$$

The balance, in words, for the crystallizers is

$$\begin{aligned} &\text{Rate of accumulation} = \text{Inflow} - \text{Outflow} \\ &+\text{Production due to collisions} \end{aligned} \tag{D.39}$$

With no crystal inflow and no production by collisions, the balance becomes

$$\tfrac{d}{dt}\int_o^V \int_r^{r+\Delta r} f\, dr\, dV = -F\int_r^{r+\Delta r} f(t,r)\, dr. \tag{D.40}$$

An application of the Reynolds transport theorem to (D.40) gives the balances below:

$$\int_r^{r+\Delta r}\left(\frac{\partial f}{\partial t}+\frac{\partial\left(f\,\frac{dr}{dt}\right)}{\partial r}-\frac{F}{V}f\right)dr = 0. \tag{D.41}$$

The velocity in (D.41) is the "slow" crystal growth rate, $\frac{dr}{dt}$. Let

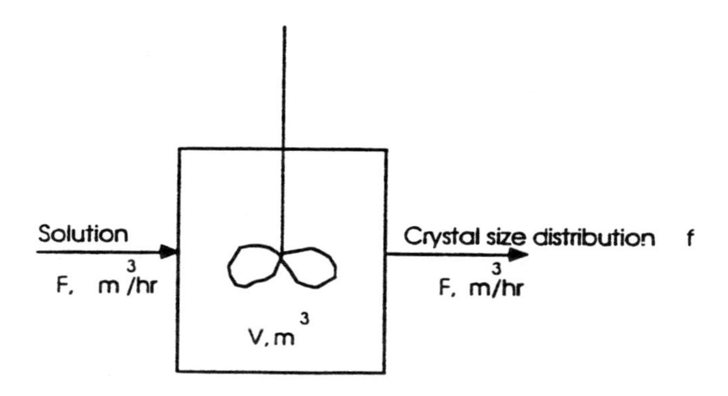

Fig. D.2 Stirred tank.

$$\dot{R} = \frac{dr}{dt}. \tag{D.42}$$

Then the population balance becomes the simple first order partial differential equation in terms of the unknown growth rate $\dot{R}$:

$$\frac{\partial f}{\partial t} + \frac{\partial(f\dot{R})}{\partial r} = -\left(\frac{F}{V}\right)f. \tag{D.43}$$

McCabe's crystal growth rate law assumes that $\dot{R}$ is a constant. For a steady state, (D.43) becomes an ordinary differential equation:

$$\dot{R}\frac{df}{dr} = -\left(\frac{F}{V}\right)f. \tag{D.44}$$

To integrate (D.44) the number of nuclei f_o must be specified:

$$f(o) = f_o. \tag{D.45}$$

Then integration of (D.44) gives the exponential frequency distribution,

$$f = f_o \exp\left(-\frac{Fr}{V\dot{R}}\right). \tag{D.46}$$

For spherical particles the cumulative weight distribution, W, is as follows:

$$W = \frac{\int_o^R r^3 \exp\left(-\frac{Fr}{V\dot{R}}\right) dr}{\int_o^\infty r^3 \exp\left(\frac{Fr}{V\dot{R}}\right) dr}. \tag{D.47}$$

It can be expressed as

$$W = \frac{1}{6}\int_o^{\frac{Fr}{V\dot{R}}} x^3 e^{-x}\, dx. \tag{D.48}$$

The frequency of the weight distribution, f_w, is the derivative of W obtained by using the Leibniz rule in Appendix C:

$$f_w = \frac{F}{6V\dot{R}}\left(\frac{Fr}{V\dot{R}}\right)^3 \exp\left(-\frac{Fr}{V\dot{R}}\right). \tag{D.49}$$

It has a maximum at

$$r_{\text{dominant}} = \left(\frac{3V\dot{R}}{F}\right). \tag{D.50}$$

The dominant crystal size depends upon the hold-up time, V/F and the crystal growth rate. A similar result holds for an ash agglomerator in a fluidized bed reactor used to make gas from coal via the U-GAS process (Vora *et al.*, 1980). Coal is fed into the reactor, as depicted in Fig. D.2, and ash agglomerates are removed, just as in the crystallizer. The air and gas production do not enter into this simplified balance.

APPENDIX E

REYNOLDS TRANSPORT THEOREM

Balances are made on some quantity that can be counted and that changes with time. Following Aris (1962), call this quantity $F(t)$. In space we define a property per unit volume $\Im(t,\mathbf{x})$, where t is time and $\mathbf{x}$ the position vector, such that

$$F(t) = \iiint\limits_{V(t)} \Im(t,\mathbf{x})\,dV. \tag{E.1}$$

In the Lagrangian representation let three parameters $(x^o, y^o, z^o) = \mathbf{x}^o$ identify the individual particle of the continuum. Thus, (x^o, y^o, z^o) are the coordinates of the particle at some fixed time t^o. Then the spatial coordinates of the particle at any time are given by functions of

$$\begin{aligned} x &= x\left(t, x^o, y^o, z^o\right), \\ y &= y\left(t, x^o, y^o, z^o\right), \\ z &= z\left(t, x^o, y^o, z^o\right), \end{aligned} \tag{E.2}$$

as shown in Fig. E.1. The functions x, y, z are taken to be single-valued and at least twice differentiable. The transformations are assumed to be one-to-one, so that the inverse transformations also exist and are twice continuously differentiable. Hence,

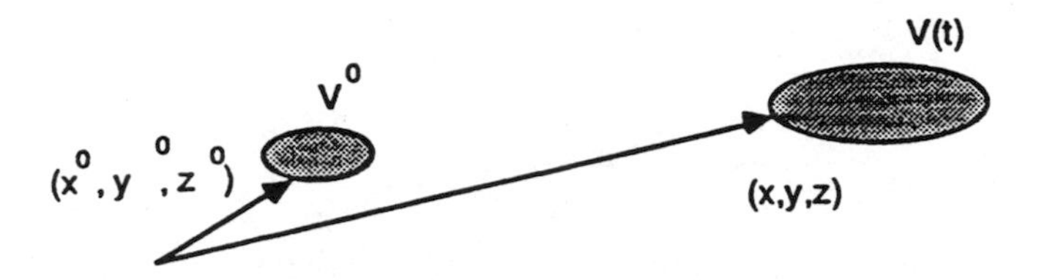

Fig. E.1 Motion of a system of constant mass.

$$\begin{aligned} x^o &= x^o(t,x,y,z), \\ y^o &= y^o(t,x,y,z), \\ z^o &= z^o(t,x,y,z). \end{aligned} \tag{E.3}$$

The functions $\mathbf{x}$ and $\mathbf{x}^o$ are inverses. The flow velocities or "particle velocities" for the continuum are then defined as

$$\mathbf{v}(t,\mathbf{x}) = \frac{d\mathbf{x}}{dt} = \frac{\partial \mathbf{x}(t,\mathbf{x}^o)}{\partial t}. \tag{E.4}$$

Differentiation of (E.1) can be done by changing the region of integration from the arbitrary volume $V(t)$ in Fig. (E.1) to the fixed initial volume. The mathematics is essentially that already described in Appendix D. The new aspect presented here is its generalization to three dimensions. Its extension to six or higher dimensions needed in the Boltzmann equation formulation should be simple.

To change the limits of integration one makes use of a formula from advanced calculus (Brand, 1955) also derived in Aris (1962):

$$dV = J dV^o, \tag{E.5}$$

where J is the Jacobian determinant. Then differentiation of (E.1) gives

$$\frac{dF(t)}{dt} = \frac{d}{dt}\iiint_{V(t)} \Im(t,\mathbf{x})\,dV = \frac{d}{dt}\iiint_{V^o} \Im\left[\left(t,\mathbf{x}(t,\mathbf{x}^o)\right)\right] J\,dV^o, \tag{E.6}$$

$$\text{where } J = \begin{vmatrix} \frac{\partial x}{\partial x^o} & \frac{\partial y}{\partial x^o} & \frac{\partial z}{\partial x^o} \\ \frac{\partial x}{\partial y^o} & \frac{\partial y}{\partial y^o} & \frac{\partial z}{\partial y^o} \\ \frac{\partial x}{\partial z^o} & \frac{\partial y}{\partial z^o} & \frac{\partial z}{\partial z^o} \end{vmatrix} = \frac{\partial(x,y,z)}{\partial\left(x^o,y^o,z^o\right)}, \tag{E.7}$$

$$\frac{dF(t)}{dt} = \iiint_{V^o} \left(\frac{d\Im}{dt}\cdot J + \Im\cdot\frac{dJ}{dt}\right) dV^o. \tag{E.8}$$

In (E.8) only the integrant was differentiated, since the limits are fixed.

Differentiation of the Jacobian

For the element of column one we have

$$\frac{d}{dt}\frac{\partial x}{\partial x^o}=\frac{\partial}{\partial x^o}\cdot\frac{dx}{dt}=\frac{\partial u}{\partial x^o}, \tag{E.9}$$

since

$$x=x\left(t,x^o,y^o,z^o\right) \text{ and } u=\frac{dx}{dt}. \tag{E.10}$$

To carry out $\frac{\partial u}{\partial x^o}$, note that

$$u=u\left[t,x\left(t,x^o,y^o,z^o\right),y\left(t,x^o,y^o,z^o\right),z\left(t,x^o,y^o,z^o\right)\right]. \tag{E.11}$$

Then

$$\frac{\partial u}{\partial x^o}=\frac{\partial u}{\partial x}\cdot\frac{\partial x}{\partial x^o}+\frac{\partial u}{\partial y}\cdot\frac{\partial y}{\partial x^o}+\frac{\partial u}{\partial z}\cdot\frac{\partial z}{\partial x^o}. \tag{E.12}$$

For $\frac{dJ}{dt}$ we obtain the sum of three determinants. The first is

$$\begin{vmatrix} \frac{\partial u}{\partial x^o} & \frac{\partial y}{\partial x^o} & \frac{\partial z}{\partial x^o} \\ \frac{\partial u}{\partial y^o} & \frac{\partial y}{\partial y^o} & \frac{\partial z}{\partial y^o} \\ \frac{\partial u}{\partial z^o} & \frac{\partial y}{\partial z^o} & \frac{\partial z}{\partial z^o} \end{vmatrix} = \begin{vmatrix} \frac{\partial u}{\partial x}\cdot\frac{\partial x}{\partial x^o}+\frac{\partial u}{\partial y}\cdot\frac{\partial y}{\partial x^o}+\frac{\partial u}{\partial z}\cdot\frac{\partial z}{\partial x^o} & \frac{\partial y}{\partial x^o} & \frac{\partial z}{\partial x^o} \\ \frac{\partial u}{\partial x}\cdot\frac{\partial x}{\partial y^o}+\frac{\partial u}{\partial y}\cdot\frac{\partial y}{\partial y^o}+\frac{\partial u}{\partial z}\cdot\frac{\partial z}{\partial y^o} & \frac{\partial y}{\partial y^o} & \frac{\partial z}{\partial y^o} \\ \frac{\partial u}{\partial x}\cdot\frac{\partial x}{\partial z^o}+\frac{\partial u}{\partial y}\cdot\frac{\partial y}{\partial z^o}+\frac{\partial u}{\partial z}\cdot\frac{\partial z}{\partial z^o} & \frac{\partial y}{\partial z^o} & \frac{\partial z}{\partial z^o} \end{vmatrix}$$

$$=\frac{\partial u}{\partial x}\underbrace{\begin{vmatrix} \frac{\partial x}{\partial x^o} & \frac{\partial y}{\partial x^o} & \frac{\partial z}{\partial x^o} \\ \frac{\partial x}{\partial y^o} & \frac{\partial y}{\partial y^o} & \frac{\partial z}{\partial y^o} \\ \frac{\partial x}{\partial z^o} & \frac{\partial y}{\partial y^o} & \frac{\partial z}{\partial z^o} \end{vmatrix}}_{J}+\frac{\partial u}{\partial y}\underbrace{\begin{vmatrix} \frac{\partial y}{\partial x^o} & \frac{\partial y}{\partial x^o} & \frac{\partial z}{\partial x^o} \\ \frac{\partial y}{\partial y^o} & \frac{\partial y}{\partial y^o} & \frac{\partial z}{\partial y^o} \\ \frac{\partial y}{\partial z^o} & \frac{\partial y}{\partial z^o} & \frac{\partial z}{\partial z^o} \end{vmatrix}}_{\text{2 identical columns, hence zero}}+\frac{\partial u}{\partial z}\begin{vmatrix} \\ \\ \text{zero} \end{vmatrix}. \tag{E.13}$$

Similarly differentiating the second and the third columns, we find

$$\frac{dJ}{dt}=\left(\frac{\partial u}{\partial x}+\frac{\partial v}{\partial y}+\frac{\partial w}{\partial z}\right)\cdot J, \tag{E.14}$$

$$\frac{\frac{dJ}{dt}}{J}=\nabla\cdot\mathbf{v}=\text{divergence of }\mathbf{v}, \tag{E.15}$$

or the relative time rate of expansion which is independent of coordinate systems. Thus, we obtain

$$\frac{dF}{dt}=\iiint_{V^o}\left[\frac{d\Im}{dt}+\Im(\nabla\cdot\mathbf{v})\right]J\,dV^o=\iiint\left(\frac{d\Im}{dt}+\Im\nabla\cdot\mathbf{v}\right)dV. \tag{E.16}$$

By chain rule,

$$\frac{d\Im(t,x,y,z)}{dt}=\frac{\partial\Im}{\partial t}+\frac{\partial\Im}{\partial x}\cdot u+\frac{\partial\Im}{\partial y}\cdot v+\frac{\partial\Im}{\partial z}\cdot w=\frac{\partial\Im}{\partial t}+\nabla\Im\cdot\mathbf{v}, \tag{E.17}$$

where u, v, w are the x, y, and z components of the velocity.

The derivative in (E.17) is sometimes referred to as the substantial derivative. It is, in general,

$$\frac{d\Im(t,\mathbf{x}(t))}{dt}=\frac{\partial\Im}{\partial t}+\mathbf{v}\cdot\nabla\Im. \tag{E.18}$$

Then using (E.16) and (E.18), the Reynolds transport theorem becomes

$$\frac{d}{dt}\iiint_{V(t)}\Im(t,\mathbf{x})dV=\iiint_{V(t)}\left(\frac{\partial\Im}{\partial t}+\nabla\cdot\Im\mathbf{v}\right)dV. \tag{E.19}$$

It is clearly applicable to an n-dimensional space.

An alternate form of the Reynolds transport theorem can be obtained by use of the divergence theorem from advanced calculus (Brand, 1955). It states that for any vector function $\mathbf{a}$ that has continuous first partial derivatives,

$$\iiint\nabla\cdot\mathbf{a}\,dV=\oint\!\!\oint\mathbf{a}\cdot\mathbf{n}\,dS, \tag{E.20}$$

where $\mathbf{n}$ is the unit normal to the surface S.

Then the Reynolds transport theorem assumes the form obtained naturally from the Eulerian view:

$$\frac{dF}{dt}=\iiint_{V(t)}\frac{\partial\Im}{\partial t}dV+\oint\!\!\oint_{S(t)}\Im\mathbf{v}\cdot\mathbf{n}\,dS. \tag{E.21}$$

Rate of accumulation of F + Net rate of outflow across the closed surface, where $\mathbf{n}$ = outward drawn normal

A special case useful in single phase balances is presented in Bird *et al.* (1960). Thermodynamics quantities are expressed per unit mass, while $\Im$ in the Reynolds transport theorem is per unit volume. If ρ is density, let $\Im$ equal ρf, where f is the function per unit mass. Then

$$\frac{d}{dt}\iiint_{V(t)} \rho f \, dV = \iiint_{V(t)} \left(\frac{\partial(\rho f)}{\partial t} + \nabla \cdot \rho f \mathbf{v} \right) dV. \tag{E.22}$$

By the conservation of mass,

$$\frac{\partial \rho}{\partial t} + \nabla \cdot \rho \mathbf{v} = 0. \tag{E.23}$$

Then using (E.23) and writing f in convective form, the special case of the Reynolds transport theorem becomes

$$\frac{d}{dt}\iiint_{V(t)} \rho f \, dV = \iiint_{V(t)} \rho \frac{df}{dt} dV. \tag{E.24}$$

In Eq. (E.24), $\frac{df}{dt}$ is the substantial derivative. It is often written as

$$\frac{Df}{Dt}.$$

Since it simply means

$$\frac{df}{dt} = \frac{df(t, \mathbf{x}(t))}{dt}, \tag{E.25}$$

which can be evaluated by differentiation by the chain rule of calculus, it seems to the author that creation of a new symbol for differentiation causes unnecessary confusion for a novice. The notation created by one of the inventors of calculus, Leibniz is suggestive and hence useful. Despite this criticism, continuum mechanics engineering scientists (Eringen, 1975) use the D/Dt notation. They refer to the substantial derivative as the "material" derivative, since we follow a path of constant mass.

APPENDICES

THE METHODS OF CHARACTERISTICS: INTRODUCTION

Conservation laws in space and time are expressed in terms of a set of first order partial differential equations. For single phase flow the equations are the conservation of mass, momentum, and energy for the dependent variables of pressure, velocity vector, and energy, or equivalent variables. In multiphase flow, there are three n such variables for n phases. Because of the appearance of volume fraction of the phase as a variable, some of the equations in the literature are not well-posed as an initial value problem. Since such equations produce numerical instabilities and non-physical behavior (Lyczkowski *et al.*, 1982) it is necessary to fully understand the concept of well-posedness and characteristics. Furthermore, the characteristic directions dictate the prescription of inlet or boundary conditions. In view of this a short, self-contained description of the method of characteristics is presented. The technique described here is based on a decoupling method. It is easier to apply to a large system of equations than the classical technique described in standard mathematics texts, such as Abbott (1969), Courant and Hilbert (1962) and Garabedian (1964). The classical method involves forming a compatibility relation for each variable and solving for the gradient of the dependent variables to test for its non-uniqueness. This process leads to high order determinants that are more difficult to evaluate than the method given here. Of course, for a full comprehension of the subject the classical method must be studied.

APPENDIX F

FIRST ORDER PARTIAL DIFFERENTIAL EQUATION

F.1 Integration Theory

Multiphase flow is described by a system of first order partial differential equations, by a set of balance laws. Sometimes one can reduce the phenomenon to be described to just one partial differential equation. Hence it is necessary to understand the mathematics of first order partial differential equations. Although many texts treat this subject (Sneddon, 1957; Abbott, 1966; Rhee *et al.*, 1986), a physically motivated description stripped of non-essentials for an elementary treatment is useful for understanding of multiphase flow.

In one dimension, multiphase flow gives rise to the following quasi-linear partial differential equation treated in mathematics courses (Sneddon, 1957):

$$P\frac{\partial z}{\partial x}+Q\frac{\partial x}{\partial g}=R, \tag{F.1}$$

where

$$P, Q \text{ and } R=f_i(x,y,z). \tag{F.2}$$

The standard equation (F.1) can be written as

$$\frac{\partial z}{\partial x}+\frac{Q}{P}\frac{\partial z}{\partial y}=\frac{R}{P}. \tag{F.3}$$

In the form (F.3) it can be expressed in the form of a substantial derivative:

$$\frac{dz(x,y,(x))}{dx}=\frac{R}{P}, \tag{F.4}$$

where the curve $y(x)$ is determined by

$$\frac{dy}{dx}=\frac{Q}{P}. \tag{F.5}$$

Differentiation of (F.4) by chain rule shows that (F.4) is equivalent to (F.3) when the derivative, a velocity in the substantial derivative nomenclature, is set to Q/P, as in (F.5). Then it was shown that the solution of the quasi-linear partial differential equation (F.1) is equivalent to the solution of the two ordinary differential equations (F.4) and (F.5). Thus it was shown that the solution to the partial differential equation (F.1) is written in terms of the ratios

$$\frac{dx}{P}=\frac{dy}{Q}=\frac{dz}{R}. \tag{F.6}$$

F.2 Unsteady Plug Flow Reactor

The method of solution of first order partial differential equations needs to be illustrated by examples. An unsteady state generalization of conservation of species in a flow reactor discussed in Appendix D can serve as the first example. For constant molar density, the first order reaction balance is

$$\frac{d}{dt}\int_{x(t)}^{x+\Delta x} C\,dx=-\int_{x(t)}^{x+\Delta x} kC\,dx, \tag{F.7}$$

where C is molar concentration and k is the rate constant. An application of the Reynolds transport theorem and the limiting process gives the first order partial differential equations:

$$\frac{\partial C}{\partial t}+v\frac{\partial C}{\partial x}=-kC. \tag{F.8}$$

By (F.6) its solution can be written as

$$\frac{dt}{1}=\frac{dx}{v}=\frac{dC}{-kC}. \tag{F.9}$$

To integrate (F.9) we need an initial and an inlet condition:

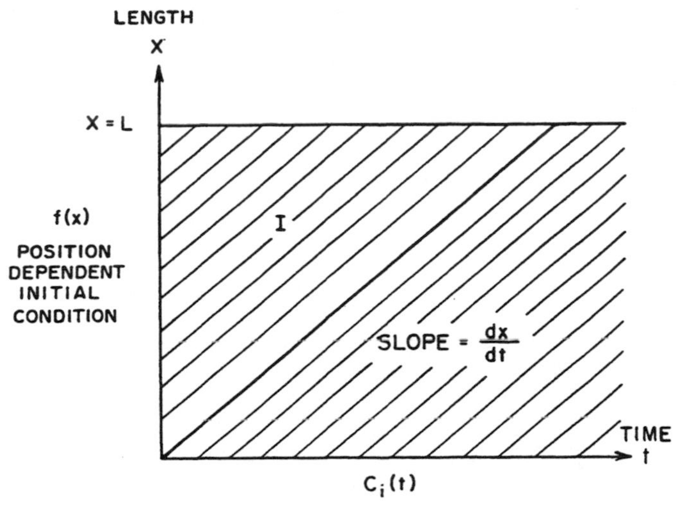

Fig. F.1 Transient plug flow reactor.

$$\text{at } t = 0, \quad C = f(x) \tag{F.10}$$

and

$$\text{at } x = 0, \quad C = C_i(t). \tag{F.11}$$

In terms of time the solution to (F.9) is

$$C = Ae^{-kt}. \tag{F.12}$$

The use of the initial condition (F.10) determines the constant A to give

$$C = f(x)e^{-kt}. \tag{F.13}$$

But the solution to

$$\frac{dx}{dt} = v \tag{F.14}$$

shows that (F.13) is valid only in the region determined by the initial condition, in Region I in Fig. F.1. For constant velocity, v, the curves generated by (F.14) are straight lines.

Equation (F.8) written as a substantial derivative or Eq. (F.9) shows that

$$\frac{dC}{dt} = -kC, \tag{F.15}$$

that is, the concentration changes as given by the right-hand side of Eq. (F.15) along the curve generated by Eq. (F.14). Thus, the concentration coming out of the reactor at $x = L$ will be determined by the initial condition in Region I. After the time

$$t = x/v, \tag{F.16}$$

the concentration leaving the reactor will be determined by the inlet condition, which may be time dependent as in (F.11). Integration of (F.9) with respect to x shows that in Region II the solution is

$$C = C_i(t)\exp(-kx/v). \tag{F.17}$$

For a constant inlet condition this is the steady state solution derived in Appendix D.

F.3 Separation in a Chromatographic Column

F.3.1 Problem Formulation

Separation in a packed chromatographic column is a good illustration of various characteristic paths. Consider the case of equilibrium sorption dynamics with no resistance to mass transfer. The problem of sorption with diffusion has been solved by Chase *et al.* (1970) using the method of Green's functions for a linear isotherm and the problem of simultaneous heat and mass transfer for non-linear isotherms, was expressed in terms of a Green's matrix by Roy and Gidaspow (1972) and applied to the design of desiccant air conditions by Roy and Gidaspow (1974) and by Ghezelayagh and Gidaspow (1982). Rhee *et al.* (1986) have given an exhaustive treatment of chromatographic separation using the methods developed for analysis of compressible fluid flow.

Let the fluid phase concentration by C_i and the concentration of species i absorbed or dissolved in the stationary phase be Γ_i. In the absence of appreciable fluid phase diffusion, all the species i flow with the fluid phase velocity v.

With these assumptions, the species i shell balance on an element of volume Δx is

$$\frac{d}{dt}\int_x^{x+\Delta x} \varepsilon C_i \, dx \quad + \quad v\varepsilon C_i \Big]_x^{x+\Delta x} \quad + \quad \frac{d}{dt}\int_x^{x+\Delta x} (1-\varepsilon)\Gamma_i \, dx \;=\; 0. \tag{F.18}$$

Rate of accumulation in the flowing phase	+ Rate of accumulation = 0 outflow in the flowing phase	Net rate of species in the stationary phase	+

An application of the mean value theorems and the limiting process gives the mixture species balance for constant porosity ε:

$$\varepsilon\frac{\partial C_i}{\partial t}+(1-\varepsilon)\frac{\partial \Gamma_i}{\partial t}+v\varepsilon\frac{\partial C_i}{\partial x}=0, \quad i=1,2,3...n \text{ phases.} \tag{F.19}$$

Equilibrium relationships with no diffusional resistances, that is, complete accessibility of reactants, complete the specification of the problem of equilibrium sorption dynamics, as shown below:

$$\Gamma_i=\Gamma_i\left(C_1,C_2...C_n\right), \quad i=1,2,3...n \text{ phases.} \tag{F.20}$$

Equilibrium constants K_{ij} are defined to be

$$K_{ij}=\frac{\partial \Gamma_i}{\partial C_j}. \tag{F.21}$$

Then, the use of the chain rule on (F.20) and substitution into (F.19) gives the system of quasi-linear partial differential equations given by (F.22),

$$\varepsilon\frac{\partial C_i}{\partial t}+v\varepsilon\frac{\partial C_i}{\partial x}+\sum_{j=1}^{n}K_{ij}\frac{\partial C_j}{\partial t}(1-\varepsilon)=0, \quad i=1,2,3...n; \tag{F.22}$$

$$K_{ij}=K_{ij}\left(C_1,C_2...\right). \tag{F.23}$$

The system of quasi-linear partial differential equations can be decoupled using the techniques to be discussed in the next appendix. Here only the case of one partial differential equation is treated. The physical assumption made is that the equilibrium is a function of its own concentration only, that is, interferences are neglected. Hence,

$$K_{ij_{(i\neq j)}}=0, \tag{F.24}$$

and let

$$K_{ii} = K_i\left(C_i\right). \tag{F.25}$$

F.3.2 Linear Isotherm: Henry's Law

In thermodynamics of solution, Henry's law is frequently applicable for dilute absorption. It is

$$\Gamma_i = K_i C_i. \tag{F.26}$$

Then the equilibrium sorption dynamics equation becomes the decoupled single partial differential equation for the concentration C_i:

$$\left(\varepsilon + (1-\varepsilon)K_i\right)\frac{\partial C_i}{\partial t} + v\varepsilon\frac{\partial C_i}{\partial x} = 0. \tag{F.27}$$

Using the theory of first order partial differential equations the solution to (F.27) can be written as a "substantial" or material derivative:

$$\frac{dC_i\left(t, x(t)\right)}{dt} = 0, \tag{F.28}$$

along the "characteristic" path

$$\frac{dx}{dt} = \frac{v\varepsilon}{\varepsilon + (1+\varepsilon)K_i}. \tag{F.29}$$

The validity of (F.28) and (F.29) also follows from using the chain rule on (F.28). For no sorption or zero K_i (F.29) becomes simply the fluid velocity,

$$\frac{dx}{dt} = v. \tag{F.30}$$

This means that species i injected at the inlet of the column simply moves with the fluid velocity as assumed in the formulation.

Chromatographic columns or sorption separation devices are designed such that K_i is large compared with one and K_1 differs from K_2, etc.:

$$K_1 << K_2 << K_3. \tag{F.31}$$

Then Eq. (F.29) shows that the velocities of propagation of species i will be much slower than the fluid velocity, and that each species will move through the column at a distinctly different velocity resulting in separation of a mixture into its components. Although the preceding discussion was not restricted to the Henry's law case, an understanding of column operation is best obtained for this simplified case. Consider the case of constant initial concentration c_f and constant inlet condition c_o. Then as in the unsteady plug flow example in Region I, up to a time

$$t_i = \frac{\left(\varepsilon + (1-\varepsilon)K_i\right)x}{v\varepsilon}, \tag{F.32}$$

the concentration leaving the column of length x will be a constant equal to c_f, which is normally zero in chromatography. At time t_i, it will suddenly jump to c_f. For a properly chosen sorbent this time t_i is different for each i to be separated. In a chromatographic column the sample is injected for only a period of Δt. Then purge fluid of zero i is injected. Thus, after a period of Δt, the concentration will jump down to zero. In this linear isotherm model, a pulse injected into the column stays as a pulse. In a real column a pulse of fluid to be separated never emerges from a column as a pulse. The pulse disperses, as depicted in Fig. F.2. This dispersion is due to the neglect of diffusion in the mobile and in the stationary phases.

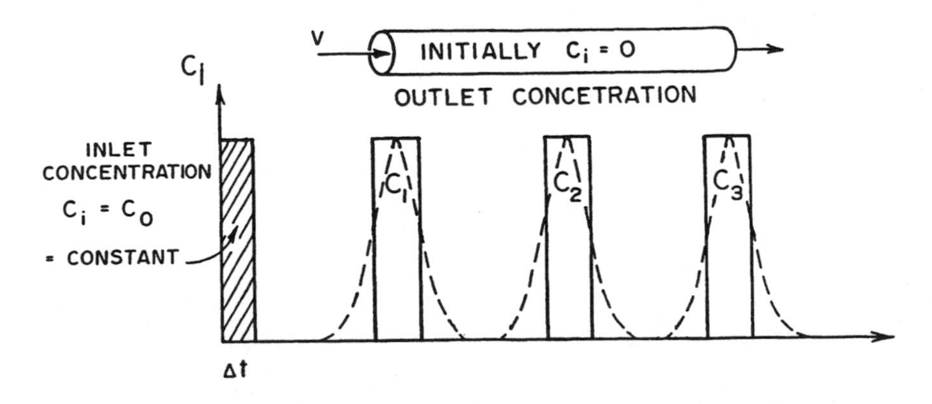

Fig. F.2 Idealized and real separation in a chromatography column with a linear isotherm.

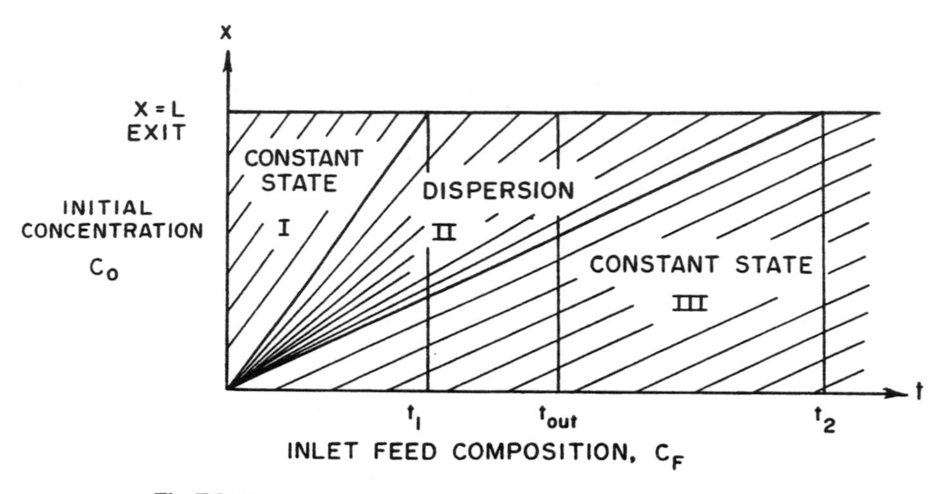

Fig. F.3 Dispersion in a chromatograph due to an unfavorable isotherm.

F.3.3 Dispersion

For the case of a nonlinear unfavorable isotherm,

$$\frac{\partial^2 \Gamma_i}{\partial C_i^2} > 0, \tag{F.33}$$

the characteristics determined by Eq. (F.29) will have different slopes when the concentration jumps from the initial condition of C_o to the inlet feed composition C_f, as illustrated in Fig. F.3. The illustration in this figure is for the case

$$\Gamma_i = kC_i^2 \text{ and } C_o < C_f, \tag{F.34}$$

$$\frac{dx}{dt} = \frac{v\varepsilon}{\varepsilon + (1-\varepsilon)2KC_i}. \tag{F.35}$$

In Region I, the concentration is a constant equal to C_o and that in Region III is C_f. In II it can be determined as follows. Pick a concentration $C_o + \Delta C$. This concentration is a constant along the path (F.35) with C_i replaced by $C_o + \Delta C$.

Using this equation the outlet concentration $C_o + \Delta C$ can then be determined by integration of (F.35) to give expression (F.36) for a column of length L and time t_{out} in Fig. F.3:

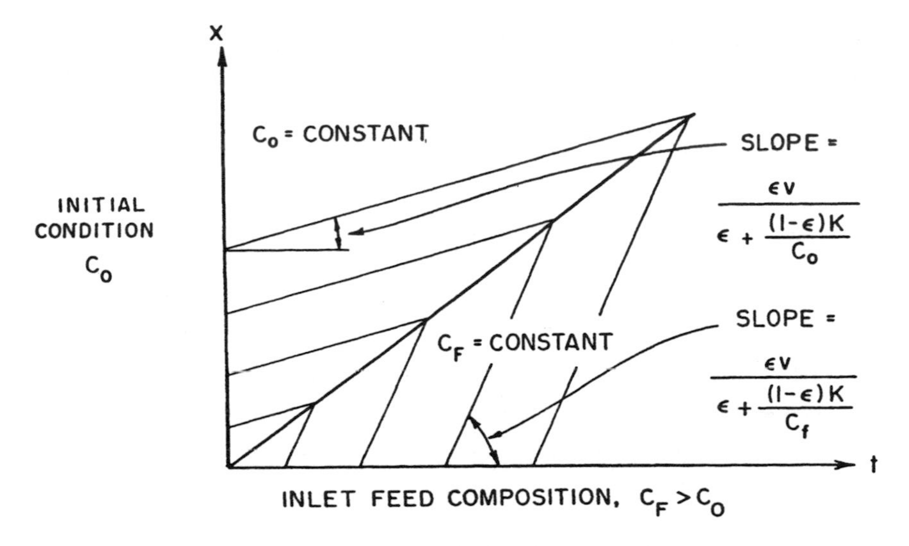

Fig. F.4 Shock formation in a chromatograph.

$$\frac{L}{t_{\text{out}}} = \frac{v\varepsilon}{\varepsilon + (1-\varepsilon)2K(C_o + \Delta C)}. \tag{F.36}$$

Then the concentration changes gradually from C_o to C_f. For a Dirac delta type input, we obtain a curve that looks like a normal distribution curve as in Fig. F.2.

F.3.4 Shock Formation

For the case of a nonlinear favorable isotherm,

$$\frac{\partial^2 \Gamma_i}{\partial C_i^2} < 0, \tag{F.37}$$

the characteristics starting from the initial and the inlet conditions for the case $C_o < C_f$ will intersect, as depicted in Fig. F.4 for the case when

$$\frac{\partial \Gamma_i}{\partial C_i} = \frac{K}{C_i}. \tag{F.38}$$

The shock velocity U_s can be obtained by the rigorous limiting process illustrated in Chapter 10 for settling. Alternatively, it can be obtained by making a species balance in which we see from Fig. F.4 that the fluid phase concentration jumps from C_o to C_f forcing the stationary phase composition to jump from Γ_i to Γ_f. Then the species balance across the shock can be written as

$$(v-U_s)\varepsilon[C]=U_s(1-\varepsilon)[\Gamma], \tag{F.39}$$

where

$$[C]=C_f-C_o, \tag{F.40}$$

$$[\Gamma]=\Gamma_f-\Gamma_o. \tag{F.41}$$

The shock velocity then takes the same form as the continuous propagation velocity for the constant porosity case:

$$U_s=\frac{\varepsilon v}{\varepsilon+(1-e)\frac{[\Gamma]}{[C]}}. \tag{F.42}$$

The dispersion and the shock analysis can be extended to the case of coupled partial differential equations given by Eq. (F.22).

APPENDIX G

SOLUTION OF A HYPERBOLIC SYSTEM OF FIRST ORDER PARTIAL DIFFERENTIAL EQUATIONS

G.1 Decoupling of Quasilinear First Order PDE

Conservation laws in one dimension usually consist of a system of partial differential equations (PDE):

$$\frac{\partial \mathbf{u}}{\partial t} + \mathbf{A}\frac{\partial \mathbf{u}}{\partial x} = \mathbf{b}, \tag{G.1}$$

where the $n \times n$ matrix $\mathbf{A}$ is a function of the n-dimensional dependent vector variable $\mathbf{u}$:

$$\mathbf{A} = \mathbf{A}(\mathbf{u}), \tag{G.2}$$

and where t is time and x is the spatial dimension. For some conservation forms, such as those involving the relative velocity, time, and space must be reversed to maintain unity in front of the first term in Eq. (G.1) (Gidaspow *et al.*, 1983). The standard method (Garabedian, 1964) to diagonalize or to decouple the matrix equation (G.1) is to multiply it by a diagonalizing matrix $\mathbf{W}$, as shown below:

$$\mathbf{W}\frac{\partial \mathbf{u}}{\partial t} + \mathbf{WA}\frac{\partial \mathbf{u}}{\partial x} = \mathbf{Wb}, \tag{G.3}$$

and to observe that for constant $\mathbf{A}$ and $\mathbf{W}$, Eq. (G.3) can be written as an ordinary differential equation if one lets

$$\psi = \mathbf{Wu} \tag{G.4}$$

and rewrites (G.3) in terms of the inverse matrix $\mathbf{W}^{-1}$ as

$$\frac{\partial \psi}{\partial t} + \mathbf{WAW}^{-1}\frac{\partial \psi}{\partial x} = \mathbf{Wb}. \tag{G.5}$$

The diagonal form is obtained by letting

$$\mathbf{WAW}^{-1} = \boldsymbol{\lambda}, \tag{G.6}$$

where $\boldsymbol{\lambda}$ is the diagonal matrix all of whose diagonal elements are one and whose non-diagonal elements are zero. With these definitions, the vector equation (G.1) is equivalent to vector ordinary differential equations:

$$\frac{d\psi}{dt} = \mathbf{Wb} \tag{G.7}$$

along the paths or characteristics generated by the *n* ordinary differential equations

$$\frac{d\mathbf{x}}{dt} = \boldsymbol{\lambda}. \tag{G.8}$$

This theory is the generalization of that discussed in Appendix F for a single partial differential equation.

Equation (G.6) can be written in the more useful form shown below by multiplying it by **W**:

$$\mathbf{W}(\mathbf{A} - \boldsymbol{\lambda}) = \mathbf{0}. \tag{G.9}$$

Equation (G.9) shows that for non-zero W, the following determinant must be zero:

$$|\mathbf{A} - \boldsymbol{\lambda}| = 0. \tag{G.10}$$

This determinant is called the characteristic determinant. It determines the characteristics, $\boldsymbol{\lambda}$. The diagonalizing matrix **W** can then be obtained by solving Eq. (G.9) for its components. Chase *et al.* (1969) give an example of this technique for simultaneous heat and mass transfer in a porous medium with bulk flow, in a chromatographic column, essentially. Although this technique is useful for illustrating the theory, it does not give the equations along the characteristic for non-constant **A**. For the matrix **A** that is of the form (G.2), the method used by Gidaspow *et al.* (1983) and earlier by Koo and Kuo (1977) is more useful than that found in standard texts.

In this method one introduces an intermediate vector function **z**,

$$\mathbf{z} = \mathbf{z}(\mathbf{u}), \tag{G.11}$$

such that the decoupling matrix **W** is the Jacobian matrix

$$\mathbf{W} = \frac{\partial \mathbf{z}}{\partial \mathbf{u}^T}, \tag{G.12}$$

where the superscript T denotes the transpose. Multiplication of (G.1) by the transpose of the vector $\mathbf{w}_i^T$ gives the equation

$$\mathbf{w}_i^T \frac{\partial \mathbf{u}}{\partial t} + \mathbf{w}_i^T \mathbf{A} \frac{\partial \mathbf{u}}{\partial x} = \mathbf{w}_i^T \mathbf{b} \tag{G.13}$$

Let

$$\mathbf{w}_i^T \mathbf{A} = \lambda_i \mathbf{w}_i^T, \tag{G.14}$$

which, on using the rule $(\mathbf{AB})^T = \mathbf{B}^T \mathbf{A}^T$, can be written as

$$\mathbf{A}^T \mathbf{w}_i = \lambda_i \mathbf{w}_i. \tag{G.15}$$

Then the following system of homogeneous equations determines $\mathbf{w}_i$:

$$\left(\mathbf{A}^T - \lambda_i\right)\mathbf{w}_i = 0 \tag{G.16}$$

Equation (G.16) also determines the characteristics λ_i. Alternatively, the characteristic determinant given by Eq. (G.10 can be obtained from Eq. (G.14).

Using (G.14) and (G.12), the partial differential equation (G.13) becomes

$$\frac{\partial z_i}{\partial \mathbf{u}^T} \cdot \frac{\partial \mathbf{u}}{\partial t} + \lambda_i \frac{\partial z_i}{\partial \mathbf{u}^T} \frac{\partial \mathbf{u}}{\partial x} = \mathbf{w}_i^T \mathbf{b}. \tag{G.17}$$

To obtain an ordinary differential equation from (G.17) let

$$\lambda_i = \frac{dx}{dt^i}, \quad i = 1,2,3 \ldots n. \tag{G.18}$$

Then, in view of the chain rule, as in Appendix F, (G.17) becomes the ordinary differential equation

$$\frac{dz_i\left(\mathbf{u}^T(t,x(t))\right)}{dt} = \mathbf{w}_i^T\mathbf{b} \tag{G.19}$$

The chain rule

$$\frac{dz_i}{dt} = \frac{\partial Z_i}{\partial u_1}\frac{du_1}{dt^i} + \frac{\partial z_i}{\partial u_2}\frac{du_2}{dt^i} + \ldots \frac{\partial z_i}{\partial u_n}\cdot\frac{du_n}{dt^i} \tag{G.20}$$

permits one to write (G.19) in the useful form shown below:

$$\mathrm{W}_{i1}\frac{du_1}{dt^i} + \mathrm{W}_{i2}\frac{du_2}{dt^i} + \ldots \mathrm{W}_{in}\frac{du_n}{dt^i} = \mathbf{w}_i^T\mathbf{b}, \quad i = 1,2,3\ldots n. \tag{G.21}$$

Thus, the system of n quasi-linear partial differential equations (G.1) has been converted into n ordinary differential equations (G.21) along the n paths or characteristics given by (G.18). The characteristics can be obtained from the determinant (G.10). The decoupling coefficients W_{ij} are obtained from the solution of homogeneous equation (G.16). In view of redundancy we can choose W_i equal to unity. The example to follow will illustrate the method.

The success of this method depends upon our ability to generate n real and distinct eigenvalues λ_i obtained from the n by n characteristic determinant (G.10). When all such values are real and distinct, the system of partial differential equations given by (G.1) is called a hyperbolic system. When two roots of (G.10) become equal, then an independent gridwork cannot be generated by Eq. (G.18). Such a degenerate system is referred to as being parabolic. We obtain an elliptic system if two or more roots of the characteristic determinant (G.10) are complex roots. This would mean that there exists no paths along which information is propagated. Such a problem is ill-posed as an initial value problem, as described in Chapter 7. A system of first order partial differential equations with complex characteristics is called elliptic. Many steady state systems, with t being dimension one, and x as the second coordinate, are elliptic.

G.2 Transient Pipe Flow

Transient one-dimensional flow is important in transmission of natural gas in long pipelines and in the analysis of thermalhydraulics of nuclear reactors where a homogeneous equilibrium model is used (Moore and Rettig, 1973; Lyczkowski *et al.*, 1982). The one-dimensional continuity equation for a compressible gas or for a homogeneous mixture can be written in terms of the flux, $F = \rho v$; the pressure, P; and the sonic velocity, c, as

$$\frac{\partial P}{\partial t} + c^2 \frac{\partial F}{\partial x} = 0. \tag{G.22}$$

For the purpose of illustrating the decoupling procedure, the velocity square term is neglected in the momentum equation. Such an approximation is valid for reasonably small flow of gas in pipelines compared with the sonic velocity, but is not valid for blow-down in a nuclear reactor where the velocity of discharge may be near the critical flow. With such an approximation, the momentum balance with frictional force per unit volume represented by f_w, usually given by means of a Fanning's type expression, as in Chapter 2, Eq. (2.49), is

$$\frac{\partial F}{\partial t} + \frac{\partial P}{\partial x} = f_W. \tag{G.23}$$

In pipe flow, the energy equation is not needed, since the temperature is constant, while in homogeneous two phase flow, the energy equation shows that one of the characteristics is the velocity of the fluid. Hence, Eqs. (G.22) and (G.23) define the problem. In matrix form they are written as

$$\begin{pmatrix} \frac{\partial P}{\partial t} \\ \frac{\partial F}{\partial t} \end{pmatrix} + \begin{pmatrix} 0 & c^2 \\ 1 & 0 \end{pmatrix} \begin{pmatrix} \frac{\partial P}{\partial x} \\ \frac{\partial F}{\partial x} \end{pmatrix} = \begin{pmatrix} 0 \\ f_w \end{pmatrix}. \tag{G.24}$$

The characteristic determinant is

$$|A - \lambda| = \begin{vmatrix} -\lambda & c^2 \\ 1 & -\lambda \end{vmatrix} = \lambda^2 - c^2 = 0. \tag{G.25}$$

Hence, the characteristics are

$$\lambda = \pm c. \tag{G.26}$$

The decoupling coefficients are obtained as follows:

$$\left(\mathbf{A}^T - \lambda\right)\mathbf{w}_i = \begin{pmatrix} -\lambda & 1 \\ c^2 & -\lambda \end{pmatrix}\begin{pmatrix} \mathrm{w}_{i1} \\ \mathrm{w}_{i2} \end{pmatrix} = \begin{pmatrix} -\lambda \mathrm{w}_{i1} + \mathrm{w}_{i2} \\ c^2 \mathrm{w}_{i1} - \lambda \mathrm{w}_{i2} \end{pmatrix} = \begin{pmatrix} 0 \\ 0 \end{pmatrix}. \tag{G.27}$$

Let $W_{i1} = 1$. Then (G.27) gives

$$\begin{pmatrix} -\lambda + \mathrm{w}_{i2} \\ c^2 - \lambda \mathrm{w}_{i2} \end{pmatrix} = \begin{pmatrix} 0 \\ 0 \end{pmatrix}. \tag{G.28}$$

The two equations from (G.28) give

$$\mathrm{w}_{i2} = \lambda \tag{G.29}$$

and

$$\mathrm{w}_{i2} = c^2/\lambda. \tag{G.30}$$

But in view of (G.26), Eqs. (G.29) and (G.30) are the same. Hence,

$$\mathrm{w}_{i2} = \pm c. \tag{G.31}$$

Let i of one be the positive values of c. The right-hand side of (G.21) is evaluated as

$$\mathbf{w}_i^T \mathbf{B} = \left(1 \mathrm{w}_{i2}\right)\begin{pmatrix} 0 \\ f_w \end{pmatrix} = \mathrm{w}_{i2} f_w. \tag{G.32}$$

Then, Eq. (G.21) gives the following ordinary differential equations:

$$\frac{dP}{dt} + c\frac{dF}{dt} = cf_w \tag{G.33}$$

along the path given by

$$\frac{dx}{dt} = c, \tag{G.34}$$

and

$$\frac{dP}{dt} - c\frac{dF}{dt} = -cf_w \tag{G.35}$$

along the path given by

$$\frac{dx}{dt} = -c. \tag{G.36}$$

Chapter 4 presents the more general case for granular flow. It can be shown that Eqs. (G.33) and (G.35) are simply a rearranged form of the original equations (G.22) and (G.23). By convective differentiation of (G.33) using (G.34), this equation as partial differential equation is

$$\frac{\partial P}{\partial t}+c\frac{\partial P}{\partial x}+c\frac{\partial F}{\partial t}+c^2\frac{\partial F}{\partial x}=cf_w. \tag{G.37}$$

But (G.37) is simply the sum of c times Eq. (G.23) plus Eq. (G.22). Equation (G.35) can be interpreted similarly.

If c^2 in Eq. (G.22) were negative, the characteristics in (G.25) would be imaginary. The system is then called elliptic. The paths given by Eqs. (G.34) and (G.36) are then imaginary, that is, no real paths exist. Fortunately, the Second Law of Thermodynamics guarantees that the sonic velocity c is real. It is the so-called stability condition (Callen, 1960).

G.3 Dimensionless Representation of Pipe Flow

Transient flow in a pipeline was described by Eqs. (G.22) and (G.23) in the previous section. Now consider a pipe of length L and a characteristic pressure P_o at the inlet of the pipe at zero time. For isothermal transmission of ideal gas, the sonic velocity c becomes simply the constant group

$$c=\sqrt{\left(\frac{\partial P}{\partial \rho}\right)_T}=\sqrt{\frac{\mathrm{R}T}{M}}, \tag{G.38}$$

where T is the absolute temperature, R the ideal gas law constant, and M its molecular weight. Then, in terms of the dimensionless pipe length $\bar{x}$,

$$\bar{x}=\frac{x}{L}, \tag{G.39}$$

Eq. (G.34) shows that the natural dimensionless time $\bar{t}$ is

$$\bar{t}=\frac{tc}{L}=\text{Time}\times\text{Frequency}. \tag{G.40}$$

This group is a ratio of the real time to the time it takes a pressure wave to move through the pipe. Since c is of the order of 500 m/s, the length of the pipe must be of the order of 500 m for the transient to be important. This can be more clearly seen from the basic Eqs. (G.22) and (G.23). For a large c^2, Eq. (G.22) shows that the flux variation with distance must be near zero and the pressure is nearly steady. Then the pressure drop is simply balanced by friction, as seen in Eq. (G.23). Thus, transients are important only for short durations or for long pipes. Alternatively, we see that transients are of a high frequency for a steady time averaged flow, where the frequency is c/L. For a gas, the frequencies are high. For a granular solid flow, where c is of the order of one, the frequencies are low.

In terms of the dimensionless variables discussed earlier, the basic Eqs. (G.33) to (G.35) become

$$\frac{d\bar{P}}{d\bar{t}}+\frac{d\bar{F}}{d\bar{t}}=\left(\frac{Lf_w}{P_o}\right) \tag{G.41}$$

along the path

$$\frac{d\bar{x}}{dt}=1, \tag{G.42}$$

$$\frac{d\bar{P}}{d\bar{t}}-\frac{d\bar{F}}{d\bar{t}}=-\left(\frac{Lf_w}{P_o}\right) \tag{G.43}$$

along the characteristic

$$\frac{d\bar{x}}{d\bar{t}}=-1, \tag{G.44}$$

where

$$\bar{P}=\frac{P}{P_o}\ ,\quad \bar{F}=\left(\frac{cF}{P_o}\right)\quad \text{and}\quad \bar{f}=\frac{Lf_w}{P_o}. \tag{G.45}$$

The set of Eqs. (G.41) to (G.44) characterize both the gas flow in transmission systems and one-dimensional transient granular flow, if c is taken to be simply a constant.

G.4 Frictionless Flow Example

It was already described in the text that frictionless flow is described by a wave equation. Such an equation for pressure or flux is equivalent to Eqs. (G.41) to (G.44) with zero f_w. As an illustration of the method of characteristics consider frictionless flow in a pipe as illustrated in Fig. G.1. Initially the flow was constant and equal to 0.1 in terms of the dimensionless flow rate given by Eq. (G.45). The pressure drop was linear, decreasing to zero at the end of the pipe. At zero time the pressure at the end of the pipe was decreased to –0.1 of its initial value at the pipe entrance. The problem is to compute the transient pressure and flow.

For frictionless flow, Eqs. (G.41) to (G.44) show that

$$\overline{P} + \overline{F} = \alpha \tag{G.46}$$

along the straight line characteristics with a slope of 45°, called α characteristics, and

$$\overline{P} - \overline{F} = \beta \tag{G.47}$$

along the 135° characteristics, called β characteristics.

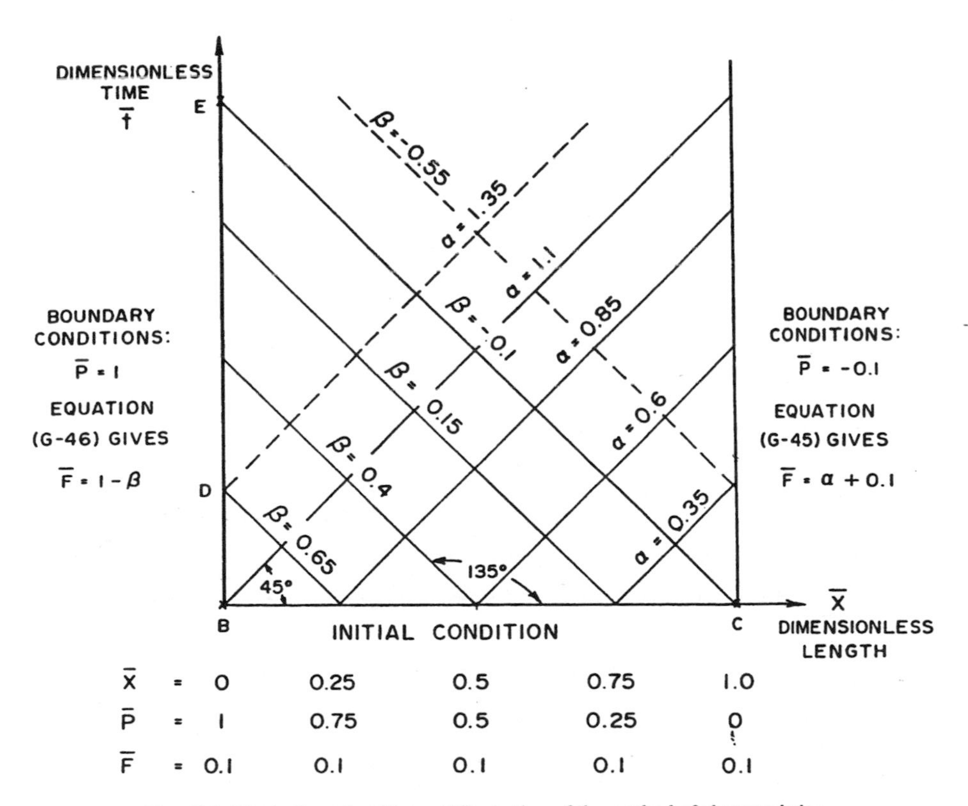

Fig. G.1 Frictionless pipe flow: an illustration of the method of characteristics.

The dimensionless flow and the pressure can be obtained in terms of the parameters α and β by adding and subtracting equation (G.46) and (G.47) to give

$$\overline{P} = (\alpha + \beta)/2, \tag{G.48}$$

$$\overline{F} = (\alpha - \beta)/2. \tag{G.49}$$

The α values in Fig. G.1 are obtained by adding the values of $\overline{P}$ and $\overline{F}$ from the initial condition as required by (G.46). Similarly, the values of β are obtained using (G.47). In the triangular region ABC the pressure and the flow rate are completely determined by the initial condition from the values of α and β shown in the figure and from Eqs. (G.48) and (G.49). To obtain the solution outside this region, boundary conditions are necessary. By specifying the pressure at the two boundaries, the flow rate can be obtained as shown in Fig. G.1. The flow rate increases with time at both ends of the pipe, as expected.

The α characteristics which begin at the pipe inlet, such as that shown through point D, are obtained from the prescribed value of the pressure at the boundary and Eq. (G.46) to be

$$\alpha = 2 - \beta. \tag{G.50}$$

Thus, the inlet and the initial conditions determine the solution in region BCE. To obtain the complete solution, a boundary condition at the pipe exit is required. With the prescribed pressure depicted in Fig. G.1, the values of β are determined using (G.41) to be

$$\beta = -0.2 - \alpha. \tag{G.51}$$

G.5 Numerical Solution by Characteristics

The numerical solution to the transient pipe flow problem given by Eqs. (G.41) to (G.45), with the wall friction f_w expressed by the Fanning's equation

$$f_w = \frac{2 f_g \rho v^2}{D_t}, \tag{G.52}$$

with the gas friction coefficient given by Eqs. (2.50) to (2.53), can be obtained by integrating Eqs. (G.41) and (G.43) along this respective α and β characteristics,

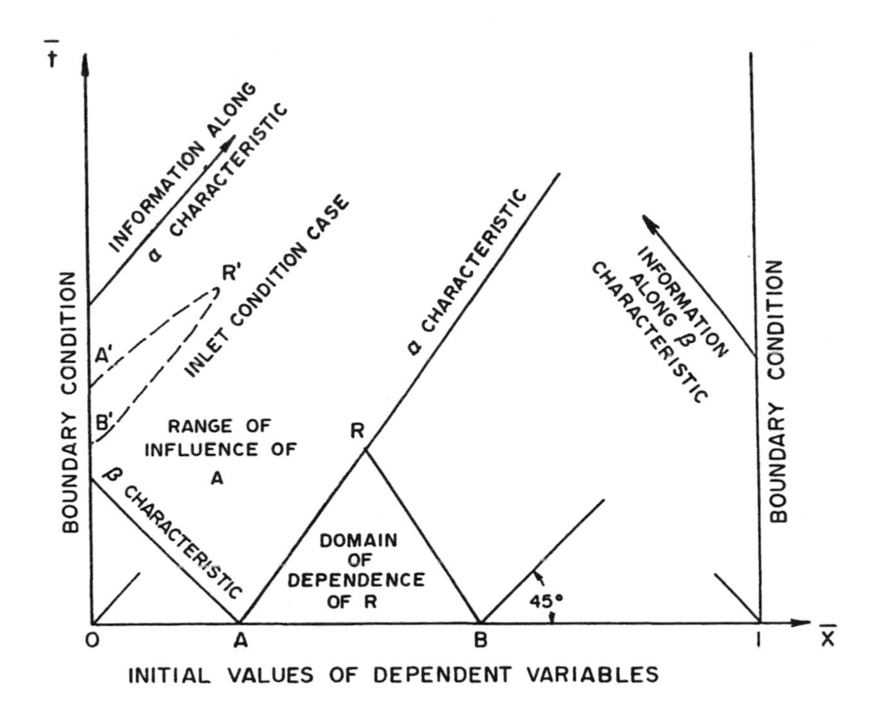

Fig. G.2 Integration using the method of characteristics.

as depicted in Fig. G.2. Integration of (G.41) from points A to R gives

$$\int_{\overline{P}_A}^{\overline{P}_R} d\overline{P} + \int_{\overline{F}_A}^{\overline{F}_R} d\overline{F} = \int_{\bar{t}_A}^{\bar{t}_R} \bar{f}\, d\bar{t} \tag{G.53}$$

The initial conditions give the values of $\overline{P}_A$ and $\overline{F}_A$. The integral on the right of (G.53) can be integrated by various quadrature formulas. The use of the trapezoidal approximation gives the average value of $\bar{f}$ at the two points times the length of the internal. Hence, (G.53) becomes

$$\overline{P}_R - \overline{P}_A + \overline{F}_R - \overline{F}_A = \tfrac{1}{2}\left(\bar{f}_A + \bar{f}_R\right)\cdot\left(\bar{t}_R - \bar{t}_A\right). \tag{G.54}$$

Similarly, along the β characteristic we obtain

$$\bar{P}_R - \bar{P}_B - \bar{F}_R + \bar{F}_B = -\tfrac{1}{2}\left(\bar{f}_B + \bar{f}_R\right)\cdot\left(\bar{t}_R - \bar{t}_B\right). \tag{G.55}$$

In view of the nonlinearity of $\bar{f}$ as given by Eq. (G.52), Eqs. (G.54) and (G.55) are two algebraic non-linear equations for $\bar{P}_R$ and $\bar{F}_R$. They can be solved simultaneously by iteration using the secant method. With the given boundary conditions, as discussed in the previous frictionless case, the complete solution can be obtained, as illustrated for the two points above. This technique was extended to unequal velocity two phase flow by Rasouli *et al.* (1983). A set of six first order partial differential equations were solved for the case when the discharge from the pipe could become critical, as illustrated in Chapter 4 for granular flow.

Figure G.2 also illustrates the concepts of the domain of dependence and the range of influence. The region ABR is called the domain of dependence of point R. The range of influence of A is the region in between the α and the β characteristics drawn through the point A. The values given at A affect all the values in such a region. Hence its name. The figure also shows how the boundary information is propagated along the characteristics for the pipe flow. For the chromatographic problem given by Eq. (F.22), the characteristics are all real and positive. Chase *et al.* (1969) show this for the case of simultaneous heat and mass transfer with a constant equilibrium constant. In such a situation, which can be easily generalized to the general variable coefficient case discussed in this section, the characteristics demand that data be prescribed at the inlet, as indicated in Fig. G.2 by the dashed line and points A', B', and R', which correspond to points A, B, and R in the pipe flow problem.

G.6 Uniqueness of Solution and Boundary Conditions

It has been shown that the n conservation laws given by the set of n quasi-linear partial differential equations (G.1) can be expressed as set of n ordinary differential equations (G.21) along the n paths generated by the n ordinary differential equations (G.18), when the characteristics given by the determinant (G.10) are all real and distinct. In chromatographic separation given by Eq. (F.22), when $K_{ij} = K_{ji}$, the matrix A is symmetric. Then the stability condition of thermodynamics (Callen, 1960) appears to require the matrix A to be positive definite. Such requirements lead to real and positive characteristics. Hence, for a chromatographic column all information must be prescribed at the inlet, as is done in actual chromatographic chemical analysis, where a sample to be analyzed is injected at the inlet of the column. Multiphase flow problems give rise to boundary value problems, as illustrated by the simple gas transmission problem discussed in the previous sections where the characteristics were positive and negative. A further complication with two phase flow problems is that the popular model A where the pressure drop is divided between the liquid and the gas phases gives rise to complex characteristics and hence to numerical instability and a violation of the principle of cause and effect, as illustrated in Fig. 7.1 A decade of numerical solution with such models has shown that they can lead to non-physical behavior in regions where stabilization with a modulus of elasticity or viscosity fails in an unexpected way in view of the strong nonlinearity. Hence, such models are not considered further.

Then for real characteristics the paths can be obtained by integrating Eq. (G.18) between points R and point i=1, 2, 3...n:

$$\frac{x(R)-x(i)}{t(R)-t(i)} = \int_{t(i)}^{t(R)} \lambda_i \, dt, \quad i=1,2,3...n. \tag{G.56}$$

Along each path i generated by (G.56) Eq. (G.21) gives the integral equation shown below:

$$\int_{t(i)}^{t(R)} \sum_{j=1}^{n} w_{ij} \frac{du_{ij}}{dt^i} = \int_{t(i)}^{t(R)} \mathbf{w}_i^T \mathbf{b} \tag{G.57}$$

For the region determined by the initial values, one can apply the contraction mapping theorem of functional analysis (Collatz, 1966) to show that when the functions under the integrals in (G.56) and (G.57) satisfy the Lipschitz condition, these equations will generate a unique solution which can be obtained by integration. The Lipschitz condition restriction is rather mild since the integrals give Δt times a Jacobian matrix evaluated at some mean values and Δt can be made arbitrarily small. Hence, starting with reasonable initial data we are guaranteed a unique numerical solution. A prediction can be made. When characteristics are positive and negative, as in pipe flow, initial values do not determine the complete solution, as illustrated in Fig. G.2. For the general non-linear problem given by Eq. (G.1), an existence and uniqueness theorem has surprisingly not been given in the applied mathematics literature. Since the values of λ_i are not arbitrary, but are obtained from the conservation equations, as illustrated in Chapter 7, such a task appears to be feasible.

Literature Cited in Appendices

Abbott, M. B. (1966). *An Introduction to the Method of Characteristics.* New York: Elsevier.

Aris, R. (1962). *Vectors, Tensors, and the Basis Equations of Fluid Mechanics.* Englewood Cliffs, New Jersey: Prentice Hall.

Bird, R. B., W. E. Stewart, and E. N. Lightfoot (1960). *Transport Phenomena.* New York: John Wiley & Sons.

Brand, L. (1955). *Advanced Calculus.* New York: John Wiley & Sons.

Callen, H. B. (1960). *Thermodynamics.* New York: John Wiley & Sons.

Carslaw, H. S., and J. C. Jaeger (1959). *Conduction of Heat in Solids,* 2nd Ed. London, U.K.: Oxford University Press.

Chase, C. A., D. Gidaspow, and R. E. Peck (1969). "Adiabatic Adsorption in a Regenerator," *CEP Symp. Series No. 96* **65**, 34–47.

Chase, C. A., D. Gidaspow, and R. E. Peck (1970). "Transient Heat and Mass Transfer in an Adiabatic Regenerator: A Green's Matrix Representation," *Ind. J. Heat and Mass Transfer* **13**, 817–833.

Collatz, L. (1966). *Functional Analysis and Numerical Mathematics,* translated from the German by H. Oser. New York: Academic Press.

Courant, R., and D. Hilbert (1962). *Methods of Mathematical Physics Vol. II: Partial Differential Equations.* New York: Wiley-Interscience.

Eringen, C. (1975). *Continuum Physics, Vol. II: Continuum Mechanics of Single-Substance Bodies.* New York: Academic Press.

Garabedian, P. R. (1964). *Partial Differential Equations.* New York: John Wiley & Sons.

Ghezelayagh, H., and D. Gidaspow (1982). "Micro–Macropore Model for Sorption of Water on Silica Gel in a Dehumidifier," *Chem. Eng. Sci.* **37**, 1181-1197.

Gidaspow, D. (1978). "Hyperbolic Compressible Two Phase Flow Equations Based on Stationary Principles and the Fick's Law," pp. 283–297 in *Two Phase Transport and Reactor Safety,* Vol. 1. Washington D.C.: Hemisphere Publishing Corp.

Gidaspow, D., and R. T. Ellington (1964). "Surface Combustion of Hydrogen, Part I: On Platinum Coated Alumina" *AIChE J.* **10** (5), 707–713.

Gidaspow, D., F. Rasouli, and Y. W. Shin (1983). "An Unequal Velocity Model for Transient Two-Phase Flow by the Method of Characteristics," *Nucl. Sci. & Eng.* **84**, 179–195.

Goldberg, S. (1958). *Introduction to Difference Equations.* New York: John Wiley & Sons.

Koo, H. and K. K. Kuo (1977). "Transient Combustion in Granular Propellant Beds Part I: Theoretical Modeling and Numerical Solution of Transient Combustion Processes in Mobile Granular Propellant Beds," Final report of Feb. 15, 1974 – Feb. 14, 1977 from BRL-CR, Ballistic Research Laboratory, AD-A044998, Aberdeen Proving Ground. Aberdeen, Maryland.

Li, J., Y. Tung, and M. Kwauk (1988). "Method of Energy Minimization in Multi-scale Modeling of Two-Phase Flow," in P. Basu and J. F. Large, Eds. *Circulating Fluidized Bed Technology II.* New York: Pergamon Press.

Li, J., M. Kwauk, and L. Reh (1992). "Role of Energy Minimization in Gas/Solid Fluidization," in O. E. Potter and D. J. Nicklin, Eds. *Fluidization VII*. New York: Engineering Foundation.

Lyczkowski, R. W., D. Gidaspow, and C. W. Solbrig (1982). "Multiphase Flow Models for Nuclear, Fossil and Biomass Energy Production," pp. 198–351 in A. S. Majumdar and R. A. Mashelkar, Eds. *Advances in Transport Process Vol. II*. New York: Wiley-Eastern Publisher.

Moore, K. V., and W. H. Retting (1973). "RELAP-4-A Computer Program for Transient Thermal-Hydraulic Analysis," Aerojet Nuclear Co., available from the National Technical Information Service.

Rasouli, F. D., D. Gidaspow, and Y. W. Shin (1983). "Unequal Velocity, Two Component, Two Phase Flow by the Method of Characteristics," *Argonne National Laboratory Report, ANL-83-98*. Argonne, Illinois; also see F. Rasouli (1981). Ph.D. Thesis. Chicago, Illinois: Illinois Institute of Technology.

Rhee, H.-K., R. Aris, and N. R. Amundson (1986). *First Order Partial Differential Equations*, Vol. I. Englewood Cliffs, N.J.: Prentice Hall.

Roy, D., and D. Gidaspow (1972). "A Cross Flow Regenerator: A Green's Matrix Representation," *Chem. Eng. Sci.* **27**, 779–793.

Roy, D., and D. Gidaspow (1974). "Nonlinear Coupled Heat and Mass Exchange in a Cross-Flow Regenerator," *Chem. Eng. Sci.* **29**, 2101–2114.

Sneddon, I. A. (1957). *Elements of Partial Differential Equations*, New York: McGraw-Hill.

Spielberg, N., and B. D. Anderson (1987). *Seven Ideas That Shook the Universe*. New York: John Wiley & Sons.

Vora, M. K., W. A. Sandstrom and A. Rehmat (1980). "Historical Development of the U-Gas Process at the IGT Pilot Plant," pp. 648–653 in *Proceedings of the 15th Intersociety Energy Conversion Engineering Conference*, Vol. 1. AIAA.

INDEX

A

Abbott, M.B. 75, 84, 95, 431, 453
Abed, R. 218, 235
Acrivos, A. 212, 236, 383, 388
Added Mass Force 35
Aldis, D.F. 26, 28
Amundson, N. 116, 126
Anderson D.H.A. 183, 192
Angle of repose 76, Fig. 4.1
Annular flow 34, 36—Table 2.1, 209—Fig. 8.1, 212—Fig. 8.4, 214—Fig. 8.6, 224, 226—Fig. 8.15, 232—Fig. 8.23
Araki, S. 299, 334
Arastoopour, H. 41, 50, 52, 58, 117, 214, 235, 344, 353
Archimedis number
Aris, R. 2, 28, 423, 425, 453
Assumptions in kinetic theory 248, 272, 339, 345
  binary collisions 247, 339
  chaos 248, 339
  inelastic particles 245, 337, 338
Attrition 112
Averaging
  ensemble 259, 339
  kinetic theory 259, 339
  operator 21
  volume 21
Averaging theorems 21, 259, 276, 339
Auzerais, F.M. 374, 388

B

Babu, S.P.S. 126, 146, 192
Bader, R. 224, 236
Bagnold, R.A. 239, 294, 304, 334
Bagnold number 304
Bahary, M. 125, 126
Ball probe 220, 223—Fig. 8.12
Batchelor, G.K. 141, 192, 361, 388
Batch settling 357
  interface history 363
  particle velocity 355, 380—Fig. 12.8
Bernoulli equation 78
Bezbaruah, R. 324, 334
Bilicki, Z.
Bird, R.W. 35, 37, 39, 58, 106, 113, 201, 236, 427, 454
Birkhoff, G. 35, 58, 280, 294
Blake, T.R. 152, 180, 192, 202, 236
Boiling 19, 53
Boltzmann equation 256, 339
Boltzmann constant 269—Table 9.5
Borgwardt, R.H. 54, 59
Boundary condition
  slip 254, 295, 298, 302, 325—Fig. 10.16
Bouillard, J.X. 82, 95, 113, 120, 126, 172, 192
Boundary layer, (problem) 234
Boure, J.A. 9, 28, 85, 95
Bowen, R.M. 3, 4, 7, 12, 15, 17, 22, 28
Brand, L. 426, 454
Breault, R.W. 58, 117, 126
Brennen, C.E.

Bubble
- computed 154, 158—Figs. 7.6, 7.7, 7.9, 7.11, 7.14, 7.21, 7.24, 7.28, 7.31; 317—Fig. 10.5, 327—Fig. 10.8
- experimental 158—Figs. 7.6, 7.7, 7.11, 7.13; 313—Fig. 10.5b
- fast 178—Fig. 7.28
- formation 161—Fig. 7.8
- growth 161—Fig. 7.8 and Fig. 7.11
- pressure distribution 176—Figs. 7.26, 7.27
- shape 158—Figs. 7.5, 7.6, 7.7, 7.9, 7.10, 7.28
- slow 149
- velocity 149, 173, 175—Fig. 7.25

Bubbling criterion 119, 120—Fig. 6.3, 121
Buscall, R. 374, 388

C

Callen, H.B. 4, 28, 80, 95, 134, 192, 452, 454
Campbell, C.S. 302, 335
Carlos, C.R. 241, 294, 301, 335
Carslaw, H.S. 187, 193, 241, 294, 375, 388
Cause and effect violation, 133—Fig. 7.1
Capillary force 374
Chapman, S. and Cowling, T.G. 88, 95, 240, 295, 339, 353
Channeling, in fluid beds 330—Fig. 10.17
Chaos 240—Fig. 9.1, 269—Fig. 9.5, 241
Charge 109, 377, 383
Characteristics 83—Fig. 4.3, 133, 137, 138, 186–192, 431
Characteristic velocity *(see Characteristics)*
Chase, C.A. 434, 454
Chemflub model Table 7-2, 153
Chemical reaction 3, 20, 54, 55, 394, 418, 434
Choking 82, 91
Churchill, S.W. 141, 193
Churn flow 124
Chute flow 307, Figs. 10.3 and 10.4
Circulating fluidized bed (CFB) 99—Fig. 5-1, 207, 210, Fig. 8-2, 324, Fig. 10.16—Fig. 10.23, 330
Circulation 330
Cloud in bubbling beds 149, 178—Fig 7.28
Cluster 214, 216, Figs. 8–10
Coal 109, 422
Coalescence 125—Fig. 6.3
Cohesive shear 76—Fig. 4.1
Collatz, L. 453, 454
Collier, J.G. 50, 59
Collins, R. 146, 193
Collisions
- binary 244
- elastic 244
- inelastic 245, 338—Fig. 11.1

Colloidal suspension 376
Combustion 57
Compression settling 366
Complex characteristics 135, 138, 191
Component equation 20, 54, 55
Computer model 149
Conductivity
- dilute 292
- dense 294
- Onsager type 380

Cook, K.T.
Condiff, D.W. 280, 295, 339, 353
Conservation laws
- energy 7
- enthalpy 9
- entropy 9
- mass 3, 217—Table 8.1, 314—Table 10.1
- momentum 3, 31, 36—Table 2.1, 217—Table 8.1, 221—Table 8.2, 314—Table 10.1
- species 20

Consolidation coefficient 375
Constitutive equations for
- conductivity 294, 315—Table 10.1

viscosity 290, 314—Table 10.1
Coulaloglu, C.A. 125, 127
Courant, R. 135, 193, 429, 456
Continuity equation 3, 412, 415
Convective heat transfer 8
Conveying, pneumatic 40, 41—Fig 2.3
Critical flow
gas 83—Fig. 4.3
kinetic theory 88
particles 86, 254—Table 9.1
vapor-liquid 94—Fig. 4.5
Crystallizer 422

D

Dahler, J.S. *(see Condiff, D.W.)*
Davidson, J.M. 104, 114, 115, 126
Davies, J.T. 378, 388
Debye, P. 376, 388
Decoupling of first order PDE
Deformation tensor
DeGroote, S.R. 379, 388
Delaplaine 76, 95
Dense suspensions
conductivity 290, 347—Table 11.2
diffusivity 255, 368
viscosity 212, 216—Fig. 8.8, 317—Fig. 10.7, 323—Fig. 10.14, 327—Fig. 10.18, 347—Table 11.2
Desalting 185, 381
Design
computer 149, 324
scale factors 201
Diffusivity 255, 368
Diffusion model 234, 375
Dilute suspension 253
viscosity and conductivity 257—Table 9.2
Ding, J. 86, 95, 240, 295, 311, 335
Dispersion 121—Fig. 6.2, 440—Fig. F.3
Distribution function
normal 241, 320—Fig. 10.10
particle 240, 263
Distribution of particle size 337
Dissipation of energy
Dobran, F. 23, 28
Donsi, G.M. 171, 193
Drag coefficient
dense suspension, Ergun 35
dilute suspension 37
sphere, standard curve 38—Fig. 2.2
Drew, D.A. 134, 193, 201, 236, 280, 295
Drift flux
bubble column 67—Fig. 3.2
definition 64
dense flow 68
graphical method 65—Fig. 3.1
riser 68
three phases 69
Drying 56
Drops 67
Dimensionless groups *(see Scale factors)*
Dimensionless numbers *(see Scale factors)*

E

Elasticity modulus 42, 80, 83, 88—Table 4.1, 121, 131, 137, 286, 368, 375
Electrophoresis 378
Electrophoretic mobility 378
Electrostatic charging 109, 376
Electrostatic separation 109
Elliptic PDE 133
Energy equation 7, 55
Entrainment 23
Entropy 10, 18, 269—Table 9.5
Equilibrium 53
Ergun equation 35
Erosion 112
Ettehadieh, B. 156, 193
Experimental study
CFB 207
pneumatic transport 40
settling 371, 370
Eulerian approach 405

F

Fan, L.S. 70, 71, 124, 126
Fanning's equation 33, 46
Fanucci, J.B. 116, 126

Fast bubble 149, 178
Fast fluidization 208
Flow regime
  bubble 154
  cluster 220
  core-annular 224
Fick's law of diffusion 350, 375
Film coefficient 56
Fitzgerald, T.J. 103, 114, 202, 236
Flow over a flat plate 198, 234
Fluidized bed
  bubble dynamics 115
  bubble formation 119, 161
  circulating fluidized bed (CFB) 207
  clusters 220
  flow patterns 220, 224, 311, 219
  flow regime 220, 224, 319
  fluctuations 204, 328
  Geldart's classification 103
  kinetic theory approach 311, 324
  minimum fluidization 98
  segregation 109
  viscosity 216, 257
  void propagation 116
Font, R. 363, 388
Force
  buoyant 37
  drag 39
Foscolo, P.U. 116, 126
Freeboard 157
Free settling 360
Friction coefficient
  definition 33
  interface 35
  Onsager 379
Froude number 205
Fuel cell 395

G

Gamwo, I. 311, 335
Galileo relativity 199
Galtier, P.A. 208, 236
G modulus 42, 80, 83, 88, 121, 131, 286, 368, 375
Garabedian, P.R. 441, 454
Gas dynamics 444
Gas-liquid flow 18, 21, 53, 190
Gasification 55
Geldard, D. 103, 114, 206, 236
Geldard's classification 106—Fig. 5.3
Ghezelayagh, H. 434, 454
Glicksmann, L. R. 202, 236
Gorin, E. 381, 388
Gough, P.S. 81, 96
Grace, J.R. 149, 173, 194
Grain inertia regime 305
Granular flow
  constitutive equation 75, 83
  critical flow 82, 86
  Jannssen equation 75
  mass balance 74
  maximum flux 86, 91
  momentum balance 74
  statics 75
  sound speed 88, 286
Granular temperature
  definitions 242, 341
  multiparticle size 341
Green's functions 145, 241
Green's theorem 142
Gupta, R. 348, 353

H

Heat exchanger 55, 187
Heat flux 7
Heat transfer
  fluidized bed 197
  riser 56, 57
Heat transfer coefficient 8, 55, 56
Hewitt, G.F. 50, 58, 61, 72
Hirschfelder, J.O. 342, 353
Hold-up
  computation 43, 44—Fig. 2.4, 61, 63, 166, 168, 227, 233, 326, 331, 371, 387
  definition 61
  in homogeneous flow 62
  in particulate flow 43, 44—Fig. 2.4, 63

in gas-liquid flow 66
in gas-liquid-solid flow 70
in a riser 69, 220, 226, 232, 326, 331
Matsen's approach 69
Homogeneous flow 53, 62
Hopper flow 83, 87
Hydrodynamic models
CHEMFLUB 153—Table 7.2
kinetic theory 314—Table 10.1
one dimensional 36—Table 2.1
two dimensional 150—Table 7.1
Viscosity Coefficient Model B 221—Table 8.2
Hyperbolic PDE 133, 441

I

Identity, for mixtures 15
Inclined plane flow 307
Information flow 83—Fig. 4.3
Ill-posed problem 135, 191, 453
Instabilities *(see Ill-posedness)*
Instability, numerical *(see Well-posedness)*
Institute of Gas Technology (IGT) correlation 51
Integration theory of PDE 441
Interfacial drag 35, 39
Intergranular stress 373
Inviscid multiphase flow
basic equations 129, 150
bubble formation 158, 170
jet flow 158, 170
Irreversibility
definition 11
in a fuel cell 402
violation 19
Irreversible process *(see Irreversibility)*
Ishida, M. 309, 335
Ishii, M. 7, 28

J

Jackson, R. 115, 126, 178, 194
Jannssen equation 75
Jayaswal, U.K. 384, 388
Jeffrey, D.J. 212, 236
Jeffreys, H. 201, 237
Jenkins, J.T. 88, 96, 131, 240, 295, 339, 354
Jenkins-Savage transport theorem 276
Jet
bubble formation 158, 161
large scale experiment 182
penetration 182
region of influence 144
Johnk, C. 324, 335
Johnson, P.C. 240, 295
Johnson, K.L. 338, 354
Jump mass balance 363, 442

K

Kalinin, A.V. 10, 28
Kaza, K.R. 79, 96
Kinetic
energy balance 107
energy dissipation (ch. 5) 107, 282, 298, 347
energy fluctuation 281, 290, 315, 347
Kinetic theory
definition of temperature
thermal 269
granular 242
multiparticle size 341, 347
pressure 267, 269, 347
Keairns, D.L. 103, 114, 180, 196, 333, 336
Klein, H.H. 116, 126
Klinzig, G. 44, 51, 59
Konno, H. 47, 51, 59
Koo, H. 442, 454
KRW bubbling bed 179
Kunii, D. 35, 59, 186, 194
Kuo, K.K. 442, 454
Kuwabura, G. 301, 335
Kynch, G.J. 363, 388

L

Lackme, C. 66, 72
Lagrangian approach 413

Lamella settler 383
Laplace's equation 133, 147
Lax, P.D. 135, 194
Leibnitz formula 411
Leung, L.S. 33, 40, 59, 61, 72
Levenspiel, O. 23–27, 28, 59, 146, 186, 422
Levich, V.G. 377, 389
Li, J. 392, 457
Lin, J. 5, 319, 335
Liquid, settling in 357
Littman, H. 178, 194
Liu, Y. 117, 126, 143, 194
Lun, C.K.K. 88, 96, 240, 295, 307, 335
Luo, K.M. 38, 47–49, 59, 65, 72, 220, 237
Lyczkowski, R.W. 10, 20, 28, 34, 59, 134, 194, 206, 236, 429, 455

M

Mace, G.R. 383, 389
Mach number 203, 205
Manometer formula 33
Mass balance 2, 74, 129, 150, 261, 264, 357, 394, 412
Matsen, J.M. 61, 68, 72
Maximum solids flux 86
Maxwellian distribution
  for gas 252, 254
  for particles 242
  Graphical representation 320—Fig. 10.10
Maxwell's transport equation 259
McCabe, W.L. 75, 96, 421
Mechanical energy balance 22, 107
Miller, A. 208, 237, 327, 335
Mixing, multiparticle 342
Mixture
  Boltzmann equation 339
  conservation of mass 3, 11, 14, 20
  energy 12, 18
  entropy 18
  identities 15
  momentum 16, 36—Table 2.1
  steady, one dimensional 31, 36—Table 2.1
Minimum
  bubbling velocity 105
  fluidization velocity 98
Mean free path
  definition 252
  for molecules 254—Table 9.1
  for particles 254—Table 9.1
Models
  A 36—Table 2.1, 150—Table 7.1
  B 36—Table 2.1, 150—Table 7.1, 314—Table 10.1
  C 36—Table 2.1
  CHEMFLUB 153—Table 7.2
  ill-posed 133—Fig. 7.1
  well-posed 133—Fig. 7.1
Molecular dynamics simulations 302
Momentum
  balances 3, 32, 36, 73, 129, 201, 279
  models *(see Models)*
  steady, one dimensional
  steady, one dimensional mixture 36—Table 2.1
Mosmelian, D. 318, 335
Muir, J.F. 94, 96
Multicomponent 20
Multisize particle 337, 360
Murray, J.D. 180, 194
MuschelKnautz 343, 354
Mutsers, S.M.P. 115, 126

N

Nakamura, K. 47, 60, 343, 354
Navier-Stokes equation
  compressible multiphase derivation 200
  incompressible multiphase derivation 199
  kinetic theory derivation 274, 284, 347
  kinetic theory summary 347—Table 11.1
Nedderman, R.M. 79, 96
Newton's drag 37

Non-Newtonian fluid behaviour 304
Nozzle flow 52
Nusselt number 55
Number density 355

O

Objectivity principle 199
Oil-water flow 22
One-dimensional flow
  critical 73
  direct contact heating 55
  gas-liquid-solid 69
  in a combustor 57
  in a nozzle 52
  in a spray column 67
  riser flow 56
  steady mixture momentum 31, 36—Table 2.1
  transport reactor 54
Open system balance 402
Oscillations
  high frequency 254—Table 9.1
  low frequency 328—Figs. 10.19 and 10.20
Osmotic pressure 370

P

Packed bed flow 35
Pair-distribution function 247
Pal, S.K. 135, 195
Particle concentration measurement 45, 156, 166, 168, 211, 212, 370, 387
Particle Reynolds number 37, 52, 205
Particle properties *(see Powder)*
Particles
  dynamics of encounter 244
  frequency of collision 240
  restitution coefficient 245, 358
Particulate flow *(see Granular flow)*
Peculiar particle velocity 258
Permeability 27, 367
Phase change 3, 57
Piepers, H.W. 115, 126, 171, 195
Pipe flow
  by characteristics 448
  experimental 38, 41, 207
  one-dimensional 32
Pneumatic conveying 32
  apparatus 41—Fig. 2.3
Poots, G. 141, 196
Porosity
  continuum definition 2
  kinetic theory derivation 261, 264—Table 9.4
  relation to bulk density 74
  relation to number density 357
  volume averaging 21
Porous medium *(see Packed bed flow, Permeability)*
Powder
  Geldart classification 106—Fig 5.3
  granular conductivity 283, 290, 315, 347
  K. Rietema 127
  kinetic theory equation of state 269—Table 9.5, 285, 347—Table 11.2
  modulus 286
  particle to particle drag 343
  pressure (Eq. 9–230, Table 9.6) 269—Table 9.5, 285, 347
  stress 262—Table 9.3, 274, 290, 314—Table 10.1, 346, 347
  viscosity 216—Fig. 8.8, 257—Table 9.2, 289, 314—Table 10.1, 347
Prandtl number 56
Predictive model 314—Table 10.1
Presssure
  collisional 277, 284, 287, 290
  definition (Tables 9.4, 9.5) 7, 264—Table 9.4, 269—Table 9.5
  kinetic 267
Pressure drop correlation 51
Pressure tensor 262, 274, 290, 314
Prigogine, I. 11, 29
Pritchett, J.W. 116, 127, 154, 195
Probstein, R. 384, 389
Pumping requirement 91
Pyroflow CFB 325

Q

Quality of steam *(see Weight fraction)*
Quantity such as mass flux, momentum flux 253

R

Radial distribution function 248, 304
Radial variation of
flux 214 (Fig. 8.6)
gas velocity 230—Fig. 8.20, 224—Fig. 8.13
granular temperature 326—Fig. 10.17
particle velocity 209, 232
solids viscosity 327—Fig. 10.18
solids volume fraction 227—Fig. 8.16, 223—Fig. 8.11, 233—Fig. 8.24, 326—Fig. 10.17, 330—Fig. 10.22
Radioactive isotopes 41—Fig. 2.3, 156—Fig. 7.3, 211—Fig. 8.3, 319—Fig. 10.9, 370—Fig. 12.4, 387—Fig. 12.12
Random motion 240—Fig. 9.1, 248, 320—Fig. 10.10
Rasouli, R. 94, 96, 452
Rate of entropy production 11, 19, 54
Reactor 54, 55, 57, 416, 432
Regime(s)
flow in vertical pipes 216, 224
fluidization 106, 119, 311, 324
grain inertia 303
Reklaites, G.V. 24, 29
Restitution coefficient
definition 245
for various materials 338- -Fig. 11.1
Reversible process,
definition 11
Reynolds number 37, 46, 52, 151, 203, 205, 222, 233, 316
Reynolds transport theorem 1, 421, 425
Reynolds stress 229, 262
Rhee, H.K. 431, 455
Rhodes, M.J. 208, 237
Richardson, J.F. 36, 60
Richner, D.W. 182, 195
Richtmeyer, R. D. 135, 195
Rietema, K. 82, 96, 115, 127
Riser 325
Roco, M. 280, 295
Rotation 280
Rowe, P.N. 37, 60, 116, 127, 171, 195
Roy, D. 434, 455
Rudinger, G. 137, 195

S

Savage, S.B. 33, 60, 88, 96, 131, 195, 240, 295, 301, 335
Sandusky, H.W. 81, 96
Scale factors for
bed inventory 207
creeping flow (Stokes flow) 39
densities 205
dilute to dense flow 233
dilute transport 38—Fig. 2.2
electrical forces 109, 381
fast particle flow (Newtonian regime) 37
fluid velocity 205
gas compressibility 203, 205
geometry 206
granular compressibility 205
minimum fluidization 102
mixture Reynolds number 233
packed bed flow 35, 63
particle inertia (Bagnold #) 304
particle size 202
surface tension (Weber number) 71
viscosities 205
zeta potential 381
Scrubber 54
Scheidegger, A.E. 367, 389
Schiffman, R.L. 371, 389
Schlichting, H. 198, 237, 297, 336
Schneyer, G.P. 180, 195
Schugerl, K. 212, 237
Second law of thermodynamics
restriction 11, 19
stability 134, 381
statement 11

violation 19
Sedimentation *(see Settling)*
Sedney, R. 135, 195
Seo, Y. 171, 195
Settling
apparatus 370—Fig. 12.4
compression settling 366
conservation of particles 355
density effect 358, 381
electric field effect 380—Fig. 12.8, 384—Fig. 12.10
experiment 371—Fig. 12.5, 380—Fig. 12.8
free settling 360
improvement methods 381
Kynch's analysis 363
lamella settler 384—Fig. 12.10
particle to particle drag 343
shock balance 363
size effect 360—Fig. 12.2
thickener 382—Fig. 12.9
zeta potential 380—Fig. 12.8
Separation
by size or density 342, 358
by shaking or vibration 342
chromotographic 434
liquid-solid 359
particles *(see also Settling)* 109, 342, 359
particle-particle 109, 342
solid-gas 326
with electric field 109, 380
Sinclair, J.C. 235, 237, 302, 336
Shahinpour, M. 240, 296
apparatus 305
bubble 119
definition 297
definition 363
in chromatography 439
in settling 363
Shale 183
Shear flow, granular
Shen, H. 299, 336
Shih, Y.T.D. 109, 114, 369, 389
Shock
Shook, C.A. 52, 60
Shrinking core model 55
Slattery, J.C. 3, 29
Slip
definition 62
fast bed 208
Sludge 370
Slug flow 99
Slurry flow 320
Sneddon, I.A. 431, 457
Sobreiro, L.E.L. 171, 196
Soil mechanics 372
Sokolovskii, V.V. 75, 96
Soo, S.L. 37, 60, 240, 296
Sound speed
homogeneous mixture 94
in gas 140, 254—Fig. 9.1
in particles 88, 254—Fig. 9.1, 286
Spielberg, N. 391, 455
Spouted bed 99
Squires, A.M. 208, 237, 330, 336
Stand-pipe 325—Fig. 10.16
Stability
colloidal 379
numerical 138
thermodynamic 134
Statistical distribution 320
Stewart, H.B. 21, 29
Stokes law 39
Stream function 142
Stress tensor 4, 199, 262—Fig. 9.3, 290, 347
Surface tension 71
Suspension
colloidal 379
gas-solid 31, 197, 324
Syamlal, M. 8, 29, 197, 237, 240, 296
System
constant mass 413, 423
constant volume 405
macroscopic 393

T

Tarmy, B.L. 125, 127
Taylor, D.W. 371, 389
Terminal velocity 51, 63, 358
Temperature
granular, definition 242, 341
granular, numerical values 254, 317, 322, 326
granular, for a mixture 341
Terzagi, D.W. 372, 389
Three phase fluidization 124
Tiller, F.M. 363, 389
Time averaged profiles 166, 168, 170, 173, 212, 214, 215, 223, 227, 319, 329, 387
Tolstoy, I. 81, 96
Trajectorics of particles 40, 143
Transport equations 1, 259, 276, 413, 423
Transport, Leibnitz formula
Transport theorems
dense 276
Jenkins-Savage 276
Maxwell 259
Reynolds 419, 423
Truesdell, C.J. 188, 196
Tsuo, Y.P. 222, 237
Turbulence *(also see Oscillations)* 228, 328
Two-fluid models

U

U-gas process 422
Uniqueness of solution 452

V

Variance 241, 320—Fig. 10.10
Variables, dependent 149
Velocity
of bubble 116, 149, 175
concentration wave 436
characteristic 83—Fig. 4.3, 84, 133—Fig. 7.1, 138, 186–192
fluidization 98
measurement 170, 213, 319, 380
profiles 170, 215, 223
relative 36
sedimentation 358, 380
superficial 35
terminal 38, 63
Velocity gradient 199
Venturi meter 52
Verloop, J. 115, 127
Vertical flow 31, 210
Vibration 343
Vigil, S.C. 130, 196
Viscosity of gases 257—Table 9.2
Viscosity of particles
bulk 289
dilute phase 256, 257—Table 9.2, 314—Table 10.1
dense phase 289, 290, 314—Table 10.1
simple model 256
measurement 216—Fig. 8.8
Viscous stress tensor *(see Stress tensor)*
Void fraction
definition 357
equation for 2, 43
measurement 44—Fig. 2.4, 65—Fig. 3.1, 166—Fig. 7.15, 168, 212, 223
Volume fraction
definition 2
equations for 2
measurement 44—Fig. 2.4, 49—Fig. 2.7, 65—Fig. 3.1, 166—Fig. 7.15, 168, 212, 371, 387
Von Mises, R. 84, 96
Vora, M.K. 422, 455
Vorticity
definition 142
equation for 142

W

Wake 161—Fig. 7.9
Wall effects 32, 75
Wall friction 32, 46, 75, 106, 218

Wallis, G.G. 61, 72
Walton, O.R. 302, 336
Wave(s)
  computation of 441
  concentration 436
  density 116, 133
  pressure 139
Weber number 71
Well-posedness 134
Wear 112
Weinstein, H. 208, 232, 237
Wen, C.Y. 36, 60, 99, 102, 114, 146, 196
Westinghouse data 170, 179
Whitehead, A.B. 311, 336
Wissa, E.Z. 371, 389
Woods, L.C. 22, 29

X

X-ray densitometer 211
X-ray system 45

Y

Yang, W.-C.D. 103, 114, 180, 196, 333, 336
Yerushalmi, J. 208, 237
Yoon, H.S. 55, 60

Z

Zenz, F.A. 50, 60
Zeta potential 379
Zone settling 360